Synthese Library

Studies in Epistemology, Logic, Methodology, and Philosophy of Science

Volume 477

The aim of *Synthese Library* is to provide a forum for the best current work in the methodology and philosophy of science and in epistemology, all broadly understood. A wide variety of different approaches have traditionally been represented in the Library, and every effort is made to maintain this variety, not for its own sake, but because we believe that there are many fruitful and illuminating approaches to the philosophy of science and related disciplines.

Special attention is paid to methodological studies which illustrate the interplay of empirical and philosophical viewpoints and to contributions to the formal (logical, set-theoretical, mathematical, information-theoretical, decision-theoretical, etc.) methodology of empirical sciences. Likewise, the applications of logical methods to epistemology as well as philosophically and methodologically relevant studies in logic are strongly encouraged. The emphasis on logic will be tempered by interest in the psychological, historical, and sociological aspects of science. In addition to monographs *Synthese Library* publishes thematically unified anthologies and edited volumes with a well-defined topical focus inside the aim and scope of the book series. The contributions in the volumes are expected to be focused and structurally organized in accordance with the central theme(s), and should be tied together by an extensive editorial introduction or set of introductions if the volume is divided into parts. An extensive bibliography and index are mandatory.

Cristián Soto
Editor

Current Debates in Philosophy of Science

In Honor of Roberto Torretti

Springer

Editor
Cristián Soto
Departamento de Filosofía
Universidad de Chile
Ñuñoa, Chile

Newton International Fellow
British Academy/CPNSS, LSE
London, UK

ISSN 0166-6991 ISSN 2542-8292 (electronic)
Synthese Library
ISBN 978-3-031-32374-4 ISBN 978-3-031-32375-1 (eBook)
https://doi.org/10.1007/978-3-031-32375-1

This Springer imprint is published by the registered company Springer Nature Switzerland AG
The registered company address is: Gewerbestrasse 11, 6330 Cham, Switzerland

In Memoriam Roberto Torretti

Roberto Torretti passed away on 12 November 2022. His life and work have received extensive recognition and endless praise from his friends and colleagues worldwide. Even though he did not get to see the final version published, he was familiar with this book. The origins of the volume go back to February 2019, when I had the fortunate chance to discuss the project with Otávio Bueno, who strongly encouraged me to proceed swiftly. From the beginning, Roberto Torretti was enthusiastically involved, suggesting names of potential contributors, and providing advice on the book's outline. He felt humbled when he knew the final table of contents, and he was grateful to the authors for their contributed chapters. This volume represents a collective effort to honor Roberto Torretti's name and place in the philosophy of science. He represents an outstanding example of life-long devotion to the field, with exceptional standards of passion, rigor and professionalism that will continue to inspire many over the time.

London, England, UK
9 March 2023

Cristián Soto

Picture 1 Roberto Torretti at the age of 24. Picture taken by Carla Cordua, Chilean philosopher and Roberto Torretti's wife

Picture 2 Roberto Torretti at the age of 70. By Pablo Hermansen

Acknowledgments

While editing this book, I amassed various debts to many friends and colleagues. First and foremost, I thank Roberto Torretti, who was incredibly generous with his time, engaging in extended e-mail correspondence and arranging online meetings while the COVID-19 pandemic was at its peak. We extensively discussed his works and this volume's goals in his honor. He facilitated an updated list of his publications and some photos we include now on these pages. Roberto Torretti's outstanding example of intellectual rigor and devotion to the philosophy of science goes almost unparalleled in the discipline.

Likewise, with his usual insight and genuine interest in making other people's works thrive, Otávio Bueno, editor-in-chief of the Synthese Library Series, encouraged this project from the very beginning. He offered his guidance throughout the process. I am deeply grateful for his ongoing support in this and other ventures.

This book was only possible thanks to the trust of the authors in this editorial project. They have shared valuable results of their investigations, and I have been fortunate to have the chance to bring them together to honor Roberto Torretti's life and work. Many thanks to Hernán Lucas Accorinti, Pablo Acuña, Jonas Arenhart, Katherine Brading, Otávio Bueno, Jeremy Butterfield, Alejandro Cassini, Aldo Filomeno, Sebastián Fortin, Roman Frigg, Henrique Gomes, Manuel Herrera, Jesús Jaimes, Qiu Lin, Olimpia Lombardi, John D. Norton, Patricia Palacios, Thomas Ryckman, and Charlotte Werndl.

I thank the Departamento de Filosofía, Universidad de Chile, for providing me with the necessary conditions to undertake this project. And to the British Academy for awarding me a Newton International Fellowship to conduct research at the Centre for the Philosophy of Natural and Social Sciences, LSE, UK, which finally allowed me to bring the volume to a good end. I am particularly indebted to Bryan Roberts, the Centre's Director, and to Roman Frigg for kindly hosting me at the Centre, as well as for his invaluable guidance in this and other projects.

For making it all worthwhile and being a constant source of joy and adventure, I thank Dominique and Damián – to them with endless love.

Contents

Contributors

Hernán Lucas Accorinti CONICET and Universidad de Buenos Aires, Buenos Aires, Argentina

Pablo Acuña Institute of Philosophy, Pontificia Universidad Católica, Santiago, Chile

Jonas R. B. Arenhart Department of Philosophy, Universidade Federal de Santa Catarina, Florianópolis, Brazil

Katherine Brading Department of Philosophy, Duke University, Durham, NC, USA

Otávio Bueno Department of Philosophy, University of Miami, Coral Gables, FL, USA

Jeremy Butterfield Trinity College, University of Cambridge, Cambridge, UK

Alejandro Cassini CONICET and Universidad de Buenos Aires, Buenos Aires, Argentina

Aldo Filomeno Institute of Philosophy, Pontificia Universidad Católica de Valparaíso, Valparaíso, Chile

Sebastián Fortin CONICET and Universidad de Buenos Aires, Buenos Aires, Argentina

Roman Frigg Department of Philosophy, Logic and Scientific Method, and CPNSS, LSE, London, UK

Henrique Gomes Trinity College, University of Cambridge, Cambridge, UK

Manuel Herrera CONICET and Universidad de Buenos Aires, Buenos Aires, Argentina

Jesús Alberto Jaimes Arriagada CONICET and Universidad de Buenos Aires, Buenos Aires, Argentina

Qiu Lin Department of Philosophy, Duke University, Durham, NC, USA

Olimpia Lombardi CONICET and Universidad de Buenos Aires, Buenos Aires, Argentina

John D. Norton Department of History and Philosophy of Science, University of Pittsburgh, Pittsburgh, PA, USA

Patricia Palacios Department of Philosophy, University of Salzburg, Salzburg, Austria

Thomas Ryckman Department of Philosophy, Stanford University, Stanford, CA, USA

Cristián Soto Departamento de Filosofía, Universidad de Chile,Ñuñoa, Chile
Newton International Fellow, British Academy/CPNSS, LSE, London, UK

Charlotte Werndl Department of Philosophy, University of Salzburg, Salzburg, Austria

Chapter 1
Editor's Introduction: Celebrating Roberto Torretti

Cristián Soto

Abstract This volume collects previously unpublished contributions to the philosophy of science. What brings them together is a twofold goal: first and foremost, celebrating the name of Roberto Torretti, whose works in this and other areas have had – and continue to have – a significant impact on the international philosophy of science community; and second, the desire of advancing novel perspectives on various issues in the philosophy of science broadly construed. In what follows, I firstly offer a brief biographical note on Roberto Torretti; second, I suggest considering Torretti's philosophical production as belonging to the tradition of natural philosophy; and third, I provide an outline of what is to come in the rest of the book.

1.1 A Biographical Note

Roberto Torretti was born in Santiago, Chile, on February 15, 1930. As per his recollections, he gained awareness of the theoretical and practical relevance of science in society after the dropping of atomic bombs on Japan in 1945. For the first time, this event directed his attention to books in physics and philosophy at the age of 15. Philosophy, however, was not a career path his family would support. Among possible alternatives, he was encouraged to consider enrolling in the more traditional undergraduate programs, such as engineering, medicine, and law. He ended up pursuing studies in law at the Universidad de Chile. However, he soon realized that his most personal, intellectual interests were in the riddles emerging from the understanding of nature we obtain from philosophical reflection upon the sciences. While in his third year of law, in 1948, he enrolled in the Philosophy program at the Universidad de Chile.

C. Soto (✉)
Departamento de Filosofía, Universidad de Chile, Ñuñoa, Chile

Newton International Fellow, British Academy/CPNSS, LSE, London, UK
e-mail: C.Soto1@lse.ac.uk; cssotto@uchile.cl

C. Soto (ed.), *Current Debates in Philosophy of Science*, Synthese Library 477,
https://doi.org/10.1007/978-3-031-32375-1_1

After completing his undergraduate studies, he moved to Germany to write a doctoral dissertation on Fichte's political system at the University of Freiburg, which he completed in 1954 under Wilhelm Szilasi. Reasons for deciding to write a dissertation on Fichte included his previous work on modern German philosophy, especially Kant and Fichte, and his specialization in jurisprudence.

Although Torretti preserves the manuscript of his doctoral dissertation in his bookshelves, it is a work he rarely returns to in later publications. As an exception, he revisited Fichte's ideas in 2014, when he delivered a keynote address at the Universidad Diego Portales in Santiago, Chile, to commemorate the 200th anniversary of Fichte's death. On this occasion, however, at the very beginning of his intervention, he nicely remarked that he is "convinced that Fichte's philosophy, just like any other fundamentalism from Plato to Husserl, failed to achieve its goal" (Torretti, 2014, in Cordua & Torretti, 2017, p. 58). It is worth mentioning this because his opposition to philosophical fundamentalism was a driving force underpinning his works. Soon after completing his doctorate, he readily shifted the focus of his investigations towards the history and philosophy of science broadly construed. From the early 1960s and for the next 60 years, Torretti's intellectual efforts were systematically devoted to natural philosophy from the ancient Greeks to the moderns, and the history and philosophy of science, focusing on physics and geometry.

Torretti held a brief appointment at the Universidad de Chile in Valparaíso between 1954 and 1955 (currently known as Universidad de Valparaíso, Chile). This was immediately before temporarily emigrating to take on a position as Translator-Secretariat at the United Nations in New York (1955–1958). Over the years 1958–1961, he was a lecturer at the University of Puerto Rico before returning to Chile to work at the Universidad de Concepción from 1961 to 1964. A crucial turn in his academic career occurred in 1964 when he moved back to the Universidad de Chile. He was a founding member of the Centro de Estudios Humanísticos, a center for the humanities engaging with the sciences and technology at the Faculty of Physical and Mathematical Sciences. The center initially granted him the time and funding to pursue research. He served as the center's Director from 1964 to 1970, years during which he sketched and wrote his *Manuel Kant: Estudio sobre los Fundamentos de la Filosofía Crítica* (*Immanuel Kant: Studies of the Groundworks of Critical Philosophy*, 1967) and his collection – including translations of his own – of manuscripts in natural philosophy from the ancient Greek to the early moderns (Torretti, 1971). Both manuscripts were published in Spanish by the Universidad de Chile press, representing unprecedented works in the recent history of Chilean academic philosophy and novel contributions to international debates.

By the late 1960s, the political situation in Chile was precarious, and it inevitably affected the quality of academic life in universities. Torretti was fortunate enough to relocate once more to the University of Puerto Rico, holding a position as professor of philosophy from 1970 to 1995. The University of Puerto Rico provided him with academic freedom and a stimulating intellectual environment that allowed him to undertake some of his most essential endeavors in the philosophy of physics and geometry. To the Puerto Rico years belong some of his key works: *Philosophy of*

Geometry from Riemann to Poincaré (1978), *Relativity and Geometry* (1983), and *Creative Understanding: Philosophical Reflections on Physics* (1990).

After the year 1995, having returned to Chile, Torretti continued his philosophical work. In 1998, he published *El Paraíso de Cantor: La Tradición Conjuntista en Filosofía Matemática* (*Cantor's Paradise: The Set Theoretical Tradition in Mathematical Philosophy*) at the Editorial Universitaria, which was the rebranding of the abovementioned Universidad de Chile press. Only a year later, in 1999, his landmark manuscript *The Philosophy of Physics* was published by Cambridge University Press. Additionally, he held a brief appointment as a professor of philosophy at the Universidad de Chile (1999–2001), this time at the Department of Philosophy, before retiring from teaching, hence securing additional time to continue his philosophical endeavors. Retirement, of course, was not in vain, since in the years to come, a series of books in Spanish were sent to the press: *Relatividad y Espaciotiempo* (2003, *Relativity and Spacetime*); *De Eudoxo a Newton: Modelos Matemáticos en Filosofía Natural* (2007a, *From Eudoxo to Newton: Mathematical Models in Natural Philosophy*); and *Crítica Filosófica y Progreso Científico: Cuatro Ejemplos* (2008, *Philosophical Criticism and Scientific Progress: Four Examples*).[1]

Roberto's philosophical production is a neat example of a lifelong devotion to the philosophy of science. In recent years, the Ediciones Universidad Diego Portales press, Chile, has collected, edited, and published an impressive number of Torretti's works. The volumes in that collection amount to an almost independent section in the bookshelves of a library (as I have done at home), providing a vivid, direct image of Torretti's extensive philosophical contributions. Five of these volumes appear under the title *Estudios Filosóficos* (*Philosophical Studies*, see Torretti, 2006, 2007b, 2010, 2013, 2014), including writings from the years 1957 to 1987, 1986 to 2006, 2007 to 2009, 2010 to 2011, and 2011 to 2014, respectively. Most of them are translations of manuscripts previously published in English. However, some are additions to Torretti's opus, especially in the philosophy of biology and the philosophy of mathematics, issues with which he engaged with renewed efforts since the early 2000s. This edition of Torretti's works comprises other titles, such as the Spanish translation of *Creative Understanding*, this time under the title *Inventar para Entender* (Torretti, 2012), whose translation was prepared by Roberto himself; the abovementioned volume *Crítica Filosófica y Progreso Científico: Cuatro Ejemplos* (Torretti, 2008, *Philosophical Criticism and Scientific Progress: Four Examples*), which addresses four issues in the history and philosophy of physics to put to the test Hasok Chang's views on history and philosophy of science; and an extensive 500-page interview by the Chilean philosopher Eduardo Carrasco from the Universidad de Chile, which delivers a unique window into both Torretti's

[1] The books mentioned in the last three paragraphs represent only a sample of Torretti's philosophical production. They provide, nevertheless, an informative impression of his intellectual interests. The reader can find a complete list of Torretti's publications, including his 34 books and numerous articles, book chapters, encyclopedia entries, and reviews, in Appendix 1 in this volume.

personal life and his views on issues such as art, history, and politics (see Carrasco, 2006).

After reading Torretti's philosophical production, one identifies certain shared features that appear here and there in most, if not all, of his writings. One such feature is that he tirelessly emphasized the relevance of the historiographical approach to science and philosophy. For this, Torretti felt the need to master the relevant languages from ancient Greek and Latin to German, French, and English (Spanish being his mother tongue), thus privileging first-hand access to the relevant sources. We see examples of this in his studies of Descartes, Newton, Maxwell, Herschel, Whewell, Mill, Mach, Plank, Einstein, and others, whose works he approached directly with a remarkable sensitivity to their respective historical context.

A second such feature is this: when studying geometry and physics, he went through the hard road of learning the relevant mathematical and physical theories, which enabled him to advance sound philosophical interpretations of Euclid, Gauss, Riemann, Minkowski, Cantor, and Einstein, among others. In this regard, he took seriously the motto of developing a philosophical reflection on the sciences paying attention to the history and practice of the relevant disciplines.

Furthermore, a third key feature of Torretti's philosophy is undoubtedly this: he conceived philosophy as a continuous discipline with scientific and mathematical practices, at least insofar as they are all concerned with the human attempt to articulate our best conception of various domains. For him, the philosophy of science brings together historical and philosophical reflections upon our socially institutionalized epistemic practices, be they theoretical or experimental, mathematical or physical.

The impact of Torretti's work on the philosophical community is substantial and multifaceted. A broad Latin American readership may be familiar with Torretti's name thanks to his scholarly contributions to Kantian transcendental philosophy and his interpretations of natural philosophy from the ancient Greeks to the early moderns. This is the case since he published his study of Kant's *Critique of Pure Reason* and his selection of writings in natural philosophy in Spanish in the late 1960s and early 1970s in Spanish. Nevertheless, in the Anglo-Saxon philosophical world, he is best known for his contributions to the philosophy of geometry and physics, particularly in relativity theory, investigations which Torretti began in the 1970s and kept him busy for most of his career. This scenario highlights a nice feature of his philosophical production. Whereas some of us are routinely trained to distinguish between the history of philosophy and the philosophy of science, Torretti sees an uninterrupted, continuous series of attempts at producing reliable knowledge of various domains.

Torretti was awarded various prizes and recognitions throughout his career. In 2001, he became Emeritus Professor at the University of Puerto Rico. In 2005, he was granted a Doctor Honoris Causa, at the Universitat Autónoma de Barcelona, Spain. Likewise, the Republic of Chile granted him the National Prize for Humanities and Social Sciences in 2011. Furthermore, he received various academic awards from the Alexander-von-Humboldt Dozentenstipendiat, Kant-Archiv, Bonn

(1964–1965); the John Simon Guggenheim Memorial Fellow in the periods 1975–1976 and 1980–1981; and a Visiting Fellowship Appointment at the Pittsburgh Center for Philosophy of Science in 1983–1984.

Apart from this book, in 2016, primarily due to the efforts of philosophers Juan Redmond and Rodrigo López Orellana, the *Revista de Humanidades de Valparaíso* published a volume in honor of Roberto Torretti, including contributions from Alejandro Cassini, Jordi Cat, Hasok Chang, José Ferreirós, Olimpia Lombardi, Carlos Ulises Moulines, Luis Pavez, Wilfredo Quezada, Hans-Jörg Rheinberger, Adán Sús, and David Teira, along with an original article by Roberto himself. Before this, in 2006, David Teira organized and published a Symposium in the journal *Teorema*, in Spain, under the title "La filosofía de la ciencia de Roberto Torretti" ("Roberto Torretti's Philosophy of Science"), which includes four reviews of Torretti's books commissioned to philosophers Juan Bautista Bengoetxea, Ricardo Parellada, José Romo, and Xavier de Donato, including a reply by Roberto himself.

1.2 Roberto Torretti, Natural Philosopher

Torretti's numerous contributions to the philosophy of science are as diverse as they are relevant to the discipline's recent developments. No single statement can neatly summarize the scope of his work's impact. The results of his philosophical thinking encompass a wide variety of research fields, among them, as we have mentioned, the history of philosophy and the history and philosophy of science, with a particular focus on geometry and physics. It is indeed challenging for a single reader to draw a detailed map of Torretti's philosophical views as they appear throughout his entire set of works. I will at least partially do so in Chap. 2 of this volume, in which I examine some of his takes on various issues in the general philosophy of science.

As will become more transparent, Roberto Torretti can best be considered a natural philosopher. Let me dwell on this shortly. He understands natural philosophy as an intellectual effort to investigate reality by implementing various empirical and theoretical tools. Although standard historical interpretations would restrict the development of natural philosophy to practices that converged in the investigation of nature over perhaps 1543, with the publication of Copernicus's *Revolutions of Heavenly Bodies*, and 1726, with the publication of the third edition of Newton's *Principia Mathematica*, Torretti holds the view that natural philosophy goes beyond these frontiers both back and forth in time, spanning from the very ancient Greek philosophers of nature to our own scientific and philosophical practices (see Torretti, 1971, 13, 1999, chapter 1).

We are inclined to see Torretti's intellectual life as belonging to the tradition of our institutionalized practices in history and philosophy of science. Think of his *Philosophy of Physics,* which occupies center stage in his production. Apart from including chapters on the history and philosophy of classical mechanics, electrodynamics, statistical mechanics, relativity theory, and quantum mechanics, it opens with the chapter "The Transformation of Natural Philosophy in the

Seventeenth Century" (1999, 1–40), where he examines conceptual and practical transformations that made possible the invention of science in the hands of Galileo, Descartes, and Newton, among those who applied geometrical reasoning to the investigation of the physical world. Nevertheless, Torretti's *Philosophy of Physics* goes beyond what we would expect from a standard textbook, including sections on philosophy, which he deems essential for comprehending the evolution of physics. He addresses concerns in Peirce's pragmatism and looks into the notion of physical laws from a structuralist perspective. We have pointed out that this represents Torretti's intellectual approach: he sees continuity and interaction where current academic, highly compartmentalized canons would find discontinuity and segregation. For Torretti, natural philosophy continues to be cultivated in philosophy, mathematics, and scientific research. Moreover, the uninterrupted efforts he put into his investigations in both the history of philosophy and the history and philosophy of geometry and physics were driven by the same force.

Torretti invites us to see current epistemic practices as seconding the spirit of natural philosophy, stressing that their relevance goes beyond scholarly matters, hence contributing to shaping many facets of human life. He believes philosophers, historians, and sociologists of science have a crucial role in reasonably interpreting science as a cultural phenomenon. In his words,

> the human genre has stopped from being an abstract idea of prophets and philosophers to become a concrete society, united – for good or bad – by a common destiny. [...] Somehow, modern natural science appears in every case as a decisive reality for current human existence, hence a reality that a 'humanist' education must teach to comprehend (Torretti, 1971, 13, my translation).

Torretti's words provide an initial glimpse into the motivations underpinning his studies in history and philosophy of science. He is thoroughly concerned with the technical details of the relevant theoretical frameworks in mathematics, geometry, physics, and philosophy, but this always goes hand in hand with a vivid concern regarding the human quest for understanding reality.

1.3 Outline of the Book

As stated above, the editorial project that motivates the present volume seeks to celebrate Roberto Torretti's place in the philosophy of science, offering novel contributions to various issues in this research field. Chapters comprising this book can be approached from that perspective. Some authors directly address views defended by Torretti, especially concerning topics in the philosophy of physics or Kantian philosophy of science. Instead, other chapters pursue investigations in areas of Torretti's interest, although without directly engaging with his views. Taken together, contributions to this volume celebrate his name in their own way by doing one of the things he enjoys the most: articulating in-depth analyses of various branches of science. Here is a summary for each of the chapters that follow, which will help the reader to navigate the volume[2]:

Chapter 2. *Cristián Soto. "Roberto Torretti's Philosophy of Science."* Abstract: "I put forward a detailed analysis of Torretti's views in the general philosophy of science. For this, I consider such issues as his Kantian take on objectivity, the creative understanding thesis, his critique of scientific realism, his elaboration of mathematical fictionalism, his structuralist conception of physical laws, and his analysis of the contribution of philosophical reflection to scientific progress. In so doing, I examine Torretti's publications from the late 1960s to the early 2010s in the hope of expounding some of his ideas that may not be well known to his readers, coming with either an undivided interest in the ancient or early modern history of ideas, or to those who read Torretti with an exclusive focus on the history and philosophy of physics and geometry. With this, I aim to fill a gap, assessing the motivations and extent of his conclusions and arguing that they deliver an attractive alternative for various issues in current debates in the philosophy of science." (See Soto, Chap. 2 in this volume.)

Chapter 3. *Katherine Brading and Qui Lin. "Du Châtelet on Absolute and Relative Motion."* Abstract: "In this chapter, we argue that Du Châtelet's account of motion is an important contribution to the history of the absolute versus relative motion debate. The arguments we lay out have two main strands. First, we clarify Du Châtelet's threefold taxonomy of motion, using Musschenbroek as a useful Newtonian foil and showing that the terminological unity between the two is only apparent. Then, we assess Du Châtelet's account in light of the conceptual, epistemological, and ontological challenges posed by Newton to any relational theory of motion. What we find is that, although Du Châtelet does not meet all the challenges to their full extent, her account of motion is adequate for the goal of the *Principia*: determining the true motions in our planetary system." (See Brading & Lin, Chap. 3 in this volume.)

Chapter 4. *Thomas Ryckman. "Effective Field Theories: A Case Study for Torretti's Perspective on Kantian Objectivity."* Abstract: "Those enlightened philosophers of physics acknowledging some manner of descent from Kant's 'Copernican Revolution' have long found encouragement and inspiration in the writings of Roberto Torretti. In this tribute, I focus on his "perspective on Kant's perspective on objectivity" (2008), a short but highly stimulating attempt to extract the essential core of the Kantian doctrine that 'objects of knowledge' are constituted, not given, or with Roberto's inimitable pungency, that "*objectivity is an achievement, not a gift*." That essential core Roberto locates in the Kantian notion of *apperception*, or self-activity, manifested in cognition in the idea of combination (*Verbindung*) or composition, which, Kant tells us, "among all ideas ... is the one that is not given through objects, but can only be performed by the subject itself, because it is an act of self-activity" (B 130). I first rehearse

[2] I thank the authors of each chapter for providing these abstracts.

Roberto's proposal for how an imaginative interplay between sensibility and understanding can be fashioned via the *productive imagination* or *power of reflective judgment* (of the 3rd *Critique*). In this way, the notion of composition in general, unfettered from needless period constraints issuing in "pure forms of sensibility" and "pure concepts of the understanding", can be seen as the intellectual motor for the "free creation" of concepts celebrated by Einstein and others, furnishing structural scaffolding required to articulate and display physical objects and processes, a conceptual panoply that "cannot be fished out of the stream of impressions". Roberto emphasizes that historical case studies are needed to evaluate his proposal, suggesting one himself, the continuous conceptual development inaugurated by Riemann's *Habilitätionsschrift* (1854) resulting, some hundred years later, in the fiber bundle formalism of modern differential geometry and topology. I sketch a related suggestion, that the gauge groups of modern particle physics are the outcome of a similar line of conceptual advance, a structural scaffolding saving the phenomena of high-energy experiment within the framework of 'effective field theory.'" (See Ryckman, Chap. 4 in this volume, and references therein.)

Chapter 5. *Olimpia Lombardi. "A Kantian-Rooted Pluralist Realism for Science."* Abstract: "After the preeminence of logical positivism/empiricism during the most part of twentieth century, during the last decades many authors began to recognize the relevance of the Kantian thought for present-day philosophy of science. This chapter follows this general trend, adopting a realist reading of Kantian teachings. On this basis, I will delineate a Kantian-rooted realism according to which the worlds of science are always the result of a synthesis between the conceptual schemes embodied in scientific theories and practices and the independent noumenal reality. However, my position takes distance from the Kantian doctrine by admitting the possibility of different conceptual schemes, both diachronically and synchronically. This view not only leaves room for abrupt and discontinuous changes in the history of science, but also leads to an ontological pluralism that allows for the coexistence of irreducible and different, even incompatible ontological domains at the same historical time. I will focus particularly on the synchronic case to reject both ontological reductionism and emergentism from a perspective that denies any priority or dependence between domains, in resonance with a non-hierarchical articulation between scientific theories and disciplines." (See Lombardi, Chap. 5 in this volume.)

Chapter 6. *Otávio Bueno. "Mathematical Fictionalism Revisited."* Abstract: "Mathematical fictionalism is the view according to which mathematical objects are ultimately fictions, and, thus, need not be taken to exist. This includes fictional objects, whose existence is typically not assumed to be the case. There are different versions of this view, depending on the status of fictions and on how they are connected to the world. In this paper, I critically examine the various kinds of fictionalism that Roberto Torretti identifies, determining to what extent they provide independent, defensible conceptions of mathematical ontology and how they differ from platonism (the view according to which mathematical objects and structures exist and are abstract, that is, they are neither causally active nor are located in spacetime). I then contrast Torretti's forms of fictionalism with a version of the view that, I argue, is clearly non-platonist and provides a deflationary account of mathematical ontology, while still accommodating the attractive features of the view that Torretti identified." (See Bueno, Chap. 6 in this volume)

Chapter 7. Jeremy Butterfield and Henrique Gomes. "Functionalism as a Species of Reduction." Abstract: "This is the first of four papers prompted by a recent literature about a doctrine dubbed spacetime functionalism. This paper gives our general framework for discussing functionalism. Following Lewis, we take it as a species of reduction. We start by expounding reduction in a broadly Nagelian sense. Then we argue that Lewis' functionalism is an improvement on Nagelian reduction. This paper thereby sets the scene for the other papers, which will apply our framework to theories of space and time. (So those papers address the space and time literature: both recent and older, and physical as well as philosophical literature. But the four papers can be read independently.) Overall, we come to praise spacetime functionalism, not to bury it. But we criticize the recent philosophical literature for failing to stress:

(i) functionalism's being a species of reduction (in particular: reduction of chrono- geometry to the physics of matter and radiation);
(ii) functionalism's idea, not just of specifying a concept by its functional role, but of specifying several concepts simultaneously by their roles;
(iii) functionalism's providing bridge laws that are mandatory, not optional: they are statements of identity (or co-extension) that are conclusions of a deductive argument, rather than contingent guesses or verbal stipulations; and once we infer them, we have a reduction in a Nagelian sense.

On the other hand, some of the older philosophical literature, and the mathematical physics literature, is faithful to these ideas (i) to (iii) – as are Torretti's writings. (But of course, the word 'functionalism' is not used; and themes like simultaneous unique definitions are not articulated.) Thus in various papers, falling under various research programmes, the unique definability of a chrono-geometric concept (or concepts) in terms of matter and radiation, and a corresponding bridge law and reduction, is secured by a precise theorem. Hence our desire to celebrate these results as rigorous renditions of spacetime functionalism." (See Butterfield & Gomes, Chap. 7 in this volume.)

Chapter 8. Patricia Palacios. "Intertheoretic Reduction in Physics beyond the Nagelian Model". Abstract: "In this chapter, I defend a pluralistic approach to intertheoretic reduction, in which reduction is not understood in terms of a single philosophical "generalized model", but rather as a family of models that can help achieve certain epistemic and ontological goals. I will argue then that the reductive model (or combination of models) that best suits a particular case-study depends on the specific goals that motivate the reduction in the intended case-study." (See Palacios, Chap. 8 in this volume.)

Chapter 9. John D. Norton. "Inductive Inferences on Galactic Redshift, Understood Materially." Abstract: "A two-fold challenge faces any account of inductive inference. It must provide means to discern which are the good inductive inferences or which relations capture correctly the strength of inductive support. It must show us that those means are the right ones. Formal theories of inductive inference provide the means through universally applicable formal schema. They have failed, I argue, to meet either part of the challenge. In their place, I urge that background facts in each domain determine which are the good inductive inferences; and we can see that they are good in virtue of the meaning of the pertinent background facts. This material theory of induction is used to assess the competing inductive inferences in the debate in 1972 between John N. Bahcall and Halton Arp over the import of the redshift of light from the galaxies." (See Norton, Chap. 9 in this volume, and references therein.)

Chapter 10. *Charlotte Werndl and Roman Frigg. "When Does a Boltzmannian Equilibrium Exist?"* Abstract: "The received wisdom in statistical mechanics (SM) is that isolated systems, when left to themselves, approach equilibrium. But under what circumstances does an equilibrium state exist and an approach to equilibrium take place? In this paper we address these questions from the vantage point of the long-run fraction of time definition of Boltzmannian equilibrium that we developed in our two papers Werndl and Frigg (2015a, 2015b) (see also Frigg and Werndl 2019; Werndl and Frigg 2017, 2020). After a short summary of Boltzmannian statistical mechanics (BSM) and our definition of equilibrium (Section 2), we state an existence theorem which provides general criteria for the existence of an equilibrium state (Section 3). We first illustrate how the theorem works with a toy example (Section 4), which allows us to illustrate the various elements of the theorem in a simple setting. After looking at the ergodic programme (Section 5) we discuss equilibria in a number of different gas systems: the ideal gas, the dilute gas, the Kac gas, the stadium gas, the mushroom gas and the multi-mushroom gas (Section 6). In the conclusion we briefly summarise the main points and highlight open questions (Section 7)." (See Werndl & Frigg, Chap. 10 in this volume, and references therein.)

Chapter 11. *Roman Frigg and Charlotte Werndl. "Boltzmannian Non-Equilibrium and Local Variables."* Abstract: "Boltzmannian statistical mechanics (BSM) partitions a system's space of micro-states into cells and refers to these cells as 'macro-states.' One of these cells is singled out as the equilibrium macro-state while the others are non-equilibrium macro-states. It remains unclear, however, how these states are characterised at the macro-level as long as only real-valued macro-variables are available. We argue that physical quantities like pressure and temperature should be treated as field-variables and show how field variables fit into the framework of our own version of BSM, the long-run residence time account of BSM. The introduction of field variables into the theory makes it possible to give a full macroscopic characterisation of the approach to equilibrium." (See Frigg & Werndl, Chap. 11 in this volume.)

Chapter 12. *Pablo Acuña. "Scientific Understanding in Astronomical Models from Eudoxus to Kepler."* Abstract: "In the following essay I present a narrative of the development of astronomical models from Eudoxus to Kepler, as a case-study that vindicates an insightful and influential recent account of the concept of scientific understanding. Since this episode in the history of science and the concept of understanding are subjects to which Professor Roberto Torretti has dedicated two wonderful books – *De Eudoxo a Newton: modelos matemáticos en la filosofía natural* (2007), and *Creative Understanding: philosophical reflections on physics* (1990), respectively – this essay is my contribution to celebrate his outstanding work and career in this volume. I dedicate this piece to Roberto, dear friend and mentor, in gratitude for all his inspirational work and personal support, which has greatly helped me, and many others, to better understand that human wonder we call scientific knowledge." (See Acuña, Chap. 12 in this volume, and references therein.)

Chapter 13. *Alejandro Cassini. "Reinterpreting Crucial Experiments."* Abstract: "Crucial experiments have been largely neglected by philosophers of science. The main reason for this predicament is that Duhem's criticism of that kind of experiments has been accepted as sound and definitive. In this article, I start by revisiting the main argument against the possibility of crucial experiments, which is based on epistemological holism. I contend that the argument rests on the confusion between crucial and decisive experiments. When crucial experiments are deprived of its supposed decisive character, the argument loses its bite. Epistemological holism applies to any experiment, whether crucial or not, but it does not imply that experiments are not possible or that they do not have epistemological import. This variety of holism simply shows that any evidence has to be interpreted and assessed within a theoretical context that includes many auxiliary hypotheses and presupposed theories, which are regarded as accepted background knowledge. This knowledge is not put to test in a given experiment, but it is rather employed in describing the experimental result and interpreting its theoretical consequences. The meaning of any crucial experiment has then to be extracted from the theoretical context in which the experimental result is interpreted. When the background of accepted knowledge undergoes a drastic change, a crucial experiment may be reinterpreted in such a way that confirms or refutes hypotheses or theories not available at the moment in which it was performed. I will illustrate this kind of reinterpretation with the historical cases of Fizeau's 1851 experiment, the Michelson and Morley 1887 experiment, and Eddington's 1919 experiment. I will conclude by vindicating crucial experiments." (See Cassini, Chap. 13 in this volume, and references therein.)

Chapter 14. *Jonas Arenhart. "Non-Reflexive Logics and Their Metaphysics. A Critical Appraisal."* Abstract: "Non-reflexive logics are systems of logic in which the reflexive law of identity is restricted or violated. The most well-known such systems are Schrödinger logics and quasi-set theory; both are related with the metaphysics of quantum mechanics, attempting to formalize the idea that quantum entities are non-individuals. We argue in this paper that non-reflexive logics may be seen as attempting to characterize two metaphysically incompatible notions of non-individuals: (i) non-individuals as violating self-identity and (ii) non-individuals as indiscernible entities. The problem is that any choice between these options brings difficult questions, making the understanding of non-individuals through the apparatus of non-reflexive logics rather implausible." (See Arenhart, Chap. 14 in this volume.)

Chapter 15. *Aldo Filomeno. "Typicality of Dynamics and Laws of Nature."* Abstract: "Certain results, most famously in classical statistical mechanics and complex systems, but also in quantum mechanics and high-energy physics, yield a coarse-grained stable statistical pattern in the long run. The explanation of these results shares a common structure: the results hold for a 'typical' dynamics, that is, for most of the underlying dynamics. In this paper we argue that the structure of the explanation of these results might shed some light – a different light – on philosophical debates on the laws of nature. In the explanation of such patterns, the specific form of the underlying dynamics is almost irrelevant. The conditions required, given a free state-space evolution, suffice to account for the coarse-grained lawful behaviour. An analysis of such conditions might thus provide a different account of how regular behaviour can occur. This paper focuses on drawing attention to this type of explanation, outlining it in the diverse areas of physics in which it appears, and discussing its limitations and significance in statistical mechanics." (See Filomeno, Chap. 15 in this volume.)

Chapter 16. *Hernán Lucas Accorinti, Sebastian Fortin, Manuel Herrera, and Jesús Jaimes. "The Case of Phonons: Explanatory or Ontological Priority."* Abstract: "Recent discussions about the microstructure of materials generally focus on the ontological aspects of the molecular structure. However, there are many types of substances that cannot be studied by means of the concept of molecule, for example, salts. For the quantum treatment of these substances, a new particle, called phonon, is introduced. Phonons are generally conceived as a pseudo-particle, that is, a mathematical device necessary to perform calculations but which does not have a "real" existence. In this context, the aim of this paper will be to analyze the ontological status of phonons. For such purposes, we will critically analyze the arguments that would account for the presumed nonexistence of phonons. Finally, having already demonstrated that there are not enough reasons to consider phonons as non-existing entities, we will explore some possibilities that allow us to elucidate their ontological status." (See Accorinti et al., Chap. 16 in this volume.)

As can be seen, the volume encompasses several issues: Torretti's philosophy of science; Du Châtelet's contribution to the investigation of absolute and relative motion in classical mechanics; discussions on objectivity and pluralism from a Kantian perspective in the philosophy of physics; mathematical fictionalism; reductionism in the philosophy of physics; the philosophy of induction in astrophysics; the philosophy of Boltzmannian statistical mechanics; scientific understanding and crucial experiments; non-reflexive logics; laws of nature; and arguments for the reality of phonons in theoretical physics. They reflect variety, and so do Torretti's interests in the philosophy of science. Chapters come from scholars worldwide, and they pursue independent investigations in one or another area of the philosophy of science broadly construed. However, they all come together as a token of recognition of Torretti's longstanding contributions to the discipline.

References

Carrasco, E. (2006). *En el Cielo Solo las Estrellas. Conversaciones con Roberto Torretti*. Ediciones Universidad Diego Portales.

Cordua, C., & Torretti, R. (2017). *Perspectivas*. Ediciones Universidad Diego Portales.

Torretti, R. (1967). *Manuel Kant: Estudio sobre los Fundamentos de la Filosofía Crítica*. Editorial Universitaria de la Universidad de Chile.

Torretti, R. (1971). *Filosofía de la Naturaleza*. Editorial Universitaria de la Universidad de Chile.

Torretti, R. (1978). *Philosophy of geometry from Reimann to Poincaré*. D. Reidel Publishing.

Torretti, R. (1983). *Relativity and geometry*. Pergamon Press.

Torretti, R. (1990). *Creative understanding: Philosophical reflections on physics*. The University of Chicago Press.

Torretti, R. (1998). *El Paraíso de Cantor: La Tradición Conjuntista en Filosofía Matemática*. Editorial Universitaria de la Universidad de Chile.

Torretti, R. (1999). *The philosophy of physics*. Cambridge University Press.

Torretti, R. (2003). *Relatividad y Espaciotiempo*. RIL editores.

Torretti, R. (2006). *Estudios Filosóficos 1957–1987*. Ediciones Universidad Diego Portales.

Torretti, R. (2007a). *De Eudoxo a Newton: Modelos Matemáticos en la Filosofía Natural*. Ediciones Universidad Diego Portales.

Torretti, R. (2007b). *Estudios Filosóficos 1986–2006*. Ediciones Universidad Diego Portales.

Torretti, R. (2008). *Crítica Filosófica y Progreso Científico*. Ediciones Universidad Diego Portales.

Torretti, R. (2010). *Estudios Filosóficos 2007–2009*. Ediciones Universidad Diego Portales.

Torretti, R. (2012). *Inventar para Entender*. Ediciones Universidad Diego Portales.

Torretti, R. (2013). *Estudios Filosóficos 2010–2011*. Ediciones Universidad Diego Portales.

Torretti, R. (2014). *Estudios Filosóficos 2011–2014*. Ediciones Universidad Diego Portales.

Chapter 2
Roberto Torretti's Philosophy of Science

Cristián Soto

Abstract In this chapter, we provide an analysis of Torretti's main views in the general philosophy of science. I shall examine his Kantian take on objectivity (Sect. 2.2), the creative understanding thesis (Sect. 2.3), his critique of scientific realism (Sect. 2.4), his elaboration of mathematical fictionalism (Sect. 2.5), his analysis of physical laws (Sect. 2.6), and his examination of the contribution of philosophical reflection to scientific progress (Sect. 2.7). For this, I will consider Torretti's publications from the late 1960s to the early 2010s in the hope of expounding some of his ideas that may not be well known to some of his readers coming exclusively from either the history and philosophy of physics and geometry, or from the history of philosophy. With this, we fill a gap in our understanding of Torretti's views in various fields in the philosophy of science, assessing their motivations and arguing that they yield an attractive alternative for various issues in current debates.

2.1 Introduction

Roberto Torretti is widely known for his contributions to the history and philosophy of physics and geometry, as well as for his work on the history of philosophy. Apart from these research fields, Torretti was a prominent contributor to various debates in the general philosophy of science. In this chapter, to provide an analysis of what I consider Torretti's main views in the general philosophy of science, I shall examine his Kantian take on objectivity (Sect. 2.2), the creative understanding thesis (Sect. 2.3), his critique of scientific realism (Sect. 2.4), his elaboration of mathematical fictionalism (Sect. 2.5), his analysis of physical laws (Sect. 2.6), and his examination of the contribution of philosophical reflection to scientific progress (Sect. 2.7). For this, I will take into consideration Torretti's publications from the late 1960s to the

C. Soto (✉)
Departamento de Filosofía, Universidad de Chile, Ñuñoa, Chile

Newton International Fellow, British Academy/CPNSS, LSE, London, UK
e-mail: C.Soto1@lse.ac.uk

C. Soto (ed.), *Current Debates in Philosophy of Science*, Synthese Library 477,
https://doi.org/10.1007/978-3-031-32375-1_2

early 2010s in the hope of expounding some of his ideas that may not be well known to some of his readers coming exclusively from either the history and philosophy of physics and geometry, or from the history of philosophy. With this, I shall aim to fill a gap in our understanding of the scope and limits of Torretti's views in various research fields in the philosophy of science, assessing their motivations and arguing that they yield an attractive alternative for various issues in current debates.

I should observe this: I will draw heavily from Torretti's works in the following sections for obvious reasons, and I will also take the liberty to quote extensively from his writings. Among the benefits of proceeding in this way are, first, that it will help the reader gain a closer-to-first-hand impression of Torretti's views regarding issues in the general philosophy of science; and second, that I will translate into English some of Torretti's works which are only available in Spanish. This will provide a more accurate picture of his ideas. Sections in this chapter will all come together in the most comprehensive survey and assessment of Torretti's views on the general philosophy of science.

2.2 Kantian Objectivity and the Philosophy of Science

According to Torretti, the most important novelty of Kant's philosophy consists in arguing that the objects of knowledge are not given but constituted. That is, "objectivity is an achievement, it is not a gift." (Torretti, 2008a, 81).[1] The targets of scientific investigation are the results of what Kant called composition or synthesis as an act of human cognition. Torretti's elaboration of the Kantian view of objectivity has two aspects. The first deals with the scholarly interpretation of Kant's writings, particularly those we find in the 1781 and 1787 versions of the *Critique of Pure Reason's* transcendental deduction, along with the bulk of manuscripts and letters that paved the way for the project of critical philosophy. The second aspect goes beyond this scholarly work, advancing an original take on the philosophical problem of the objectivity of scientific knowledge, hence providing a live alternative to answer various issues in the current philosophy of science. Let us look at them in turn.

[1] Spanish-speaking readers will be delighted to know that Torretti's (2008a) "Objectivity: A Kantian Perspective", which appears as chapter 4 in Massimi, Ed., (2008, 81–95), *Kant and Philosophy of Science Today*, was later on translated into Spanish as "La objetividad en el sentido de Kant" (Torretti, 2010, 13–32).

2.2.1 Kantian Preliminaries

Torretti's extensive studies of Kantian philosophy led his philosophical thought along various roads. He is not alone in believing that Kant's conception of the objectivity of phenomenal knowledge delivers the groundwork for an approach to the philosophy of science. Here is an initial motivation that Torretti finds in Kant's philosophy:

> Roughly stated, the argument in KrV A proceeds as follows: absolutely nothing is simple in the stream of *Erlebnisse*; therefore, fragments of the stream can be sensed only if they are run through and held together in 'the synthesis of apprehension.' To bring this about in the midst of the stream's incessant flow, apprehension must be assisted by recollection. Thus, for a melody to be heard, every segment of it must somehow be remembered at the time the last note comes forth. Yet, as Kant pointedly remarks, if there is no awareness that what is being noticed now is one and the same thing that was noticed a moment ago, the entire reproduction in the sequence of *Erlebnisse* will be in vain. The requisite identification is achieved by 'the synthesis of recognition in a concept'. 'Every cognition – says Kant– requires a concept, no matter how imperfect or unclear it might be. By nature, a concept is always something (. . .) that serves as a rule.' Indeed, a concept can only work for the recognition, reproduction and apprehension of *Erlebnisse* insofar as it grasps the regular patterns of the *Erlebnis* manifold and thereby represents 'the synthetic unity in the awareness thereof'. On the other hand, recognition, that is, awareness of the identity of that which is being successively grasped, cannot take effect without awareness of the identity of the very act of grasping. This form of awareness, that is, self-awareness, is called by Kant apperception (after Leibniz). It is the key to Kant's conception of objectivity. (Torretti, 2008a, 83)

According to the Kantian transcendental subject, the necessary awareness of one's identity acts as the necessary unity of the composition of phenomena. The categories of mind make the representation of phenomena reproducible with (Kantian) necessity, making it possible to determine an object for their intuition in sense experience. In Kant's picture, the human mind would not be able to think of its own identity in the manifold of its representations (*Vorstellungen*) should it "not have before its eyes the identity of its act, which subordinates all synthesis of apprehension (. . .) to a transcendental unity and makes its regular interconnection a priori possible" (Kant, KrV A 108). A few lines below this passage, Kant contends:

> The pure concept of this transcendental object (which in effect is always in all our cognitions the same unspecified X) is what can generally procure a relation to an object, that is, objective reality, to all our empirical concepts. This concept cannot contain any definite intuition, and therefore concerns exclusively that unity which must be found in a manifold of cognition insofar as it stands in relation to an object. (Kant, KrV A 109)

Torretti (2008a, 87) highlights a brief passage from the 1787 edition of the *Critique of Pure Reason*, which throws light on the categories of understanding and how they enable us to achieve objectivity. In that passage, Kant claims: "Among all ideas, combination is the only one that is not given through objects, but can only be performed by the subject itself, because it is an act of its self-activity" (Kant, KrV, B 130). Furthermore, in a letter to Tieftrunk, Kant makes this clearer: "The concept of the composite in general is not a particular category, but is contained in all categories" (Kant, Ak 20, 275 f., 11 December 1797, as quoted in Torretti, 2008a, 88). Thus, the concept of composition cannot be accessed through the intuitions of space and time, in Kant's nomenclature. Instead, it precedes perception to make the manifold of experience a single, composite representation of an object.

2.2.2 Kantian Objectivity for the Philosophy of Science

Examining Kant's conception of the objectivity of phenomenal knowledge, and its related notions of the unity of consciousness and the act of composition, provides the tools for elaborating a Kant-inspired view of objectivity in the philosophy of science. One way of depicting this view is by comparing it with the ready-made world of traditional scientific realism. In its most general form, scientific realism maintains that we should interpret scientific theories at face value as accurate descriptions of both observable and unobservable domains. The scientific realist's world is, Torretti submits, full of "perfectly definite objects waiting for the scales to fall from our eyes so that we get to see things in their own light as they are in themselves" (Torretti, 2008a, 81). The Kantian notion of objectivity opposes realist inclinations, seeking to provide a "partial, approximate construction of objects by ever imperfect but endlessly perfectible modeling according to our lights" (Ibid).

An attractive feature of the Kantian approach to objectivity consists in this: advocates of scientific realism postpone the achievement of objectivity to the end of the research process (until reaching convergence with a final, true theory, if ever), whereas the scientist inspired by a Kantian notion of objectivity "can take pride in their daily achievements of contextual, improvable, incomplete, but reasonably working and passably stable objective truths" (Ibid.). In this regard, by acknowledging the role that human understanding plays in the attainment of objectivity in our phenomenal knowledge, "Kant opens a wide door to intellectual pluralism, which indeed has thrived in his wake" (Torretti, 2008a, 87; see Lombardi, Chap. 5 in this volume).

We shall return to Torretti's rejection of scientific realism in Sect. 2.4 below. For now, it is clear that the Kantian approach has its own limits. Furthermore, Torretti does not mince his words in this respect. An overall constraint is imposed by the architectural pretension of embodying a closed, fixed set of categories within a conception of human understanding, which would determine not only the kinds of thoughts we can have, but also the sorts of phenomena we may individuate or conceptualize in our conscious cognitive activity. The closed, fixed categorical framework or conceptual scheme does not suit our understanding of concept formation in the sciences and everyday life, which is broadly unruly. Torretti asserts against the Kantian architecture of mind: "the very idea of intellectual closure [as presented in the 12 categories of understanding] is unpalatable to us, especially if it goes with the idea that the complete set of primary concepts of the understanding is equinumerous with that of the signs of the Zodiac and that of the apostles of Jesus." (Torretti, 2008a, 87).

Torretti aims to overcome this constraint. What is needed is to ensure that the Kantian productive imagination has maximal freedom at its disposal when intertwined with the pure forms of sensibility and the categories of understanding. The objectivity of phenomenal knowledge does not have to be grounded on unnecessary constraints to lay out knowledge's foundations. By contrast, the free undertakings of the human mind routinely stimulate the progress in knowledge of the physical world:

> Kantians who adopt my proposal will no longer be put to shame by the growth of geometry or the turnabouts of mechanics. Thanks to it, Kant's pure reason can take pride in the open-endedness that actual human reason so glaringly displays. More significantly perhaps, we renounce the claim to completeness and invariability that Kant made on behalf of the categories and principles of the understanding. Thereby, we extend to reason itself Kant's intrepid rejection of unconditioned totalities, instead of allowing her to stand as the ultimate transcendental illusion. (Torretti, 2008a, 93)

Furthermore, the destruction (or deconstruction, as you will) of the Kantian strategy continues as follows:

> Open-endedness and detotalization yield two philosophical benefits. First of all, if we admit that scientific concept formation is ever in progress and there is no closed system of basic notions by which it must abide, we can in good faith dismiss the widespread and yet highly unlikely view of the history of science as a series of quantum jumps between incommensurable intellectual systems. The glass-and-steel towers of theory can always communicate with each other across the quicksand of ordinary discourse on which they repose. The second philosophical benefit I see is this: if even the formation of basic concepts remains incomplete, the very idea of a full inventory of predicates by comparison with which every single object might be exhaustively determined makes no sense at all, and the thoroughgoing determination of things asserted by modern pre-critical metaphysics is unthinkable. (Torretti, 2008a, 93)

Elsewhere, Torretti refers to the paragraph 20 of Kant's *Prolegomena* to stress that, whereas the categories of the understanding would help us decipher appearances in order to read them as experience, "the spelling changes deeply if the orthography of experience is not prefixed by the eternal nature of human reason, but it is emerging from its own free inventiveness" (Torretti, 2014, 19).

Should we take Kant's proposal in its original shape, we would see that his "system of 'forms,' categories, and principles could not stem the tide of conceptual innovation but was swept away by it" (Torretti, 1990, 36). Torretti's introduction of free inventiveness within the Kantian framework aims at responding to this problem, enabling his conception of objectivity to accommodate the advancement of physical science from Newton's mechanics to twentieth-century relativity theory and quantum theory. The development of space, time, and mass concepts illustrate the inalienable freedom of thinking in our ongoing investigation of various aspects of reality. Presuppositions about determinism and objective simultaneity, crucial for Kant's worldview, went through significant transformations from the early formulations of Newtonian mechanics to the rise of contemporary physical theory. Undoubtedly, Torretti's efforts oppose those of the clergy of scholars seeking to keep the scriptures of their messiah alive. Instead, he is interested in accounting for the construction of objectivity overall.

2.3 The Creative Understanding Thesis

Torretti submits that the understanding of natural phenomena that we obtain through the development of mathematical physics amounts to one of humankind's most outstanding achievements. Theoretical results and technological applications derived from mathematical physics have directly impacted human life, and its methods have set a paradigm for undertaking scientific investigation in a wide range of domains (Torretti, 1990, ix).

Torretti proposes his views on the creative understanding following his elaboration of a Kant-inspired account of the objectivity of scientific knowledge. He begins with an analysis of observation, but not to pay "lip service to the philosophy of inductivism, but to underscore its inadequacy" (Torretti, 1990, x–xi). He argues that observation without understanding is blind. By contrast, we come to understand phenomena "under universal concepts in order to make them out, and so make them into facts" (Ibid.). To elaborate on his creative understanding thesis, Torretti reminds us of the following passage from Einstein:

> But if experience is the beginning and end of all our knowledge about reality, what role is there left for reason in science? A complete system of theoretical physics consists of concepts and basic laws to interrelate those concepts and of consequences to be derived by logical deduction. [...] We have assigned to reason and experience their place within the system of theoretical physics. Reason gives the structure to the system; the data of the experience and their mutual relations are to correspond exactly to consequences in the theory. On the possibility alone of such a correspondence rests the value and the justification of the whole system, and especially of its fundamental concepts and basic laws. But for this, these latter would simply be free inventions of the human mind which admit of no a priori justification either through the nature of the human mind or in any other way at all. (Einstein, 1934, 164–165; see Torretti, 2014, 13–14; see also Torretti, 1990, ix)

Observation alone would not suffice to articulate a theory about phenomena. Human understanding plays an active role in creating concepts and laws laying out the core structure of scientific theorizing. They are the free inventions of the human mind and are dictated neither by sense experience nor by the eternal structure of understanding.

A remarkable feature of the creative understanding thesis has to do with its historical, contingent character. Torretti embraces the influence of historicism in the philosophy of science, broadly predominant from the 1960s to the 1980s. Scientists do not deal with scientific realism's ready-made world. By contrast, they face an experience of the world that is permeated by the conceptual and technological development of their own time, which inevitably sets a perspective on the kinds of phenomena they investigate. More than two decades after the publication of *Creative Understanding* in 1990, Torretti (2014, 26) revisits his views on the matter, this time quoting a passage from Kuhn:

> Finally, what replaces the one big mind-independent world about which scientists were once said to discover the truth is the very variety of niches within which the practitioners of these various specialties practice their trade. Those niches, which both create and are created by the conceptual and instrumental tools with which their inhabitants practice upon

> them, are solid, real, resistant to arbitrary change as the external world once was said to be. But, unlike the so-called external world, they are not independent of mind and culture, and they do not sum to a single coherent whole of which we and the practitioners of all the individual scientific specialties are inhabitants. (Kuhn, 2000, 120)

Our creative understanding does not freely float without touching the actual world. By contrast, we account for the workings of creative understanding keeping in mind the historical context. A theory responds to the challenges of its own time. It makes substantial use of previously available theoretical systems and technology, if relevant, to provide the best possible response granted certain conditions. Within this framework, concerning the nature of theories, Torretti maintains:

> When speaking about physical theories in general one ought to keep in mind that they issue from a peculiar mode of thought pertaining to a particular tradition. Surely, mathematical physics is the most impressive enterprise of human knowledge, and this is indeed a philosopher's chief motivation for reflecting on it. But it would be silly to expect that its methods can be fruitfully brought to bear on every conceivable subject of empirical research [. . .] Not sharing the faith in *unam sanctam catholicam scientiam* we have no business in reaching for a conception of scientific theorizing so broad that it is relevant to every field of study. To work on a single comprehensive notion of "scientific theory," capable of fitting such diverse creatures as Darwin's theory of evolution, Freud's theory of neuroses, and Schumpeter's theory of capitalist development, as well as the Salam-Weinberg theory of the electroweak interactions, seems pointless to me – except, perhaps, for the sake of invidiously pronouncing unscientific any purported theory which does not comply with it. (Torretti, 1990, 99)

Once more, as with the reading of Kantian objectivity, pluralism is a relentless feature of the creative understanding thesis and the view of theories it involves. We conceive of parcels of reality in diverse, sometimes incompatible ways. They result from a complex intertwining of theoretical, technological, and historical considerations, which are contingent to particular cultures and epochs. From this perspective, scientific practices take different forms, given the theories that each puts forward and the concomitant methodological assumptions they imply about how best to investigate the world. Although mathematical physics sets an example, it does not provide an exclusive style or threshold for thinking about the targets of other epistemic practices. The creative understanding thesis is inherently pluralist in this regard, hence partially anticipating what recent literature on scientific pluralism has shown in various ways (see Longino, 1990; Douglas, 2009; Chang, 2012, 2022; Massimi, 2022). We do as best as we can, provided the means we have and the historical context that motivates the problems we deem interesting.

2.4 Against Scientific Realism

Apart from his leaning towards pluralism and historicism in the philosophy of science, Torretti's reflection on the nature of science pays close attention to the details of scientific practice as much as to the historiographical context of specific disciplines, especially those of geometry and physics. As mentioned above,

Torretti's perspective on Kant's perspective on objectivity (to draw from Ryckman's expression in Chap. 4 in the present volume) directly opposes scientific realism. For the latter view, a key question is whether scientific theories refer to actual entities or to figments of the scientific imagination. In a sense, the scientific realist's faith suggests a belief in the targets of scientific discourse as though they were "real in a stronger sense than the things of everyday life" (Torretti, 2000, 113). His proximity to historicism and the turn to practice in the philosophy of science motivated Torretti to distance himself from traditional scientific realism. Even though scientific realists usually call themselves materialists, being closer to Hobbes than to Descartes, Torretti believes they remain frozen in the prejudices rooted in the seventeenth century. Their talk about the external world makes them appear as if they were disembodied spirits reflecting on reality from the outside, aiming at achieving God's eye point of view or at deciphering the true laws of nature that God imposes on the world. Likewise, scientific realists often assume that reality's existence is well defined independently from our cognitive activities, as if we could think about physical domains beyond our epistemic limitations. In this picture, one of the critical goals of science would be to provide a true theory that neatly depicts reality as a whole, cutting it at its joints, as Plato dreams in the *Phaedrus* (see Torretti, 2000, 114) – thus supposedly avoiding any traces of human intellectual activity.

Contrary to this, Torretti explicitly opposes scientific realism in some of his works. Nevertheless, he did not spend much time elaborating a systematic attack on scientific realism nor articulating an anti-realist conception of science. His views appear in a few of his publications, some of whose relevant passages I quote below. Nevertheless, hints of his opposition to scientific realism can be found throughout his extensive, detailed studies of the history and philosophy of physics and geometry, as much as in his historiographical approach to Cantor's set theory and his reflections on mathematical physics. In brief, these studies provide a consistent picture illustrating that an adequate understanding of science ultimately differs from what scientific realism has traditionally suggested. Here is a first passage in this direction:

> I contend, moreover, that scientific discourse is just the verbal aspect of scientific practice and has no serious justification apart from it; that this verbal aspect is no closer to the aim of science than, say, its manipulative aspects; that science does not have a primary aim but that with her, as with any other form of human activity, the distinction between means and goals is continually shifting from one context to another. [. . .] From this perspective, science is, if I may say so, a continuation of common sense by other means, as required by the overall pragmatic situation. (Torretti, 2000, 114–115)

Over the last two decades or so, Torretti has developed his own pragmatic realism in more detail, which, he submits, "best expresses the real facts of human knowledge and the working scientist's understanding of reality" (2000, 115). This leads to a particular form of pragmatic realism, which distances from philosophical preconceptions encapsulated in standard defenses of scientific realism, avoiding at once falling prey to constructivist inclinations. Torretti examines two standard arguments against scientific realism: the underdetermination of theory by evidence and the pessimistic meta-induction. He provides for each argument an example taken from the history of physics. The second of these arguments is particularly interesting, given Torretti's allegiance to pragmatism:

> The anti-realist tendencies among recent historians and sociologists of science draw their sustenance [...] from the *actual* succession of *real* theories most of which one already has decided against because they are not so good as the last. Inevitably the point is made that even the currently accepted theory is not good enough to be final and will in all likelihood be superseded later. The accurate theoretical representation of reality would thus be deferred forever. Against this conclusion it has been argued that, although science never quite hits the truth about reality, it comes closer and closer to it. This idea of truth by approximation can be held in two versions: either we assume that reality stands there ready-made outside the process of scientific research, which aims at finding the truth about it; or we conceive the true articulation of reality as the limit to which the succession of scientific theories converges, and which is constituted by this succession in the sense in which a real number is constituted by a Cauchy sequence of rationals. The second version is perfectly acceptable to the pragmatist [...] but the scientific realist can only countenance the first. (Torretti, 2000, 116)

In a similar line, an argument frequently employed by advocates of scientific realism points out the purported continuity between science and common sense. Such continuity takes, nevertheless, two different forms, namely: practical and theoretical. Firstly, when it comes to practical continuity, Torretti embraces the fact that scientific interaction with the environment is continuous with our ordinary interaction with medium-sized, ordinary objects, with the caveat that the former takes place in highly artificial, controlled scenarios, where measurements of various sorts and a range of technological devices are employed in order to obtain relevant information. Continuity from science to common sense can be accepted in this case. But secondly, concerning theoretical continuity, scientific realism suggests that scientific discourse is a continuation of commonsense, ordinary language. We talk about electrons just like we talk about trees. Hence, granted that we believe in the reality of the targets of ordinary language, we should likewise believe in the reality of the unobservable posits of scientific theorizing. Torretti (1999, 398–405) questions this move, highlighting the violence that mathematical physics does to commonsense conceptions of the world:

> Modern mathematical physics began in open defiance of common sense. Galileo declared – through his spokesman Salviati– that he could not "sufficiently admire the outstanding acumen" of the heliocentrist astronomers, who, "through sheer force of intellect," had "done such violence to their own senses as to prefer what reason told them over that which sensible experience plainly showed them to the contrary" (EN VII, 355; Drake translation). Furthermore, he judged color and sound, heat and cold to be mere affections of the human senses, like the tickling one feels when a feather is introduced into one's nose, which, of course, lies not on the feather but on the nerves stimulated by it (1623, §48). (Torretti, 1999, 398)

Physical discourse may retain some terms shared with commonsense expressions, mainly when they can be expressed with elementary arithmetic and geometric notions. Yet, modern mathematical physics has entailed a "wholesale dismissal of core ingredients of ordinary language right at the start of modern physics as being only a first step, a preparation for and anticipation of what would come later."

(Torretti, 1999, 398–399) The rejection of scientific realism is clear: advocates of this view would believe that reality is well-defined or ready-made, as implied in a monotheistic conception of God (Torretti, 2000, 14). However, it is still worth doubting that human agents can fully articulate such a view, should there be any. Torretti dubs such pretension "acute provincialism" (Ibid.). By contrast, epistemic humility encourages us to adopt a pragmatic stance. Pragmatists, Torretti contends, "do not need to feed on fanciful expectations and can make do from day to day with what is really there, because they readily accept that physics, like every other major human enterprise, is a patchy, makeshift affair" (Torretti, 2000, 120).[2]

2.5 Mathematical Fictionalism

Torretti's critique of scientific realism has a natural continuation in his elaboration of mathematical fictionalism (see Bueno, Chap. 6 in this volume). Particularly, in his analysis of Bunge's mathematical fictionalism, Torretti distinguishes three forms that this view can take, which, for the sake of clarity, he calls *mathematical fictionalism*$_1$, *mathematical fictionalism*$_2$, and *mathematical fictionalism*$_3$. The first holds that mathematical entities are thoroughly fiction. In Bunge's words, as quoted by Torretti (1981, 400):

> We do not claim that they exist in themselves but only that it is often convenient (for example in mathematics but not in metaphysics) to feign or pretend that they do. We do not accept that the Pythagorean theorem exists anywhere except in the world of phantasy called 'mathematics,' a world that will go down with the last mathematician. (Bunge, 1974–, II, 85).

Let us call this *radical mathematical fictionalism*, which denies mathematical existence and makes mathematics dependent on mathematicians' imagination.

The second form, mathematical fictionalism$_2$, is derived from what Bunge (1974–, II, 166) calls *literalists* as follows:

> If someone says that we feign that there are constructs, to which our mathematical statements refer, he will most naturally be understood to mean that we produce objects in our fancy, which thereafter constitute definite, stable, albeit ghostly referents of our discourse. It is in this sense that some atheists maintain that there is a fantastic being, created by men in their own likeness, to whom the majority of mankind refer in their persistent talk about God. It is in this same sense that literary authors are usually said to create the characters about which they write in their books. (Torretti, 1981, 400)

[2] Those who are familiar with Peirce's early elaborations of pragmatism, which recognize the role of scientific communities in the fixation of beliefs about the world, and Putnam's rejection of the scientific realist's pretension to achieve a God's eye point of view, will readily note their influence on Torretti's views. An additional, late influence is Hasok Chang's elaboration of pluralism and pragmatism in *The Invention of Temperature* (2004), which stirred up Torretti's most recent investigation on philosophy's contribution to scientific progress. See Sect. 2.7 below.

Moreover, the third form is mathematical fictionalism$_3$, which is particularly apt to account for applied mathematics and the roles of mathematical idealizations in physical sciences. Here is Torretti's characterization:

> When asked to calculate the period of a given pendulum at a given place, we feign that the pendulum hangs from a weightless, inextensible string, and that the sinus of the angle of displacement is equal to the angle itself, and rounding up the values of the pendulum's length and the local acceleration of gravity, we compute 2π times the square root of their quotient to an agreed decimal. From our fictitious assumptions we derive a result which is admittedly false, but which will differ from the measured values of the period by less than an agreed decimal. (Torretti, 1981, 401)

According to mathematical fictionalism$_3$, "one 'approximates' a real situation by an unreal one that is mathematically more manageable" (Torretti, 1981, 412). It amounts to a separate form of mathematical fictionalism since it provides both a way of understanding mathematical representations in terms of fictions and a way to make sense of the usefulness of mathematics in physical sciences. Hence, mathematical fictionalism$_3$ provides a range of possibilities for the view to take a more specific form. Considering this early exploration in mathematical fictionalism, one may be inclined to conjecture that Torretti leans towards some form of this view, hence opposing Platonist and Pythagorean projects. However, evaluating which of the three forms best represents his own view is not a straightforward task.

More than 30 years later, in 2014, Torretti resumes his reflections on mathematical fictionalism.[3] In this new manuscript, he confirms his rejection of mathematical Platonism (and for this matter, we should add mathematical Pythagoreanism as well), offering a nuanced articulation of his view combining features of mathematical fictionalism$_2$ and mathematical fictionalism$_3$. He distinguishes between a maximalist (strong) and a minimalist (weak) sense of existence. The former exclusively applies to physical bodies, and it leads the Platonist to lucubrate about the supposed reality of mathematicalia. The minimalist conception, however, appeals to the way in which language, both natural and formal, fixes the referents of the terms. According to Torretti, this happens equally in literary narratives and mathematical theorizing, where language is employed to individuate and describe the properties and relations of objects. Torretti suggests that the minimalist sense of existence allows us to submit that mathematical objects exist, but just in the sense that we claim that literary fictions exist.

Underpinning this contention, we find this: for something to exist in the relevant, minimal sense, *it suffices for us to be able to say that it exists*. Here is Torretti:

> Unfortunately, the inevitable presence of bodies, and of physical objects and processes in general, and our justified fixation with them –after all, their behavior usually represents a matter of life or death for us– incline us to believe that the general features of their way

[3] This manuscript from 2014 was originally written in Spanish, and no English translation is available as of now (2023). The manuscript is particularly relevant since it yields a substantive account of Torretti's views on the philosophy of mathematics, especially concerning problems about the ontological status of mathematical posits. I shall translate a few passages in the remainder of this section.

> of being belong to every existence; and hence, that what does not exist in this 'strong' sense, properly does not exist. Such inclination is, I would say, the root of a philosophical requirement stating that everything that exists needs to fulfill criteria that are similar to those that are applied to certify that a body, which we appear to have seen, is neither a mirage nor an illusion; or that a physical object that we believe we have discovered is not an artifact of our own experimental procedure or an improper interpretation of observations. This maximalist interpretation of the verb 'to exist' messes up literary works and mathematical structures into a bunch of philosophical problems, which under the suggested broad and generous interpretation – which I will call minimalist – would not be more than pseudo-problems. (Torretti, 2014, 125–126 – my translation)

Accordingly, the minimal requirement for something to exist is that it is effectively possible to mention it. After all, without mentioning the object, it is not possible to say that it exists. With this, Torretti seeks to bypass standard criteria for existence, which are broadly restrictive, such as those claiming that for something to exist, it must be independent of our discourse and mental life. By contrast, he argues that the existence we attribute to mathematical entities when we reason about them is neither less certain nor less stringent than everyday, commonsense existence. As he suggests: think of linguistic objects, such as characters and words. They do not have by themselves causal powers in the maximalist sense of existence. But they do play a role in our mental and social life. Accordingly, Torretti (2014, 130) asks: why should we assume that mathematical objects have causal powers? They do not, and we still recognize them and pay attention to them even though they are impassible and inactive.

Torretti outlines one challenge that the minimalist conception of existence is deemed to face:

> for the ontological minimalism I propose, it suffices that something is mentionable for it to be acknowledged as existing, and hence being possible to reason about it reasonably. Nevertheless, this liberality, which at first sight is so big, may not be enough for modern mathematics, which posits the existence of collections of objects which are by definition so immense that not even an immortal intelligence could generate names for every single one of them. (Torretti, 2014, 140 – my translation)

There are more mathematical posits than those we can mention. The scope of mathematical language finds a limit in our ability to display them linguistically. If we focus on the case of mathematical practice, what we encounter is a complex net of theorizing that includes an infinity of objects which we will never be able to mention, given the finitude of our lives and restrictions in cognitive skills; likewise, we may never be interested in mentioning some others due to the irrelevance of so doing in view of mathematical practice. Should we grant Torretti's minimalist criterion for existence, a distinction would need to be introduced in the world of mathematics, separating those mathematical posits that we have come, or will come, to mention from those that have not yet deserved (or will never deserve) this honorific treat. However, we should refrain from settling such distinction as a matter of principle.

2.6 Physical Laws

A crucial aspect of Torretti's views on the philosophy of science concerns his reflections on physical laws and necessity, which appear in various places throughout his works. Among others, chapter 5 of his *Creative Understanding* is entitled "Natural Necessity"; and a separate contribution is entitled "Mathematical structures and physical necessity" (Torretti, 1992). His insights into the character of physical laws and necessity develop along several lines, but they all contribute to shaping an articulated approach to core issues in the philosophy of laws of nature's recent developments.

In what follows, I shall refer to "Laws and Patterns" (Torretti, 1999, 405–420), where Torretti systematically articulates his take on physical laws and necessity. As usual in his philosophical style, he begins with a historiographical investigation aiming at tracing the path of laws of nature in the classical Greek texts and then in the early modern natural philosophy. He swiftly moves through the nineteenth century, hence arriving at what he takes to be the views of necessity and laws developed by the three main conceptions of scientific theories, namely: positivism, historicism, and structuralism. I shall briefly highlight some of the main points.

Here is a beginning in Plato's hands:

> We have all heard that the aim of physics is the discovery and formulation of the *laws of nature*. The phrase 'law of nature' is first attested in Plato's dialogue *Gorgias* (483e), where it is pointedly used as an oxymoron by Callicles, an otherwise unknown young Athenian rightwinger. Fifth-century sophists regularly opposed 'nature' (φύσις) and 'law' (νόμος), the latter being the product of human consensus and the defining mark of civilized society. [...] After asking rhetorically on what principle of justice Xerxes campaigned against Greece and Xerxes's father against the Scythians, Callicles offers this answer: 'They acted according to the nature of what is just and indeed, by Zeus, according to the very law of nature'. (Torretti, 1999, 405)

Torretti contends that the founders of modern physics would have taken Plato's metaphor literally in their quest for nature's law-like principles. The contention is interesting, although it is hard to follow a continuous path throughout historiographical documents. Indeed, the historiography about the origin of the metaphor *law of nature* rarely refers to Plato's *Gorgias* as a source for the concept of law of nature in natural philosophy. By contrast, jurisprudence, geometry, and theology are widely recognized among the cultural influences that blended to foster the emergence and consolidation of laws in the seventeenth and eighteenth centuries. In this regard, the role that Christianity played in the formation of this notion is well known to us. As Torretti points out,

> In the writings of these Christian authors the word 'law' does not signify the universal scope of the prescribed regularities, but rather the legislative authority of their divine source; thus, 'Kepler's Laws' became the standard name for what at most might qualify as local traffic regulations. However, all the founders shared the quaint belief that God, although infinitely powerful, is a paradigm of good husbandry and therefore has used the thriftiest means to achieve the richest variety and abundance of effects. They understood this to imply that all the lowly local laws of nature must follow from a few all-embracing ones. (Torretti, 1999, 406–407)

Descartes is among the first natural philosophers who provided us with a written document that formulates three laws of nature, whose scope of application is universal for every time and every place, and whose necessity is grounded in the immutability of God's will. And so did Newton as well, a few decades after Descartes. However, the Newtonian strategy took the consolidation of laws a step further, showing that Kepler's laws could be obtained from his second dynamical law and the law of universal gravitation. Just like Descartes, Newton appeals to the Judeo-Christian God, at once drawing from the deductive structure of demonstrations in Euclid's *Elements*.

Things, nevertheless, changed throughout the nineteenth century with the works of Herschel, Whewell, and Mill in Great Britain. Torretti rightly stresses this:

> For John Herschel (1830) a law of nature is either "a general proposition, announcing, in abstract terms, a whole group of particular facts relating to the behavior of natural agents in proposed circumstances," or "a proposition announcing that a whole class of individuals agreeing in one character agree also in another" (p. 100). [...] To find the laws of nature we must rely entirely on experience, on "the observation of facts and the collection of instances" (p. 118). (Torretti, 1999, 407)

According to this, the Newtonian spirit of Herschel's approach is close to the hypothetic-deductive method that moves from facts and observations to laws, and then derives propositions from these laws. Here is a passage from Herschel that nicely highlights this:

> The analysis of phenomena, philosophically speaking, is principally useful, as it enables us to recognize, and mark for special investigation, those which appear to us simple; to set methodically about determining their laws, and thus to facilitate the work of raising up general axioms, or forms of words, which shall include the whole of them; which shall, as it were, transplant them out of the external into the intellectual world, render them creatures of pure thought, and enable us to reason them out a priori. (Herschel, 1830, 97; as quoted in Torretti, 1999, 407)

Herschel's passages anticipate J. S. Mill's view of laws in a few years. Originally published in 1842, Mill's *System of Logic* submits:

> the question, What are the laws of nature?, may be stated thus: What are the fewest and simplest assumptions, which being granted, the whole existing order of nature would result? [...] thus: What are the fewest general propositions from which all the uniformities which exist in the universe might be deductively inferred?" (Mill, 1874, 230). (Torretti, 1999, 407–408)

Reflections along the lines of Herschel and Mill prepared the road for twentieth-century developments of the expression law of nature. But Torretti considers this development through the lenses of the structuralist tradition. A landmark contribution in this regard comes from Hilbert, who in personal correspondence writes to Frege:

> Of course every theory is only a scaffolding or schema of concepts together with their necessary mutual relations, and the basic elements can be conceived in any way you wish. [...] In other words: every theory can always be applied to infinitely many systems of basic elements. One needs only to apply an invertible one-one transformation and to stipulate that

> the axioms for the transformed things are respectively the same. (Hilbert to Frege, 29-12-1899; in Frege KS, pp. 412–13; see Torretti, 1999, 409)

In this view, a system of axioms does not correspond to a system of statements about something, but to a formal system of conditions that specifies a concept of a particular type. It does not, by itself, delimit a region of reality that would satisfy the requirements of the relational system. It only provides the conditions any system could fulfill, instantiating the relevant structural features. Hence, the structuralist view comes to the fore to replace the received view of laws as statements partaking in an axiomatic presentation of scientific theories. Here is Torretti's nuanced characterization of structuralism's central tenet:

> At the core of a physical theory there is always a coherent piece of mathematics that is intended to throw light on the processes and states of affairs in the theory's chosen field of study. Any coherent piece of mathematics can be articulated –as Hilbert did for Euclidean geometry– as the conception of a relational system or structure. Such a conception throws light on a physical domain when this is grasped as an instance or family of instances of the structure. (Torretti, 1999, 412)

According to this, laws play the role of the axioms of a deductive system: they are the principles that enable us to arrange the theory in a deductive fashion and explain phenomena capturing their relevant structure. Yet, the misleading sense of cleanliness and systematicity is a weakness of the structuralist construction of theories and laws. Constructing scientific theories is a formal game with no interest in the physical realm for the structuralist. Of course, the structuralist would claim that that is not the case, hence being forced to provide the tools for making sense of the actual construction of scientific concepts and theories, which is usually murky, full of rugged edges, and contingent. Torretti highlights this point and then goes on: "Physical theories do not fall like ripe fruits from a 'place beyond heaven' (Plato, *Phaedrus* 247c), but are gropingly fashioned by physicists on Earth. Their concepts are not all found ready-made on the smorgasbord of extant mathematics, but have often been created by the physicists themselves, or, at any rate, adapted to their needs" (1999, 414). Structuralism offers an idealized image of mature scientific theories to shed light on their scope and limits, but not an account of actual (i.e., historically contingent) theory construction processes.

In a sense, structuralism represents substantial progress from both theology-based speculations about the character of laws of nature and their place in scientific theories, and the received view claiming that scientific theories are sets of statements, whose laws were just the axioms of the system expressing a selection of regularities. On the face of it, structuralism invites us to look at the human face of the scientific endeavor, namely: we construct theories; we abstract from actual physical domains; and we build up idealizations in order to articulate representations, explanations, and predictions of phenomena. In this scenario, laws and necessity are not divine commands, nor are they the unique way of grasping generalizations about physical domains. We make up theories and systems, as much as we make up laws for spelling patterns of our interest in specific domains:

> Since the seventeenth century, physicists have been focusing on particular patterns that they isolate and grasp by means of suitably contrived mathematical concepts. Surely, they have thrived not only by their modesty, but also and chiefly by their uncanny eye for patterns. Thanks to it they ably discerned those features in actual phenomena that make together a distinctive pattern from the rest, which only mess it up, and they recognized structural affinities and identities where formerly one had seen only abysmal differences (e.g., between the fall of apples and the circulation of the moon). (Torretti, 1999, 415)

Patterns in nature came to be represented in terms of mathematical structures, and similarities between patterns came to be expressed in terms of morphisms of various sorts. Mathematical models replaced physical target systems, making the mathematics involved in high-level theorizing tractable. Moreover, to make possible contact, so to speak, between mathematical theory and the world, two things are in order: first, corrections are added to mathematical structures; and second, margins of error are admitted in our representation of physical domains. Only this makes possible the successful investigation of natural patterns in terms of mathematical structures expressing laws. Models instantiate mathematical structures, aiming at capturing the relevant structure of patterns in the world.

2.7 HPS and Scientific Progress

Motivated by Hasok Chang's (2004) landmark contributions to recent HPS, Roberto Torretti explores the problem of whether philosophical reflection can contribute to scientific progress. Over the years 2005 and 2006, Torretti wrote a draft of the manuscript "Can science advance effectively through philosophical criticism and reflection?" which he made public on the *PhilSci Archive,* University of Pittsburgh, in 2006. The manuscript is complex in many ways. To begin with, it sets out what Torretti conceives of as Chang's challenge:

> Chang claims that HPS can actually *generate* scientific knowledge in at least two ways. On the one hand, through the recovery of forgotten scientific knowledge, HPS can reopen neglected paths of inquiry. On the other hand, by applying the philosopher's scalpel to the thick and often opaque tissue of scientific discourse, HPS can positively contribute to clarifying or eliminating the confused, ambiguous or downright inept notions that bedevil innovative scientific thinking and appear to blunt its cutting edge. (Torretti, 2006, 2)

To address the challenge, Torretti's manuscript encompasses four separate investigations from an HPS perspective: first, the role of absolute space in Newtonian dynamics; second, rod contraction and clock retardation in special relativity; third, the reality of the electromagnetic ether over the eighteenth and nineteenth centuries, until Einstein's 1905 publications; and fourth, the problem (or set of problems) of the arrow of time. In each of these cases, Torretti examines whether philosophical reflection could have effectively enhanced the advancement of relevant scientific theories. Allow me to look into Torretti's analyses of the third case study.[4]

Here is the question: could philosophical reflection carried out by HPS practitioners have led the scientific community to the abandonment of the ether before Einstein's 1905 dictum? As is known, in "Zur Elektrodynamic bewegter Körper," Einstein (1987, 277, originally from 1905) claims: "The introduction of a luminiferous ether will prove to be superfluous." The case, in this regard, would run as follows. Even though the ether was an ontological presupposition that was not backed up by hard facts, it played crucial roles over the development of theories of electricity, magnetism, and astronomy, from the seventeenth to the very beginning of the twentieth century. Contrasting the lack of hard facts supporting the presupposition, the ether had a great history since the ancient Greek philosophy of nature, for which it meant the stuff making up heaven. Aristotle, for one, deemed it to be the fifth, immutable element that helped explain heavenly appearances. Furthermore, this presupposition smoothly passed into the hands of Judeo-Christian beliefs in the sixteenth century, and from then onwards, as we will recount below. The point is this: such cultural forces would have made it strongly unlikely that the sole philosophical reflection could have effectively contributed to do away with the ether before 1905.

Torretti revisits the history of ether in physics, collecting a series of indispensable passages for a comprehensive understanding of the ether's place in the construction of the world from Tycho and Descartes to Lorentz and Poincaré. Torretti argues:

> The conception of ether as a fluid, transparent, extremely subtle form of matter that thoroughly fills the vast interstellar and interplanetary spaces is usually attributed to Descartes. In the *Principles of Philosophy*, Descartes consistently refers to this form of matter as 'the second element' (1644, III, §§ 52, 70, 82, 123, 1996, 8: 105, 121, 137, 172). However, in the same year in which this book was published, Sir Kenelm Digby used 'aether' in English to designate the transparent interstellar matter (*Oxford English Dictionary*, s.v. ether, 5.a), and the term was later employed in this sense, as a matter of course, both in England and in the Continent. According to Descartes, this form of matter is immensely more abundant than the two other kinds acknowledged by him: the opaque matter of the Earth and the other planets, and the radiant matter of the Sun and the stars. (Torretti, 2006, 18–19)

According to Descartes and Huygens, the ether was supposed to be the carrier of light either as a rigid solid or as an elastic fluid, respectively, involving instant or finite velocities in each case. But it also occupies a central place in explaining planetary motions and free fall. Torretti highlights:

> In Newton's *Principia*, much of Book II was designed to prove that the phenomena of the Solar System cannot be accounted for by ether vortices (not, at any rate, under Newton's Laws of Motion). And Newton's well-known declaration that he does not feign hypotheses, coming as it does right after his admission that he has been unable to assign a cause to

[4] This section of the PhilSci Archive manuscript was later published as a separate article in Torretti (2007c). Likewise, the PhilSci Archive document was translated into Spanish in the book (Torretti, 2008b).

> gravity, can surely be read as a gibe at the ether-based Cartesian theories. In Part III, §44, of his *Principia,* Descartes declared that he did not claim to have found the "genuine truth" concerning the physical questions he dealt with, and that what he would henceforth write about them should be understood "as an hypothesis" (the French translation adds: "that is perhaps very far from the truth"; as I pointed out in footnote 24, this is probably from Descartes own hand). Nevertheless, if every consequence inferred from such an hypothesis fully agrees with experience, 'we shall gather from it no less utility for life than from the knowledge of truth itself' (1644, III, §44; 1996, vol. 8, p. 99). In unwitting (or was it deliberate?) opposition to Descartes' words, Newton's First Rule of Philosophy prescribes that no causes of natural things should be admitted *unless they are true* (Newton, 1726, p. 387; 1999, p. 794). (Torretti, 2006, 20)

Torretti maintains that Newton's methodology undermined the ether, which could only exist hypothetically. The predominance of Newtonian mathematical physics led to calling into question its presupposition. "By 1771 the enlightened founders of *Encyclopedia Britannica* thought it appropriate to explain 'ether' as "the name of an imaginary fluid, supposed by several authors [. . .] to be the cause [. . .] of every phenomenon in nature" (2006, 21). Along the same lines, Torretti reminds us of Priestley's words:

> Here the imagination may have full play, in conceiving of the manner in which an invisible agent produces an almost infinite variety of visible effects. As the agent is invisible, every philosopher is at liberty to make it whatever he pleases, and ascribe to it such properties and powers as are most convenient for his purpose. (Priestley, 1775, vol. 2, p. 16; quoted by Laudan, 1981, p. 159)

However, the ether hypothesis was resilient enough to continue to play a central role in various branches of the physical sciences developing throughout the eighteenth and nineteenth centuries:

> In his second paper, "On Physical Lines of Force" (1861/62), Maxwell valiantly proposes a mechanical model of the ether. Electricity and magnetism depend on the presence of molecular vortices in it. The model allowed the existence of transverse waves propagating in the ether with a speed equal to the ratio between the electrostatic and the electromagnetic unit. Back to town from the country house where he worked out this result, Maxwell verified that this value, as established by Weber and Kohlrausch (1857) differed by less than 1.5% from the speed of light, as it was then known. (Torretti, 2006, 30)

Maxwell's relevant passage reads as follows: "The velocity of transverse undulations in our hypothetical medium [. . .] agrees so exactly with the velocity of light [. . .] that we can scarcely avoid the inference that *light consists in the transverse undulations of the same medium which is the cause of electric and magnetic phenomena.*" (Maxwell, 1890, 1, 500) And Torretti observes:

> There was, however, one significant philosophical reason for Maxwell's retention of the ether hypothesis. He expected to account for electric currents and electrostatic charge distributions as epiphenomena of ether dynamics. This would spare physicists the need to postulate one or two special electric fluids (as they did in the 18th century) or to acknowledge electric charge as a primitive property of matter (as they have been doing since the 1890s). Maxwell's ether eludes our senses and was endowed by him and other researchers with either a far-fetched or an altogether unperspicuous mechanical structure, but it was only assigned properties or relations that could be conceived in classical mechanical terms. (Torretti, 2006, 32)

Poincaré (1968, 215, originally from 1889) is among those who had foreseen the possibility that the ether would be rejected at some point. Yet, the ether's resilience by the end of the nineteenth century was stronger than the philosophers' ideas about the possibility of doing away with it. Poincaré himself adduces a reason for holding our belief in the ether in his lecture at the Paris Congress of Physics of 1901. Here is Torretti's (2006, 34–35) translation of Poincaré:

> We know where our belief in ether comes from. If we receive light from a distant star, for several years that light is no longer at the star and is not yet on the Earth. It must therefore be somewhere, sustained, so to speak, by some material support. The same idea can be expressed in a more mathematical and abstract way. What we record are the changes suffered by material molecules; we see, for example, that our photographic film displays the consequences of phenomena staged many years earlier in the incandescent mass of the star. Now, in ordinary mechanics, the state of the system under study depends only on its state in an immediately preceding state; the system therefore satisfies differential equations. But if we did not believe in the ether, the state of the material universe would depend not only on the immediately preceding state (*l'état immédiatement antérieur*), but on much older states; the system would satisfy finite difference equations. To avoid this derogation of the general laws of mechanics we have invented the ether. (Poincaré, 1968, 180–181)

Let us go back to Torretti's take on Chang's challenge, which asks whether HPS practitioners can effectively contribute to the progress in scientific knowledge. Remember that this can take two non-exclusive forms: first, the cultivation of HPS can help us reopen (or keep open) lines of research that were abandoned in scientific practice; and second, the philosophical and historiographical analyses developed by an HPS practitioner may yield reasons for eliminating or clarifying concepts that undermine cutting edge research due to their obscurity or inanity. Torretti's response to this twofold challenge differs from Chang. While the latter positively builds an optimistic case for the interaction between scientific practice and HPS, Torretti suggests a more cautious interpretation considering two different aspects.

First, concerning the role of HPS, Torretti observes that an HPS practitioner, fully familiarized with the tools of philosophical thinking, could have perceived by the mid-nineteenth century that the presupposition of the ether was idle. Nevertheless, "William Thomson, Lord Kelvin, who certainly was such a thinker, remained during the next half-century ether's most adamant advocate." (Torretti, 2006, 39). The limit that such critical thinkers faced was imposed by various cultural dimensions of their era, such as the long-standing Greek tradition in Western culture, the association between the ether and the celestial realm of Christianity, and other intellectual forces predominating throughout the eighteenth and nineteenth centuries. The sole presence of HPS departments over that period could hardly have debunked the belief in the reality of some form of ether. In brief, Torretti submits a somewhat cautious attitude towards the possibility that "a greater presence of HPS in 19th century academic life could have significantly speeded the abolition of ether." (Torretti, 2006, 39).

Moreover, second, while acknowledging that the rejection of the ether represented a significant improvement in the advancement of physics by 1905, Torretti points out that many historians of science could still show that the presupposition of the ether had a beneficial influence on the growth of various physical theories about, among other things, planetary motion, light, electricity, magnetism, and else, hence representing a fortunate, though dated and soon-to-be-eliminated addition to the scientific ontology for over a few hundred years. In this sense, "it was perhaps just right that it survived as long as it did" (Torretti, 2006, p. 39). Although spurious, the ontological presupposition of ether was intellectually stimulating and came to be crucial for addressing a plethora of physical conundrums.

2.8 Conclusion

Let us summarize Torretti's key tenets in the philosophy of science as follows:

- First, beginning with his scholarly work on Kant's philosophy, Torretti develops a Kant-inspired conception of objectivity, enabling him to address a range of issues in the philosophy of science in new ways.
- Second, the Kantian background places Torretti in the right position to develop his creative understanding thesis, which provides a framework for addressing concerns about observation, the structure of theories, necessity, etc.
- Third, Torretti's philosophy of science involves a rejection of scientific realism, particularly of the thesis that scientific investigation cuts nature at its joints and that we inhabit a ready-made world that we can objectively know independently of our epistemic endowment. Contrary to this, Torretti primarily endorses pluralist and pragmatist stances that grant priority to the peculiarities of scientific practice.
- Fourth, along with his rejection of scientific realism, Torretti articulates a form of mathematical fictionalism that pays attention to our use of linguistic discourse and how we refer to things in ordinary and formal language.
- Fifth, Torretti addresses the concept of physical laws in his characteristic style, tracing the expression back to its roots in ancient Greece, passing through its developments over the seventeenth, eighteenth, and nineteenth centuries, and ending with a critical approach to the structuralist interpretation of physical laws as mathematical formalisms capturing patterns of phenomena.
- And sixth, in his most recent work, Torretti engages with Chang's contributions to HPS, assessing whether science can effectively advance through philosophical reflection by recovering abandoned epistemic practices in the history of particular sciences or by eliminating obscure, idle concepts.

Torretti's philosophical work is vast and complex. His studies of Kant's philosophy continue to be of interest to Kantian scholars, and his investigations in the history and philosophy of geometry and physics have rightfully earned a central place in current debates. In this chapter, our analyses have provided a broader window into his philosophical understanding of science, focusing on elements that help us gain further insight into Torretti's philosophical views.

The following image emerges from our analyses throughout the previous sections, viz.: Roberto Torretti truly exemplifies a philosopher who forcefully opposes any form of dogmatic philosophical fundamentalism. As he maintains,

> In philosophy, things are said to be as we understand them to be, but we are well aware that they might not be that way. Such awareness, however, does not result from our transcending our understanding and glimpsing at things beyond it. It simply expresses our discontent with our views and thoughts, which we feel to be incomplete, murky, or plainly inconsistent. But improvement can only be had by thinking harder, and we alone must see to that. (Torretti, 1990, 7)

These lines provide a direct entry into Torretti's philosophical world. His philosophy is thoroughly secular in every sense: no strong metaphysical presuppositions are introduced to make his worldview stable and pleasant, and his ideas duly pay attention to the details of the relevant sciences in each case, hence being tirelessly crafted through the combination of mathematical and scientific knowledge, along with a deep commitment to historiographical research and philosophical thinking. He advances a philosophy with a human face, which always illustrates that we only have access to a restricted perspective of what things are and how they work. At every step, Torretti is more interested in expounding his ideas in a clear, analytic fashion rather than defending them from external criticism. This chapter has conveyed an impression of such central features of Torretti's philosophy of science.

References

Bunge, M. (1974–). *Treatise on basic philosophy.* D. Reidel.

Chang, H. (2004). *Inventing temperature: Measurement and scientific progress.* Oxford University Press.

Chang, H. (2012). *Is water H_2O? Evidence, realism, and pluralism.* Springer.

Chang, H. (2022). *Realism for realistic people. A new pragmatist philosophy of science.* Cambridge University Press.

Descartes, R. (1644). *Principia Philosophiae.* Elzevier. (Reprinted in Descartes, 1996 *Œuvres.* Publiées par Charles Adam & Paul Tannery. Vrin, vol. 8, pp. 1–348).

Douglas, H. E. (2009). *Science, policy, and the value-free ideal.* University of Pittsburgh Press.

Einstein, A. (1934). *Mein Weltbild.* Querido Verlag.

Einstein, A. (1987) *The collected papers of Albert Einstein.* Princeton University Press.

Frege, G. (KS/1967). *Kleine Schriften* (Herausgegeben von I. Angelelli). Wissenschaftliche Buchgesellschaft.

Herschel, J. F. W. (1830). *A preliminary discourse on the study of natural philosophy.* Longman, Rees, Orme, Brown, 6t Green, and John Taylor (facsimile reprint: University of Chicago Press, 1987).

Kant, I. Ak. Kant's gesammelte Schriften, Deutschen Akademie der Wissenschaften zu Berlin (volumes 1–23) and Akademie der Wissenschaften in Göttingen (volumes 24, 25, 27–29), Berlin 1902–.

Kuhn, T. (2000). *The road since structure: Philosophical essays, 1970–1993, with an autobiographical interview.* University of Chicago Press.

Laudan, L. (1981). The medium and the message: A study of some philosophical controversies about ether. In G. N. Cantor & M. J. S. Hodge (Eds.), *Conceptions of ether: Studies in the history of ether theories, 1740–1900* (pp. 157–185). Cambridge University Press.

Longino, H. (1990). *Science as social knowledge. Values and objectivity in scientific enquiry*. Princeton University Press.
Massimi, M. (2022). *Perspectival realism*. Oxford University Press.
Maxwell, J. C. (1890). *The scientific papers of James Clerk Maxwell* (W. D. Niven, Ed.). Cambridge University Press. 2 vols. (unaltered reprint: Dover, 1965).
Mill, J. S. (1874). *A system of logic ratiocinative and inductive, being a connected view of the principles of evidence and the method of scientific investigation* (8th ed.). Harper & Brothers.
Newton, I. (1726). *Philosophiae naturalis principia mathematica*. Editio tertia aucta *&C* emendata. Londini: Apud Guil. & Joh. Innys, Regiae Societatis typographos.
Poincaré, H. (1968). *La science et l'hypothèse*. Flammarion. (First edition: Flammarion, 1902).
Priestley, J. (1775). *The history and present state of electricity: With original experiments* (3rd ed., corrected and enlarged). C. Bathurst and others. 2 vols.
Torretti, R. (1981). Three forms of mathematical fictionalism. In J. Agassi & R. S. Cohen (Eds.), *Scientific philosophy today* (pp. 399–414). D. Reidel.
Torretti, R. (1990). *Creative understanding: Philosophical reflections on physics*. The University of Chicago Press.
Torretti, R. (1992). Mathematical structures and physical necessity. In J. Echeverría, A. Ibarra, & T. Mormann (Eds.), *The space of mathematics: Philosophical, epistemological, and historical explorations* (pp. 132–140). Walter de Gruyter.
Torretti, R. (1999). *The philosophy of physics*. Cambridge University Press.
Torretti, R. (2000). »Scientific realism« and scientific practice. In E. Agazzi & M. Pauri (Eds.), *The reality of the unobservable: Observability, unobservability and their impact on the issue of scientific realism* (pp. 113–122). Kluwer.
Torretti, R. (2006). *Can science advance effectively through philosophical criticism and reflection?*http://philsci-archive.pitt.edu/archive/00002875/
Torretti, R. (2007c). Getting rid of the ether: Could physics have achieved it sooner, with better assistance from philosophy. *Theoria: An International Journal for the Theory, History and Foundations of Science, 60*, 253–274.
Torretti, R. (2008a). Objectivity: A Kantian perspective. In M. Massimi (Ed.), *Kant and the philosophy of science today* (pp. 81–95). Cambridge University Press. See the Spanish translation in Torretti 2010, pp. 13–32.
Torretti, R. (2008b). *Crítica Filosófica y Progreso Científico*. Ediciones Universidad Diego Portales.
Torretti, R. (2010). *Estudios Filosóficos 2007–2009*. Ediciones Universidad Diego Portales.
Torretti, R. (2014). *Estudios Filosóficos 2011–2014*. Ediciones Universidad Diego Portales.
Weber, W., & Kohlrausch, R. (1857). *Elektrodynamische Maassbestimmungen insbesondere Zurückührung der Stromintensitäts-Messungen auf mechanische Maass*. K. Sächsische Gesellschaft der Wissenschaften zu Leipzig; math.-phys. Cl. Abhandlungen. Reproduced in Wilhelm Weber, Werke. Springer, 1892–1894, 3: 609–676.

Chapter 3
Du Châtelet on Absolute and Relative Motion

Katherine Brading and Qiu Lin

Abstract In this chapter, we argue that Du Châtelet's account of motion is an important contribution to the history of the absolute versus relative motion debate. The arguments we lay out have two main strands. First, we clarify Du Châtelet's threefold taxonomy of motion, using Musschenbroek as a useful Newtonian foil and showing that the terminological affinity between the two is only apparent. Then, we assess Du Châtelet's account in light of the conceptual, epistemological, and ontological challenges posed by Newton to any relational theory of motion. What we find is that, although Du Châtelet does not meet all the challenges to their full extent, her account of motion is adequate for the goal of the *Principia*: determining the true motions in our planetary system.

Keywords Du Châtelet · Absolute motion · Relative motion · True motion · Musschenbroek · Newton

3.1 Introduction

Émilie Du Châtelet's principal work, her *Foundations of Physics*, was first published in 1740: fourteen years after the third edition of Newton's *Principia*; four years after Euler's *Mechanica*; three years before d'Alembert's *Treatise on Dynamics*; and eight years before Euler's "Reflections on Space and Time". The central theme of all these texts is the motion of bodies. More specifically, these texts intersect in the philosophical space associated with the following problem of bodily motion: given the initial motions of a collection of bodies, what will their motions be at a later time? This apparently simple problem in physics was, at the time, inextricably embedded in a web of metaphysical, epistemological, and conceptual difficulties. Among these difficulties lies the debate over absolute space, time and motion, with

K. Brading (✉) · Q. Lin
Department of Philosophy, Duke University, Durham, NC, USA
e-mail: katherine.brading@duke.edu; Qiu.Lin@Duke.edu

C. Soto (ed.), *Current Debates in Philosophy of Science*, Synthese Library 477,
https://doi.org/10.1007/978-3-031-32375-1_3

the Newtonians on one side, advocating an "absolute" conception of space, time and motion, and the Leibnizians on the other, advocating a "relational" one. In this chapter, we situate Du Châtelet's account of motion in the context of the absolute versus relative motion debate. In our view, Du Châtelet's account is an important contribution to the history of this debate in the eighteenth century.[1]

One of us has argued elsewhere (Brading, 2019) that Du Châtelet modelled her *Foundations* on the textbooks of such figures as 's Gravesande (1720), Musschenbroek (1734), and Pemberton (1728). Against this background, the most striking thing about the book is its non-Newtonian elements, and especially the Leibnizian themes. As noted in the literature, these themes include Du Châtelet's versions of the principle of sufficient reason and the law of continuity, her non-extended simples ("monads"), and her Leibnizian conceptions of force.[2] What has not been studied, however, are the less obvious ways in which Du Châtelet deviated from the Newtonian textbooks that were her model, and what these tell us about her own broader philosophical position. On the topic of motion, she made essential use of resources she found in Musschenbroek. Yet, as we will see, while Musschenbroek accepted Newtonian absolute motion, Du Châtelet did not.

Du Châtelet's rejection of Newtonian absolute motion comes as no surprise to those familiar with her views on space. In Chapter 5 of the *Foundations*, "On Space", she sides with Leibniz in rejecting absolute space and endorsing a relational view of space. But those who reject absolute space must deal with Newton's arguments as to why such a notion is necessary in order for the project of the *Principia* to proceed. For this project, Newton argued, we need a distinction between absolute and relative motion. We assess the extent to which Du Châtelet has the resources to meet the demands of the *Principia* without appeal to absolute space, and therefore without adopting Newtonian absolute motion. Spoiler: she is surprisingly successful.

---

[1] The history of space, time, and motion in the eighteenth century plays an important role in Torretti's work in philosophy of physics (see Torretti, 1999, and references therein). Situated between Newton and Kant, both temporally and philosophically, Du Châtelet should be of especial interest to philosophers of physics interested in this time period.

[2] See Iltis (1977) and Janik (1982) for the view that what Du Châtelet seeks to provide in the *Foundations* are Leibnizian foundations for Newtonian physics, and Brading (2019) for a different assessment, according to which the basic foundational problem Du Châtelet attempts to address is not the lack of metaphysical foundation of Newtonian physics, but the lack of an epistemically secure basis for physical theorizing. See Stan (2018) for a useful discussion of Du Châtelet's metaphysics of substance, which emphasizes its Wolffian ingredients against the received view that Leibniz is the decisive influence. See Janiak (2018) for a discussion of how Du Châtelet utilizes the resources of her metaphysics to provide a treatment of the force of gravity, which she regards Newton as failing to offer. Also see Brading (2018) for a reconstruction of Du Châtelet's solution to the problem of bodies, which is a version of a Leibnizian solution that begins with non-extended simple beings. For discussions of Du Châtelet's views on *vis viva*, see Iltis (1977, pp. 38–45), Hutton (2004, pp. 527–29), Hagengruber (2012, pp. 35–8), Suisky (2012, pp. 144–6), Reichenberger (2012, pp. 157–71), Terrall (1995, pp. 296–8), Kawashima (1990), and Walters (2001). For a discussion of Du Châtelet's exchange with Mairan on the topic of *vis viva* in relation to Kant's early philosophy of matter and body, see Massimi and De Bianchi (2013) and Lu-Adler (2018).

3.2 In Search of True Motion

The principal aim of Newton's *Principia* is to determine the system of the world: Newton sought the true motions of the bodies comprising our planetary system, and thereby to adjudicate once and for all between the geocentric and heliocentric hypotheses. A prior question required attention: what is the appropriate definition of true motion? Famously, Newton argued in favor of *absolute* motion (motion with respect to absolute space and time) and against relative motion.[3] In particular, he thought that Descartes's definition of motion as relative to other bodies must be rejected. In the scholium to the definitions in Book 1 of his *Principia* (Newton, 1999, pp. 408–15), Newton distinguished absolute from relative time, space, place, and motion, and argued that absolute rather than relative motion is needed for a physics of bodies in motion. He did so by comparing the properties, causes and effects of absolute and relative motion.

In *The Leibniz-Clarke Correspondence* (Alexander, 1956), Leibniz pushed back, rejecting Newton's conception of absolute motion and arguing for a relational conception instead. The exchange concerning absolute versus relative motion in these letters remains a source for ongoing debates today, with the balance of opinion weighing strongly in favor of absolute motion: Leibniz simply did not understand the requirements on a concept of motion adequate for the purposes of a theory of bodies in motion. This is the context for eighteenth century discussions of space, time and motion.

The focus of the debate over space and time has been primarily ontological: are space and time absolute or relative? However, as one of us has shown,[4] Du Châtelet shifts the debate into a different key. This forces us to parse Newton's arguments

[3] We distinguish true from absolute motion. In his discussion of Newton's scholium, Huggett (2012) argues that the terms "true motion" and "absolute motion" differ in meaning. We agree with Huggett that "absolute motion" means motion with respect to absolute space and time, but we disagree that the meaning of the term "true motion"—as distinct from "absolute motion"—is implicitly (partially) defined by the laws. True motion, in our view, is that motion which is proper to a body, and to assert that a body has a true motion is to assert that there is a unique motion proper to it. The next question is then whether that motion is absolute (i.e. with respect to absolute space and time) or relative (e.g. with respect to some unique privileged body or set of bodies). And so, in our view, it is motion simpliciter that is implicitly (partially) defined by the laws (for something to move just is for it to move in accordance with the laws of motion); the open questions of the *Principia* are whether that motion is true (whether there is a unique motion proper to a body), and if so, whether it is absolute. Newton's assertion in the scholium is that it is both. For further discussion of the interpretation of "absolute, true, and mathematical" see Brading (2017). Schliesser (2013) offers an alternative interpretation of the terminology for the case of time. While we do not have space to address these proposals in detail here, one advantage of the approach to the terminology that we are proposing is its consistency. Instead of "true" and "absolute" being treated differently for time as compared to motion, as they would be if we accepted both Schliesser's (2013) account for time and Huggett's (2012) account for motion, the terminology as we interpret it is uniform across time, space, place and motion.

[4] Lin, "Du Châtelet on the Representation of Space", ms.

against relational motion into three: conceptual, epistemological, and ontological. First, Newton sought to show that absolute motion is superior to relative in providing the conceptual resources necessary for a theory of true motion. Second, Newton used these resources to pursue the epistemological project of determining true motions (and, in particular, the true motions of the bodies in our planetary system). Third, Newton used the ontological status of absolute space and time to underwrite the conceptual distinctions that make the epistemological project possible.

In what follows, we discuss Du Châtelet's definitions of motion in light of this context. As we will see, she offers a threefold taxonomy of motion—"absolute motion", "common relative motion" and "proper relative motion"—using terminology she seems to have adopted from Musschenbroek. However, whereas Musschenbroek endorsed Newtonian absolute space, Du Châtelet did not, and this leads to important differences between their treatments of motion, as we shall see. We use Musschenbroek as a useful foil for explicating Du Châtelet's account of motion.[5]

With Du Châtelet's account of motion on the table, we then turn our attention to the conceptual (Sect. 3.3), epistemological (Sect. 3.4), and ontological (Sect. 3.5) challenges posed by Newton. Ultimately, the test of Newton's account of motion is its success in delivering on the main goal of the *Principia*: determining the true motions of the bodies in our planetary system. With our examination of Du Châtelet's account of motion in hand, we assess whether she has the resources to meet this demand.

3.2.1 Motion and Change of Place

Du Châtelet opens her chapter on motion (Chapter 11 of the *Foundations*) with the following definition (§211): Translations are from Du Châtelet 2009 and Du Châtelet 2018.

> Motion is the passage of a Body from the place that it occupies into another place.

By itself, this definition is neutral between absolute and relative motion; we need also a definition of "place". In the *Principia*, Newton distinguished between absolute

[5] Musschenbroek used this terminology in a series of texts in the 1730s (see, for example, Musschenbroek, 1734 and 1739). We use his *Elementa Physicae* of 1734 as our source. Our quotations and references are to the 1744 English translation, which is a translation of a later, expanded, version of the 1734 Latin original. Multiple versions of Musschenbroek's text, which are based on his lecture notes, were published under a variety of different titles. We have compared the relevant passages from the 1744 English translation to the 1734 Latin edition of *Elementa Physicae*, and also to a 1739 French translation of a similar Musschenbroek text, to ensure that the Musschenbroek materials we cite would indeed have been available to Du Châtelet during the time she was writing her *Foundations*, if not exactly as quoted here, then as close as is necessary for the points that we wish to make.

and relative place,[6] that distinction in turn being parasitic on the distinction between absolute and relative space. If Du Châtelet had adopted Newton's account of space, and thereby of place, then her definition of motion would have yielded Newtonian absolute motion. But she did not.

In Chapter 5 of the *Foundations*, immediately after her rejection of absolute space, Du Châtelet defined "place" as follows (§88):

> We call the location or the place of a Being its determined manner of coexisting with other Beings.

This is a relational definition of location or place, in which the place of a being depends (in some way) on its relations to other beings. She explains as follows (§88, continued):

> Thus, when we pay attention to the manner in which a table exists in a room with the bed, the chairs, the door, etc., we say that this table has a place; and we say that another Being occupies the same place as this table when it obtains the same manner of coexisting that the table had with all the Beings.
>
> This table changes place when it obtains another situation with respect to the same things that we regard as not having changed place at all.

This relational approach to place is consistent with her rejection of absolute space and her endorsement of a relational conception.[7,8]

Given Du Châtelet's relational definition of place, it seems we should understand her definition of motion (§211, see above) to be relational too. And this is right. But things turn out to be more complicated—and more interesting—than this simple claim suggests, as we shall now see.

3.2.2 Absolute Motion

Immediately following her definition of motion, Du Châtelet distinguishes motion into three kinds (§212): absolute motion, common relative motion, and proper

---

[6] "Absolute space, of its own nature without reference to anything external, always remains homogeneous and immovable. Relative space is any movable measure or dimension of this absolute space", and "Place is the part of space that a body occupies, and it is, depending on the space, either absolute or relative" (Newton, 1999, p. 409).

[7] Du Châtelet also distinguishes between location and place (§92), defining the place of a thing as the location of all its parts. She further defines situation (§93) as "the order that several coexistent but non-contiguous things maintain through their coexistence".

[8] Du Châtelet's account of space (see her Chapter 5) is extremely interesting in its own right, see Lin, "Du Châtelet on the Representation of Space" ms. Here, our interest is in her account of motion (in Chapter 11), and so we note her rejection of absolute space (as well as of absolute time, see her Chapter 6) and move on. See Hutton (2012) for a focused treatment of Du Châtelet's disagreements with Samuel Clarke, including the disagreement on the issue of space; see Jacobs (2020) for a comparative study of Du Châtelet's views on the ontology of space, extension, and bodies.

relative motion. In this, she is departing from Newton's own twofold distinction and is, we suggested above, following Musschenbroek (see his 1744, for example) in adopting a threefold terminology. However, in Musschenbroek's case, the corresponding distinctions have Newton's conceptions of absolute and relative motion as their source, for Musschenbroek endorses Newtonian absolute space.[9] He defines absolute motion as follows (§101):

> Absolute motion is the successive existence of a body in different parts of the space of the immovable universe.

Clearly, Musschenbroek is adopting a Newtonian conception of absolute motion.

At first sight, Du Châtelet seems to simply adopt Musschenbroek's definition, with the latter part of it modified to reflect her endorsement of a relational conception of space (§213):

> Absolute motion is the successive relation of a Body to different Bodies considered as immobile, and this is real motion, and properly so called.

Notice that this modification introduces terminology familiar from Descartes's definition of proper motion in his 1644 *Principles of Philosophy* II.25 (1991, p. 51):

> What movement properly speaking is. ... it is the transference of one part of matter or of one body, from the vicinity of those bodies immediately contiguous to it and considered as it rest, into the vicinity of others.

In particular, both Descartes and Du Châtelet offer us a definition of "proper" motion in which the standard of rest is provided by bodies that are "*considered* as immobile" or "at rest". However, notice too this important difference between Du Châtelet and Descartes: Du Châtelet's definition relaxes the contiguity condition on the bodies that provide the standard of motion (i.e. which are considered to be at rest). Both of these points will be important later on.

It seems that Du Châtelet has offered a definition of absolute motion in terms of relative motions among bodies, rather than with respect to absolute space. How is this anything other than an abuse of words? In the *Principia*, Newton distinguished absolute from relative motion precisely because he believed that no

---

[9] In the chapter preceding his discussion of motion, Musschenbroek argued for absolute space, independent of and distinct from any body or bodies, concluding in words that echo Newton's discussion of absolute and relative space in his *Principia* (Musschenbroek, 1744, §90, p. 55):

> The space of the universe is one, invisible, intangible, extended, of infinite amplitude, nor confined by any limits, homogeneous, always similar to itself, continuous, immovable, indivisible; and in which are no actual parts, but there may be accidental, which are intercepted between surfaces of bodies, and constitute relative space. Yet these cannot be seen, nor distinguished by our senses: therefore in their stead we use sensible measures, taken from the distances of bodies; and thus the parts are mensurable, though immoveable. The order of the parts is immutable, because space is one, immovable and indivisible. Moreover, it is penetrable by bodies without any resistance, containing all bodies within it, allowing them motion in and by itself.

relative motion among bodies was adequate for the purposes of physics: hence the need for introducing absolute motion as motion with respect to absolute space. Du Châtelet looks to be confused: she seems to use the words "absolute motion" to define a relational type of motion, not realizing that this defeats the whole purpose of introducing the terminology of absolute motion in the first place. In order to address this puzzle, we first need to take a closer look at what Du Châtelet has to say about relative motion.

3.2.3 Relative Motion

Du Châtelet persists with Musschenbroek's terminology, distinguishing absolute motion from two different types of relative motion: common relative motion and proper relative motion.

Consider first common relative motion. Musschenbroek writes (§102):[10]

> That is called motion relatively common, when a body carried on together with others, in respect of them keeps the same situation, and so seems to be at rest, yet together with those bodies passes through the several parts of universal space. With such a motion as this a mariner is carried, who sits at rest in his ship under sail. Or with such all things are moved that adhere to the surface of the earth, while it revolves about its own axis, and is carried around the sun. Or lastly, with such a motion a dead fish moves, which is rolled along with the stream.

Similarly, Du Châtelet writes (1740, §214):

> Common relative motion is that which a Body experiences when, being at rest with respect to the Bodies that surround it, it nevertheless acquires along with them successive relations, with respect to other Bodies, considered as immobile, and this is the case in which the absolute place of Bodies changes, though their relative place remains the same; and it is what happens to a Pilot, who sleeps at the tiller while his Ship moves, or to a dead fish carried along by the current of water.

Once again, she seems to have adopted Musschenbroek's definition, modifying it to reflect her rejection of absolute space and making explicit reference to the surrounding bodies.

In addition to common relative motion, Musschenbroek also introduces proper relative motion, writing (1744, §103):

> Motion relatively proper is a successive application of a body to the different parts of the bodies that immediately surround or touch it. With this motion all things seem to us to be carried, which in our earth we perceive to be moved.

For Musschenbroek, proper relative motion is with respect to the *immediately* surrounding bodies, and insofar as these bodies are taken to be at rest in evaluating

[10] The different word order is an artefact of the English translations being used here. Musschenbroek (1739) and Du Châtelet (1740) both use the two phrases "mouvement relatif commun" and "mouvement relatif propre".

the proper relative motion of a body, Descartes's "movement properly speaking" corresponds to Musschenbroek's proper relative motion. Yet again, Du Châtelet follows suit in adopting the terminology of "proper relative motion" while changing the content of the definition (1740, §215):

> Proper relative motion is that which one experiences when, being transported with other Bodies in a relative common motion, one nevertheless changes one's relations with them, as when I walk on a Ship that is sailing; for I change at every moment my relation with the parts of this Ship, which is transported with me.

Notice that she makes no reference to the immediately surrounding bodies and so, unlike for Musschenbroek, her definition of proper relative motion does not correspond to Descartes's "movement properly speaking".

Thus, notwithstanding the similarities in terminology, Du Châtelet's taxonomy of motion is very different from that of Musschenbroek, and the two views can be summarized as follows.

In Musschenbroek there is a primary distinction between absolute motion (which is the motion of a body with respect to absolute space and absolute time) and relative motion (which is the motion of a body with respect to other bodies). Within relative motion, there is a further distinction between common and proper. The relative motion that a body shares with some group of bodies, when moving with that group of bodies with respect to some other body or bodies, is their common (i.e. communal) relative motion. For example, the kernel and the shell of a nut may move together through the air when the nut falls from a tree, and this is their common relative motion (with respect to the air), and the kernel may also move within the shell (perhaps it has come loose and rotates within the shell), in which case the kernel has a proper motion relative to the shell, in addition to the common relative motion that it shares with the shell.

Like Musschenbroek, Du Châtelet claims a distinction between absolute and relative motion, as well as one between common and proper relative motion, but she defines all three types of motion in relational terms. In absolute motion, the reference bodies are considered immobile. In common relative motion, several bodies move together in absolute motion. In proper relative motion, a body not only moves together with other bodies in absolute motion, but also changes its relations with respect to those bodies. Therefore, despite the use of Musschenbroek's terminology, Du Châtelet has a very different account of motion. In particular, her account is thoroughly relational. What, then, is the true motion of a body, and how are we to find the true motions? In the remainder of this chapter, we examine the extent to which Du Châtelet's account is capable of addressing the challenges to a relational theory of motion posed by Newton.

3.3 The Conceptual Challenge: Properties, Causes and Effects

In his *Principia*, in the scholium to the definitions, Newton wrote (1999, p. 411):

> [A]bsolute and relative rest and motion are distinguished from each other by their properties, causes, and effects.

He then offered a series of arguments intended to show the superiority of his concept of absolute motion for the purposes of constructing a theory of matter in motion. Since Du Châtelet's account seems to admit only relative motion, despite her use of the term "absolute motion", our first question is whether her account allows her to make the conceptual distinctions that Newton argues for in his discussion of "properties, causes, and effects". With this in hand, we will then be in a position to assess whether Du Châtelet has the conceptual resources needed to carry out the epistemological and ontological work for which Newton appealed to absolute motion.

3.3.1 The Properties of Absolute and Relative Motion

We begin with the properties. It is here that Newton offers his famous nut example. He writes (1999, p. 411):

> It is a property of motion that parts which keep given positions in relation to wholes participate in the motion of such wholes. ... Therefore, when bodies containing others move, whatever is relatively at rest within them also moves. And thus true and absolute motion cannot be determined by means of change of position from the vicinity of bodies that are regarded as being at rest. ... For containing bodies are to those inside them as the outer part of the whole to the inner part or as the shell to the kernel. And when the shell moves, the kernel also, without being changed in position from the vicinity of the shell, moves as a part of the whole.

Newton's target here (as has been convincingly argued by Belkind (2007), see especially pp. 285–6) is Descartes, and the conflict Newton perceives between Descartes's definition of motion (as motion with respect to the immediately surrounding bodies themselves considered to be at rest) and the quantity of motion (as the product of bulk and speed) that he associates with a body (as needed for his rules of collision). In the case of the nut falling from the tree, only the shell moves relative to its immediately surrounding bodies, yet the total volume or bulk of the nut (the shell plus the kernel) contributes to the quantity of motion. How can something that is at rest (the kernel, which is at rest with respect to its immediately surrounding bodies) contribute to the quantity of motion of the nut? Newton's response is that if we define motion with respect to absolute space, rather than the immediately surrounding bodies, then the entire nut (the kernel plus the shell) is in motion, and both the kernel and the shell contribute to the quantity of motion of the nut. In short, according to Newton, a necessary condition on an adequate definition of motion is that the parts of a body in motion contribute to the quantity of motion of the whole.

Musschenbroek, in adopting Newton's definition of absolute motion, adopts a definition that meets this condition. Moreover, he makes the point about the relationship between the motion of a body and its quantity of motion explicitly (§§. 120–122, p. 65), asserting that for an extended body its motion is "equally distributed into all its parts" such that "the whole quantity of motion may be conceived alike divisible as the body, and in every part of the body it will be proportional to the magnitude of that part".

Interestingly, Du Châtelet is also able to meet Newton's condition. All parties grant that the nut is in motion (with respect to the air surrounding it, for example); the issue is the motion of the parts. Given Descartes's definition of motion, the kernel is at rest since it is at rest with respect to the immediately surrounding bodies, and so Descartes fails Newton's test concerning the motion of the parts. For Du Châtelet, however, the absolute motion of a body is not defined with respect to the immediately surrounding bodies, so she does not immediately fail Newton's test. Moreover, the kernel and the shell may be in common relative motion, even when the kernel is at rest with respect to the shell (and therefore has no proper relative motion). So Du Châtelet's definition of common relative motion allows her to evade Newton's objection. One might respond that unless Du Châtelet tells us *which* bodies we are supposed to take as our standard of rest, she cannot tell us the quantity of motion associated with the nut; this is true, but it is not the thrust of the nut example. Newton's example is intended to show that, if the immediately surrounding bodies provide the standard of rest, then the kernel must be considered as at rest even when the shell is in motion. By relaxing the condition on which bodies are used as the standard of rest, and by invoking common relative motion, Du Châtelet's relational conception of motion evades the immediate force of the nut example. In short, she has the conceptual resources to meet Newton's challenge.

It is not just the properties of motion, but also the properties of rest, that are important for Newton. He writes (1999, p. 411):[11]

> It is a property of rest that bodies truly at rest are at rest in relation to one another.

While Musschenbroek follows Newton in asserting the above property of rest (see Musschenbroek, 1744, §104) Du Châtelet once again goes her own way. She first defines rest in general, as she did for motion, before defining relative rest and then absolute rest (*Foundations*, §§220–222):

> 220. Rest is the continuous existence of a body in the same place.

[11] This claim harks back to his rejection in "De Gravitatione" (Newton, 2004) of Descartes's definition of motion. Descartes's definition allowed him to say both (1) that the Earth is at rest properly speaking (since it is at rest with respect to the immediately contiguous bodies of the surrounding fluid), and yet (2) that when considered with respect to the Sun it is in orbit around the Sun. Newton found this problematic as a basis for developing an account of planetary motion, as he argued there at length.

> 221. Relative rest is the continuation of the same relationships of the body being considered to the bodies which surround it, though these bodies move with it.
>
> 222. Absolute rest is the permanence of a body in the same absolute place, this is to say, the continuation of the same relationships of the body being considered to the bodies that surround it, considered as immobile.

This is parasitic on her definition of absolute place, which (as we saw above, and as she notes here) is a relational definition. As such (at least pending further consideration of her account of absolute place), it does not deliver the Newtonian result that bodies truly at rest are at rest with respect to one another. Du Châtelet lacks the resources by which to obtain this result.

Does this matter? In the methodology we are following here, it does so only insofar as it presents an obstacle to pursuing the project of the *Principia*: of finding the true motions of the bodies in our planetary system and thereby determining the system of the world. Do we need Newton's property of rest for this purpose? As it turns out, this condition is a sufficient condition for Newton to be able to carry through the argument of the *Principia*, but it is not necessary. As corollary VI to his laws of motion, and the twentieth century developments associated with General Relativity, make clear, the evidence Newton was working with requires a distinction between free fall and non-gravitationally forced motion, yet systems in free fall may be in accelerated motion with respect to one another. Therefore, it would be premature to reject Du Châtelet's account on the grounds that it lacks this aspect of the Newtonian account. The conceptual distinction that Newton makes turns out not to be necessary for his purposes and so, pending further investigation, it is no criticism of Du Châtelet's definition that it fails to allow for this distinction. We will not pursue this further here. Our preliminary conclusion is that Du Châtelet's failure to replicate Newton's criterion of rest is not, in itself, a problem for her definition of motion.[12]

3.3.2 The Causes of Absolute and Relative Motion

Newton writes (1999, p. 412):

> The causes which distinguish true motions from relative motions are the forces impressed upon bodies to generate motion. True motion is neither generated nor changed except by forces impressed upon the moving body itself, but relative motion can generated and changed without the impression of forces upon this body. ... Therefore, every relative motion can be changed while the true motion is preserved, and can be preserved while the true one is changed, and thus true motion certainly does not consist in relations of this sort.

[12] Rather than prematurely rejecting Du Châtelet's account for its failure to meet Newton's criterion, we should first revise Newton's criterion such that it is necessary, and then assess the adequacy of Du Châtelet's definition with respect to that. We do not pursue this here.

Musschenbroek seems to follow suit, writing (1744, §113, p. 63):

> Though true and absolute motion requires that forces should be impressed upon the bodies moving, yet relative motion may be generated and changed without force impressed immediately upon the body. It is enough if it be impressed upon such other bodies, to which the relation is made, that by their motion that relation may be changed, in which the relative rest or motion of the other consists.

Du Châtelet, though, says something different. We find a clue in her definition of absolute rest. The first part of this definition (§222) was quoted above. The second part is as follows (§223):

> When the active force or the cause of motion is not in the body which can move, this body is at rest, and this is, strictly speaking, real rest.

This indicates that absolute and relative rest and motion are distinguished by their causes. For absolute motion, the cause must be in the body itself. That this is, indeed, Du Châtelet's view, is confirmed by her treatment of the motion of bodies throughout the *Foundations*. Moreover, she is explicit about it in her discussion of place, in the same paragraph in which she defines location. She writes that for a thing to "really" change its place, the cause of that change must lie in the being itself (§88).[13] This position follows Leibniz in *The Leibniz-Clarke Correspondence* (Alexander, 1956). In the fifth letter, Leibniz re-iterates his view that Newton has not shown "the reality of space in itself", and he then says (L5: 53):

> However, I grant there is a difference between an absolute true motion of a body, and a mere relative change of its situation with respect to another body. For when the immediate cause of the change is in the body, that body is truly in motion; and then the situation of other bodies, with respect to it, will be changed consequently, though the cause of that change be not in them.

Therefore, absolute and relative rest and motion are indeed distinguished from one another, but very differently for Leibniz as compared to Newton. For Newton, changes in the state of rest or uniform motion are absolute when brought about by a force impressed on the body in question, and relative when brought about by forces impressed on other bodies. Such causes are therefore impressed (i.e. arising from outside the body rather than being internal to the body in question), and the presence and absence of impressed forces is correlated with a distinction between non-uniform and uniform motion. For Leibniz, all true motion of a body (be it uniform or otherwise) requires a force in that body. Causes of motion are therefore internal to the body in question, and the presence or absence of such forces is correlated with a distinction between motion and rest.

Musschenbroek may also have been a source for Du Châtelet, for he too follows Leibniz in asserting that when a body moves there must be a real force

[13] She writes: "Thus, in order to make certain that a Being has changed its place, and in order for this change to be real, the reason for its change, that is to say the force that produced it, must be in the Being at the moment at which it moves, and not in the coexisting Beings. This is because if we ignore where the true reason of change lies, we also ignore the reason why these Beings changed place."

in the body.[14] This may come as a surprise given that, as we have emphasized, Musschenbroek's account of motion has been standardly Newtonian up to this point. However, Musschenbroek's view on the force of bodies in motion reflects the ongoing difficulties with Newton's Definition 3 in the *Principia*, in which "inherent force of matter"—also called "force of inertia"—is introduced. The postulation of this force precedes, and in Musschenbroek's case justifies, Newton's first law of motion (see Musschenbroek, 1744, §§129–130, p. 67). It was only later that Euler (1752) insisted on reserving the word "force" for impressed force, and moved away from thinking of inertia as a force.

So for Musschenbroek, as for Leibniz, there is a real cause of motion in any body in motion, and Du Châtelet's own position is in line with this approach. Where Du Châtelet goes beyond Musschenbroek is in attempting to theorize this inherent force of body in terms of active and passive force, which she does in her *Foundations* in Chapter 8. She then puts this to use in Chapter 11 to move from her theory of motion to her laws of motion, and from there to the later chapters on the behaviors of bodies (especially Chapters 20 and 21 on statics, the equilibrium of forces, and the famous problem of *vis viva*).[15]

These concerns seem orthogonal to Newton's purposes in discussing the causes of true motions in the *Principia*. If, by changing our standard of rest, we are able to change whether or not a body moves uniformly, then the absence/presence of impressed forces is no longer a means by which to distinguish uniform from non-uniform motions, and thereby to identify true motions. So the issue of causes concerns whether or not there is a non-arbitrary standard adequate for distinguishing uniform from non-uniform motions. Newton proposes absolute space. Du Châtelet, in rejecting absolute space, must offer an alternative.

Du Châtelet's theory of absolute and relative motion, as we have explored it so far, does not provide an alternative. This is for two reasons. First, her definitions of motion are all relational, and so (pending further guidance on our choice of reference bodies) an appropriate change of reference bodies would suffice to change the motion of our target body from uniform to non-uniform. Second, her account of the force of motion internal to a body does not distinguish between uniform and non-uniform motions of that body. Instead, it distinguishes between motion and rest

---

[14] Here is Musschenbroek (§110, p. 62): "A moved body is transferred from one part of space into another. This transference is a real effect, which requires a real cause in the body. This must be some force moving the body. This passes from one body into another. It penetrates from the external to the internal parts of the body, not through its pores, but through the solid substance itself, and is received into every atom, though otherwise immutable, in quantities infinitely diversified from one another." He goes on (§111, p. 62): "Now we may conclude that force passes from body to body, because whatever force is lost by one, just so much is gained by the other body." And (§112, p. 62): "Is force therefore an ens physicum? Or a substance of its own kind? Or is it an idea first produced in an intelligent mind, then communicated to bodies, and passing out of one into another? None of all these can be demonstrated. It is better to acknowledge our ignorance, and that the mind is not capacitated to form a clear idea of it."

[15] For a systematic engagement with Du Châtelet's theory of forces, see Brading (2019), in particular Chapters 3 and 4.

(§225).[16] However, given her account of how one body acts on another, she *can* say at least this much: when a body changes its state of motion, its internal quantity of active force changes.

Where does this leave Du Châtelet? For the Newtonians, absolute space together with absolute time provide the resources for a conceptual distinction between uniform and non-uniform motion: a body moves uniformly when it traverses equal intervals of space in equal intervals of time. Moreover, since absolute places retain their identity over time, Newtonian absolute space provides the resources for a distinction between rest and motion. Therefore, Newtonian absolute space and time provide the resources for a distinction between the presence and absence of causes because, as will be important in the next section, non-uniform absolute motions are the *effects* of impressed forces. However, when considering the *causes* themselves, Du Châtelet has a means to distinguish, conceptually, between the causes of rest, uniform motion, and non-uniform motion.

3.3.3 The Effects of Absolute and Relative Motion

We turn our attention now to the effects of absolute motion. This has long been thought to contain the strongest argument demonstrating the superiority of absolute motion as providing the conceptual resources for a theory of bodies in motion, and so it is here that we expect to find Du Châtelet's most difficult test. Newton writes (1999, p. 412):

> The effects distinguishing absolute motion from relative motion are the forces of receding from the axis of circular motion. For in purely relative circular motion these forces are null, while in true and absolute circular motion they are larger or smaller in proportion to the quantity of motion.

There follows Newton's famous bucket example, in which he demonstrates a correlation between rotation with respect to absolute space and the shape of the surface of the water (as it recedes from the axis of circular motion), and the failure of such a correlation between the rotation of the water with respect to the immediately surrounding body (the bucket) and the shape of the surface of the water.

More specifically, the conceptual challenge being posed to the relationist is as follows. The bucket stands for any scenario in which the relative motions—no matter which body or bodies you choose as your reference body—are the same, while the observable consequences are different. These observable consequences can be thought of in two ways. First, Newton himself describes the effects of absolute rotation as the *forces* of receding from the axis of rotation. We can label this a *dynamic* reading of the bucket experiment. One can also read this scenario kinematically, i.e. without explicit reference to forces: the observed shape of the

[16] She writes (1740, §225): "the only real motion is that which operates by a force residing in the body that moves, and the only real rest is the absence of that force."

water differs when it is at absolute rest (flat) from when it is in absolute motion (curved) even though (once the water is moving at the same angular speed as the bucket) the relative motions are the same in both cases. The relationist is being challenged to show that her account of motion has sufficient resources to make these distinctions.

The bucket argument shows that the postulation of absolute space is sufficient to allow a definition of motion that supports the above correlation between forces and motions, but it does not show that it is necessary. Even if we accept that the argument succeeds against Descartes's definition of motion, which appeals to the immediately surrounding bodies for the standard of rest, we still need to investigate whether Du Châtelet, who offers a different definition of motion, has the resources to tackle Newton's bucket example.[17]

In *The Leibniz-Clarke Correspondence*, Leibniz offers only this (Alexander, 1956, L5: 53):

> 'Tis true that, exactly speaking, there is not any one body, that is perfectly and entirely at rest; but we frame an abstract notion of rest, by considering the thing mathematically.

Du Châtelet gives us just a little more (§89):

> We ordinarily distinguish the location of a body into absolute location and relative location; the absolute location is the one that suits a Being insofar as we consider its manner of existing with the entire universe considered as immobile; and its relative location is its manner of coexisting with some particular Beings.

What does it mean to consider the "entire universe" as immobile? Without an answer to this question, we cannot evaluate whether Du Châtelet has the resources to meet the challenge of Newton's bucket. We shall have to return to it below.

3.4 The Epistemological Challenge

In the final section of the scholium to the definitions in his *Principia*, Newton posed the following epistemic problem (1999, p. 414):

> It is certainly very difficult to find out the true motions of individual bodies and actually to differentiate them from apparent motions, because the parts of that immovable space in which the bodies truly move make no impression on the senses.

The problem is that the motion of a body with respect to absolute space is unobservable, because absolute space itself is unobservable. What we actually observe are the apparent motions—the motions of bodies as they appear to us, from our vantage point—and from this we can determine the relative motions. The problem we are then faced with is how to arrive at the absolute motions, since these are, for Newton, the true motions. The solution, Newton tells us, is "to draw

[17] It is widely held that Newton's absolute space posits *too much* structure (see Torretti, 1983, ch. 1, for example), but that is not the issue here.

evidence, partly from the apparent motions, which are the differences between the true motions, and partly from the forces that are the causes and effects of the true motions" (1999, p. 414). Musschenbroek too makes note of this very problem (1744, §101).

The *Principia* is a spectacular demonstration of how to solve the epistemological problem. We begin with a guess—we assume we have some sort of rough epistemic access to the presence or absence of impressed forces, and to whether motion is uniform or non-uniform, for at least some cases. We then move, using a sophisticated interplay between theory and observation, through a series of successive approximations.[18] In this way, we are able to arrive at the absolute and true motions.

Du Châtelet does not have this epistemic problem, for she does not equate true motion with Newtonian absolute motion. Nevertheless, she faces the problem of determining the true motions.

For Du Châtelet, the true (or "real") motions are those that arise from the internal force of a body (§225): "the only real motion is that which operates by a force residing in the body that moves, and the only real rest is the absence of that force." And she is explicit that it is only by discovering these forces in the bodies themselves that we can adjudicate on the problem of the system of the world; knowledge of the apparent motions alone are insufficient (see §88).

The true motions of bodies coincide with the "absolute motions", or so she seems to suggest (§213):

> Absolute motion is the successive relation of a Body to different Bodies considered as immobile, and this is real motion, and properly so called.

Similarly, for absolute rest, she writes (§222):

> Absolute rest is the permanence of a body in the same absolute place, this is to say, the continuation of the same relationships of the body being considered to the bodies that surround it, considered as stationary.

And for absolute location (§89):

> absolute location is the one that suits a Being insofar as we consider its manner of existing with the entire universe considered as immobile...

Therefore, to find the true motions it suffices to find the "absolute motions", thus conceived. How are we to proceed, and what would justify the claim that the resulting "absolute motions" are indeed the *true* motions?

Consider first her assertion that we should consider the "the entire universe" as immobile when assigning an absolute location to a Being. It is tempting to suggest that the immobile universe posited here is supposed to somehow play a role akin to absolute space in Newton, providing the immobile places to which all motions ultimately refer. However, we do not think that this was Du Châtelet's intention. Rather, we interpret her as offering an epistemic analysis of the means by and extent

---

[18] For in-depth discussions of Newton's scientific methodology, see Harper (2011) and Smith (2014, pp. 262–345).

to which we are able to arrive at true motions. The role of the bodies "considered as immobile" is not to approximate Newtonian absolute space, but to provide a material frame of reference useful for the problem at hand. To explain what we mean by this, we return to the main problem of determining the true motions for the system of the world.

In astronomical theorizing, the preferred material frame had long been the fixed stars: they are called the fixed stars because, as viewed from Earth, they appear to us to be mutually at rest in the night sky. Du Châtelet is clear that in practice we use the fixed stars as the standard of rest to measure the location of other celestial bodies—the Moon, the "wandering stars" (the planets), and so forth—even though the fixed stars may not be truly immobile (§91):

> We perceive that a Being has changed location when its distance from other Beings, which are immobile (at least for us), is changed. Thus, we made the catalogs of fixed stars in order to know whether a Star changes location, because we regard the others as fixed, and indeed they effectively are relative to us.

Note the phrases "at least for us" and "effectively". What these each emphasize is that, as observers on Earth, our epistemic situation is such that the fixed stars appear to be at rest relative to each other, and so we can ascribe rest to them. In other words, we use the apparent rest of the fixed stars with respect to one another for the practical purpose of providing us with a standard of rest, even though we do not know whether they are truly at rest. With the benefit of hindsight, we know that using the fixed stars as a standard of rest is well-suited for the task of determining the changing locations of celestial bodies in our planetary system. Thus, while our lack of epistemic access to the true state of the fixed stars may sound discouraging at first, as it turns out, the limitation does little harm to our theorizing. Is it just a matter of epistemic luck, one might ask, that we happen to inhabit a particular part of the universe from which so many stars appear as mutually at rest? The answer is yes: this is one instance of serendipity in the history of astronomy, one that we have been able to put to good epistemic use.[19]

Du Châtelet defines absolute motion in terms of the relation to "different bodies considered as immobile", and draws attention to the epistemic significance of the fixed stars for astronomy, which are "effectively" at rest relative to us. We suggest that these two points could be linked in a useful way by taking the motion of celestial bodies relative to the fixed stars as their *effective absolute motion*. Different from Newtonian absolute motions, which refer to unobservable absolute space, effective absolute motions refer to the fixed stars. Now we are in better place to engage with the following question: what justifies the claim that effective absolute motions are

[19] Barbour's (2001) magnificent history of the discovery of dynamics makes vivid the role of luck (both good and bad) in the observations that were available from our vantage point on Earth in the development of astronomy and the clues they provided (or masked) concerning the system of the world. See also Smith (2012) for an insightful discussion of how the method of what Smith calls "successive approximations", which lies at the heart of Newton's methodology, meets the challenge presented by the likely parochialism of our observational situation.

the true motions arising from the internal forces? In order to address this, we return to the bucket experiment.

In our view, a Du Châtelean response to Newton's bucket experiment would be as follows. First, we can infer from the different observed effects displayed by the water (including its changing shape and endeavor to recede from the axis of rotation) to the presence or absence of forces within the water. The origins of these forces lie in the bodies themselves, according to Du Châtelet's theory of forces. Second, we compare the inferred presence or absence of internal forces to the effective absolute motions of the water and bucket, using the fixed stars as our standard of rest. Finally, insofar as the forces and motions correlate appropriately, we say that the effective absolute motion (defined in terms of relations to the fixed stars) *just is* the true motion (defined in terms of the presence of forces in the bodies) whose effects we observe. Until the correlation fails, we continue to trust the fixed stars for providing us with an adequate standard of rest for the purpose of physical theorizing. However, where we find discrepancies that we cannot resolve, this may indicate the need for modifying our standard of rest.

This process is, of course, true to the *practice* of physics, for whether or not we endorse Newtonian absolute space, the apparent motions are all that we have to work with. From the Newtonian perspective, the continual modification of our standard of rest is a process of ever closer approximation to absolute space. From the Du Châtelean perspective, this continual modification brings us ever closer to the forces of bodies, from which the true motions arise, but there is no background "absolute space" relative to which those motions are "true".

In our opinion, this is a compelling analysis of the epistemic situation. However, there is a further layer to the challenge posed by the bucket experiment. The Newtonian explains the results of this experiment by appeal to the ontology of absolute space and time: absolute rotation has observable effects. More generally, absolute space and time provide the Newtonian with the resources for an ontological distinction between uniform and non-uniform motion, and this in turn both underwrites the corresponding conceptual distinction, and provides justification for the means by which the epistemological challenge is met (that is, for the claim that the observable effects of absolute motion are a guide to the true motions of bodies). Du Châtelet lacks absolute space and time, and so can appeal to no such ontological resources to back up her conceptual and epistemological analyses. We call this the "ontological challenge"; we explain it in more detail in the next section, and offer a response on behalf of Du Châtelet.

3.5 The Ontological Challenge

For Descartes, the material world is to be explained in terms of parts of matter moving around: the shapes, sizes and motions of the parts of matter are the explanatory resources to which natural philosophers may appeal. Particularly important for our purposes is the claim—widely shared, especially among those advocating "mechanical philosophy"—that motion does explanatory work.[20] As a consequence, a definition of motion will be inadequate if it yields the result that *different* outcomes are associated with the *same* motions. The bucket experiment illustrates this point: it shows that, if we begin with Descartes's relational definition of motion, we have cases where the *same* state of motion (e.g. the water at rest with respect to the bucket) yields *different* shapes for the surface of the water (flat when both water and bucket are at absolute rest; curved when both water and bucket are rotating in absolute space, as Newton would say). Therefore, Descartes's theory of motion is unable to explain the results of the bucket experiment.

Newton's claim is that, if we adopt absolute motion, then the same states of motion are correlated with observable outcomes that are the same, and when the observable outcomes differ the state of motion is different too. So, his definition of motion provides the appropriate correlations between states of motion and observations. More importantly, if we adopt the ontological commitments that correspond to his definition, so that for a body to move is for it to move with respect to absolute space and time, then different states of motion can be used to *explain* different observable outcomes. When the surface of the water is flat, this is because the water is at rest with respect to absolute space; when the surface is curved, this is because the water is rotating with respect to absolute space. This is the kinematic reading of the bucket experiment (see above, Sect. 3.3.3). We can also give a dynamical reading, in which we describe the different observable outcomes in terms of the presence and absence of impressed forces, such that the different states of motion are correlated with the presence and absence of forces. Specifically, uniform motion is correlated with the absence of impressed forces, whereas non-uniform motion involves their presence (again, see Sect. 3.3.3, above). Either way, what explains the observed effects in the bucket experiment (the shape of the water, the endeavor to recede from the axis of rotation), is the motion of the water with respect to absolute space.

For Newton, there is a real difference between uniform and non-uniform motion, and this difference, ontologically, lies in true motion being absolute: it is motion with respect to absolute space. Absolute space and time provide the ontological resources that underwrite the conceptual distinctions on which Newton relies in his pursuit of true motion.

Lacking these ontological resources, the relationist is hard-pressed to explain the results of the bucket experiment. We can summarize the challenge thus: give me a theory of motion that differentiates the scenarios in the bucket experiment, so that different states of motion *explain* the observed effects.

Du Châtelet, as we have seen, chooses the fixed stars to provide her with "effective absolute motion". This suggests a response to the bucket experiment along the following lines. We take the rest frame of the fixed stars to have not just *epistemic*

[20] This motion, as Descartes was at pains to emphasize, is not the richly varied "motion" of the Aristotelians, encompassing many different kinds of change, but strictly "local motion", that is changed of place.

significance (see Sect. 3.4), but also ontological significance. When the water rotates with respect to the rest frame of the fixed stars, the changing spatial relations result in an endeavor to recede from the axis of rotation, and the observed change in the shape of the surface of the water follows. This is a puzzling suggestion. If motion is truly relational, could we not equally use the bucket as our standard of rest, and expect the fixed stars to recede from their axis of rotating around the bucket? And even if that relational consequence is rejected, why should we take motion with respect to the distant stars as *explanatory* of such localized effects in the bucket? Is this a *causal* action of the stars on the water? Given Du Châtelet's rejection of action-at-a-distance, it seems unlikely that she would have embraced this attempted response to the bucket experiment.

An alternative response would be an endorsement of an ether theory, in which a background ether provides a standard of rest, and accounts locally for the observations in the bucket experiment. Since Du Châtelet endorsed the plenum, this might seem a more promising approach. But such a view has the following consequence: Newton's laws, by which we predict the outcome of the bucket experiment, do not hold unless an ether—to which we make no reference in applying the laws and deriving our predictions—exists. At best, this leaves the supposed explanatory role of the ether mysterious.

Neither of these options for providing an ontological underpinning, by which to explain the results of the bucket experiment, looks promising. And indeed, as later developments have shown, constructing a fully relational theory of motion is an elusive task.

We submit that Du Châtelet would have rejected the ontological challenge as misguided. Du Châtelet focuses our attention on the epistemology of the theory of motion, and in particular on the challenge of how to determine the true motions. The ontological explanation for these motions lies in the forces of bodies, and indeed ultimately in the forces of the simples from which bodies arise. It is not *motion* that is explanatory of the presence/absence of forces, but the forces of bodies that explain the apparent motions. En route to discovering the forces of bodies, we proceed via the effective absolute motions, and we are epistemically cautious: we may not have a way to arrive at a perfect correlation between effective absolute motions and the presence of forces, but Newton's *Principia* has shown us that the methodology is promising and worth pursuing, at least for now.

In Newton's *Principia*, absolute space and time underwrite the conceptual structure of true motion: they distinguish rest from motion, yield quantity of speed (as a determinate distance travelled in a determinate amount of time) and quantity of acceleration (as rate of change of speed and/or direction), and distinguish uniform from non-uniform motion. Newton's laws of motion require some, but not all, of these resources. The first law states that every body continues in its state of rest or uniform motion unless acted upon by an external force. The second law states that the quantity of deviation from uniform motion is correlated to the magnitude of the external force. Non-uniform motions of a body indicate that an impressed force is involved, the magnitude of which is correlated with the quantity of acceleration, and the source of which must be located in another body. This is the basis on which

Newton undertakes the project of determining the true motions of the bodies in our planetary system. True acceleration requires an impressed force, and the correlation between accelerations and impressed forces is the key by which to unlock the puzzle of determining the true motions. Anyone who appeals to Newton's laws can do so only to the extent that they have the resources to distinguish between uniform and non-uniform motion, and to quantify acceleration. For Newton, this is done with the ontology of absolute space and time.

The Du Châtelean response is straightforward and pragmatic: she can make these distinctions effectively, for the purposes of theorizing, and she does not require that they are underwritten ontologically in order to proceed. Indeed, to commit to an ontology of absolute space, time and motion would exceed limits of that which is epistemically warranted by the methods and results of either the *Principia* itself, or of her own methodology for scientific theorizing (see especially Chapter 4 of her *Foundations*).[21] We do not pretend that Du Châtelet herself offered this response to the bucket experiment, but we do maintain that it is consistent with her approach, and that she has the resources to meet the demands of the *Principia* without adopting Newtonian absolute motion.

3.6 Conclusions

The history of space-time theory since Newton indicates that no relational theory of space and time can provide appropriate structure for ontologically underwriting the distinction between inertial and non-inertial motion.[22] Relational attempts to explain the bucket experiment (or rotation more generally) fail because relationists lack the spatiotemporal structure to say whether or not a body truly accelerates. Since Du Châtelet offers a relational account of motion, it would seem at first sight that she is in the same tough spot as all the other relationists. Closer inspection reveals that this is not the case. Rather, she changes the focus of the debate away from ontology and to epistemology (and methodology). In so doing, she successfully meets all of the conceptual and epistemic demands placed on an account of motion by Newton's *Principia*, while also rejecting absolute space, time and motion. In our opinion, this makes her account of motion a most interesting contribution to the absolute-relative motion debate in the eighteenth century.

[21] For more discussion on Du Châtelet's methodology for scientific theorizing, see Brading (2019), Chapter 2, which argues that the problem of method lies at the heart of the *Foundations*. Also see Detlefsen (2019) for a useful study comparing Du Châtelet and Descartes's views on the use of hypothesis in science, which finds Du Châtelet's attitude toward hypothesis "considerably more modern" than Descartes's.

[22] See, for example, Torretti (1983, pp. 9–11) and Earman (1989). For a twentieth-century attempt at relational mechanics, see Barbour and Bertotti (1982).

References

Alexander, R. G. (1956). *The Leibniz-Clarke correspondence: together with extracts from Newton's Principia and Optiks*. Manchester University Press.

Barbour, J. B. (2001). *The discovery of dynamics: A study from a Machian point of view of the discovery and the structure of dynamical theories*. Oxford University Press.

Barbour, J. B., & Bertotti, B. (1982). Mach's principle and the structure of dynamical theories. *Proceedings of the Royal Society of London. A. Mathematical and Physical Sciences, 382*(1783), 295–306.

Belkind, O. (2007). Newton's conceptual argument for absolute space. *International Studies in the Philosophy of Science, 21*(3), 271–293.

Brading, K. (2017). Time for empiricist metaphysics. In *Metaphysics and the philosophy of science: New essays*. Oxford University Press.

Brading, K. (2018). Émilie Du Châtelet and the problem of bodies. In *Early modern women on metaphysics* (pp. 150–168). Cambridge University Press.

Brading, K. (2019). *Émilie Du Châtelet and the Foundations of Physical Science*. Routledge.

Detlefsen, K. (2019). Du Châtelet and descartes on the roles of hypothesis and metaphysics in natural philosophy. In *Feminist history of philosophy: The recovery and evaluation of women's philosophical thought* (pp. 97–127). Springer.

Du Châtelet, E. (1740). *Institutions de physique*. Paris: Prault.

Du Châtelet, E. (2009). *Selected philosophical and scientific writings*. University of Chicago Press.

Du Châtelet, E. (2018). *Foundations of physics*, (trans. K Brading et al.), Available at www.kbrading.org.

Earman, J. (1989). *World enough and spacetime*. Cambridge, MA: MIT Press.

Euler, L. (1752). Recherches sur l'origine des forces. *Mémoires de l'académie des sciences de Berlin, 6*, 419–447.

Hagengruber, R. (2012). Émilie du Châtelet between Leibniz and Newton: The transformation of metaphysics. In *Émilie Du Châtelet between Leibniz and Newton* (pp. 1–59). Springer.

Harper, W. L. (2011). *Isaac Newton's scientific method: Turning data into evidence about gravity and cosmology*. Oxford University Press.

Huggett, N. (2012). What did Newton mean by 'Absolute Motion'? In *Interpreting Newton: Critical essays* (pp. 196–218). Cambridge University Press.

Hutton, S. (2004). Émilie Du Châtelet's *Institutions de Physique* as a document in the history of French Newtonianism. *Studies in History and Philosophy of Science Part A, 35*(3), 515–531.

Hutton, S. (2012). Between Newton and Leibniz: Emilie du Châtelet and Samuel Clarke. In *Emilie du Châtelet between Leibniz and Newton* (pp. 77–95). Springer.

Iltis, C. (1977). Madame Du Châtelet's metaphysics and mechanics. *Studies in History and Philosophy of Science, 8*, 30.

Jacobs, C. (2020). Du Châtelet: Idealist About Extension, Bodies and Space. *Studies in History and Philosophy of Science Part A, 82*, 66–74.

Janiak, A. (2018). Émilie du Châtelet: Physics, Metaphysics and the Case of Gravity. In E. Thomas (Ed.), *Early modern women on metaphysics*. Cambridge: Cambridge University Press.

Janik, L. G. (1982). Searching for metaphysics of science: the structure and composition of Madame du Chatelet's Institutions de physique, 1737–1740. *Studies on Voltaire, 201*, 85–113.

Kawashima, K. (1990). La participation de Madame Du Châtelet à la querelle sur les forces vives. *Historia scientiarum, 40*, 9–28.

Lu-Adler, H. (2018). Between Du Châtelet's Leibniz exegesis and Kant's early philosophy: A study of their responses to the vis viva controversy. *History of Philosophy and Logical Analysis 21*, 177–194.

Massimi, M., & De Bianchi, S. (2013). Cartesian echoes in Kant's philosophy of nature. *Studies in History and Philosophy of Science Part A, 44*(3), 481–492.

Musschenbroek, P. V. (1734). *Elementa physicae*. Leyden: Samuel Luchtmans.

Musschenbroek, P. V. (1739). *Essai de physique*. Leyden: Samuel Luchtmans.

Musschenbroek, P. V. (1744). *The elements of natural philosophy. Chiefly intended for the use of students in universities*. J. Nourse, at the Lamb without Temple-Bar.
Newton, I. (1999). *The Principia: Mathematical principles of natural philosophy*. University of California Press.
Newton, I. (2004). *Newton: Philosophical writings*. Cambridge University Press.
Pemberton, H. (1728). A view of Isaac Newton's philosophy. Dublin: John Hyde, and John Smith and William Bruce.
Reichenberger, A. (2012). Leibniz's quantity of force: A 'Heresy'? Émilie du Châtelet's *Institutions* in the context of the *Vis Viva* controversy. In *Emilie du Châtelet between Leibniz and Newton* (pp. 157–171). Springer.
's Gravesande, W. J. (1720). *Mathematical elements of physicks*, Book I, translated into English by John Keill from the 1720 Latin edition.
Schliesser, E. (2013). Newton's philosophy of time. In *A companion to the philosophy of time* (pp. 87–101). Wiley-Blackwell.
Smith, G. E. (2012). How Newton's *Principia* changed physics. In *Interpreting Newton: Critical essays* (pp. 360–395). Cambridge University Press.
Smith, G. E. (2014). Closing the loop. In *Newton and empiricism* (pp. 262–352). Oxford University Press.
Stan, M. (2018). Emilie Du Châtelet's metaphysics of substance. *Journal of the History of Philosophy, 56*(3), 477–496.
Suisky, D. (2012). Leonhard Euler and Émilie Du Châtelet. On the post-Newtonian development of mechanics. In *Émilie Du Châtelet between Leibniz and Newton* (pp. 113–155). Springer.
Terrall, M. (1995). Émilie Du Châtelet and the gendering of science. *History of Science, 33*(3), 283–310.
Torretti, R. (1983). *Relativity and geometry*. Pergamon Press.
Torretti, R. (1999). *The philosophy of physics*. Cambridge University Press.
Walters, R. L. (2001). La querelle des forces vives et le rôle de Mme Du Châtelet. *Studies on Voltaire and the Eighteenth Century, 11*, 198–211.

Chapter 4
Effective Field Theories: A Case Study for Torretti's Perspective on Kantian Objectivity

Thomas Ryckman

Abstract Those enlightened philosophers of physics acknowledging some manner of descent from Kant's 'Copernican Revolution' have long found encouragement and inspiration in the writings of Roberto Torretti. In this tribute, I focus on his "perspective on Kant's perspective on objectivity" (2008), a short but highly stimulating attempt to extract the essential core of the Kantian doctrine that 'objects of knowledge' are constituted, not given, or with Roberto's inimitable pungency, that "*objectivity is an achievement, not a gift*." That essential core Roberto locates in the Kantian notion of *apperception*, or self-activity, manifested in cognition in the idea of combination (*Verbindung*) or composition, which, Kant tells us, "among all ideas ... is the one that is not given through objects, but can only be performed by the subject itself, because it is an act of self-activity" (B 130). I first rehearse Roberto's proposal for how an imaginative interplay between sensibility and understanding can be fashioned via the *productive imagination* or *power of reflective judgment* (of the third *Critique*). In this way, the notion of composition in general, unfettered from needless period constraints issuing in "pure forms of sensibility" and "pure concepts of the understanding", can be seen as the intellectual motor for the "free creation" of concepts celebrated by Einstein and others, furnishing structural scaffolding required to articulate and display physical objects and processes, a conceptual panoply that "cannot be fished out of the stream of impressions". Roberto emphasizes that historical case studies are needed to evaluate his proposal, suggesting one himself, the continuous conceptual development inaugurated by Riemann's *Habilitätionsschrift* (1854) resulting, some hundred years later, in the fiber bundle formalism of modern differential geometry and topology. I sketch a related suggestion, that the gauge groups of modern particle physics are the outcome of a similar line of conceptual advance, a structural scaffolding saving the phenomena of high-energy experiment within the framework of 'effective field theory.'

T. Ryckman (✉)
Stanford University, Stanford, CA, USA
e-mail: tryckman@stanford.edu

C. Soto (ed.), *Current Debates in Philosophy of Science*, Synthese Library 477,
https://doi.org/10.1007/978-3-031-32375-1_4

Philosophers of physics acknowledging intellectual descent from Kant's Copernican Revolution have long found both inspiration and encouragement in the writings of Roberto Torretti. Here I focus on his "perspective on Kant's perspective on objectivity" (Torretti, 2008), a brief but highly stimulating challenge to comprehend contemporary physical theory through recognition and application of "the great and lasting novelty of Kant's approach",

> the notion that the objects which are the referents of epistemic discourse, be they things and their attributes, or objective situations and objective developments, are not *given* but *constituted,* articulated in the stream of becoming by our own regulated activity of composition or 'synthesis'. (2008, p. 81)

This would be to show that the objects of discourse of contemporary fundamental physical theory – *inter alia*, energy-momentum tensors, quarks, gauge bosons, scalar fields – are synthesized or composed in a to-be-articulated sense within cognitive activity as *objects of knowledge* rather than the "ready-made ... "perfectly definite objects" that physicists manage to discover "once the scales fall from their eyes". In brief, "objectivity is an *achievement,* not a *gift*". Certainly in order to accommodate concepts and laws of contemporary physical theory involves considerable pruning of Kant's eighteenth century architectonic to extract what is deemed essential to his "approach". Minimally this must decouple Kant's general views on physical objectivity from the *a priori* physics ("pure natural science") of the *Metaphysical Foundations of Natural Science* and its claim of a necessarily Newtonian nature. In the first half of the twentieth century this was the driving impulse of several neo-Kantian philosophers, in particular Ernst Cassirer (of whom more below) and maverick interlopers in philosophy of physics, especially mathematician Hermann Weyl (more below). Bypassing mention of such like-minded precursors of his project, Torretti seeks to loosen the fetters of the "closed system of reason" of Kantian doctrine[1] to accommodate post-Newtonian physical theory, proposing to rehabilitate Kant's conception of objectivity by locating its "key" in the notion of self-awareness, "called by Kant *apperception* (after Leibniz)"(2008, p. 83). Naturally, particular case studies from recent physics are required to support this proposal with the goal of exhibiting.

> how a new concept or structured panoply of concepts issues from those that were previously available – and ultimately perhaps from pre-conceptual intuitions – through an intellectual effort resembling that of Kant's productive imagination or reflective judgment. (2008, p. 93)

Torretti advances several examples, and two are thematically linked: the continuous conceptual development from intuitive notions of space and quantity to the concepts of differentiable manifold and Riemannian metric inaugurated in Riemann's 1854 *Habilitätionsschrift*, continuing still further a century later with the fiber bundle formalism of modern differential geometry and topology. Stepping back a minute to consider: the astonishing burden of Torretti's project is to make

[1] E.g., (2008, p. 89): "the 4 × 3 table of the categories ... a millstone around Kant's neck".

a plausible case that the Kantian notion of apperception is somehow involved or implicated in "all the grand conceptual structures employed by mathematical physics or yet to be created by its practitioners". (2008, p. 93)

I share Torretti's optimism that this can be done, outlining below another case study broadly supporting his bold project though within the limitations of this paper it can be no more than a preliminary sketch. In doing so, I am assisted by the circumstance that Ernst Cassirer, building on insights from Richard Dedekind, previously advanced a revitalized account of Kantian objectivity remarkably similar to that proposed by Torretti. I begin with a brief overview of several of the relevant Kantian texts. Section 4.2 then summarizes how Cassirer, inspired by Dedekind's account of mathematical cognition, likewise found the ground of Kantian objectivity in the spontaneity of thought that is the "productive imagination". Section 4.3 outlines the broadly neo-Kantian philosophical context in which the gauge principle emerged, a half a century later to become the unifying principle of the Standard Model. Section 4.4 briefly reviews the problem of renormalization in quantum field theories, while Section 4.5 introduces the new view of renormalization under the rubric of "effective field theory". I sketch here how gauge groups within the framework of quantum field theories of the Standard Model, understood as "effective field theories", provide the contemporary structural scaffolding for the phenomenology of high-energy experiment. I conclude by suggesting several aspects of these conceptually-linked developments that manifest the spirit of Torretti's "perspective on Kant's perspective on objectivity".

4.1 A Triad of Notions: Apperception, Productive Imagination, Reflective Judgment

As a matter of textual fact, the Kantian notion of apperception is not entirely univocal; the A Deduction (A 107) distinguishes "inner sense or empirical apperception" from "transcendental apperception"; the B Deduction (B 132) "pure" from "empirical" apperception.[2] Torretti himself does not plunge into the deep waters of the respective merits, demerits, or distinctive roles of the A and B Deductions, wisely situating the notion in Kant's response to the primacy to subjective experience (*Erlebnis*) in empiricist (British) epistemology to which Kant opposed to the notion of objective experience (*Erfahrung*). Torretti's brief remarks on the A Deduction then tell us that, in order to read subjective appearances as objective experience (*Erfahrung*), that is, to have cognition via concepts ("the synthesis of recognition in a concept") requires two distinct acts of awareness: a "synthesis of apprehension" whereby fragments of the manifold of *Erlebnisse* are

[2] Kant (1997), a translation of the 1781 ('A') and 1787 ('B') editions.

grasped as regular patterns, and simultaneously a higher order act of recognition, which is just that

> awareness of the identity of that which is being successively grasped, cannot take effect without awareness of the identity of the very act of grasping. (2008, p. 83)

It is the latter act, identified as the form of awareness called apperception, that serves as "the key to Kant's conception of objectivity". Now the A Deduction advances three acts of synthesis comprising the order of empirical genesis of representations ("synthesis of apprehension in intuition", "synthesis of reproduction in imagination", and "synthesis of recognition in a concept"); each is a distinguished moment that depends on *transcendental apperception*, that "unity of consciousness" without which the mind, in cognition of a sensible manifold, "could not become conscious of the identity of the function by means of which this manifold is synthetically combined into one cognition." (A 108) And the A Deduction introduces a distinct active faculty of mind combining different perceptions and exhibiting them as a series: the "reproductive faculty of imagination". (A 120) As a "merely empirical" faculty, it naturally has an objective ground, *viz.*, "pure imagination ... a fundamental faculty of the human soul grounding all cognition *a priori*". The latter's transcendental function is to bring the necessary unity of apperception ("the standing and lasting I ... constituting the correlate of all our representations") into combination with the manifold of intuition; this is the "transcendental function of the imagination" necessarily connecting the faculties of sensibility and understanding. (A 124) The end result in the A Deduction is that the three moments of synthesis yielding representation of an object by formation of concepts depend ultimately on this fundamental faculty which, insofar as it pertains to *the form of an experience in general*, are the categories grounding "all formal unity in the synthesis of the imagination" (A 125).

The B Deduction drops mention altogether of a fundamental faculty of "pure imagination" and its transcendental function of mediation between sensibility and understanding. It begins (§15, B 130) by simply underscoring the notion of combination: of all ideas *combination* (*conjunctio, Verbindung*) "is the only one that is not given through objects but can be executed only by the subject itself, since it is an act of its self-activity". Combination comes to us not from the senses, but "is an act of the spontaneity of the power of representation" which, in the B Deduction, is an action of the understanding. Apperception is then the consciousness of the act of thinking itself, consciousness of the "spontaneity of my thought". (B 158n). However, in §24 of the B Deduction Kant distinguished *reproductive* imagination (whose synthesis is entirely subject to empirical laws") from *productive* imagination, which is imagination regarded as the spontaneity of thought. (B 152) And as Torretti observes, Kant elsewhere (A 163/B 204) described the emergence of geometrical thought in the interplay of understanding and sensibility as "the successive synthesis of productive imagination". It is "productive imagination" or "reflective judgment"[3] for which Torretti seeks "maximal freedom" from the constraints of the pure forms of sensibility and those of the pure concepts of the understanding (2008, p. 92).

We cease further discussion of these textual intricacies but focus instead on the B Deduction's broad yet promising identification of "spontaneity of (discursive) thought" with the notion of combination itself. In this way, the notion of combination in its various modes (as a concept uniting sensible representations according to a rule; as combining concepts into complex concepts; as composition of concepts in judgment) might be seen as the intellectual motor for "free creation" of concepts celebrated by Einstein, i.e., for furnishing an evolving framework for the articulation and display of physical objects and processes, a conceptual panoply that "cannot be fished out of the stream of impressions". Employing the "productive imagination" (now unfettered from needless constraints stemming from "pure forms of sensibility" and "pure concepts of the understanding") in fashioning an "imaginative interplay" between sensibility and understanding in order to allow "free invention in Einstein's sense", and hence "all the grand conceptual structures employed by mathematical physics or yet to be created by its practitioners" (2008, p. 93) is then Torretti's proposal for re-vitalizing the Kantian story of constitution of objects for contemporary physical theory.

4.2 Cassirer's "Dedekindian" Account of Concept Formation Via the Productive Imagination

As early as 1907 and into the 1940s, Ernst Cassirer advanced a functional doctrine of concepts inspired in significant ways by Dedekind. Despite its title, the 1910 monograph *Substanzbegriff und Funktionsbegriff* provided no precise definition of either "substance concept" or "function concept". Rather, the intended contrast is one of competing accounts of *Begriffsbildung. Concept formation* today is a topic for psychology and cognitive science, while *conceptual change* is allocated to history and philosophy of science. But to period neo-Kantians, *Begriffsbildung* belonged to *Wissensschaftslehre*, distinct from both formal logic and psychology. As it happens, Cassirer's most comprehensive treatment of his intellectual debt to Dedekind is buried away in the first two chapters ("*Theorie des Begriffs*", "*Begriff und Gegenstand*") of Part III (concerning theoretical natural science and mathematics) of the third volume of the *Philosophy of Symbolic Forms*.[4] At the beginning of the volume Cassirer presented a brief hint of the "functional theory

[3] Torretti (2008, p. 90) cites Kant's 1797 letter to Tieftrunk which refers to the "creative interplay between the sensibility and understanding" in the *Critique of Judgment* as "the schematism of the power of judgment".

[4] Cassirer (1929). All translations are my own. The corresponding text in the published translation by R. Manheim (1957) follows the semi-colon.

of objectivity" to be developed there: "the function of knowledge is to build up and constitute the object ... as a phenomenal object, conditioned by this very function" which involves "a 'spontaneous' act of the understanding ... a specific mode and direction of *formation* leading to the world view (*Weltbild*) of theoretical knowledge." (1929, pp. 5–6; 1957, p. 5).

We pick up Cassirer's discussion at the point where he observed that the meaning of the term 'analogy' in Kant (particularly in the Analogies of Experience, see concluding remarks) retains the sense of Greek linguistic usage as "the general term for the concept of relation":

> It expresses therefore a basic orientation (*Grundrichtung*) of *relational thinking* ... equally indispensable for apprehending the 'meaning' of number and the 'meaning' of relational thought formulated in language, of a linguistic 'metaphor'. (1929, pp. 297–8; 1957, p. 257)

Cassirer then immediately turns to Dedekind's statement of this basic trend of relational thinking in the Preface to the second edition of *Was sind und was sollen die Zahlen*, famously tracing the concept and system of natural numbers back to.

> the capacity of the mind to relate things to things, to allow one thing to correspond to another, or to represent (*abzubilden*) one thing through another, without which capacity in general no thinking is possible. (1893, p. VII)

The idea of mapping (*Abbildung*), a notion broader than that of function, lies at the core of Dedekind's account of concepts as "free creations of the human mind" and in no uncertain terms, Dedekind believed "free creation" to be essential to science, as the "greatest and most fruitful advances in mathematics and the other sciences have invariably been made by the creation and introduction of new concepts". Cassirer's own account of concepts then draws from Kantian doctrine as revitalized by Dedekind. A streamlined paraphrase of A 98–110 laid out the broad Kantian parameters:

> In the construction of 'objective' knowledge, the 'synthesis of recognition in the concept' must be added as the actual capstone to the synthesis of 'apprehension in intuition' and that of 'reproduction in the imagination'. Recognizing an 'object' means nothing other than subjecting the manifold of intuition to a rule that determines it in reference to its order. However, consciousness of such a rule, and the unity posited through it, this and nothing else is the concept. (1929, p. 362; 1957, p. 315)

But then for Cassirer Dedekind's *Abbildung* is the "basic form of relation" (*Grundform der Beziehung*); it manifests "the unity of *relation* by virtue of which a manifold is innerly (*innerlich*) determined as belonging together". While the core of the number concept and of mathematical thinking in general, "it is by no means restricted to this domain" but "dominates the entirety of cognizing (*Gesamtheit des Erkennens*)" since "precise logical and epistemological analysis in general shows that "to grasp" ('*Begreifen*') and "to relate" ('*Beziehen*') are correlates, interchangeable concepts (*Wechselbegriffe*)". (1929, p. 243; 1957, p. 298) The notion of *Abbildung* thus delineates "the essence of the *general form* of the concept" (*das Wesen dieser* ***allgemeinen Form*** *des Begriffs*, (1929, p. 328, 1957, p. 285), a form that

> not only holds for mathematical concepts but represents an essential trait of all genuine conceptual structures ... by establishing a new, ideal *reference point*" [whereby] particulars which had previously tended apart, *are directed* towards this point of reference, and through this unity of direction are stamped with a unity of 'essence' – though this essence is not to be taken ontically but logically, as a pure determination of meaning.[5]

In effect, Cassirer recognized Dedekind's notion of mapping as going proxy for nearly all of the gears and cranks of Kantian conceptual doctrine. What is essential to this doctrine, and according to Torretti, "opens a wide door to intellectual pluralism, which indeed has thrived in [Kant's] wake" (2008, p. 87), is merely what is stated in the B Deduction. Underling all concept formation is "the capacity to connect a priori" (B 135) yielding a "functional theory of objectivity" according to which an object "is that in whose concept the manifold of intuition is united" (B 137). Liberated from the narrow constraints of what Torretti termed Kant's "in effect, Aristotelian – view of concept formation" (2008, p. 90), in Cassirer's Dedekindian treatment, the function of concepts is to scout out the "free realm of the 'possible'", i.e., to project "the direction of inquiry ... not so much as a ready-made path along which thought progresses, but as a method, a process of pathfinding." A newly acquired concept, "long before it is itself exactly 'defined'", strikes out new hypothetical connections, and merely "sets up a definite direction and norm of *discursus* ... indicating the point of view under which a manifold of contents ... are apprehended and 'seen together'". (1929, pp. 346–7 & 353–4; 1957, pp. 298–9 & 305–6) And indeed, the underlying impetus of this revitalized Kantian view of concepts inspired by Dedekind is traced back to the "spontaneity of thought" that is the "productive imagination":

> In the concept this achievement of the productive imagination (*diese Leistung der produktiven Einbildungskraft*) stands before us in more intensified and potent form. ... The achievement of the concept ... is a productive constructive achievement ... a presupposition of experience and hence a condition of the possibility of its objects. (1929, p. 353 & p. 362; 1957, p. 306 & p. 315)

4.3 The Gauge Idea

The idea of gauge has an unusual 'context of discovery': invoked on *broadly* Kantian ("transcendental phenomenological") grounds by mathematician Hermann Weyl, it emerged in the context of general relativity in 1918 prior to quantum mechanics

[5] (1929, p. 349): "*Und dies gilt nicht nur für die mathematischen Begriffen, sondern es stellt einen Wesenszug aller echten begrifflichen Strukturen dar. ... daß ein neuer ideeller* ***Bezugspunkt*** *für dasselbe auggestellt wird. Indem das Besondere, das zuvor Auseinanderstrebende sich nach diesem Bezugspunkt* ***richtet***, *wird ihm in dieser Einheit der Richtung eine neue Einheit des "Wesens" aufgeprägt – wobei ebendieses Wesen selbst nicht ontisch, sondern logisch, al seine reine Bestimmung der Bedeutung, zu nehmen ist.*" Cf. (1957, p. 303)

as a stipulated invariance of gravitational and electromagnetic field laws under a generator of coordinate scale change acting at each space-time point. As a local function of space-time coordinates this generator is mathematically identical to the four-vector potential of electromagnetism, giving Maxwell's theory a basis in space-time geometry.[6] But to Weyl, the primary charm of the idea of local symmetry lay in that it simultaneously satisfied two desirable philosophical goals. First, it mandates that the epistemic reach of an experiencing/cognizing subject is initially restricted to what is "evident", i.e., to what can be linearly constructed within the tangent space to each manifold point.[7] Secondly, Weyl took from Riemann and Lie the metaphysical command of "*Nahewirkungphysik*", that "the true lawfulness of nature is expressed in laws of nearby action, connecting only values of physical quantities at space-time points in the immediate vicinity of one another". (1927, p. 61; 1949, p. 86). Weyl's idea didn't work in the context of general relativity but rather directly prompted his purely mathematical work in 1925–6 on Lie theory (i.e., on representations of semisimple Lie groups and 'Lie algebras').[8] Weyl (1929) carried the gauge principle over to quantum theory where he (and others) showed it to be a phase invariance in quantum electrodynamics (QED) . Finding Lie group structure in both general relativistic manifolds and in quantum electrodynamics convinced Weyl that objectivity in physical theory is constituted as an invariance "for a subject with its continuum of possible positions", and that it arises in step-by-step symbolic construction from a (linear) basis of what is *aufweisbar* (evident), "something to which we can point to *in concreto*" as demonstrably evident to the constituting mind.[9] "Symbolic construction" would become Weyl's term of

---

[6] Weyl (1923) is the comprehensive presentation; for discussion see Ryckman (2005) and the refences cited there.

[7] Torretti (2008, p. 92) offers a similar suggestion: In place of "the pure forms of sensibility" the constituting subject has available "the intuitive 'form' in every *Erlebnis* [that] merely adumbrates the mathematical notion of a – should we say four-dimensional? – coordinate patch".

[8] See Hawkins (2000, Chapters 11 & 12) and Eckes (2013). 'Lie algebra' is the term coined by Weyl in 1934 lectures at the Institute of Advanced Study for what Lie, Killing and Cartan referred to as an "infinitesimal group". Weyl showed that it designates a linear vector space structure at the identity of the Lie group from which most, not all, of the information of the group can be derived.

[9] Weyl (1954, p. 628 & p. 627): "(T)he constructions of physics are only a natural prolongation of operations [the] mind performs in perception, when, e.g., the solid shape of a body constitutes itself as the common source of its various perspective views. These views are conceived as appearances, for a subject with its continuum of possible positions, of an entity on the next higher level of objectivity: the three-dimensional body. Carry on this 'constitutive' process in which one rises from level to level, and one will land at the symbolic constructions of physics. Moreover, the whole edifice rests on a foundation which makes it binding for all reasonable thinking: of our complete experience it uses only that which is unmistakably *aufweisbar*." "... The words 'in reality' must be put between quotation marks; who could seriously pretend that the symbolic construct is the true real world?" In fact, Weyl uses the term 'symbolic construct' to encompass not merely the symbolic universe in which physical systems, states, transformations and evolutions are mathematically defined in terms of manifolds, functional spaces, algebras, etc., but also the symbolic specification of idealized procedures and experiments by which the basic physical quantities or observables of the theory are related to observation and measurement. It reflects an insistence, reinforced by

art for epistemological reflection upon the enterprise of theoretical natural science, revealing a step-by-step process of constitution of objects *as known*, beginning from the evidentially privileged local standpoint of the constituting subject. Reference to a mind-transcendent world purportedly portrayed in physical theory is then to be 'critically' understood as reference to this symbolic construction.

Revived and generalized to non-Abelian groups by Yang and Mills (1954), the gauge idea's mandate of *radical locality* of all interactions plays a central unifying role in the Standard Model (SM) of elementary particles. In the modern view the gauge idea is the idea of local symmetry as a dynamical principle in field theory. In general, a field is a dynamical object that has a value at every point of the space it occupies. In quantum fields, these are values in a so-called "internal" or "configuration" space. A local symmetry is the invariance of the field theory's laws (and predictions) if a global symmetry of the theory's Lagrangian (in which the dynamical variables have the same value everywhere) is "gauged", i.e., is required to be a local symmetry. Thus the field laws remain invariant as the dynamical variables are allowed change independently at each point of space (or space-time).[10] Remarkably, the imposition of local symmetry requires the existence of interactions; in quantum fields, i.e., it introduces an arbitrary vector function signifying a local symmetry in the internal dynamical space associated with each space-time point.[11] These degrees of freedom serve to represent interactions as proceeding through the exchange of spin 1 bosons such as the photon of QED, or the gauge bosons of the electro-weak interaction ($W^{\pm}, Z^0$). Both Lie groups and Lie algebras play prominent roles in the development of the gauge principle leading up to the SM; the latter play a particularly important role inasmuch as nearly all the information of interest about the structure of a gauge field theory can be ascertained by focusing on infinitesimal gauge transformations, i.e., group elements continuously connected to the group identity. However, to arrive at the SM, a fundamental difficulty with quantum field theories had first to be mastered: to show that the interactions described by theory are 'renormalizable', i.e., mathematically tractable and predictive. It turned out that the massless (QED) and massive gauge theories (provided the masses are generated by spontaneous symmetry breaking via the Higgs mechanism[12]) are renormalizable.

---

quantum mechanics, that physical quantities (beginning with 'inertial mass') are not simply given, but "constructed" See especially Weyl (1931, p. 76).

[10] The terms "local" and "global" are a bit misleading since the fields and their transformations ostensibly "live in" an internal dynamical space but the local gauge transformations are functions of the space or space-time coordinates at the given point.

[11] Gauging a global symmetry mandates (to restore invariance of the field Lagrangian) introduction of a gauge-covariant derivative; the new derivative is required to transform in a manner that introduces a new (gauge) field; the gauge field provides the form of the interaction forces of a matter field. The same mathematical expressions appear with only minor changes in the different quantum fields of the SM, e.g., in place of the phase of the electron field in QED, there are generalized phases associated with the wave functions of multicomponent matter fields.

[12] A symmetry of a system is said to be "spontaneously broken" if its lowest energy state is not invariant under the operations of that symmetry. This is an extremely important concept in the weak interaction as the bosons introduced by gauge symmetries are massless, like the photon; their

4.4 Quantum Field Theory and the Problem of Renormalization

Since the earliest days of quantum electrodynamics in 1929–1930, quantum field theory (was plagued by the problem of infinities, a legacy of modeling particles and their interactions as point-like. Like the classical electron, the point-like relativistic Dirac electron had an infinite self-energy (energy of the electromagnetic field generated by the electron, in addition to the energy of interaction of the electron with this field) understood as an electrostatic repulsion manifested by the electron's continual emission and re-absorption of a photon. In particular, processes in which electrons continually emit and reabsorb photons apparently shift the energy of the electron state by an infinite amount. These infinities were discovered in 1930 by Oppenheimer, calculating these effects for the bound electron in a hydrogen atom, and by Waller, calculating the effects for the free electron. Adding up these energies in perturbation theory in powers of the unit of electric charge e, the next term of the perturbation series, proportional to e^2, is already infinitely large. Thus, when Dirac electron theory is pushed beyond lowest approximation (the range of the Dirac equation), it yields ultraviolet divergent results for radiative self-energies. Other formal infinities also cropped up e.g., 'polarization of the vacuum', as Dirac called the process of an electron's attraction of virtual positrons and repulsion of virtual electrons; taken together, the infinities are the "self-energy problem."

In order for the quantum electrodynamic theory (QED) to make sense, the formal infinities had to be removed or circumvented. However, by the experimental standards of the time, the Dirac equation was empirically adequate; for an electron in a Coulomb field there were no discrepancies from the predictions derived from the Dirac equation until after WWII (the so-called 'Lamb shift'). In the late 1940s Tomonaga, Schwinger, and Feynman resurrected QED by independently developing methods to deal with the infinities. Taken as a whole, they provided an unambiguous recipe to calculate to any desired accuracy phenomena resulting from coupling of electrons to the electromagnetic field. For this they received the Nobel Prize in 1965. Dyson, synthesizing their results, showed that in a certain class of QFTs the infinities are of such a kind that they can be eliminated by renormalization, i.e., by redefining fundamental parameters of the theory (such as masses and charges) are by identifying them with the measured values. The observed finite values of electron mass and charge are not identified with the parameters m_e and e appearing in the QED Lagrangian, but with the mass and charge calculated when the clouds of virtual photons and electron-positron pairs are taken into account. The mass of the electron in the Lagrangian is then theoretically the sum of its bare mass and the mass associated with the energy of interaction with its own electromagnetic field. To render the measured mass finite, the bare mass must be

masses arise from the "spontaneous breaking" of the $SU(2) \times U(1)$ symmetry through couplings to the scalar Higgs field.

given an infinite negative value that cancels the positive infinity of the self-energy. In short, the infinities could be removed by introducing infinite counter-terms. Renormalized QED became the ideal model for physical theory even though it required defining QED by a limiting procedure. An energy or momentum cut-off Λ is introduced so that parameters of the renormalized theory can then be expressed in terms of physically measurable quantities (e.g., mass and charge of particles, scattering probabilities). This modifies physics at momenta or distances $< \Lambda$ so that calculations make sense (eliminate divergences). Physics at scales higher than the cut-off (or, at shorter distances than the corresponding cut-off length scale) can then be redefined, so that all calculated quantities become finite but cut-off dependent. However, the renormalization procedure in principle allowed $\Lambda \rightarrow \infty$,seemingly permitting valid description of nature up to arbitrarily high energies.

Physicists subsequently learned that QED is atypical of QFT; renormalization theory, as developed in context of QED, does not apply to all interactions. Attempts to construct renormalizable quantum field theories of other elementary particle interactions failed, for some two decades leading to demise of field theory and the rise of alternative approaches to the weak and strong interactions (S-matrix, current algebra). Influential theorists (Dirac, Landau, even Dyson and Schwinger) became critics of renormalization procedure. In the late 1960s, field theory re-emerged triumphant when Weinberg, and independently Salam, showed that Glashow's $SU(2)$ theory of weak nuclear forces could be made to work by borrowing the idea of broken symmetry from condensed matter theory. Glashow's basic idea was that just as electromagnetic forces are transmitted by exchange of photons, hypothetical "intermediate" bosons transmit the weak nuclear force. These particles are massive (they act at short distances) but are related to the photon by a continuous global $SU(2)$ symmetry, where '2' indicates that the symmetry acts differently on left-handed and right-handed components of the different particles. This posed something of a dilemma. The different types of bosons meant that the global symmetry was broken. A known result, Goldstone's theorem, stated that a broken global symmetry would give rise to massless particles yet these were not observed. In fact, the hypothetical bosons were not massless, indeed they were so heavy they had still to be revealed in high energy experiments. Weinberg and Salam then explained that the symmetry of $SU(2)$ is energetically unstable and is spontaneously broken by a massive scalar (Higgs) field that gives mass to the heretofore massless $W^{\pm}$ and neutral Z^0.

At accessible energies the $SU(2)$ theory is unified with QED into a "electroweak" theory, whose gauge group $SU(2) \times U(1)_Y$ is "spontaneously broken" to $U(1)_{EM}$. [13]It initially seemed that the theory of weak interactions could not be renormalized since its 'intermediate bosons' have large masses. However G. 't Hooft and M.

[13] Technical detail: Symmetry breaking allows the full electroweak gauge symmetry $SU(2) \times U(1)_Y$ ('Y' for *weak hypercharge*) be replaced by the electromagnetic subgroup $U(1)_{EM}$ at low energies. The "broken" $U(1)$ gauge group is a linear combination of the original $U(1)$ and a subgroup of $SU(2)$.

Veltman in 1971 and subsequently others showed that the various multiparticle exchanges involving photons, charged intermediate vector bosons, neutral intermediate vector bosons, and other particles add up so as to cancel all non-renormalizable infinities. The new bosons with the predicted masses were discovered in the mid-1970s and Glashow, Weinberg and Salam received the Nobel Prize for the electroweak theory in 1979. The last remaining piece of the Standard Model came in 1973 when D. Gross, K. D. Politzer and F. Wilczek produced an $SU(3)$ gauge field theory of the strong interaction, the theory of so-called quantum chromodynamics ('3' refers to the triplet of quark 'colors', a species of charge). Because the strong force is strong, all possible virtual exchange processes had to be taken into account at each stage of calculation. Theoretical treatment seemed hopeless until Gross, Politzer and Wilczek showed that the strong interaction among quarks is asymptotically free (the strength of the strong interaction decreases as distances between interacting quarks grows shorter). At high enough energies (sufficiently short distances) the force was weak enough to be amenable to the approximate methods of renormalization theory. Gross, Politzer and Wilczek received the 2004 Nobel Prize for the discovery of "asymptotic freedom". By 1975 or so, the Standard Model of the strong, weak and electromagnetic forces was considered complete. The fundamental dynamical interactions of all known matter fields are theoretically cast by the SM as a direct product of the local semi-simple Lie symmetry groups $SU(3) \times SU(2) \times U(1)$.

4.5 Effective Field Theory: A New View of Renormalization and of QFT

At present, *known* physical laws currently regarded as fundamental are the quantum field theoretic Standard Model of the three non-gravitational interactions and the classical Einstein theory of gravity, General Relativity. Their unification, if it exists, may be the locus of *truly* fundamental laws, in the sense of being the ultimate stopping point, the unknown bedrock from which (to use Steven Weinberg's (1987) metaphor) all arrows of explanation originate. Few, if any, theoreticians consider either the Standard Model or General Relativity to be *truly fundamental* in this sense.[14] Soon after its inception in the mid-1970s, many quantum field theorists and cosmologists have tended to view the Standard Model, the $SU(3) \times SU(2) \times U(1)$ gauge theory of the strong and electroweak interactions, as merely a provisional stage in the descent to ever smaller distance scales corresponding to an ascent to the ever-higher energies of the early universe. Correspondingly, it is known that General Relativity generically gives rise to singularities, and theoretically breaks down at

[14] Besides the limitations to be mentioned, the measured values of some 26+ free parameters in the SM lack theoretical explanation, nor does the SM tell us why it contains three families of fermions.

smaller and smaller scales; below Planck scale $\left(Gh/c^3\right)^{\frac{1}{2}} \approx 10^{-35}$m the effects of quantum gravity are believed to be significant. But in the current physical world picture, the laws of fundamental interactions within the Standard Model together with Einstein's non-quantum theory of gravity describe a low-energy regime of the universe out of which we and our apparent surroundings are made; bracketing effects of so-called dark matter and dark energy, they comprise what can be called *known physics*. These laws might legitimately be termed fundamental since they can be invoked to explain all (in principle, and with the provisos about dark matter, etc.) physical phenomena over the truly impressive range from 10^{-13}cm (the scale of quark interactions) to 10^{29}cm. (the extent of the observable universe).

Still, from the haughty perspective of an unknown and possibly non-existent future "theory of everything", laws currently deemed fundamental may have a parochial or environmental character even as certain conceptual structures and principles such as locality and symmetry underlying the Standard Model and General Relativity may be explanatorily deeper than their implementation in either. At both smaller (down to Planck scale 10^{-33} *cm*) and larger scales (the Multiverse?), the physics remains unknown, and perhaps unknowable into the distant future due to the high energies and vast (spacelike) distances that render experiment and observation beyond the reach of current (14 TeV at the LHC; 1 TeV $= 10^3$ GeV) or foreseeable technologies. One senses here a dialectical situation kindred in spirit to the storied debate of regularity and necessitarian theorists. Known fundamental laws are believed to be not truly fundamental; they seem contingent regularities in the sense that their explanatory validity is restricted to low-energy epochs of the universe. The *pro tem* objectivity rendered by these laws, if such it is, may originate only in conditions that are accidental or environmental according to both string theory and most models of inflationary cosmology. It is instructive that theoretical physicists themselves have come to terms with this provisional status of fundamental laws by developing, since around 1980, the idea of an "effective field theory" (EFT). In this view, the SM is but an "effective field theory" valid only up to the level of accessible energies but ill-defined at higher energies. This view of the SM is coupled with one where GR, governing largescale phenomena, is merely a low-energy approximation to a quantum theory of gravity. If philosophical consideration of contemporary physical theory is brought to bear on the fundamental theories in known physics, the surprising finding is that the validity of theoretical representations governing physical phenomena extends only to within the bounded energy domains of the respective theory. Clearly we see here need for the philosophical virtues of "open-endedness and detotalization" sought by Torretti. (2008, p. 93)

The idea of an "effective field theory" (EFT) originated with Steven Weinberg (1979) and has been especially advocated by Harvard theorist Howard Georgi (2009).[15] A convenient way to think about the concept of an EFT is that it is a new

[15] Bain (2013) is an overview written for philosophers.

viewpoint on renormalization whereby it is nothing more than a parameterization of the sensitivity of renormalizable low-energy physics to non-renormalizable high-energy physics. (Cao, 1993) In the original perspective, renormalization is a "constitutive" principle of QFT requiring couplings to remain small so that calculations are reliable; any non-renormalizable theory is intrinsically defective by giving rise to meaningless mathematically infinite results. Recall that in the earlier view, a renormalized QFT has no fixed built-in mass or energy scales; though momentum or energy cut-offs Λ are employed in the term by term renormalization group procedure, in principle, the validity of a renormalized theory can be taken to the limit $\Lambda \rightarrow \infty$. The theory can then be regarded as both mathematically consistent and predictive at any energy scale. But the new 'philosophy' of renormalization stresses instead that non-renormalizable theories can be tamed by adopting energy bounds, since the physical problem with non-renormalizability after all concerns not so much mathematical consistency but predictivity. This change of emphasis is important for it anchors the idea that the Lagrangians of quantum field theory are first, and foremost, phenomenological; they are devised to permit systematic calculation of renormalizable interactions for successful prediction of the phenomenology of experiment at higher and higher energies. In brief: the low-energy world accessible to current experiment is described by renormalizable theories because these are the only theories relevant at low-energies.

As an illustration, consider QED. As previously understood, renormalized QED could be seen as, in principle, a valid description of nature up to arbitrarily high energies. But as incorporated into the SM, QED is an effective description of electromagnetic interactions at energies low compared to scales where new physics can be expected, e.g., at grand-unified (GUT) scale of 10^{15} GeV. QED has approximate validity at 'long distances', that is, at distances comparable to the Compton wavelength of the electron $\lambda_e = h/m_e c$.[16] At this distance, energies are not available to produce more massive particles and they can be excluded from the theory (the QED Lagrangian). But it is known experimentally that at mass scales of ~100 GeV, these particles are definitely around; as described above, heavy charged $W^{\pm}$ and neutral Z^0 bosons that mediate the weak nuclear reaction, allowing protons and neutrons to convert into one another in radioactive decays, appear in the $SU(2) \times U(1)$ electroweak theory of Glashow, Salam and Weinberg. Thus, at mass scales of ~100 GeV, QED is assimilated into the electro-weak theory, part of the Standard Model that contains QED as a low-energy approximation. Viewed as an effective field theory, QED is then a theory involving only the electron and the photon. Since all observable phenomena of QED involve only these interactions, nothing is lost in omitting heavier particles (and their fields) from the Lagrangian. Though now regarded as an EFT whose validity is now bounded by a given energy scale, QED's predictivity has not been altered one whit.

[16] The Compton wavelength is the wavelength of the quantum wave associated with a particle of mass m;the mass of the electron is m_e.

More generally, although quantum field theory in principle concerns particles of very disparate masses, the physics that can be studied in experiments obviously is manifest only at energies accessible by existing technologies. Rather than restricting the theoretical description to the physics below the cut-off as in renormalization, an EFT can contain terms in the Lagrangian for the non-renormalizable interactions of heavier particles. Indeed, the requirement of renormalizability is replaced with a condition on the non-renormalizable terms. Of course, including non-renormalizable interactions has a price: the theoretical description becomes much more complex as the EFT must contain *all* interactions allowed by the symmetries (Lorentz and gauge) of its Lagrangian, symmetries that might be presumed to provide information about the laws of nature at higher, though at present experimentally inaccessible, energies. The condition on the non-renormalizable interactions is along the following lines.[17] Since the masses and energies of non-renormalizable interactions are very large (say of order M) compared to the masses and energies of the effective particles (say, at energy scale $\mu \ll M$) the effects of the non-renormalizable interactions will be small as they are represented in the Lagrangian by terms that are suppressed by powers of μ/M. It is then possible to ignore heavy particles of mass M and to consider the effective theory as renormalized to scale μ. On the other hand, if effects of heavier particles of mass M_i are detected at accessible energies, a new effective theory is required, with the new particle's Compton wavelength $\lambda_i = h/M_i c$ taken as the boundary between the two theories. At distances larger than $\lambda_i = h/M_i c$ or energies $\mu < M_i$, the lower energy theory omits the particle and has a cut-off $\Lambda = M_i$. For distances smaller than $\lambda_i = h/M_i c$ or energies $\mu \approx M_i$, the higher energy theory must include it so that its interactions, non-renormalizable in the lower energy theory, become renormalizable. Thus there is a kind of "matching" so that the two theories describe essentially the same physics in the boundary region between them. In this way, a "tower" of effective field theories in principle can be constructed to possibly arbitrarily high energies or shorter distances. There may or may not be a termination to this process (an "ultraviolet complete" theory), but within the limits imposed by its cut-off, an effective field theory is completely empirically adequate within the limits set by that cut-off while remaining ill-defined beyond it. While it is to be expected that the effects of non-renormalizable interactions grow and become more significant as one goes up in energy scales, 'integrating out' the super high energy/mass degrees of freedom in a fundamental theory occasions no loss of descriptive or predictive capacity since the energies at which these degrees of freedom become relevant remain inaccessible.

[17] Georgi (2009) provides a rigorous treatment, informally exposited here.

4.6 Conclusion

I submit that even this brief overview of gauge symmetries of the Standard Model and the effective field theory approach to renormalization is sufficient to point out several significant resonances within Torretti's "perspective on Kant's perspective on objectivity". First, as prescribed by Torretti (and following Cassirer), we pruned Kant's "productive imagination" or "power of reflective judgment" to broadly connote the spontaneity of thought manifested in the idea of *combination* (B 130). It is this act of the subject's self-activity that underlies the cognitive freedom to form new mathematical concepts from previous concepts and "ultimately perhaps from pre-conceptual intuitions". In the above narrative this is to consider objectivity in physical theory as an invariance "for a subject with its continuum of possible positions", each locus symbolically constructed as the domain of validity of an infinitesimal (hence, "evident") Lie algebra. Unverifiable global symmetries of a theory are then reconstructed under the idea of "radical locality": by gauging a global symmetry, the field Lagrangian remains invariant even as the smooth Lie group actions are allowed to differ from point to point. We then have here a contemporary "historical" case illustrating.

> how a new concept or structured panoply of concepts issues from those that were previously available – and ultimately perhaps from pre-conceptual intuitions – through an intellectual effort resembling that of Kant's productive imagination or reflective judgment. (2008, p. 93)

Secondly, the effective field theory approach is above all devised as a program for predicting, through step-by-step construction, the phenomenology of high-energy experiment as new technologies may evolve to extend the range of exploration to higher energies. As seen above, the explanatory validity of an effective field theory is restricted within a certain energy regime by the fact that the theory is not well-defined beyond it. In fact, although it is possible to imagine a tower of effective field theories extending to arbitrarily higher energies, no one can have the confidence to assert that this ascent will actually be achieved, for that depends on two unknowns: the relevant technologies for required experimental test of theory, and, perhaps more fundamentally, whether or not such an ascent would require a sharp break from the principles of quantum field theory. We have here, quite in accord with Torretti's gloss on Kant's Transcendental Dialectic,

> [a] conception of the world as an *Idee* or program for the local, partial, approximate construction of objects by the ever imperfect but endlessly perfectible modeling according to *our* lights. ... whereas the scientific realists must relegate the achievement of objectivity to the end of the road, the Kantian scientists can take pride in their daily achievements of contextual, improvable, incomplete, but reasonably working and passably stable objective truths. (2008, p. 81)

Finally, on a more general note, Torretti is rightly critical of Kant's essentially Aristotelian view of concept formation and, while recognizing that Kant "certainly knew" that "concept formation in mathematical physics does not proceed in this way", observes that "Kant remains beholden to the medieval conception of mathematics as the science of quantity and blissfully unaware of the geometrical significance of non-quantitative properties and relations". (90–92) Recall that in the Analogies of Experience (B 110) the concepts of the understanding are divided into two divisions, *mathematical* and *dynamical.* This distinction re-appears in the table of principles (A160-2/B199-200) and in the "proof" of the general principle of the analogies (A178-81/B220-24). Mathematical concepts or principles (quantity and quality) are concerned with objects of intuition, either as constructed in pure intuition or as possible in empirical intuition (as a synthesis). They pertain to appearances as regard to their mere possibility, and determine them as numerical magnitudes; as such they are constitutive. On the other hand, dynamical principles (pertaining to the categories of relation and modality) bring the *existence* of appearances under rules *a priori.* But as existence cannot be constructed, dynamical principles are to be merely regulative. Indeed, the three analogies (principles of persistence of substance, of temporal sequence according to the law of causality, and of simultaneity, according to the law of interaction or community) each pertain to a dynamical principle and so,

> An analogy of experience will therefore be only a rule in accordance with which unity of experience is to arise from perceptions (not as a perception itself, as empirical intuition in general, and as a principle it will not be valid of the objects (of the appearances) *constitutively* but merely *regulatively*. (A180/B222)

Ironically perhaps, the distinction between mathematical and dynamical principles may serve for a superior account of concept formation in mathematical physics in the contemporary framework of theoretical representation of nature. This is a framework (possibly Lagrangian, possibly Hamiltonian) within which particular empirical laws of nature can be set, the necessity of temporal determination of the framework infusing the particular laws. But the latter systematize particular empirical phenomena. To do this, they require theoretical construction of concepts such as fields that are qualitative: flexible enough to accommodate the wide range of target phenomena while being rigid enough not to be reducible to them. Because of their tempered rigidity, the concepts themselves are continually developed and refined, both formally and empirically, through further application to wider and wider ranges of phenomena. As the Analogies intimate, the necessity of laws of particular interactions imputed by theory is regulative because it pertains to existences, to specific empirical phenomena. As *regulative*, attributions of necessity stem from what the Transcendental Dialectic terms the "hypothetical employment of reason", the injunction to seek systematic unity that is a presupposition of any physical law. The *pro tem* necessity of particular physical laws is then a prime example of what Buchdahl (1969, p. 511) termed "Kant's method of converting metaphysical principles into something possessing purely methodological force". The known fundamental laws of nature, as viewed through the lens of the concept of EFT, appear to be a remarkable confirmation of a broadly Kantian view of laws. By specifically limiting the domain of validity of particular laws pertaining to specific energies, the laws of fundamental interactions within an effective field theory can be seen as superposing rationalism and empiricism. At the same time, the effective field theory program is yoked to a methodological and regulative injunction to seek a general "unity of nature" by producing effective theories extending to higher and higher energies. The framework does not guarantee that the "tower" of EFTs will terminate in a unique ultraviolet complete theory at the highest theoretical energies; to do so would be to succumb to transcendental illusion, to the demand of reason for an unconditioned totality, that by definition, can never be an object of possible experience. The concept of effective field theory serves to philosophically illuminate of the cognitive role and status of contemporary fundamental physical laws represented in the quantum field theories of the Standard Model as well as in the cosmological application of Einstein's relativistic theory of gravity.

References

Bain, J. (2013). Effective field theories. In R. Batterman (Ed.), *The Oxford handbook of philosophy of physics* (pp. 224–254). Oxford University Press.

Buchdahl, G. (1969). *Metaphysics and the philosophy of science: The classical origins Descartes to Kant.* Basil Blackwell.

Cao, T. Y. (1993). New philosophy of renormalization: From the renormalization group equations to effective field theories. In L. Brown (Ed.), *Renormalization from Lorentz to Landau (and beyond)* (pp. 87–133). Springer.

Cassirer, E. (1929). Philosophie der symbolischen Formen, *Dritter Teil.* Berlin: Bruno Cassirer. Pagination according to reprint by Felix Meiner Verlag: Hamburg, 2010.

Cassirer, E. (1957). *The Philosophy of Symbolic Forms* (R. Manheim, Trans. (1929), Vol. 3). Yale University Press.

Dedekind, R. (1893). *Was sind und was sollen die Zahlen ? Zweite unveränderte Auflage*. Friedrich Vieweg und Sohn.

Eckes, C. (2013). *Les groups de Lie dans l'oeuvre de Hermann Weyl.* Presses Universitaires de Nancy/Éditions Universitaires de Lorraine.

Georgi, H. (2009). *Weak interactions and modern particle theory: Revised and updated.* Dover Publications.

Hawkins, T. (2000). *Emergence of the theory of lie groups; an essay in the history of mathematics.* Springer.

Kant, I. (1997). Critique of pure reason. (P. Guyer & A. W. Wood, Trans.). In *The Cambridge Edition of the Works of Immanuel Kant.* New York and Cambridge: Cambridge University Press.

Ryckman, T. A. (2005). *The reign of relativity: Philosophy in physics 1915-25. (Oxford studies in the philosophy of science).* New York, Oxford University Press.

Torretti, R. (2008). Objectivity: A Kantian perspective. In M. Massimi (Ed.), *Kant and the philosophy of science today : Royal Institute of philosophy supplement: 63* (pp. 81–94). Cambridge University Press.

Weinberg, S. (1979). Phenomenological Lagrangians. *Physica, 96A*, 327–340.

Weinberg, S. (1987). Towards the final Laws of physics. In R. P. Feynman & S. Weinberg (Eds.), *Elementary particles and the Laws of physics: The 1986 Dirac memorial lectures* (pp. 61–110). Cambridge University Press.

Weyl, H. (1923). *Raum-Zeit-Materie*. 5 Auflage. J. Springer.

Weyl, H. (1927). *Philosophie der Mathematik und Naturwissenschaft.* R. Oldenbourg.

Weyl, H. (1929). "Elektron und Gravitation." *Zeitschrift für Physik***56** (1929), 330–352; reprinted in K. Chandrasekharan (ed.), *Hermann Weyl: Gesammelte Abhandlungen* Bd. III (pp. 245–267). Springer Verlag, 1968.

Weyl, H. (1931). *The theory of groups and quantum mechanics.* (H.P. Robertson, Trans. 2nd Ed.) Methuen and Co., Ltd. Reprinted. Dover, 1950.

Weyl, H. (1949). *Philosophy of Mathematics and Natural Science.* (O. Helmer, Trans.). Princeton University Press. Revised and augmented version of Weyl (1927).

Weyl, H. (1954). "Address on the Unity of Knowledge Delivered at the Bicentennial Conference of Columbia University."; reprinted in K. Chandrasekharan (ed.), *Hermann Weyl: Gesammelte Abhandlungen* Bd. IV (pp. 623–629). Springer Verlag, 1968.

Yang, C. N., & Mills, R. L. (1954). Conservation of isotopic spin and isotopic gauge invariance. *Physical Review, 90*, 191–195.

Chapter 5
A Kantian-Rooted Pluralist Realism for Science

Olimpia Lombardi

Abstract After the preeminence of logical positivism/empiricism during the most part of twentieth century, during the last decades many authors began to recognize the relevance of the Kantian thought for present-day philosophy of science. This chapter follows this general trend, adopting a realist reading of Kantian teachings. On this basis, I will delineate a Kantian-rooted realism according to which the worlds of science are always the result of a synthesis between the conceptual schemes embodied in scientific theories and practices and the independent noumenal reality. However, my position takes distance from the Kantian doctrine by admitting the possibility of different conceptual schemes, both diachronically and synchronically. This view not only leaves room for abrupt and discontinuous changes in the history of science, but also leads to an ontological pluralism that allows for the coexistence of irreducible and different, even incompatible ontological domains at the same historical time. I will focus particularly on the synchronic case to reject both ontological reductionism and emergentism from a perspective that denies any priority or dependence between domains, in resonance with a non-hierarchical articulation between scientific theories and disciplines.

5.1 Introduction

The idea that science moves forward by reaching truths as it unravels the veil of reality is strongly present in many areas. In addition, it is usually supposed that certain disciplines and theories count with the privilege of describing the fundamental level of reality in itself. The fact that this view pervades the scientific community is not surprising, to the extent that it is an intuitive, commonsense realism. What is unexpected is that this pre-Kantian position is still powerful among philosophers of science, when they try to explain how scientific knowledge evolves by approaching to closer-to-truth theories, or when they are concerned with

O. Lombardi (✉)
University of Buenos Aires and CONICET, Buenos Aires, Argentina

C. Soto (ed.), *Current Debates in Philosophy of Science*, Synthese Library 477,
https://doi.org/10.1007/978-3-031-32375-1_5

explaining how secondary disciplines and phenomenological theories are related to the fundamental descriptions of the world.

The aim of this article is to recall Kant's insights, and to apply them in the philosophy of science. But learning from a philosopher does not amount to glossing and repeating her works: complete fidelity is not necessary to recognize their value. I will find my inspiration in Kantian philosophy to face present-day problems, although at some point I will be forced to depart from the philosopher's doctrine. In fact, Kant's philosophy is built upon two main ideas, which were very novel at his time: the role of the subject in the constitution of the object, and the transcendental character of categories that makes them the *a priori* conditions of any possible knowledge. I will definitely accept the first idea, but at the same time I will reject the second one in its original version, turning it into the view of a relative *a priori* that underlies any theoretical body of knowledge. This not-completely Kantian but Kantian-rooted perspective will allow me to argue for a pluralist realism that, I will claim, is highly fruitful to understand the phenomenon of science.

5.2 The Constitution of the Phenomenal World

In his short story "Funes, the memorious" ("*Funes, el memorioso*") the Argentinean writer Jorge Luis Borges tells us about Ireneo Funes, who suffered a horse-riding accident after which his sensibility and his memory became *absolute*:

> He knew by heart the forms of the southern clouds at down in the 30th of April, 1992, and could compare them in his memory with the mottled streaks on a book in Spanish binding he has only seen once and with the outlines of the foam raised by an oar in the Rio Negro the night before the Quebracho uprising. (Borges, 1944a: 68–69)
>
> Funes remembered not only every leaf on every tree of every wood, but also every one of the times he had perceived or imagined it. (Borges, 1944a: 70)
>
> Not only was it difficult for him to comprehend that the generic symbol *dog* embraces so many unlike individuals of diverse size and form; it bothered him that the dog at three fourteen (seen from the side) should have the same name as the dog of the three fifteen (seen from the front.) (Borges, 1944a: 70)

Funes not only remembers everything, but also perceives everything: his memory and his sensibility are *perfect*. However, Borges explains us that this capacity is far from being a positive gift: Funes lives a sad life in a "multiform, instantaneous and almost intolerably precise world". And finally Borges admits:

> I suspect, however, that he was not very capable of thought. To think is to forget differences, generalize, make abstractions. In the teeming world of Funes, there were only details, almost immediate in their presence. (Borges, 1944a: 71)

I think that Borges proposes to us a kind of *reductio ad absurdum* game: let us suppose that Funes exists with his absolute capacities and extract all the consequences of this assumption; we will conclude that Funes cannot exist as so described, he could not think and talk as Borges tells us in his story.

Funes's sad life makes us to recall that the perfect access to reality is neither a possible nor a desirable ideal. We always access the world through our own frameworks. In Kantian terms, the object of knowledge is always the result of a synthesis between our conceptual schemes and the independent noumenal reality. This fact is at odds with metaphysical realism and its commitment with a "ready-made" world, inhabited by self-subsistent and self-identifying objects. A kind of realism that also presupposes the possibility of knowing that world as it is in itself, at least approximately, and which becomes a scientific realism when the thesis that science is the privileged means to reach that knowledge is added. As clearly expressed by Roberto Torretti:

> «Scientific realists» believe that reality is well-defined, once and for all, independently of human action and human thought, in a way that can be adequately articulated in human discourse. They also believe that the primary aim of science is to develop just the sort of discourse which adequately articulates reality – which, as Plato said, «cuts it at its joints» –, and that modern science is visibly approaching the fulfilment of this aim. (Torretti, 2000: 114)

However, this kind of realism is disproved by the scientific activity itself. Not only the properties and the behavior of the entities studied by science, but also the very existence of those entities depend on the categorical-conceptual framework implicit in a theory. As framework-dependent is the response to the question 'how is it?' as is the answer to the question 'what is it?' As Hilary Putnam claims: "'Objects' do not exist independently of conceptual schemes. We cut up the world into objects when we introduce one or another scheme of description" (Putnam, 1981: 52). In this sense, Putnam's internalist realism recognizes himself as a debtor of Kantian philosophy. Does this mean that we should embrace idealism?

5.3 Why Kantian *Realism*?

Of course, we have all learnt that Kantian philosophy is a form of idealism. And this is so because, although the existence of something independent of the subject is conceded, the access to the subject-independent reality is always permeated by the active participation of the subject. The object of knowledge is the result of a synthesis between the subject-independent reality and the framework introduced by the knowing subject. But, if knowledge essentially requires the two elements, why insist on calling this view 'idealism'?

In fact, using the term 'idealist' to describe Kantian philosophy is a realist prejudice. It seems that whenever the subject participates in the constitution of knowledge, the pure essence of realism breaks down. It seems that the only philosophical position that deserves the name 'realism' is what Putnam calls 'metaphysical realism', the "God's Eye" point of view, according to which "the world consists of some fixed totality of mind-independent objects" and "[t]here is exactly one true and complete description of 'the way the world is'." (Putnam, 1981: 49).

I want to resist that prejudice. As a consequence, I will talk of a *Kantian-rooted realism* for two reasons. First, due to the role played by noumenal reality in Kantian philosophy. Although, against pre-critical realism, Kant insists on the unknowability of noumena, he also stresses the idea that a phenomenon involves in itself the reference to something that is not phenomenon and necessarily takes part in the constitution of knowledge:

> the purely phenomenal character of the objects of experience does not exclude, but rather implies a transcendental reality that serves them as a basis, and that, although unknowable, is not for this less *effective*. [...] phenomenal objects are not mere insubstantial ghosts, [...] the perception in which their presence is manifested reveals an *effective existence*. (Torretti, 2005a: 676–677; italics added)

Although Kant compares our access to reality to that of a judge, who compels the witnesses to reply to those questions which he himself thinks fit to propose, the Kantian view does not force the responses: the independent reality must answer in the same language as that in which the questions were asked, but it can always respond negatively to those questions, making manifest its active presence.

The second reason for applying the term 'realism' to qualify a Kantian-rooted perspective is the possibility of retaining a correspondence theory of truth. Since metaphysical realism is traditionally linked to the adoption of truth as correspondence, it is usually assumed that the rejection of metaphysical realism necessarily implies giving up any correspondence view of truth in favor of some kind of coherentist or pragmatist conception. However, I think that this is not the case (see Lombardi & Pérez Ransanz, 2012). According to the correspondence view, a proposition is true if it corresponds to a fact. In general, the debates about this view turn around how it can be applied to different languages and what correspondence consists in. But in the relationship language-world, the pole "world" is usually not analyzed, under the assumption that it is the domain of the independent reality. Even in the case of the semantic view, the discussions regarding Tarski's Convention T,

'p' is true if and only if P

(where 'p' is the name of the proposition p of a language L and P is the translation of p into the metalanguage), usually focus on the relation between language and metalanguage, how to prove biconditionals in the metalanguage, and other formal matters, disregarding the discussion about how the fact referred to by p should be conceived. But such a fact might correspond to any domain (empirical, formal, fictional, etc.), whenever it is correctly defined. Therefore, nothing prevents p from referring to facts that are constituted in a Kantian sense. In other words, the active role of the subject in the constitution of the ontic domain to which language refers only leads us to reject the link between language and an independent reality, but does not force us to drop the intuition of truth as correspondence, central to realism, in favor of coherentist or pragmatist approaches.

5.4 The Plurality of Human Patterns

As shown above, Kantian resonances can be found in Borges's story about Funes. However, many clues about Borges's estrangement from the Kantian idea of transcendentality can also be found in his works. For instance, in "The analytical language of John Wilkins" ("El lenguaje analítico de John Wilkins"), he describes Wilkins's curious language and compares it with those of a Chinese encyclopedia and of the Bibliographic Institute of Brussels; from that comparison he concludes that

> it is clear that there is no classification of the Universe not being arbitrary and full of conjectures. The reason for this is very simple: we do not know what thing the universe is. (Borges, 1952: 103)

Nevertheless,

> The impossibility of penetrating the divine pattern of the universe cannot stop us from planning human patterns, even though we are conscious they are not definitive. (Borges, 1952: 103)

The human being must resign herself to be the "imperfect librarian" of "The Library of Babel": vain are the hopes of those who believe that "*there must exist a book which is the cipher and perfect compendium* of all the rest" (Borges, 1944b: 62); such a book would turn us into God, but it is not possible to locate it among the infinite number of volumes.

By contrast to Kantian philosophy, Borges tells us that it is not only that we cannot access to reality independently of our categorical-conceptual frameworks, but also that there is no privileged and definitive framework. Torretti clearly stresses the limitation of Kant's view: although Kant

> rightly compares the scientist with an investigating judge who directs his questions to nature and fixes the terms in which the answer has to be conceived, he never considers the case that, frustrated because the answers contradict with each other, the inquiries get complicated and stuck and, in general, do not seem to lead anywhere, the investigating judge rethinks his questions, modifies the categories in which the answers must be framed, and even changes the goals of the research or the criteria to evaluate their results.[1] (Torretti, 2005b)

Therefore, in order to account for this kind of situations, common in scientific research, Kantian views about the constitution of the object of knowledge must be complemented by *categorical-conceptual relativity*. According to this thesis, no concept, not even the most basic categories, need to be included in our categorical-

[1] "*compara atinadamente al científico con un juez instructor que dirige sus preguntas a la naturaleza y fija los términos en que tiene que venir concebida la respuesta, no se pone nunca en el caso de que, frustrado porque las respuestas se contradicen, las indagaciones se complican y atascan, y en general no parecen estar llegando a nada, el juez instructor repiense sus preguntas, modifique las categorías en que deben encuadrarse las respuestas e incluso cambie las metas de la investigación o los criterios para evaluar sus resultados.*" (Torretti, 2005b)

conceptual frameworks: there is no privileged concept of object or of existence. As Putnam claims:

> the phenomenon of conceptual relativity [...] turns on the fact that the logical primitives themselves, and in particular the notions of object and existence, have a multitude of different uses rather than one absolute 'meaning'. (Putnam, 1987: 19)

In other words, in their access to reality humans develop different frameworks, which are neither convergent nor reducible to a single objective one. Hence, the fundamental criticism of externalist realism is directed against its commitment to absolute categories and concepts. The idea that there is a unique correspondence between our words and frame-independent items, as if there were a metaphysical glue sticking language and world together, is an illusion.

Accepting the possibility of different categorical-conceptual frameworks leads to the thesis of *ontic pluralism*: *pace* Kant, there are different ontic domains, which are equally objective in different contexts and in function of certain interests and purposes. Ontic pluralism points in the same direction as the underdetermination of theories by evidence: both theses have been used against scientific externalism, which assumes that science converges to the true description of the independent reality. Since the privileged viewpoint of God's Eye does not exist, there is not a single "true" ontic domain: all the domains have the same metaphysical status if all of them are constituted by equally objective descriptions. It is worth emphasizing that they are not mere "epistemologized" domains as opposed to the "real" world: when there is no metaphysically objective ontic domain, the very expression 'epistemologized domain' loses any content.

According to this ontically pluralist view, the question 'What objects does the world consist of?' can only be meaningfully posed in the context of a particular framework. Only when we have adopted a system of categories and concepts, we can assume that certain facts and objects are there to be discovered. In other words, ontological questions only make sense from the perspective of knowledge. Supposing otherwise would amount to putting the cart of metaphysics before the horse of epistemology.

5.5 What Is a Categorical-Conceptual Framework?

On the basis of the above remarks, the Kantian-rooted pluralist realism just introduced heavily relies on the notion of *categorical-conceptual framework*, a notion clearly related with what several philosophers call 'conceptual scheme'. However, the move of adding the term 'categorical' to the notion is not a mere terminological whim, but carries very relevant consequences.

A categorical-conceptual framework is a system of *categories* and *concepts* that, in synthesis with the noumenal reality, constitute an ontic domain as something essentially new, in which the original components cannot longer be disjoined. In Kantian terms, a categorical-conceptual framework is condition of the possibility of knowledge, and although it is expressed through a language, it is not a linguistic entity itself: the same framework can be expressed in different languages. However, a categorical-conceptual framework can neither be identified with the mental structure of individual subjects, since it is a system of categories and concepts shared by a community: it is shaped and stabilized through social practices, not only linguistic but also material practices of manipulation and transformation, which suppose values, interests and common objectives. In turn, nothing prevents the same person from adopting different frameworks in different situations, being aware of the differences among those frameworks as well as between their corresponding ontic domains.

In order to understand the notion of categorical-conceptual framework, the first step is to distinguish between categories and concepts. According to Aristotle, there are ten categories: substance and nine types of properties. Kantian categories are twelve and are more abstract than those of Aristotle. Independently of the differences regarding the notion of category in these authors, it is quite clear that categories are not class concepts, such as "dog" or "blue", which apply to previously identified objects; categories are not *taxa*, such as "cat", "feline" and "mammal", which classify pre-existing individuals. Categories are the most basic structuring elements of both ontic and linguistic realms, logically prior to any ordering or classification (see Lewowicz, 2005). As a consequence, a system of categories does not establish a mere division into classes between items that are "out there", waiting to be classified. It provides a first identification of the items that populate the ontic domain, to the extent that it introduces the ontic categories to which such items belong. For example, the system of categories will tell us if the domain is inhabited by individuals, properties and relations, or if there are no individuals *stricto sensu* but only bundles of properties. It will tell us if possibility is an ontically irreducible feature of reality or it can be reduced to actuality. On the basis of the categories of the adopted framework we will able to say if there are causal links in the domain, as well as if the ontic items can be categorized as one or multiple, and if events are temporally arranged according past, present and future.

Besides categories, which introduce the most basic structure of the ontic domain, the framework can also include certain very generic concepts that refer to items whose existence and/or features cannot be denied. For example, in the framework of thermodynamics, the concept of heat is essential in the sense that thermodynamics is precisely the branch of physics that deals with heat and temperature, and their relation to energy, work, radiation, and certain properties of matter. Analogously, Newtonian mechanics cannot dispense with the concept of force, since it is an essential element in Newtonian laws.

The relevant difference between categories and concepts has been not sufficiently taken into account by certain contemporary philosophers. For instance, Thomas Kuhn (1983, 1993) conceives his "taxonomic categories" as a condition of possibility of knowledge; however he describes them as introducing different classifications that block complete inter-paradigm translation. In turn, although stressing his Kantian filiation, Putnam (1981) does not supply a satisfactory elucidation of his notion of conceptual scheme, and tends to talk about the constitutive role of "concepts". This assimilation between categories and concepts of class or *taxa* is what allows Ian Hacking (1983, 1993) to interpret both authors in a nominalist key: conceptual schemes would only introduce different classifications or taxonomies on a fundamental and independent reality of self-subsistent individuals. This nominalist reading of pluralism, although friendly to metaphysical realists, does not agree with the practice of science. When space and time of non-relativistic theories are replaced by spacetime, the reconfiguration of the ontic domain cannot be conceived as a mere reclassification of preexisting items (see, e.g. Sklar, 1974; Earman, 1989). The fact that quantum mechanics lacks the philosophical category of individual, at least in its traditional sense, can hardly be explained in terms of a different way of a new grouping of individual objects in classes (see, e.g. French, 1998; Lombardi & Dieks, 2016; da Costa & Lombardi, 2014).

When the philosophical notion of category is recovered to explain the role played by a categorical-conceptual framework in the constitution of an ontic domain, the discontinuity between different domains becomes manifest. Such a discontinuity implies an incommensurability that does not mean that the domains constituted by different frameworks are merely different "worlds of classes": they might not even share their most basic structures since populated by categorically different items. And it is precisely this strong form of incommensurability that sustains a genuine ontic pluralism, which rejects the metaphysical commitment with a unique reality inhabited by ultimate components interrelated according to the only real structure. There is not a unique real structure, to which the many "epistemologized" domains referred to by science will converge or will finally be reduced.

Now, not all the elements of a categorical-conceptual framework inhabit the same level of the system: the categories and sometimes some generic concepts are placed at the most basic level. Two frameworks may well share a part of their basic elements, as usually happens in the history of science. By contrast, historical examples of successive frameworks that radically differ in all their categories and concepts could hardly be found. Nevertheless, the fact that two frameworks partially agree does not prevent an ontological breakdown due to the disagreements in the rest of their structures. A simple case of this situation is that of theories that, although empirically equivalent, are still incompatible because their frameworks diverge at the non-observable level.

On the other hand, since categories are the most basic elements of the frameworks, they tend to be endowed with most generality and stability through the development of knowledge. For example, although the category of individual object is one of the most entrenched and extended across the entire history of science, it has gone into crisis in certain physical theories. Moreover, it is a fact that the historical-social context can favor – or hinder – the incorporation of very general categories and concepts in most of the frameworks, scientific or not, of a certain era. For example, the concept of evolution, practically ignored before the nineteenth century, developed into a fundamental element of the categorical-conceptual frameworks of very different scientific disciplines from the second half of that century, such as biological Darwinism and macroscopic thermodynamics. Another interesting case is the concept of probability, which, virtually absent from the picture of reality until well into modern times, became unavoidable in practically all areas of knowledge also towards the mid-nineteenth century, expanding its scope with its relation to other notions such as those of possibility, indeterminism or indetermination. Those stable categories and generic concepts play the role of the "pragmatic conception of a priori" proposed by Clarence Irving Lewis (1923), later retaken by Thomas Kuhn (1993), and are also related to the "historical a priori" of Michel Foucault (1969), which crosses the boundaries of different scientific disciplines to constrain what science can talk about at a given historical time. They have also interest affinities with Hacking's idea of "historical ontology" as the result of the "style of reasoning" prevailing at a certain historical time: "although styles may evolve or be abandoned, they are curiously immune to anything akin to refutation" (Hacking, 2002: 192), because they introduce "new ways of being a candidate for truth or falsehood" (*ibid.* 189).

5.6 Categorical-Conceptual Framework, Language, and Praxis

The difference between categories and concepts finds its counterpart in language. In fact, general terms express concepts, not only of classes or properties, but also of relations. Categories, by contrast, are manifested by the structure of language itself. Following Ludwig Wittgenstein's distinction between *saying* and *showing*, whereas language *says* something about reality through its concepts, it *shows*, by means of its own structure, the categories that inform and organize the ontic domain referred to by it. Categories are *said* neither by nouns or predicates nor by any other type of word: "*What can be shown, cannot be said*" (Wittgenstein, 1921, Proposition 4.1212). For instance, the ontic categories of object, property, relation, fact, causation, and quantity are manifested by the linguistic categories of noun, monadic predicate, *n*-adic predicate, proposition, causal connectors, and grammatical number, respectively.

Nevertheless, as already stressed, although a categorical-conceptual framework is expressed through a language, it is not a linguistic entity itself. In fact, not all the relevant import of a categorical-conceptual framework can be identified by the analysis of language. This claim can be easily explained by a concrete example: let us consider two theories whose frameworks agree in incorporating the categories of individual and property, and let us suppose that in both theories the sentence 'electrons have positive charge' is true. However, not everything is determined yet. In one of the frameworks, the concept of electron includes the property of having negative charge as one of its definitory features; in this case, the sentence 'electrons have positive charge' is analytically true since only partially expresses the content of the concept 'electron'. But in the other framework, the sentence 'electrons have positive charge' is a synthetic truth; since it is logically possible that electrons are electrically neutral or negative, the truth value of the sentence should be determined by experimental means. The difference between the two cases is beyond what language can express by itself: it is the categorical-conceptual framework that introduces them and, with this, configures the structure of the ontic domain to which the languages of the two theories refer. Therefore, the identification of the categorical-conceptual framework underlying a scientific theory requires a philosophical analysis that is not confined to the methods of traditional analytic philosophy, but rather incorporates pragmatic matters related with the effective practice of science.

In a critical comment to the application of the Kantian-rooted ontic pluralism to the case of the relationship between physics and chemistry (Lombardi & Labarca, 2005, 2006), Alexander Manafu (2013) considers that the existence of different categorical-conceptual frameworks is not sufficient to conclude that they constitute different ontic domains. According to him, in order to draw this conclusion, one needs to argue that the so-constituted domains have, in Quine's terms, the same "ontological rights". In other words, it is necessary to show that the various frameworks are equally legitimate regarding to their epistemic virtues, none being privileged over the others. This is completely right, and it is precisely in this point that the practice of science enters the scene. In fact, any realist conception of science that seeks to do justice to the way in which scientists effectively work must incorporate the pragmatic dimension of science: such a dimension embodies not only representing the world, but also and essentially intervening on it (see Hacking, 1983). The practices of manipulation, control, transformation and production of new phenomena are crucial in the way in which scientists conceive their own ontic domains: the categorical-conceptual frameworks adopted by the scientific community are those that are remain "stabilized" by successful practices. For instance, the discussions about the autonomous existence of chemical entities should no longer be grounded exclusively on considerations about intertheoretical relationships; from a genuinely pragmatic viewpoint,

> molecular chemistry holds the winning card: its astonishing success in the manipulation of known substances and in the production of new substances is the best reason for accepting the existence of the entities populating its realm. In other words, we are entitled to admit the reality of the molecular world – inhabited by, among other items, chemical orbitals, bonding, chirality, molecular shapes – on the basis of the impressive fruitfulness of molecular chemistry itself, independently of what physics has to say about that matter. (Lombardi & Labarca, 2011: 74)

5.7 Diachronic Pluralism: Scientific Change

As argued above, a Kantian-rooted view allows incorporating the realistic intuition that truth involves the correspondence between a language and an extra-linguistic domain. However, from this perspective, such a domain is not conceived as a metaphysically independent world, but as an ontic realm partially constituted by a categorical-conceptual framework. But since such a constitution is logically prior to the formulation of a theory, nothing prevents the truth value of its sentences from being established in terms of its adequacy with the facts of the categorically and conceptually constituted world.

It is worth stressing again that the ontic domain does not depend exclusively on the categorical-conceptual framework: it is not a mere creation of the mind, but arises from the synthesis between the noumenal realm and our framework. Therefore, in line with an essential element of realism, the independent reality plays an unavoidably role in the constitution of the ontic domain to which our knowledge refers, a role that is clearly manifested through the scientific practices of observation and experimentation. Taking up Kant's metaphor, the scientist is like a judge who "interrogates" the noumenal reality from the perspective of a certain theory, in particular, from the categorical-conceptual framework that this theory presupposes; in turn, such a reality must "answer" in the same language in which the question was asked, that is, with the same system of categories and concepts that theory imposes on it. However, the independent reality reserves its right to respond *negatively* to the received questions, with the result that the theory will be modified or rejected for decidedly empirical reasons. But the consequences may be even deeper: the accumulation of anomalies, unfulfilled expectations and failed predictions may also lead to modifying the framework of categories and generic concepts that the theory presupposes. In both cases, the experiential material obtained through observation and experimentation practices plays an essential role in the evaluation of scientific theories.

The operation of science, according to this perspective, can be better understood by means of a well-known historical example. In the categorical-conceptual framework of the late nineteenth century physics, physicists were able to wonder about the fixed relationship between the mass and the electric charge of the electron, since the ontic domain at the time included an entity conceived as "electron" and endowed with the properties of mass and electric charge. In that case, the received answer was positive. Indeed, J. J. Thomson designed an experiment – conceivable in terms of the framework then in force – which supplied the approximate result of -1.76×10^8 Coulombs/gram. But at that time nobody could have posed a question about the curvature of space-time, since such an item was not part of the available ontic domain. Approximately around the same decades, physicists tried to detect the motion of the Earth by measuring the difference between the speed of light traveling in different directions, parallel and perpendicular to the Earth's motion, a difference predicted by the physical theories accepted at that time. However, in this case the answer was negative, since no difference was measured. In the face of this situation, two possible strategies appeared as available. The conservative one preserved the categorical-conceptual framework and, with it, the ontic domain, and modified the theory so as to account for the new experimental result. This was the alternative adopted by Lorentz and Fitzgerald. The breakthrough strategy, by contrast, revised the categorical-conceptual framework and, consequently, reshaped the ontic domain; in this new context, a completely different theory was formulated, which explained the experimental results, but now reidentified from the new framework. This was Einstein's strategy.

This well-known historical case shows that, when the predictions of a theory are not corroborated by experience, two broad, conceptually different paths become available. In the first one, the theory is modified but without touching its underlying framework. In this case, the old and the new theories can be compared since both refer to the same ontic domain, and it makes perfect sense to say that a prediction of the previous theory contradicts a prediction of the later theory. The second path, by contrast, leads to a revolutionary change: the new theory not only rejects some of the claims of the previous theory, but also modifies, at least partially, its categorical-conceptual framework and, with it, the ontic domain which that theory referred to. It is in this sense that the new framework constitutes a "new world", a world where some of the items of the previous world no longer exist, and new items enter to inhabit the new ontic domain. In this situation, theories cannot be compared only in terms of their empirical adequacy: the replacement of the old theory by the new one is usually due to the fact that the former ceases to be effective in solving problems, especially those problems that most of the scientific community considers relevant at the time.

Summing up, the Kantian-rooted pluralist realism offers definite answers to certain traditional problems regarding scientific change. On the one hand, it allows understanding in what sense a categorical-conceptual framework is never refuted: it is not something capable of being true or false. In the situation of revolutionary scientific change, the framework is simply abandoned when a new one provides a context for the formulation more fruitful theories regarding relevant problems. On the other hand, since a kind of realism, it explains why scientists evaluate their beliefs and representations by appealing to empirical adequacy: in their daily work, they test the predictions of their theories, but without calling into question the categorical-conceptual framework underlying them and the ontic domain referred to by their languages. In other words, the framework, together with the ontic commitments that it entails, are conditions of the possibility of empirical testing. But, empirical predictions may turn out to be false because the independent reality plays an essential role in the constitution of the ontic domains to which scientific theories refer. If the evaluation of scientific knowledge did not at least partially depend on a realm independent of the knowing subjects and their categorical-conceptual frameworks, any trace of realism would fade away and science would become – as Kant warned us – a discourse that revolves around itself, a mere game of coherence among our representations, and we would enter the realm of radical relativism.

5.8 Synchronic Pluralism: A Web-Picture of Science

As stressed above, ontic pluralism points in the same direction as the underdetermination of theories by evidence. Although different forms of pluralism have been widely discussed in the context of the problem of scientific change, they have been scarcely thematized in a synchronic sense. However, when we realize that different theories and disciplines, embodying different categorical-conceptual frameworks, are accepted and successfully used at the same historical time, we must also admit that different ontic domains may coexist. The challenge is, therefore, explaining how those different domains are related to each other in the picture of reality offered by science.

The idea of fundamental entities of which everything is made was the trademark of Western philosophy since its origins in Ancient Greece, went through the thought of Modern Times, and arrived to the present-day physical sciences in its many manifestations. This idea of reduction was formalized in the twentieth century philosophy in terms of an inter-theory relationship, finding its classical locus in Ernst Nagel's (1961) model, according to which reduction is a logical relationship of deduction between theories. The difficulties in the application of the Nagelian model to real-science situations (see, e.g., Primas, 1983; Scerri & McIntyre, 1997) opened two alternatives in the discussion. Some authors insisted on the idea of reduction, confining it to the epistemological domain and adjusting the deductive model with formal resources that make it much more flexible (see, e.g., Rohrlich, 1990; Dizadji-Bahmani et al., 2010; Needham, 2010; Hettema, 2012). Other authors, assuming the failure of reduction, preferred to embrace the idea of emergence, according to which reality is organized in different levels not completely reducible to each other (see, e.g., Beckermann et al., 1992; Emmeche et al., 1997; Humphreys, 1997; Crane, 2001; Cunningham, 2001; Clayton & Davies, 2006). Of course, reductionism and emergentism are very different philosophical views; nevertheless, when endowed with ontic import, both agree in an essential feature: they both introduce an *asymmetric relationship* between levels, so that *if A emerges from/reduces to B, then B does not emerge from/does not reduce to A*. Such an asymmetry involves a kind of *ontic dependence*, which means that each level depends on the lower level for its own existence – or it is even identified with the lower level in the strongest forms of reductionism. In other words, if the lower level didn't exist, the upper level would not exist either. This assumption is so ingrained in the philosophical thought that is almost never discussed. But when a Kantian-rooted pluralism is consistently adopted, the following questions cannot be ignored: Why should this assumption be accepted? What reasons support it?

Reductionist claims usually search for support in the appeal to inter-theory relationships: for example, temperature is nothing but molecular kinetic energy because the relationship $T = (2/3k)\,\overline{E}_K$ holds. This claim is implicitly based on conceiving the symbol "=" as representing a *logical identity*, which establishes that two names denote the same extralinguistic item. But it is not difficult to see that this interpretation is misguided. When the symbol "=" is used in a mathematical intratheoretic expression as '$\boldsymbol{F} = m\boldsymbol{a}$', we are not saying that the term '$\boldsymbol{F}$' denotes the same physical item as '$m\boldsymbol{a}$', since 'm' denotes the property mass and '$\boldsymbol{a}$' denotes the property acceleration, and the "product of properties" makes no sense. '$\boldsymbol{F} = m\boldsymbol{a}$' means that the *numerical value* of the force (measured in certain measurement units) is identical to the *product of the numerical values* of the mass and the acceleration (both measured in certain measurement units). In other words, the symbol "=", as used in mathematical expressions representing scientific relationships, does not indicate logical identity but only identity of numerical values. Therefore, '$T = (2/3k)\,\overline{E}_{\mathrm{K}}$' only expresses the identity between the numerical values of the temperature and of the mean molecular kinetic energy in a gas, and not the identity between the items denoted by the terms 'T' and '$\overline{E}_{\mathrm{K}}$'. Temperature and mean molecular kinetic energy are two different extralinguistic items, each one belonging to the ontic domain referred to by its corresponding theory: temperature is defined and measured with the theoretical and experimental resources of thermodynamics, and does not need to be related to mechanical items for being so defined and measured.

Functional relationships, such as the case of $T = (2/3k)\,\overline{E}_{\mathrm{K}} = f\left(\overline{E}_{\mathrm{K}}\right)$, are the simplest inter-theory links. Other cases involve different mathematical operations, non reducible to functional dependence. For instance, the mathematical limit, as used in the relationship between classical mechanics and special relativity, does not amount to nor can be expressed as a mathematical function. This means that the mathematical limit is an operation that establishes a heterogeneity in the mathematical domain itself. As Fritz Rohrlich points out, the link between two theories through a mathematical limit introduces a breakdown that does not favor ontological reduction; on the contrary, this kind of link shows that the concepts so linked refer to entities belonging to different ontologies, and that "the ontologies of theories in different levels are *incommensurable*" (Rohrlich, 2001: 200; emphasis in the original). From a similar viewpoint, Hans Primas insists on the fact that molecular structure is a concept that can be obtained from quantum mechanics only "as an asymptotic pattern in the singular limit of infinite molecular masses" and, therefore, it is a "classical concept" that "can be discussed by using an unrestricted Boolean language" completely alien to quantum mechanics (Primas, 1983: 335).

These considerations point against the idea that mathematical relationships between theories favor the assumption of the reduction of an ontic domain to another. By contrast to that idea, ontic reductionism is a metaphysical assumption not supported by mathematical inter-theory links but "added" to them, in order to depict a reality that can be completely defined in terms of an underlying and ontologically privileged level to which our fundamental theories refer.

But if the ontic priority of a certain domain cannot be read off from the formalism of science, perhaps there are other clues to invoke such a priority. The asymmetry embodied in the idea of emergence is usually expressed in counterfactual terms as 'if A emerges from B, then if B didn't exist, A wouldn't exist either'. Of course, there are no empirical methods to verify ontological claims, nor ways to decide the truth value of a counterfactual proposition beyond any doubt. Nevertheless, the commitment to an ontological thesis can be justified by indirect arguments; in particular, what happens in the epistemological domain can offer good arguments for drawing ontological conclusions. In this case, the counterfactual can be assessed on the basis of the acceptability of its epistemic counterpart: 'if the theory describing B turned out to be wrong or unacceptable, then the theory describing A would turn also to be wrong or unacceptable'. This is no longer an ontological claim, but an assertion about what effectively happens in science, whose truth value depends on scientific practice. But historical evidence offers good arguments to reject ontic dependence by showing how certain inter-theory links, conceived in terms of reduction or emergence, were later modified due to the replacement of the "basal" theory. The paradigmatic example is the link between macroscopic thermodynamics and the theory describing the supposedly underlying domain. In this case, the "fundamental" theory changed – from caloric theory (Carnot), to classical mechanics (Boltzmann and Gibbs) and, finally, to quantum mechanics (since the advent of the theory) – and the intertheoretic links changed along the theories; however, the "phenomenological" theory – thermodynamics – remained unmodified throughout the historical process. Another relevant example is given by the different levels of description in chemistry; as Jaap van Brakel explicitly asserts:

> If quantum mechanics would turn out to be wrong, it would not affect all (or even any) chemical knowledge about molecules (bonding, structure, valence and so on). If molecular chemistry were to turn out to be wrong, it wouldn't disqualify all (or even any) knowledge about, say, water. (van Brakel, 2000: 177)

These examples show that there is no reason to expect that our best inter-theory relationships will not be replaced in the future because of the eventual replacement of the supposedly fundamental theory. And if the fate of a "phenomenological" theory may be immune to the fate of the supposedly "fundamental" theory, there are no good philosophical reasons to assume the ontic dependence of the domain described by the first on the domain described by the second. From a consistent Kantian-rooted view, there are not privileged theories that can be conceived as describing, even in an approximate way, reality as it is in itself, since the noumenal reality does not constitute an ontic domain to be described by science.

It is true that both the idea of reduction and that of emergence are directed to depicting a unified view of reality. Perhaps for this reason, some authors find in ontic pluralism the threat of the disintegration of reality, resulting from a fragmentary science whose different disciplines, and even different theories, are completely disconnected from each other (see Needham, 2006; Hettema, 2012). Certainly, the belief that reality is not an incoherent plurality has guided scientists throughout the history of science, and the ideal of unification acted as a powerful engine for scientific research. But this does not mean that such an ideal must be conceived in terms of ontic dependence. A Kantian-rooted ontic pluralism retains the aim of unification under a more flexible view, which follows the perspective inaugurated by Otto Neurath when asserting that science is not oriented towards a single whole, but proceeds by means of local systematizations and, consequently, preserves a plural and always incomplete character. On this basis, Neurath favors an idea of unification based not on hierarchical links, but rather on a picture of science as an encyclopedia, where the connections between theories adopt very different, stronger or weaker, forms (Neurath, 1935). From this perspective, inter-theory relationships do not express asymmetric relations between ontic domains: they are *symmetric links* working as "bridges" between theories, which can be "crossed" in the two directions, and do not impose relationships of priority or dependence between the corresponding ontic domains.

In turn, the relationship between two theories and their corresponding ontic domains is a single local node in a plural and complex structure. From the Kantian-rooted pluralist perspective, inter-theory relations, if can be established, do not lead necessarily to a hierarchy of levels, that is, to a "chain" where each "shackle" is connected only with the two immediately adjacent ones. On the contrary, the theories simultaneously accepted by the scientific community form a web, where each theory may be connected with more than two other theories, and through different links with each one of them. For instance, classical mechanics is related with classical statistical mechanics, with special relativity and with quantum mechanics by means of completely different links. As Gordon Belot and John Earman make the point,

> we have a web of independent theories, each of which is thought to be empirically adequate within its own domain of applicability [...] 'Web' rather than 'hierarchy' here because theories often have more than one limit: for instance, special relativity is the curvature $\rightarrow 0$ limit of general relativity, while the curved spacetime formulation of Newtonian gravity is its $c \rightarrow \infty$ limit. (Belot & Earman, 1997: 162)

Scientific theories thus form a web structure on the basis of the symmetric "bridges" connecting them; those bridges are usually much more subtle and varied than what certain traditional perspectives suppose: they involve limits, coarse-graining, approximations and other mathematical techniques far more complex than simple logical links or mathematical functions. Moreover, in this lattice, disciplinary boundaries become less important than usually conceived. By recognizing the variety of inter-theory relationships possible in science, this *non-reductive unification* transcends the conventional boundaries that separate – rather than bring closer – the different disciplines of science.

5.9 Frameworks, Theories and Models

Up to this point, the notions of categorical-conceptual framework and of theory have been the focus of our discussion. However, the notion of model has attracted a great interest in the philosophy of science of the last decades, by challenging the leading role played by theories during the twentieth century. Many philosophers of science have found in models a fundamental methodological resource for modern science that deserves to be analyzed in detail (see, e.g., Morgan & Morrison, 1999; Suárez, 2009; Weisberg, 2013; Morrison, 2015).

According to a traditional *theory-driven* perspective (e.g., van Fraassen, 1989; Bueno et al., 2002; da Costa & French, 2003), common to both the syntactic and the semantic view of scientific theories, models depend on theories in the sense that a model is a "truth-maker" of the theory. Models are mediators between theories and target systems, since theories can only be applied to specific situations through particular models. According to this view, scientific knowledge is encoded in theories, while models would be their mere application instances. This traditional view is usually combined by a *representationalist* conception of models: "a theory is true if and only if one of the members of the set of possible worlds allowed by the theory is the «real» world" (da Costa & French, 2003: 32). Otherwise, the phenomenon represented by the model cannot be conceived as evidence in favor of the truth of the theory.

In explicit confrontation with that traditional view, the idea that models depend on theories and, consequently, are its truth-makers, has been strongly challenged by the so-called *tool-box* conception of theories (Cartwright et al., 1995; Suárez, 1999; Suárez & Cartwright, 2008). From this perspective, a theory is no longer an entity to be confirmed by its models, because there is nothing that can be conceived as *its* models: models are not obtained from theories. In turn, scientific theories are not abstract structures amenable to being true or false, but useful instruments for model building. Unsurprisingly, this instrumentalist conception of theories comes together with a deflationary view about scientific models, according to which the purpose of models is not representation but manipulation of phenomena on the basis of their inferential power (Suárez, 2003).

This debate about the links between theory, model and reality seems to force us to buy a package-deal: realism about theories and representationalism about models, or instrumentalism in general. But there are more things in heaven and earth than are dreamt of in those philosophies. In fact, science works in many different and subtle ways that cannot be subsumed under single and rigid characterizations. And a Kantian-rooted realist pluralism offers conceptual tools to manage with that complexity by distinguishing three levels: categorical-conceptual framework, theory, and model. The categorical-conceptual framework is not amenable to being true or false because it is what intervenes in the constitution of the ontic domain to which theories refer. Theories refer to that logically previously constituted ontic domain: there may exist different theories, even contradictory to each other, referring to the same domain. Models, in turn, are abstract systems that are conceptually constructed for certain particular purposes; therefore, a single scientific theory is multiply applied through very different models. The distinction between these three levels makes possible to retain a realist view about theories without commitments with a strong representationalist conception of models that makes representation a definitional note of scientific models. In fact, in many cases models have the aim of partially or approximately representing the target system; but science is also full of situations in which models are only instrumental devices without any representational aspiration. However, this fact does not necessarily lead to an instrumentalist view of scientific knowledge as a whole.

The distinction between framework, theory, and model also supplies clues to understand the different ways in which scientists react to negative empirical results. Due to their local and sometimes instrumental nature, models are the first elements to be modified in the case of empirical inadequacy. Since scientific theories are more general and stable than models, only when the repeated action on models does not lead to the expected results, scientists consider the need of introducing modifications to their theories. And given that the categorical-conceptual framework is usually even more stable than the multiples theories that can be formulated on its basis, its modification tends to be the last resource scientists appeal to. These different ways in which science operates remain obscured when framework, theory, and model are not properly distinguished.

On the other hand, talking about 'science' in general also hides the fact that many different styles of scientific research coexist. Theoretical physics was implicitly adopted as the paradigm for the philosophy of science of twentieth century, and this fact led to a theory-centered perspective, according to which scientific knowledge is primarily encoded in theories. However, recent philosophical research oriented to other scientific disciplines is offering a different picture about scientific activity. For instance, chemistry, since its origins in the artisanal world (see Klein, 2008), cannot be conceived in terms of the traditional opposition between realism and instrumentalism. According to Bernardette Bensaude-Vincent (2008), the ontic domain of chemistry arises not only from cognitive frameworks, but also from action driven by caution, utility, and efficiency, since it is the result of the very nature of the discipline – science and technology. In this case theories play a less relevant role than in physics, whereas models carry out the task of supplying the approximate and/or idealized representation of the chemical world.

A still newer dimension is that introduced by the rapid and ubiquitous growing of a multiplicity of interdisciplinary fields that integrate knowledge coming both from different traditional disciplines and from novel strategies of scientific research. An example of this situation is quantum chemistry, characterized by Kostas Gavroglu and Ana Simões (2015) as an "in-between" discipline, developed in new computational laboratories where chemists, physicists, and mathematicians work together with the purpose of designing quantum mechanical models of molecular systems. According to these authors, quantum chemists have progressively abandoned any worry not only about the reference of quantum mechanics, but also about the representativity of their models. In fact, the molecular models used in quantum chemistry commonly integrate elements coming from incompatible theoretical bodies of knowledge – quantum mechanics and structural chemistry – in a constructive and empirically successful manner (see Accorinti & Martínez González, 2016). This strategy cannot be considered as a historically provisory limitation that will be overcome with future theoretical development, but it is constitutive of the discipline itself. Therefore, quantum chemistry does not count with an autonomous ontic domain as its reference, but it is a scientific field whose validity rests on its practical success in calculation and prediction (Lombardi & Martínez González, 2012).

Summing up, a Kantian-rooted pluralist realism as that proposed here should be interpreted neither as a normative approach nor as a model unrestrictedly applicable to all forms of scientific activities. By contrast, the distinctions involved in this proposal provide conceptual tools for the understanding of how both the traditional fields and the most novel areas of science work in the practice of science.

5.10 Final Remarks

In the present paper I have argued that a Kantian-rooted perspective, as sketched here, is highly fruitful to understand the phenomenon of science. On the one hand, it does not involve a metaphysical realism committed with discovering the real structure of reality in itself. Nevertheless, on the other hand, it is neither an idealist doctrine, since the noumenal independent reality plays a decisive role in the constitution of scientific knowledge. Despite belonging to a Kantian tradition, the present proposal admits the relativity of the categorical-conceptual frameworks of science, which in turn leads to an ontic pluralism that finds both diachronic and synchronic manifestations: it accounts for the different forms of scientific change, and describes a web-picture of science without dependencies or priorities. Moreover, it endows scientific praxis with a high relevance, not only through general claims about the importance of practice, but by explaining how effective theoretical and empirical activity intervenes in the different moments of the construction and consolidation of knowledge. And instead of standardizing scientific activity, this view allows understanding the diverse relationships among frameworks, theories and models in different disciplines and fields of science. At least for these reasons, the proposed Kantian-rooted pluralist realism deserves to be considered as a viable and interesting way to approach the phenomenon of science.

References

Accorinti, H., & Martínez González, J. C. (2016). Acerca de la independencia de los modelos respecto de las teorías: un caso de la química cuántica. *Theoria. An International Journal for Theory, History and Foundations of Science, 31*, 225–245.

Beckermann, A., Flohr, H., & Kim, J. (Eds.). (1992). *Emergence or reduction? Essays on the prospects for nonreductive physicalism*. De Gruyter.

Belot, G., & Earman, J. (1997). Chaos out of order: Quantum mechanics, the correspondence principle and chaos. *Studies in History and Philosophy of Modern Physics, 28*, 147–182.

Bensaude-Vincent, B. (2008). Chemistry beyond the «positivism vs realism» debate. In K. Ruthenberg & J. van Brakel (Eds.), *Stuff. The nature of chemical substances* (pp. 45–54). Königshauen & Neumann.

Borges, J. L. (1944a). Funes, el memorioso. *Ficciones*. Editorial Sur. Translated by James Irby in Yates, D., & Irby, J. (Eds.). (1962). *Labyrinths*. New Directions.

Borges, J. L. (1944b). El jardín de los senderos que se bifurcan. *Ficciones*. Editorial Sur. Translated by James Yates, D., & Irby, J. (Eds.). (1962). *Labyrinths*. New Directions.

Borges, J. L. (1952). El idioma analítico de John Wilkins. *Otras Inquisiciones*. Editorial Sur. Translated by Ruth L. C. Simms in Borges, J. L. (1971). *Other inquisitions. 1937–1952*. University of Texas Press.

Bueno, O., French, S., & Ladyman, J. (2002). On representing the relationship between the mathematical and the empirical. *Philosophy of Science, 69*, 452–473.

Cartwright, N., Shomar, T., & Suárez, M. (1995). The tool box of science. In W. Herfel, W. Krajewski, I. Niniiluoto, & R. Wójcicki (Eds.), *Theories and models in scientific processes* (Poznań studies in the philosophy of the sciences and the humanities) (Vol. 44, pp. 137–149). Rodopi.

Clayton, P., & Davies, P. (2006). *The re-emergence of emergence. The emergentist hypothesis from science to religion*. Oxford University Press.
Crane, T. (2001). The significance of emergence. In B. Loewer & G. Fillet (Eds.), *Physicalism and its discontents* (pp. 207–224). Cambridge University Press.
Cunningham, B. (2001). The reemergence of 'emergence'. *Philosophy of Science, 68*, S62–S75.
da Costa, N., & French, S. (2003). *Science and partial truth: A unitary approach to models and scientific reasoning*. Oxford University Press.
da Costa, N., & Lombardi, O. (2014). Quantum mechanics: Ontology without individuals. *Foundations of Physics, 44*, 1246–1257.
Dizadji-Bahmani, F., Frigg, R., & Hartmann, S. (2010). Who's afraid of Nagelian reduction? *Erkenntnis, 73*, 393–412.
Earman, J. (1989). *World enough and space-time*. The MIT Press.
Emmeche, C., Koppe, S., & Stjernfelt, F. (1997). Explaining emergence. Towards an ontology of levels. *Journal for General Philosophy of Science, 28*, 83–119.
Foucault, M. (1969). *L'Archéologie du Savoir*. Gallimard. English translation, Sheridan Smith, A. M. (Trans.). (1972). *The archaeology of knowledge*. Tavistock.
French, S. (1998). On the withering away of physical objects. In E. Castellani (Ed.), *Interpreting bodies. Classical and quantum objects in modern physics* (pp. 93–113). Princeton University Press.
Gavroglu, K., & Simões, A. (2015). Philosophical issues in (sub)disciplinary contexts: The case of quantum chemistry. In E. Scerri & G. Fisher (Eds.), *Essays in the philosophy of chemistry* (pp. 60–81). Oxford University Press.
Hacking, I. (1983). *Representing and intervening*. Cambridge University Press.
Hacking, I. (1993). Working in a new world: The taxonomic solution. In P. Horwich (Ed.), *World changes. Thomas Kuhn and the nature of science* (pp. 275–310). MIT Press.
Hacking, I. (2002). *Historical ontology*. Harvard University Press.
Hettema, H. (2012). *Reducing chemistry to physics. Limits, models, consequences*. Rijksuniversiteit Groningen.
Humphreys, P. (1997). How properties emerge. *Philosophy of Science, 64*, 1–17.
Klein, U. (2008). A historical ontology of material substances: c. 1700-1830. In K. Ruthenberg & J. van Brakel (Eds.), *Stuff. The nature of chemical substances* (pp. 21–44). Königshauen & Neumann.
Kuhn, T. (1983). Commensurability, communicability, comparability. In P. D. Asquith & T. Nickles (Eds.), *PSA 1982, Volume 2* (pp. 669–688). Philosophy of Science Association.
Kuhn, T. S. (1993). Afterwords. In P. Horwich (Ed.), *World changes. Thomas Kuhn and the nature of science* (pp. 311–341). MIT Press.
Lewis, C. I. (1923). A pragmatic conception of the a priori. *The Journal of Philosophy, 20*, 169–177.
Lewowicz, L. (2005). *Del Relativismo Lingüístico al Relativismo Ontológico en el Último Kuhn*. Departamento de Publicaciones de la Facultad de Humanidades y Ciencias de la Educación, Universidad de la República.
Lombardi, O., & Dieks, D. (2016). Particles in a quantum ontology of properties. In T. Bigaj & C. Wüthrich (Eds.), *Metaphysics in contemporary physics* (Poznan studies in the philosophy of the sciences and the humanities) (pp. 123–143). Brill-Rodopi.
Lombardi, O., & Labarca, M. (2005). The ontological autonomy of the chemical world. *Foundations of Chemistry, 7*, 125–148.
Lombardi, O., & Labarca, M. (2006). The ontological autonomy of the chemical world: A response to Needham. *Foundations of Chemistry, 8*, 81–92.
Lombardi, O., & Labarca, M. (2011). On the autonomous existence of chemical entities. *Current Physical Chemistry, 1*, 69–75.
Lombardi, O., & Martínez González, J. C. (2012). Entre mecánica cuántica y estructuras químicas: ¿a qué refiere la química cuántica? *Scientiae Studia Revista Latinoamericana de Filosofia e História da Ciencia, 10*, 649–670.

Lombardi, O., & Pérez Ransanz, A. R. (2012). *Los Múltiples Mundos de la Ciencia*. UNAM-Siglo XXI.

Manafu, A. (2013). Internal realism and the problem of ontological autonomy: A critical note on Lombardi and Labarca. *Foundations of Chemistry, 15*, 225–228.

Morgan, M., & Morrison, M. (Eds.). (1999). *Models as mediators*. Cambridge University Press.

Morrison, M. (2015). *Reconstructing reality: Models, mathematics, and simulations*. Oxford University Press.

Nagel, E. (1961). *The structure of science*. Harcourt, Brace & World.

Needham, P. (2006). Ontological reduction: A comment on Lombardi-Labarca. *Foundations of Chemistry, 8*, 73–80.

Needham, P. (2010). Nagel's analysis of reduction: Comments in defense as well as critique. *Studies in History and Philosophy of Modern Physics, 41*, 163–170.

Neurath, O. (1935). Pseudorationalism of falsification. In R. Cohen & M. Neurath (Eds.), *Philosophical papers 1913–1946* (pp. 121–131). Reidel.

Primas, H. (1983). *Chemistry, quantum mechanics and reductionism*. Springer.

Putnam, H. (1981). *Reason, truth and history*. Cambridge University Press.

Putnam, H. (1987). *The many faces of realism*. Open Court.

Rohrlich, F. (1990). There is good physics in theory reduction. *Foundations of Physics, 20*, 1399–1412.

Rohrlich, F. (2001). Cognitive scientific realism. *Philosophy of Science, 68*, 185–202.

Scerri, E., & McIntyre, L. (1997). The case for the philosophy of chemistry. *Synthese, 111*, 213–232.

Sklar, L. (1974). *Space, time, and spacetime*. University of California Press.

Suárez, M. (1999). The role of models in the application of scientific theories: Epistemological implications. In M. Morgan & M. Morrison (Eds.), *Models as mediators* (pp. 168–196). Cambridge University Press.

Suárez, M. (2003). Scientific representation: Against similarity and isomorphism. *International Studies in the Philosophy of Science, 17*, 225–244.

Suárez, M. (2009). *Fictions in science: Philosophical essays on modeling and idealization*. Routledge.

Suárez, M., & Cartwright, N. (2008). Theories: Tools versus models. *Studies in History and Philosophy of Modern Physics, 39*, 62–81.

Torretti, R. (2000). 'Scientific realism' and scientific practice. In E. Agazzi & M. Pauri (Eds.), *The reality of the unobservable* (pp. 113–122). Springer.

Torretti, R. (2005a). *Manuel Kant*. Ediciones Universidad Diego Portales.

Torretti, R. (2005b). Conocimiento discursivo. *Lección Inaugural* dictada en la Universidad Autónoma de Barcelona el 28 de Abril de 2005, en ocasión de la investidura del autor como Doctor Honoris Causa de dicha Universidad.

van Brakel, J. (2000). The nature of chemical substances. In N. Bhushan & S. Rosenfeld (Eds.), *Of minds and molecules. New philosophical perspectives on chemistry* (pp. 162–184). Oxford University Press.

van Fraassen, B. (1989). *Laws and symmetry*. Oxford University Press.

Weisberg, M. (2013). *Simulation and similarity: Using models to understand the world*. Oxford University Press.

Wittgenstein, L. (1921). Logisch-Philosophische Abhandlung. *Annalen der Naturphilosophie, XIV*(3/4). English version, Ogden, C. K. (Trans.). (1922). *Tractatus Logico-Philosophicus*. Routledge/Kegan Paul.

Chapter 6
Mathematical Fictionalism Revisited

Otávio Bueno

Abstract Mathematical fictionalism is the view according to which mathematical objects are ultimately fictions, and, thus, need not be taken to exist. This includes fictional objects, whose existence is typically not assumed to be the case. There are different versions of this view, depending on the status of fictions and on how they are connected to the world. In this paper, I critically examine the various kinds of fictionalism that Roberto Torretti identifies, determining to what extent they provide independent, defensible conceptions of mathematical ontology and how they differ from platonism (the view according to which mathematical objects and structures exist and are abstract, that is, they are neither causally active nor are located in spacetime). I then contrast Torretti's forms of fictionalism with a version of the view that, I argue, is clearly non-platonist and provides a deflationary account of mathematical ontology, while still accommodating the attractive features of the view that Torretti identified.

Keywords Fictionalism · Mathematics · Platonism · Objectivity · Ontology

6.1 Introduction

Over the decades, Roberto Torretti has addressed issues in the philosophy of mathematics in interesting and insightful ways. In particular, he considered the ontological issue of the existence of mathematical objects and connected it to related issues regarding the application of mathematics. He has shown sympathy toward some form of fictionalism about mathematics, identifying various formulations of this view. In this paper, I examine critically the different kinds of fictionalism that Torretti has considered, and after raising some concerns, I suggest an alternative formulation that seems to achieve what one may want from the view without the costs associated with it. I also consider whether the approach I recommend is

O. Bueno (✉)
Department of Philosophy, University of Miami, Coral Gables, FL, USA

C. Soto (ed.), *Current Debates in Philosophy of Science*, Synthese Library 477,
https://doi.org/10.1007/978-3-031-32375-1_6

compatible with the overall proposal that Torretti has, and argue that, to a certain extent, it is.

6.2 Mathematical Fictionalism: Three Types

6.2.1 *Fictionalism$_1$*

Three varieties of mathematical fictionalism are identified by Torretti. The first kind (fictionalism$_1$), as Torretti himself acknowledges, was not expressed by him but rather by Mario Bunge. Speaking of propositions, but the point easily extends to other abstract objects, Bunge notes:

> We do not claim that they exist in themselves but only that it is often convenient (for example in mathematics but not in metaphysics) to feign or pretend that they do. We do not assert that the Pythagorean theorem exists anywhere except in the world of phantasy called 'mathematics', a world that will go down with the last mathematician. (Bunge 1974-, volume II, p. 85; quoted in Torretti 1981, p. 400)

This form of fictionalism denies the ontological independence of abstract entities: they are not taken to exist independently of the intentional actions of mathematicians, which are embedded in "the world of phantasy called 'mathematics'". A form of pretense is adopted whenever it turns out to be convenient, in mathematics, to consider (in fact, pretend) that mathematical objects exist. Bunge highlights that such a pretense does not extend to metaphysics, presumably because metaphysics deals with the basic structures of reality and (if we assume for the sake of argument that they exist) these structures are ontologically independent from us and not something we can just pretend that they exist.

In contrast, acceptance of the existence of mathematical objects and propositions expressing mathematical results, such as the Pythagorean theorem, is always an option, given the lack of ontological independence of such objects and results. But there is no real commitment to their existence. Mathematicians just pretend to have such commitment when they convey the content of the results and describe the relevant mathematical objects.

Although this form of fictionalism has the advantage of avoiding, at least in principle, commitment to the existence of mathematical entities, the view incurs a significant cost: pretense talk is highly artificial and, given its intensional character, it fails to preserve inferential relations among the relevant contents. Although it follows from the axioms of Peano arithmetic that there are infinitely many natural numbers, once the axioms are entertained within the scope of a pretense operator (*It is pretended that* ...), there is no guarantee that the inferential relation will go through: nothing prevents one from pretending that there are only finitely many natural numbers and run into an inconsistent fiction. It is a pretense, after all! In other words, it is not only the existence of mathematical objects that is bypassed by the pretense operator, but part of the content of the relevant mathematical results may also be lost along the way. Moreover, in those cases in which there is no clean way of distinguishing the objects one only pretends that they exist from those that actually do exist, the pretense view becomes unworkable. If mathematics is ultimately indispensable to the formulation of scientific theories, and there is no mathematics-free formulation of scientific theories, it is unclear how to pretend that only the mathematical content of the relevant theories is false while their physical content is true, given that there is no way of independently specifying each content (for further discussion, see Azzouni, 2004, pp. 75–77 and elsewhere).

The result is that a pretense approach to fictionalism about mathematics turns out to be highly artificial. Mathematical practice does not seem to involve such a pretense. Mathematicians engage directly with the relevant mathematical principles not with their pretend counterparts. They use the characterizations of groups, metric spaces, and topologies to draw consequences from the relevant principles, which they would not be able to do properly within the intensional context of the pretense operator. They also determine logical connections among the relevant results, prove theorems, find counterexamples, and refine concepts in light of the latter. This process, especially the dynamic interaction between proofs and refutations in mathematics, is vividly described by Imre Lakatos (1976), with no pretense operators involved. None of these central aspects of mathematical practice can be properly implemented once the mathematical content in question is just pretended to be so: one can pretend all kinds of things, including impossible ones, and there is no guarantee that what is pretended to be so actually is so. Under the pretense operator, mathematical practice ends up making little sense. But without such an operator this form of fictionalism is unable to avoid commitment to mathematical objects and other abstract entities, and ultimately flounders.

6.2.2 *Fictionalism$_2$*

The second form of fictionalism that Torretti considers (fictionalism$_2$) also finds its inspiration in Bunge. This is based on Bunge's "literalists" (Bunge, 1974-, volume II, p. 166). As Torretti points out:

> If someone says that we feign that there are constructs, to which our mathematical statements refer, he will most naturally be understood to mean that we produce objects in our fancy, which thereafter constitute definite, stable, albeit ghostly referents of our discourse. It is in this sense that some atheists maintain that there is a fantastic being, created by men in their own likeness, to whom the majority of mankind refer in their persistent talk of God. It is in this same sense that literary authors are usually said to create the characters about which they write in their books. (Torretti 1981, p. 400)

Rather than relying on a pretense operator, this form of fictionalism insists that reference to mathematical objects is, in fact, very straightforward: all that it takes to secure reference to the relevant objects is to provide a suitable description of them. The objects can then be taken as referents of the relevant discourse. (A version of this view is worked out for fictional objects in Thomasson (1999).) The view is especially adequate to accommodate literary practice since positing fictional characters is something very easy. All that it takes is to describe the characters and their deeds.

The curios feature is that this form of fictionalism acknowledges that fictional objects *do exist* rather than deny their existence. It, thus, becomes inflationary in light of the resulting ontological commitments, despite the apparently straightforward way in which reference to fictional entities is maintained.

But is reference to fictional objects so easily secured? One may intend to refer to a certain object but end up referring to another instead. This may happen whenever different objects satisfy the same descriptions. This is clearly the case especially in mathematics. For instance, one may intend to refer to standard natural numbers when characterizing them in first-order Peano arithmetic. But precisely the same description is equally satisfied by nonstandard natural numbers. Nothing in the characterization of such objects allows one to secure reference to standard numbers alone. Intending to pick out the latter by specifying that the reference is meant to single out standard numbers fails since both kinds of numbers satisfy the same descriptions: any description that standard numbers satisfy is similarly satisfied by nonstandard ones. The description 'standard number', when unpacked, picks out the same conditions satisfied by nonstandard numbers.

Perhaps one could try to resist this sort of indeterminacy by formulating Peano arithmetic in second-order logic. Since in this case all models of the theory are isomorphic, nonstandard models are not forthcoming and reference to standard numbers is, in principle, secured. The problem, however, is that second-order logic allows for two different semantics: one, standard semantics, yields proper second order logic: the Löwenheim-Skolem theorem does not hold, the logic is incomplete, and all models of arithmetic are categorical; the other, Henkin semantics, has the same properties of first-order logic: the Löwenheim-Skolem theorem goes through, the logic is complete, and there are nonstandard models (for discussion, see Shapiro, 1991; Bueno, 2010). As it turns out, nothing in second-order logic allows one to pick out one semantics over the other, and with the indeterminacy of the semantics, the indeterminacy between standard or nonstandard numbers also returns. The upshot: it cannot be guaranteed that reference to mathematical objects can be so easily achieved as the literalist form of fictionalism presupposes.

6.2.3 *Fictionalism*$_3$

The third form of fictionalism (fictionalism$_3$) is especially apt to accommodate applied mathematics. As Torretti presents it:

> When asked to calculate the period of a given pendulum at a given place we feign that the pendulum hangs from a weightless, inextensible string, and that the sinus of the angle of displacement is equal to the angle itself, and rounding up the values of the pendulum's length and the local acceleration of gravity, we compute 2π times the square root of their quotient to an agreed decimal. From our fictitious assumptions we derive a result which

> is admittedly false, but which will differ from the measured values of the period by less than a prescribed number. At a less modest level, the cosmologist feigns that matter is homogeneously distributed in the universe and figures out the broad patterns of its evolution, although he knows all too well that matter is not equally distributed inside a pulsar and in intergalactic space. As the saying goes: one 'idealizes' reality in order to obtain an approximate yet manageable picture of it. (Torretti, 1981, p. 401)

Idealizations are, of course, crucial to any account of the use of mathematics in the sciences (for additional discussion, see Bueno & French, 2018). In many instances, as Torretti rightly highlights, in order to obtain a mathematically tractable account of certain empirical phenomena, one needs to distort some of their features, whether by assuming that the string of a pendulum is weightless and inextensible or by stipulating that matter is homogeneously distributed throughout the universe. The resulting account is fictional in the sense that it offers strictly false descriptions of the phenomena, by positing items or situations that do not exist. The loss of truth is hopefully compensated by the gain in tractability.

Fictionalism$_3$, nevertheless, is not restricted to the application of mathematics, as pure mathematics is also under its purview. Significantly, the form of fictionalism advanced by the account seems to shift when the focus is on pure domains. As Torretti points out:

> Turning to pure mathematics, I find that fictionalism$_3$ may have inspired some of the mathematicians who in the nineteenth century carried out the arithmetization of analysis. As a result of their labors anyone speaking, say, of a complex number could henceforth be regarded as actually referring not to some new-fangled esoteric individual entity, but 'merely' to an ordered pair of infinite classes of classes of classes of natural numbers. But then, of course, a truly hardnosed fictionalist would observe that even the natural numbers do not really exist, at any rate, not the larger ones, nor the so-called set of them. Fictionalism$_3$ was probably also at the back of Hilbert's mind when he proposed to show that the mathematical infinite was just a manner of speaking, something to which the mathematician merely seemed to refer, when he was really reasoning only about the finite. (Torretti, 1981, p. 402)

As this passage makes clear, in pure mathematics, fictionalism$_3$ is concerned not with idealizations but with strategies of reduction and elimination of certain objects. In this way, for instance, complex numbers are characterized in terms of, and ultimately dispensed with by, suitable pairs of iterated classes of natural numbers. Rather than being fictionalist about all natural numbers, however, the fictionalist$_3$ seems to reject only sets of them. In other words, at stake here are not idealization devices, but instead reformulation strategies to dispense with some unwanted entities. For some reason, though, natural numbers seem still acceptable, given that they are not actively eliminated. But it is unclear what grounds their preservation, given that natural numbers are just as abstract, non-spatiotemporally located, causally inactive objects as any other mathematical entity. One may assume that natural numbers are just given, but this still requires some account of why one is entitled to do so.

In the end, fictionalism$_3$ does not seem to provide a coherent approach to mathematics. Idealizations are fundamentally different from reductions and eliminations. One idealizes in order to simplify the descriptions of the phenomena, often making them mathematically tractable, albeit introducing deliberately false accounts along the way. In contrast, entities are reduced to others in order to be eliminated by them. However, the resulting elimination need not be constrained by an increase in mathematical tractability. Furthermore, as opposed to idealizations, reductions need to be true, or, at least, true preserving, otherwise they will fail spectacularly: without an equivalence between the reducing and the reduced, no reduction can go through. Without a proper reconciliation of these devices, or an account of how they are supposed to work in tandem, it is unclear that a coherent approach is forthcoming. The fact that fictionalism$_3$ fails to provide such an account raises doubts about the viability of the proposal.

Moreover, idealization is not a strategy that avoids ontological commitment to entities. Objects still need to be posited so that idealizations can be implemented, such as weightless and inextensible strings of a pendulum and homogeneously distributed chunks of matter throughout the universe. The fact that these items do not exist in the concrete world does not absolve the fictionalist$_3$ from having to provide an account of them.

Perhaps the most straightforward way of accommodating these items would be by deeming them to be abstract. This would explain why none of the posited entities are detected in the world. Nevertheless, if idealized objects are characterized as being abstract, they are just as problematic as the mathematical entities that the fictionalist$_3$ was trying to avoid in the first place. The fact that, in the context of pure mathematics, fictionalism$_3$ fails to provide a comprehensive elimination strategy raises questions about the ability of the view to accommodate abstract objects more generally.

6.2.4 Different Kinds of Fictionalism Compared

As will become clear, Torretti's (2014) own considered fictionalism is importantly different from the three kinds of fictionalism just examined. Crucial to his approach is the emphasis on what it takes for something to exist, as well as on the linguistic devices required to refer to an object.

The distinctive element of fictionalism$_1$, as noted above, was the use of pretense as a device to avoid commitment to the existence of mathematical objects. Given that pretense does not figure anywhere in Torretti's (2014) considered account, his fictionalism is clearly distinct from fictionalism$_1$. As will be seen, on this account, commitment to mathematical objects is, in principle, avoided by distinguishing two kinds of existence and exploring the linguistic devices common to fictional practices (although it is unclear whether the resulting commitments are successfully resisted in the end).

Interestingly, fictionalism$_2$, as discussed, also relies on the creative practices of literary authors. At least in spirit, it shares some of the features also found in Torretti's fictionalism. *Via* their intentional stances, literary authors create characters in stories, and in doing that, they refer to such characters. This referential process is central to both fictionalism$_2$ and Torretti's (2014) form of fictionalism. But, as will become clear, the latter advances an explicit conception of existence (weak existence) that clarifies the conditions under which fictional objects exist. This conception is just not part of fictionalism$_2$. Although Torretti's fictionalism is ultimately compatible with the latter, it goes beyond it with the positive account of existence that, at best, is left implicit in fictionalism$_2$. Nonetheless, as will be discussed below, in light of this conception of existence, Torretti's fictionalism ends up being ontologically inflationary, since, on this view, everything that is referred to does exist. The fact that this is a weak, easily satisfied, kind of existence does not take away from the fact that the resulting fictional objects still exist.

Fictionalism$_3$, as pointed out above, emphasizes the role played by idealization as well as by reduction and elimination in the introduction of fiction. Since none of these theoretical devices is part of Torretti's (2014) considered conception of fictionalism, his preferred form of the approach—ontological minimalism—provides a distinct philosophical view.

The upshot seems to be that none of the three forms of fictionalism considered so far seems to work. This does not pose a problem for Torretti as he is not committed to the truth of any of them. In entertaining them, he is just exploring the logical landscape of possibilities. Torretti does return to fictionalism in a later essay (Torretti, 2014), though. I will turn to it now.

6.3 Existence and Mathematical Fictionalism

6.3.1 Fictionalism: Strong and Weak

On his considered view, Torretti (2014) resists platonism, which he takes to be an inadequate approach to the ontology of mathematics and favors instead a form of fictionalism. As indicated above, the fictionalism he favors is interestingly different from the three kinds of fictionalism just discussed. Central to Torretti's approach is the articulation of a conception of existence in terms of which the issue of the existence of mathematical objects can be settled. He offers, in fact, two such accounts: a strong and a weak one.

The strong conception generalizes features made salient in the interaction with physical objects and processes around us. It requires the elimination of possible illusions when seeing something or the avoidance of artifacts in experimental procedures. Without invoking these familiar categories, this conception seems to characterize existence ultimately in terms of causal activity and spatiotemporal location. To exist is to be causally active and located in spacetime. Since mathematical objects are neither, the argument goes, they do not exist.

Platonists, who insist that abstract objects exist, will understandably complain that the use of this conception blatantly begs the question against their view. After all, they take mathematical objects to exist despite their lack of causal activity and spatiotemporal location.

The weak conception, in turn, avoids this objection. For it only demands that existent objects be referred to (in a given language). Since one is arguably able to refer to mathematical objects, such objects do exist. Interestingly, the same goes for fictional entities, which are also clearly referred to in the context of literary works. The weak conception is a form of fictionalism in the sense that it adopts resources to engage with objects that are articulated and employed in fictional discourse. Typically, in the context of fiction, it is enough to refer to something for the object in question to (be taken to) exist (see also Thomasson [1999]).

Torretti formulates the two conceptions explicitly. Starting with the first, which is the strong, maximalist account, he notes:

> [...] the inevitable presence of bodies, and of physical objects and processes in general, and our justified fixation with them—after all, their behavior usually is a matter of life or death for us—incline us to realize that the general features of their way of being belong to every existence; and hence, that what does not exist in this 'strong' sense, properly does not exist. (Torretti, 2014, p. 126)

The irresistible pull of physical objects offers a paradigmatic case of what it takes for something to exist. Physical objects are so central that anything that lacks their general features simply is not taken to exist. Location in spacetime and causal accessibility are two of these features, I take it. Epistemic criteria can then be advanced to help the determination of whether certain objects exist or not. In order to assure that they do, one needs to rule out hallucinations in perceptual experiences and the presence of artifacts in experimental procedures (two clear epistemic criteria). After all, the occurrence of these defeaters undermines the grounds to maintain the existence of the objects in question. As Torretti highlights:

> Such inclination [to realize that the general features of physical objects and processes are typical of what exists] is, I would say, the root of a philosophical requirement that everything that exists fulfills criteria that are similar to those that are applied to certify that a body, which we appear to have seen, is not a mirage nor a hallucination, or that a physical object that we believe we have discovered is not an artifact of our own experimental procedure or an improper interpretation of observations. (Torretti, 2014, p. 126)

There are costs, however, with this strong, maximalist understanding of existence. After all, neither fictional objects nor mathematical structures seem to satisfy the required conditions for something to be taken to exist. As noted, platonists about mathematics and realists about fictional discourse, who insist on the existence, respectively, of mathematical structures and fictional entities, resist the claim that physical objects are the only kind of things in existence. On their view, causally inert, non-spatiotemporally located entities, such as those in the ontologies they favor, do exist. To require spatiotemporal location and causal activity in order for something to exist is to assume the primacy of the concrete; but it is unclear, they insist, why this should be so.

These concerns motivate a different, weak, minimalist conception of existence that, by demanding less from existence, is able to accommodate more ontologically, including fictional objects and mathematical entities. Torretti emphasizes the point:

> This maximalist interpretation of the verb 'to exist' entangles literary works and mathematical structures into a skein of philosophical problems, which under the broad and generous interpretation just suggested—which I will call minimalist—would be no more than pseudo-problems. (Torretti, 2014, p. 126)

The objection is that the view that requires more in order for objects to exist (the maximalist, strong conception) has fewer resources to account literally for the entities that are typically invoked in fictional discourse and mathematical practice. The minimalist, weak alternative is allegedly able to accommodate this difficulty, though. By rejecting the primacy of the concrete, which typically also requires ontological independence from psychological processes and linguistic practices for something to exist (see Azzouni, 2004), the minimalist approach simply includes a broader range of objects in the resulting ontology.

Speaking on behalf of the minimalist, Torretti clearly remarks that if the requirement of ontological independence from mental operations and linguistic procedures is waived, a weaker conception of existence emerges, according to which it is enough, but also required, for an object to exist that it be possible to mention it (to refer to it). On his view:

> It seems clear to me that if we speak of plain existence without restrictive attributes—that is, if we do not require that the existence of what we mention be, for instance, independent from discourse or from our mental life—the minimal requisite is that *it is effectively possible to mention it*. This requisite, I maintain, is sufficient, but it is also necessary: without mentioning the object in question—by naming it or describing it—it cannot be *said* that the object *exists* either. (Torretti, 2014, p. 127)

Torretti is certainly right in noting that one cannot *say* that an object exists without referring to it. In fact, this is the case not only of objects that plainly exist "without restrictive attributes" but also of those that are ontologically independent of our linguistic practices and psychological processes (Azzouni, 2004). Arguably, one cannot *claim* that an object exists without referring to the relevant object; otherwise, it would indeterminate whose existence one is asserting. However, is it *sufficient* to refer to an object, whether by naming it or describing it in some way, in order for the object to exist (or for one to say that it does)? This is less clear. After all, one may refer to fictional objects, as we do as part of our literary practices, without thereby assuming their existence. Arguably, from the fact that Sherlock Holmes was a detective who lived in London, it does not follow that Holmes existed.

Amie Thomasson's (1999) abstract artifactual account of fictional objects denies this claim and offers a different approach to the issue by insisting that all it takes for a fictional object to exist is for someone to describe the object, for instance, by telling a story about it. However, no matter how detailed such descriptions are, they can never fully specify all the properties of the objects in question and uniquely individuate them. In contrast with concrete objects, fictional entities are ultimately incomplete. As a result, it is unclear which of a multitude of fictional objects one refers to since, for any description of characters explicitly offered in a fictional work, a huge number of distinct objects satisfy the explicit description while having different properties relative to what was *not* overtly said about them. For example, a character may be described as having brown hair. But the exact number of hairs, precise hair length, specific shade of hair color, being left unspecified, allow for a plurality of different characters to satisfy the description. And this concerns only the color and length of the fictional character's hair! In other words, it is unclear that descriptions of a fictional character, no matter how thorough they are, are sufficient for the characterization of a unique object and, thus, sufficient to guarantee the existence of *that* object rather than some other similar to, but distinct from, it.

Torretti's point, nonetheless, is more subtle. We cannot speak without using linguistic items (words, sentences, gestures). The existence of these objects is arguably presupposed when using language, and language requires objects, such as letters, that are invariant under a variety of transformations (for instance, regarding their specific shapes on a page). These objects, being abstract, Torretti notes, lack causal powers. The same goes for mathematical entities. On this view, all that is required for mathematical objects to exist is that we reason about them. This reasoning, according to Torretti, presupposes the existence of mathematical objects; it implicitly attributes such existence to the relevant entities. There is no need for the objects to be causally active, though. In this respect, the argument goes, language use and mathematical reasoning are very similar. On Torretti's view:

> I claim that the existence implicitly attributed to mathematical objects when we reason about them is not less certain nor more stringent than that the fact itself of speaking confers on [. . .] linguistic objects. If these do not require causal powers to judge the role that they play in our mental and social life, why would mathematical objects need to have them? We have seen that the possibility of language presupposes the repeated presentation of invariant objects that we recognize and pay attention to, even though they are impassible and inactive. (Torretti, 2014, p. 130)

It is important to note, nevertheless, that even though the same letter may have different shapes on a page (different tokens of the same type may be at display), what is read or spoken is never an abstract type, but always a concrete token. One needs to be able to read the statements or hear the utterances to understand what is being said. And for that to happen, statements and utterances need to be concrete (even if the types that allegedly ground the repeated presentation of invariant objects can be interpreted as being abstract). Similarly, in the case of mathematics, the notation and symbols used on a page to carry out a particular piece of mathematical reasoning are also concrete (even if what the symbols and notation stand for can be interpreted as being about abstract entities).

However, is it the case that to reason about mathematical objects one needs to attribute, albeit implicitly, existence to them? Could we not reason about these objects without taking them to exist even if to do so we use language that can only function by apparently recognizing the abstract types that code its symbols? Two issues need to be distinguished at this point: (a) Is it the case that language can function only by presupposing some abstract objects, namely, the abstract types that characterize the symbols of the language? (b) Is it possible to reason about something without implicitly attributing existence to what is reasoned about? More generally, is it possible to reason about the nonexistent without turning it into something that is presumed to exist? I will resist the implicit argument for platonism found in a positive answer to (a) and support a nominalist reading of (b), suggesting that we can, and regularly do, reason about the nonexistent.

6.3.2 Reasoning About the Nonexistent

There is no question that to use language we need to be able to recognize variations of the same letters. If different inscriptions of the letter 'a' were confused with 'e' and 'o', we would not be able to distinguish properly 'mat', 'met', and 'mot'. Clearly, the same letter can admit different inscriptions, including variations in font and writing style or multiple occurrences of the same letter in the same word or linguistic context, such as the double occurrences of 'e' and 't' in one of the words of this sentence. There are those who insist on the need for a platonist description of this fact. The situation, they insist, is characterized as a matter of the instantiation of a given universal, an abstract type (the letter), which admits several tokens (specific concrete inscriptions that instantiate the type in question).

Is this way of conceptualizing the situation really needed or even properly motivated? It is unclear that it is despite its ubiquity in philosophical circles. After all, rather than positing an ontology of multiply instantiated universals, all that is need for users to employ a language is their capacity to recognize the same letters on a page, despite variations in font and writing style. All that users need is the capacity to recognize particular inscriptions of 'e', resisting any inclination to ascent to a universal of which such inscriptions are alleged instances. It is enough to note (with suitable gestures) something along the following lines: "This is an 'e' and so is this one, albeit in a different place and in a different font", rather than "This inscription of 'e' and this other inscription of 'e' are instances of the same universal, namely, the letter 'e'". Recognition of specific inscriptions suffices. To characterize them as instantiated universals is to add an unnecessary metaphysical gloss on something that is far simpler and more straightforward than the philosophical interpretation that is offered.

It may be argued that it is only because there is a suitable universal that the similarities between different inscriptions of the same letter can be recognized. Nevertheless, this simply inverts the proper explanatory order: a universal is posited because both objects are an 'e', which gives priority to the specific inscriptions: "this is an 'e' and so is this one", the nominalist would insist. By resisting the unnecessary reification of a universal, our own capacity to use language can be accommodated without needlessly increasing the ontology.

Furthermore, we should be able to reason about the nonexistent without turning it into something that exists. Otherwise, we would lose the very content of what we are reasoning about (the nonexistent). It is important to recognize our ability to talk about and quantify over objects that do not exist, such as ghosts, witches,

Sherlock Holmes, among other fictional characters, as well as phlogiston and Vulcan (a putative planet between Mercury and the Sun), among other scientifically posited but ultimately nonexistent objects.

One way of accommodating reasoning about these entities is by adopting quantifiers that lack ontological commitment, that is, which do not conflate quantification with existence (see Azzouni, 2004; Bueno, 2005). In the domain of quantification, objects are collected, and quantifiers range over them. The universal quantifier ranges over all such objects whereas the existential quantifier ranges over some of them.

But to perform this quantificational role, such quantifiers need not have the additional function of marking the existence of the objects under consideration. In fact, it is important not to conflate these two roles, otherwise one is unable to express significant points. Consider, for instance, a classical set theorist who is discussing the Russell set: the set R of all non-self-membered sets, that is, $R = \{x: x \notin x\}$. Thinking of that set, the set theorist notes, correctly, that: "Some sets are too big to exist". In a classical set theory, whose underlying logic is classical, no such set, in fact, exists. Note that the set-theorist has not uttered a contradictory statement but only asserted that, among the various sets, some did not exist. However, if quantification and existence are conflated, her statement becomes a contradiction, a statement to the effect that there exist sets that do not exist.

No such difficulty emerges if existence is not expressed *via* quantification but instead by an existence predicate. This predicate specifies the conditions satisfied by existing objects. The determination of these conditions is no straightforward task and is fraught with challenges. Existence, in principle, can be characterized in a variety of conflicting ways, in terms of, for example, ontological independence (from one's psychological processes and linguistic practices), spatiotemporal location, causal activity, verifiability, or observability, to mention just a few conditions (see Azzouni, 2004 for an insightful discussion and a defense of the ontological independence conception). Part of the difficulty is that it is very difficult for a conception of existence not to beg the question against another if the conditions that characterize existence are formulated as necessary and sufficient conditions. Platonists will complain, against those who posit causal activity as a requirement for existence, that numbers, functions, sets, and other abstract objects exist, even though they are causally inactive. Scientific realists will complain, against those who posit observability or verifiability as requirements for existence, that photons, electrons, and quarks exist, even though they are neither verifiable nor observable (certainly not with one's naked eyes). Realists about universals will complain, against those who posit spatiotemporal location as a requirement for existence, that uninstantiated universals, despite not being spatiotemporally located, do exist.

Even if somehow no questions were begged in these ontological debates, it is often unclear how ontological disputes can be settled given that the same criterion, when applied, can yield diametrically opposed results. Consider, for instance, ontological independence. According to Jody Azzouni (2004), since mathematical objects are made up by humans, they are *not* ontologically independent from particular linguistic practices and psychological processes. Platonists agree that ontological independence is a condition for existence, and this is precisely the reason why they insist that mathematical objects *do* exist. After all, these objects would have existed even if no humans ever did. The same criterion, thus, leads to conflicting ontological conclusions.

Nevertheless, if the existence predicate specifies only *sufficient* conditions (rather than necessary conditions as well), these difficulties are often avoided, given that the determination of what is sufficient for existence is typically far less controversial than what is necessary. Objects that are spatiotemporally located or that are causally active arguably exist. Usually, these are not controversial matters.

The upshot is that debates about what exists are not, and should not, be framed as a specific logical debate, namely, a debate about quantification. Rather, they are, and should be, understood as addressing a metaphysical issue. By separating quantification and existence, the uses of ontologically neutral quantifiers and the existence predicate allow us to conduct these debates without prejudging the existent or the nonexistent. With the adoption of these quantifiers, there is no difficulty to talk about the nonexistent without presupposing that it exists (see also Azzouni, 2010; I will return to ontologically neutral quantifiers below).

6.3.3 Fictionalism: A Challenge

Torretti concludes his discussion of mathematical fictionalism by considering a formidable challenge to the ontological minimalism he favors. As he notes, mathematics posits uncountably many objects (such as all the real numbers, not to mention even larger cardinalities studied in set theory), but any linguistic representation, being formulated in terms of symbols, has at best countably many expressions. This may seem to pose a difficulty for the view that takes as a necessary condition for the existence of something that it can be referred to (mentioned), since this requires the ability to name (or, at least, refer to) the objects in question in order that they exist. After all, one will run out of names of objects (or, at least, referential devices) before every object has been referred to. Should we then conclude that there are at most countably many objects in reality? According to Torretti:

> I close these reflections by examining an issue relative to the existence of mathematical objects that could be judged problematic. It can be briefly described thus: for the ontological minimalism that I propose it suffices for something to be mentionable for it to be acknowledged as existing and for it to be reasonably reasoned about. But this liberality, which at first sight is so big, may not be enough for modern mathematics, which posits the existence of collections of objects by definition so immense that not even an immortal intelligence could generate names for every single one of them. (Torretti, 2014, p. 140)

In response to this objection, Torretti briefly considers three possible solutions (Torretti, 2014, pp. 142–143). (a) One could abandon the set-theoretic approach altogether. (b) One could adopt, at least in mathematics, collective referential devices that do not require a proper name for each object that is referred to. (c)

One could jettison the doctrine that existence requires the possibility of referring to what is taken to exist. As it turns out, there are costs associated with each option.

(a). To abandon set theory and set-theoretic devices more generally is something that constructivists and category theorists have recommended, as Torretti himself (2014, p. 142) points out. Despite the independent interest of this maneuver, it is unclear that it actually solves the problem at hand. To the extent that uncountable infinities are invoked and recognized, there will be objects that cannot be referred to, which clearly undermines the referential requirement on the existent at issue. If, in turn, the existence of uncountable infinities is simply rejected, for whatever mathematical or philosophical reasons, a price needs to be paid relative to what is available in classical mathematics, where such infinities are regularly acknowledged. Note that even if set theory is rejected, the uncountability of real numbers is a straightforward fact about analysis, and thus it still needs to be accommodated. This questions the adequacy of the requirement that only what can be referred to exists.

(b). The development of alternative collective referential devices would bypass the need for proper names to refer to arbitrary mathematical objects. As Torretti (2014, p. 142) acknowledges, it also demands a change in formal semantics for mathematical discourse, since the usual semantics invokes a universe of discourse whose items are typically taken to exist. (It is unclear, nevertheless, whether this assumption about semantics is justified, as neutral quantifiers could be adopted throughout.) However, even if collective referential devices were developed, it is not obvious that they would fully resolve the issue. After all, what is needed is to secure reference to *each* individual object of a set of uncountably many entities. Collective referential devices, although suitable for reference of a generic sort, would not be enough to secure a unique referent to each object that is referred to. Without such uniqueness, the problem under consideration is left unsolved.

(c). In light of these difficulties, one could simply reject the reference requirement on existence and admit that there may be objects that one is unable to refer to uniquely. Of course, this amounts to a rejection of the ontological minimalism recommended by Torretti. To resist this rejection, he offers two related concerns that would need to be addressed if the reference requirement were eliminated (Torretti, 2014, p. 143): (i) One would need to accommodate unmentionable existent objects. How exactly could this be done? (ii) One would need to devise an account of the way to reason about what cannot be referred to. How can this be achieved? These are significant challenges, which clearly need to be considered if the reference requirement is rejected. In what follows, I take them on and argue that, with suitable resources, it is possible to accommodate both of them.

6.4 Mathematical Fictionalism: A Neutralist Approach

6.4.1 Ontological Minimalism

It is clear by now that Torretti advances an intriguing form of ontological minimalism about mathematics, according to which it is enough, and also necessary, that an object be mentionable for it to be recognized as existing (Torretti, 2014, p. 140). I indicated earlier some concerns for the three kinds of fictionalism discussed in the beginning of this work. But there are also difficulties for the two (strong and weak) forms of fictionalism examined in the previous section. The maximalist, strong form of the view, as I noted, clearly begs the question against platonism, since it takes physical (concrete) objects as what ultimately exists and complains that since abstract objects lack the features that concrete entities have (such as spatiotemporal location and causal activity), they do not exist.

In turn, the minimalist, weak form of fictionalism faces the difficulty of recognizing as existing things for which there is no reason to think that they exist. One can refer to and talk about objects that are known *not* to exist, such as witches and ghosts, Vulcan, phlogiston, and luminiferous ether, not to mention the Russell set in classical set theory or infinitesimals as originally formulated in the early calculus. As a result, the minimalist's weak requirements on existence ends up with a severely bloated ontology. Thus, the consequences of both weak and strong fictionalism are not attractive. An alternative is called for.

Crucial to this task is the development of an account that is able to accommodate the various challenges faced by ontological minimalism. As a result, (i) a strategy needs to be devised to make room for unmentionable existent objects, and (ii) an account of how to reason about what cannot be referred to should be articulated. Moreover, this has to be achieved (iii) without bloating the ontology, otherwise no clear gains over ontological minimalism would ultimately be obtained. I will start by considering (iii) and then move to (i) and (ii).

6.4.2 Neutral Quantification

Ontologically neutral quantifiers provide a useful device for these tasks (for additional details, see Azzouni, 2004; 2010, 2017; Bueno, 2005). By allowing one to distinguish quantification from existence, these quantifiers provide the tools that are needed to quantify over things without presupposing their existence. The domains of the quantifiers need not (and should not) be thought as corresponding to existent entities. To assume that domains of quantification require the existence of the objects in them is just to presuppose the ontological import of the quantifiers in the metalanguage (Azzouni, 2004). It is unclear, nonetheless, that this ontological presupposition is justified. After all, sets of nonexistent objects, such as sets of unicorns, superheroes, golden mountains, frictionless planes, and infinitely deep oceans, can all be formed and defined unproblematically, given suitable descriptions of the objects at hand. The fact that the domains can be specified despite the nonexistence of the objects in question clearly indicates that quantification and existence come apart.

More generally, even in the case of indispensable objects (reference to which cannot be eliminated *via* suitable paraphrases and that need to be referred to in explanatory contexts), one need not assume that these objects exist. Suppose we do not know the exact number of mothers in a country, and the only way of referring to the relation between their number and the number of their kids is by talking about average moms, based on certain samples of the population. Clearly, when it is found out that the average mom has 2.4 kids, one does not conclude that there exists some mother who has precisely that number of kids (see Melia, 1995). Of course, average moms do not exist, despite being indispensable in a context like this.

In light of these considerations, quantification and existence should be distinguished, leaving the former only with the role of specifying whether the entire domain is quantified over (universal quantification) or a portion of the domain is (existential quantification). Existence is then understood, as noted above, as a predicate, whose satisfaction requires suitable (metaphysical) assumptions. This is as it should be since whereas quantification is a logical device, existence raises primarily metaphysical issues. As also noted above, provided that the satisfaction conditions for the existence predicate offer only sufficient conditions, and not necessary ones, it is possible to avoid begging the question on controversial and substantive metaphysical debates. For current purposes, the important point is that neutral quantification allows one to quantify over objects without any commitment to their existence, quite independently of the kind of objects under consideration, whether they are concrete or abstract, real or fictional, actual or made-up.

6.4.3 Unmentionable Existent Objects

How could one accommodate unmentionable existent objects? On the face of it, the expression may harbor paradox. In fact, in referring to "unmentionable existent objects", these objects have just been mentioned. So, how could there be genuinely unmentionable existent objects if in the process of specifying them, we end up mentioning what is supposedly unmentionable? Given the inconsistency, one may conclude that no such objects could possibly exist (assuming classical logic), which would then support Torretti's conception to the effect that the existent and the mentionable go hand in hand.

Of course, from the point of view of linguistic practice, the concern is of a different order: the requirement that existence be constrained by reference seems to preclude the possibility that things that cannot be referred to do exist, for no other reason than sheer linguistic limitations on the speakers' part, such as not having enough names available to mention all objects. But why should existence be tied to a linguistic trait rather than to what there is in the world? Admittedly, as the inconsistency above suggests, one cannot consider whether something exists or not without referring to the very objects whose existence is under consideration. But it does not follow that the objects in question thereby exist. After all, the existence issue is precisely what is in question.

Ontologically neutral quantifiers allow one to make sense of this situation. In quantifying over certain objects, one need not assume that the objects in question exist. Quantification only specifies the scope of the objects under consideration: some of them or all of them. Whether these objects exist or not, however, is a separate issue altogether. As a result, unmentionable objects can be quantified over, independently of their existence.

6.4.4 Reasoning About What Cannot Be Referred to

Finally, the issue of how it is possible to reason about what cannot be referred to needs to be addressed. Central here, once again, are the resources provided by neutral quantifiers.

First, it is important to highlight that neutral quantifiers are objectual quantifiers, that is, they operate on objects, namely, those in the domain of quantification (see also Azzouni, 2004). Whether such objects exist or not is independent from the fact that they can be quantified over. Objectual quantification does not require ontological commitment to the objects under consideration. It is a particular interpretation of quantifiers that requires the existence of objects in the quantification domain. Neutral quantifiers demand no such additional interpretative gloss.

Moreover, objectual quantification involves no requirement that objects that are quantified over be nameable, mentionable or referable. Compare substitutional quantification on this point. Such quantification does seem to require, as suitable substitutional instances, appropriate names for the objects that are quantified over. This imposes a significant constraint on the scope of substitutional quantifiers. In contrast, objectual quantifiers involve no such demands. Given their focus on objects (rather than on their names), it does not matter how such objects are formulated, referred to or mentioned for them to be quantified over.

In this way, neutral quantifiers allow us to reason about objects independently of how, or even whether, they are referred to. This plasticity is central to the generality involved in mathematical reasoning, which allows for the study of objects in a completely general setting. One can quantify over an arbitrary real number without specifying anything about that particular number other than that it is a number of this kind; even though it is not possible to give a name for each one of these numbers. Moreover, as part of mathematical practice, there is no need to name, mention or refer to them individually in any way other than by fact that they are real numbers. The specification of their individuality, which a name would tend to ratify, typically is not demanded by the way in which mathematicians quantify over real numbers and other mathematical objects. Hence, with neutral quantifiers, it is possible to reason about objects independently of one's ability to refer (individually) to the objects under consideration.

As an illustration, consider the archimedian property of the set of real numbers *R*, namely: if x and y are real numbers and x is strictly positive, then there is a positive integer n such that $nx > y$ (Rudin, 1976, p. 9). The result is established, *via reductio*, by relying on the least-upper-bound property, according to which there is a least upper bound of a non-empty, bounded above set (often called the *supremum* of that set). Assume the negation of the archimedian property, and note that y then becomes an upper bound for the set of all nx, which, by the least-upper-bound property, has a least upper bound a. Then derive a contradiction by showing that there is a number in the set of all nx that is greater than a. Clearly, this is impossible since a is an upper bound for that set.

Two important points should be noted: (i) Throughout the reasoning above, reference to no individual features of real numbers is made. Quantification is perfectly arbitrary; no mention or identification of any particular real number is required. Moreover, (ii) commitment to the existence of real numbers—understood, for instance, as ontological independence, causal activity, or spatiotemporal location—is nowhere to be found either. The argument relies on the fact that there is a least upper bound (a supremum), but no claim is made that this number is ontologically independent from mathematicians' linguistic practices or psychological processes, that it is causally active or located in spacetime. In other words, metaphysical issues associated with conditions of existence are not raised. The argument goes through completely independently of these metaphysical considerations. This is as it should be, given the ontological neutrality of mathematical practice. This fact becomes especially salient with ontologically neutral quantifiers.

6.4.5 Mathematical Fictionalism Revisited

These considerations suggest that perhaps there is a coherent form of fictionalism close to Torretti's considered view but which avoids the difficulties that his proposal seems to face. If quantification is understood as being ontologically neutral and, thus, as not having any ontological commitment, when Torretti notes that "it suffices for something to be mentionable for it to be acknowledged as existing and for it to be reasonably reasoned about" (Torretti, 2014, p. 140), the passage can be understood as stating that "it suffices for something to be mentionable for it to be *quantified over* and for it to be reasonably reasoned about". All that is needed here is neutral quantification over the relevant objects; there is no need to require the existence of that over which one quantifies, thus generating a proper form of fictionalism about mathematics (when the objects quantified over are mathematical).

As we saw, Torretti promptly worries that "[. . .] this liberality, which at first sight is so big, may not be enough for modern mathematics, which posits the existence of collections of objects by definition so immense that not even an immortal intelligence could generate names for every single one of them" (Torretti,

2014, p. 140). This remark, in turn, can then be understood as stating: "But this liberality, which at first sight is so big, may not be enough for modern mathematics, which *quantifies over* collections of objects by definition so immense that not even an immortal intelligence could generate names for every single one of them". Once again, since quantification does not require naming the objects that are quantified over, there is no difficulty in quantifying over any of the various kinds of objects posited by modern mathematics, from real numbers to inaccessible cardinals. And, once again, since neutral quantification does not require existence, whether any of these objects exist or not is a separate issue. In this way, by using neutral quantifiers, a coherent form of fictionalism can be articulated. It has much in common with Torretti's considered view, even if he has not invoked such quantifiers.

6.5 Conclusion

Mathematical fictionalism has a number of different formulations, some problematic, others appealing. In this work, I considered several of them, guided by Torretti's own engagement with and examination of this philosophical research program. After considering a number of difficulties that the various articulations of the proposal face, including Torretti's own preferred version of it, I sketched an alternative conceptualization of mathematical fictionalism that seems to preserve a number of the salient features that made the proposal attractive, albeit *via* resources very different from those employed by Torretti himself.

Acknowledgements It is a pleasure to dedicate this article to the towering figure that is Roberto Torretti. His writings have illuminated a huge number of central issues in philosophy, which, over decades of work, he examined so carefully with his unique combination of philosophical acumen and historical sensitivity. His work provides a continuous source of inspiration and insight—as the papers in this volume also make clear.

In a work dedicated to celebrate Torretti's own contributions to philosophy, non-philosophers may find it odd that those contributions are critically engaged with. Perhaps philosophy is odd in this respect, but it remains the case that criticism is one of the most respectful ways of showing sincere appreciation and admiration for the work of those who we esteem the most.

My thanks go to Cristián Soto for extremely insightful discussions and correspondence on the issues examined in this paper, for his feedback on an earlier version of it, and for his detailed comments on and suggestions about Torretti's work. I used his unpublished translation into English (with some added adjustments of my own) of the parts of Torretti (2014) that I quoted above. His patience, help and support throughout are sincerely acknowledged.

References

Azzouni, J. (2004). *Deflating existential consequence: A Caso for nominalism*. Oxford University Press.

Azzouni, J. (2010). *Talking about nothing: Numbers, hallucinations, and fiction*. Oxford University Press.

Azzouni, J. (2017). *Ontology without Borders*. Oxford University Press.

Bueno, O. (2005). Dirac and the dispensability of mathematics. *Studies in History and Philosophy of Modern Physics, 36*, 465–490.

Bueno, O. (2010). A defense of second-order logic. *Axiomathes, 20*, 365–383.

Bueno, O., & French, S. (2018). *Applying mathematics: Immersion, inference, interpretation*. Oxford University Press.

Bunge, M. (1974-). *Treatise on basic philosophy* (Multiple volumes.). D. Reidel.

Lakatos, I. (1976). In J. Worrall & E. Zahar (Eds.), *Proofs and refutations: The logic of mathematical discovery*. Cambridge University Press.

Melia, J. (1995). On what there's not. *Analysis, 55*, 223–229.

Rudin, W. (1976). *Principles of mathematical analysis* (3rd ed.). McGraw-Hill.

Shapiro, S. (1991). *Foundations without foundationalism*. Oxford University Press.

Thomasson, A. (1999). *Fiction and metaphysics*. Cambridge University Press.

Torretti, R. (1981). Three kinds of mathematical Fictionalism. In J. Agassi & R. S. Cohen (Eds.), *Scientific philosophy today* (pp. 399–414). Reidel.

Torretti, R. (2014). Conceptos y Objetos Matemáticos. In R. Torretti (Ed.), *Estudios Filosóficos, 2011–2014* (pp. 109–143). Ediciones Universidad Diego Portales. (All of the quotations from this essay are from an unpublished English translation of portions of it by Cristián Soto, with some changes I made.)

Chapter 7
Functionalism as a Species of Reduction

Jeremy Butterfield and Henrique Gomes

Abstract This is the first of four papers prompted by a recent literature about a doctrine dubbed *spacetime functionalism*. This paper gives our general framework for discussing functionalism. Following Lewis, we take it as a species of *reduction*. We start by expounding reduction in a broadly Nagelian sense. Then we argue that Lewis' functionalism is an *improvement* on Nagelian reduction.

This paper sets the scene for the other papers, which will apply our framework to theories of space and time. (So those papers address the space and time literature: both recent and older, and physical as well as philosophical literature. But the four papers can be read independently.)

Overall, we come to praise spacetime functionalism, not to bury it. But we criticize the recent philosophical literature for failing to stress:

(i) functionalism's being a species of reduction (in particular: reduction of chrono-geometry to the physics of matter and radiation);
(ii) functionalism's idea, not just of specifying a concept by its functional role, but of specifying several concepts *simultaneously* by their roles;
(iii) functionalism's providing bridge laws that are mandatory, not optional: they are statements of identity (or co-extension) that are conclusions of a deductive argument, rather than contingent guesses or verbal stipulations; and once we infer them, we have a reduction in a Nagelian sense.

On the other hand, some of the older philosophical literature, and the mathematical physics literature, *is* faithful to these ideas (i) to (iii)—as are Torretti's writings. (But of course, the word 'functionalism' is not used; and themes like simultaneous unique definition are not articulated.) Thus in various papers, falling under various research programmes, the unique definability of a chrono-geometric concept (or concepts) in terms of matter and radiation, and a corresponding bridge law and reduction, is secured by a precise theorem. Hence our desire to celebrate these results as rigorous renditions of spacetime functionalism.

J. Butterfield (✉) · H. Gomes
Trinity College, University of Cambridge, Cambridge, UK
e-mail: jb56@cam.ac.uk

C. Soto (ed.), *Current Debates in Philosophy of Science*, Synthese Library 477,
https://doi.org/10.1007/978-3-031-32375-1_7

7.1 Introduction

This is the first of four papers prompted by a recent literature about a doctrine dubbed *spacetime functionalism*. This paper gives our general framework for discussing functionalism. Following Lewis, we take it as a species of reduction. In this, our views are close to the 'Canberra Plan'.[1]

We first expound reduction in the broadly Nagelian sense of deduction enabled by extra premises, usually called 'bridge laws' (Sects. 7.2 and 7.3). We start with the general idea of reduction and problems it faces (Sects. 7.2.1 and 7.2.2). Then we specialize to Nagelian reduction (Sect. 7.3). Then we describe how Lewis' functionalism is a cousin of Nagelian reduction: and, we argue, an *improvement* on it (Sects. 7.4 and 7.5). Finally, we discuss connections with Torretti's writings especially about reduction and physical geometry (Sect. 7.6). Section 7.7 concludes.

This paper thereby sets the scene for the other papers, which will apply our framework to theories of space and time. So those papers address the literature about space and time—both recent and older, and physical as well as philosophical; (but they can be read independently). In the second paper, we present four older examples, drawn from geometry and dynamics (including special relativity), that give attractively precise examples of spacetime functionalism—in our reductive and Lewisian sense of 'functionalism'. Namely: results by Lie (building on Helmholtz), by Malament (building on Robb), by Mundy, and by Barbour and Bertotti. So these examples span some hundred years, from the 1880s to the 1980s. Happily for us, with our wish to honour Roberto Torretti: he has written in detail about the first two of these examples, and our verdicts about them mesh with his discussions. In Sect. 7.6 below, we will get a glimpse of this meshing.

The third and fourth papers address other more recent literature. The third paper discusses some recent philosophical literature (by authors such as Belot, Callender and Knox). The fourth discusses some physics literature, about the foundations of general relativity and kindred theories. (There, our first example will be a classic 1976 paper by Hojman et al. which, unfortunately, the philosophical literature has ignored—but which, happily for us, Torretti commends.)

Our overall position is that we come to praise spacetime functionalism, not to bury it. But we criticise the recent philosophical literature: criticisms which, it will be clear, happily do *not* apply to Torretti's views.

In short: our criticism is that although this literature picks up on the idea of *functional role*, it fails to stress three significant ways that functionalism, especially in Lewis' hands, develops that idea. Namely:

[1] This programme (surveyed in Braddon-Mitchell and Nola, 2009) is familiar in metaphysics, philosophy of mind and ethics; but unfortunately little-known in philosophy of science. But our views are only 'close to Canberra': in particular, we will not require that the functional role of a concept, extracted from the given theory, gives a conceptual analysis of it.

(i) functionalism's being a species of reduction (in this context: reduction of chrono-geometry to the physics of matter and radiation);
(ii) functionalism's idea, not just of specifying a concept by its functional role, but of specifying several concepts *simultaneously* by their roles;
(iii) functionalism's providing bridge laws that are mandatory, not optional: they are statements of identity (or co-extension) that are conclusions of a deductive argument, rather than contingent guesses or verbal stipulations; and once we infer them, we have a reduction in a Nagelian sense; i.e. the reduction is a deduction.

These lacunae are a missed opportunity, in two ways.

First, it turns out that some of the older philosophy of physics literature, and the mathematical physics literature, *is* faithful to the ideas (i) to (iii)—as are Torretti's writings. (But of course, the word 'functionalism' is not used; and themes like simultaneous unique definition are not articulated.) Thus in various papers, falling under various research programmes, the unique definability of a chrono-geometric concept (or concepts) in terms of matter and radiation, and a corresponding bridge law and reduction, is secured by a precise theorem. Witness the four older examples in our second paper, and the physics examples in our fourth paper. In short: the recent literature has missed some *golden oldies* that illustrate *real* functionalism, and not merely the idea of functional role.

The second missed opportunity concerns the wider literature in philosophy of science about reduction. We believe it has missed the power and plausibility of the ideas (i) to (iii). The reason is that the literature works predominantly with a *binary* contrast between "reduced" and "reducing": either of vocabularies or of theories, or both. (Of course, this contrast derives in part from the traditional positivist project of reducing theory to observation, as a matter of defining theoretical terms and then deducing theoretical claims.) On the other hand, functionalism especially in Lewis' hands—and more generally, the Canberra Plan—works with a *ternary* contrast. It is this ternary contrast that yields (i) to (iii).

For there is, first, within a *single* theory (the 'higher-level' theory), a binary division of its terms, with each of the terms of the first sort having a functional definition—i.e. a unique specification—using all the terms of the second sort: idea (ii) above. Here, 'functional definition' will mean 'specification' (i.e. 'uniquely picking out'), but need not give the term's conceptual analysis or meaning.

But there is also a *second* ('lower-level') theory that specifies each *definiendum*—the unique *realizer* or *occupant* of each functional role—*independently* of the first theory. Here, 'independently' means using terms (or concepts or properties) that are not in the first theory. So a single entity (extension) is picked out in two independent ways:

(a) as the unique occupant of a functional role extracted from the first theory, and
(b) as specified by the second theory.

This gives a statement of identity (coextension) that is a *derived bridge law*. It is mandatory, not optional, because it is deduced from the two theories. Besides, one can show that the collection of such statements gives a deduction of the *first* theory. The reason, in short, is that the terms in the bridge laws implicitly contain the rich information of the functional roles. Thus we obtain idea (iii) above.

So we stress that here, one must distinguish three items:

(a) the specification extracted from the first theory (using a binary division of its terms); from
(b) the specification using the second theory (so invoking a ternary contrast); and from
(c) the derived bridge law, that comes from combining (a) and (b).

Thus this paper is mostly about the second missed opportunity. We want to expound and defend what we will label as *(Lewisian) functionalist reduction*. It is another *golden oldie* that much of the literature in philosophy of science has sadly missed. Though simple, we will argue that it is not too simple. It is powerful and flexible enough to address issues that beset reduction: such as multiple realizability, meaning variance, and most real-life reductions being, in one or more ways, partial.

But even if you end up rejecting our advertisement as regards reduction in general, we in any case commend functionalist reduction for our several examples of the physics of matter and radiation contributing to determining, or explaining, some chrono-geometric concepts. As we will see in our other papers, it fits those examples very well.

In the rest of this Introduction, we will add some details about: functionalism in general (Sect. 7.1.1); functionalism about spacetime (Sect. 7.1.2); and the connection to Torretti (Sect. 7.1.3).

7.1.1 Introducing Functionalist Reduction

In philosophy, the word 'functionalism' is mostly associated with proposals in the philosophy of mind by Armstrong, Lewis and Putnam in the mid-1960s. It will be clearest to introduce functionalism, and Lewisian functionalist reduction, via this example (Sect. 7.1.1.1); and then make three general comments (Sect. 7.1.1.2).

7.1.1.1 ... in the Philosophy of Mind

These authors' initial idea was that each mental state, or concept, could be *uniquely specified* by its characteristic pattern ('web') of relations to various other states, or concepts, both mental and physical. Such a characteristic pattern, dubbed *functional role*, was spelt out in terms of nomic and-or causal relations: relations that some appropriate body of knowledge or belief (often called a 'theory'), whether everyday or scientific, claims to hold between the various states or concepts.

This idea of functional role has indeed been taken up by recent articles on spacetime functionalism (cf. Sec 7.1.2). They focus on the idea that spacetime, or specific spatiotemporal concepts such as the metric, or connection, or inertial frame, are specified by such a pattern.[2]

But there is much more to functionalism than just the idea of a state or concept having such a characteristic pattern (functional role). For after all: it is plausible that many states and concepts each have a pattern of relations to other states or concepts that is characteristic, i.e. idiosyncratic, enough, that the state or concept can be uniquely specified by that pattern. So in this minimal sense, almost every concept is functional. Cf. Footnote 2.

To say what more there is to functionalism, we begin by recalling that functionalism, like its forebears, viz. logical behaviourism and mind-brain identity theory, accepted a basic contrast between 'mind' and 'body'; or better, between 'mental' and 'material' states and concepts. The general theme, of both these forebears and

[2] We will mostly use both 'state' and 'concept' throughout, although: (i) they are used more in philosophy of mind and metaphysics than in philosophy of physics (whose main analogues are 'physical state' and 'quantity' i.e. 'physical magnitude'); and (ii) they are terms of art, without widely agreed criteria of individuation. But we shall not need to be precise about such criteria. Indeed, almost all our points are unaffected by what exactly these criteria are; and we will signal the exceptions.

It suffices for us to note that: (a) for most authors, a state is a localized fact or state of affairs; (b) some authors distinguish type-states, e.g. being in pain, or being in love, or seeing yellow in the top-left of the visual field, from token-states e.g. Fred's being in pain at t, John's loving Mary now, and these authors often go on to individuate token-states as n-tuples of the objects and times involved and the properties or relations attributed e.g. $\langle$John, Mary, now, ... loves ...$\rangle$ (such n-tuples are often called 'Russellian propositions'); (c) for most authors, a concept is the same, or pretty much the same, as a type-state. It is also the same, or pretty much the same, as a property or relation. For it is general, not localized, and is individuated more finely than by its set of instances; for example, it is individuated by a Fregean sense or Carnapian intension rather than by extension.

We can go along with (a) to (c), both in this paper and the others. We will simply note that some of the debates, e.g. in philosophy of mind about whether the mental state or concept of being in pain is reducible to material states or concepts, turn on controversies about the criteria of identity of properties—indeed a misty topic.

We can be sanguine in this way because, happily, for our examples for philosophy of physics, the state and concepts in question—the properties specified by their functional roles— are much less vague and controversial than e.g. pain. For they are physical properties: often, familiar physical quantities. Here are examples from our second paper: being freely mobile; being simultaneous with (as a relation between events); being congruent (as a relation between spatial intervals); and being isochronous, i.e. of equal duration (as a relation between temporal intervals).

of functionalism, was of course that mental discourse is *problematic*; and is to be vindicated or legitimized by showing how it is *suitably related to unproblematic* material discourse. Besides, the 'suitable relation' was, in effect, 'being a part of'. That is, mental discourse was to be *reduced* to material discourse.

But functionalism was distinctive in having this reduction proceed in two stages: stages that reflect the *ternary* contrast we stressed in this Section's preamble. In other words, functionalism makes two proposals. Or as we advocates would say: it has two insights about the reduction of mental to material.

The first is about the binary division, mental vs. material, among the terms used in everyday knowledge and belief about the mental and material—in what came to be called 'folk psychology'. The proposal is that this body of knowledge and belief is sufficiently rich that each of the many mental states and concepts can be uniquely specified by its functional role. And this can be done *simultaneously* for all of them, even though a mental state's or concept's functional role invariably mentions, not just material states and concepts, but also other mental ones. A standard pedagogic example is that the functional role of the mental state, belief that it is raining, cannot mention only bodily and-or behavioural states and concepts, such as being disposed to pick up an umbrella when going outdoors. For if someone believes it is raining, they have that disposition *only if* they also desire to stay dry—a mental state. That the functional role of a mental state needs to mention other mental states apparently implies a threat of a logical circle. But functionalism shows that the threat can be overcome. We will see later how Lewis consistently extracts simultaneous unique specifications of many terms from a theory—here, folk psychology.

The second proposal is that each mental state, each unique occupant of a corresponding functional role, is also specified by *another* body of doctrine—a second theory—that we accept, independently of (perhaps after) our acceptance of the first, i.e. folk psychology. For Armstrong, Lewis and Putnam, this second theory was of course neurophysiology.

Then the two proposals taken together yield deductions of bridge laws as statements of identity (or co-extension of predicates). By specifying a mental state in two independent ways, we can infer such a statement. These bridge laws, taken together, then yield a reduction, i.e. a deduction of the first theory from the second: details in Sect. 7.5.

This is vividly illustrated by the time-honoured example of inferring that the mental state of pain *is* a neurophysiological state. As an inference, it is trivially simple. For it is merely a case of the transitivity of identity. But since this inference will be a *leitmotif* in what follows, and will have striking parallels in our spacetime examples in other papers, it is worth rehearsing it at the outset of our discussion, albeit briefly. The classic, crystal-clear expositions are Lewis (1966, 1972, Introduction and Section 3).

Thus consider:

(i) Accepting the characterization of pain given by folk psychology, we endorse the premise: pain is the unique occupant of so-and-so role.

(ii) Accepting neurophysiology, we endorse the premise: C-fibre firing is the unique occupant of so-and-so role. (Here, 'C-fibre firing' is the ignorant philosopher's catch-all for a technical, maybe long, specification in the language of neurophysiology.)
(iii) So by the transitivity of identity, we must infer: pain *is* C-fibre firing.

Thus we see, in summary form, how the functionalism of Armstrong, Lewis and Putnam clarified and unified its forebears, logical behaviourism and mind-brain identity theory. In terms of the inference just rehearsed:

(a) logical behaviourism—roughly speaking: conceptual analysis of mental terms as they occur in folk psychology—gives the warrant for premise (i);
(b) neurophysiology—i.e. contingent empirical discoveries—gives the warrant for premise (ii);
(c) so we infer (iii), a derived so-called "bridge law": a statement of identity between pain as specified by folk psychology and pain as specified by neurophysiology.

This statement of identity is an instance of mind-brain identity theory. But the identity is not merely recommended as a hypothesis that is attractive because ontologically parsimonious (as earlier advocates of mind-brain identity theory, such as Place and Smart, had said). It is the conclusion of a valid argument from accepted premises: viz. premises that describe the unique occupant of a functional role in two ways.

Note also that we must of course expect this statement of identity, and other such identities of mental and neural states, to be relative to a *kind*. That is: which neural state is identified with a given mental state of course varies between organisms. Feeling pain might well be a very different state of a mollusc brain than of a human brain. Besides, the kind relative to which such a so-called "type-type" identity holds might well be narrower than a species (cf. Lewis, 1969, p. 25). So this kind of mind-brain identity theory has no conflict with the multiple realizability of mental states.

So overall, we have a reduction of 'mental' to 'material'. But thanks to the inference's premise (ii), 'material' now involves more than just behavioural and non-technical concepts, as in the example of being disposed to pick up an umbrella when going outdoors. Agreed, each mental state or concept has a unique functional role in terms of such behavioural and non-technical concepts. That is what we learn from behaviourism, i.e. from conceptually analysing folk psychology. But each mental state or concept is *also* uniquely specified in the language of neurophysiology. *That* is what we learn from contingent empirical enquiry. In short, as we said above: there is a *ternary*, not binary, contrast of vocabulary: 'mental', 'behavioural/non-technical' and 'neurophysiological'.

So much by way of using the familiar example from the philosophy of mind to sketch functionalist reduction. We now complete this sketch with three further comments, not tied to this example.

7.1.1.2 ... in General

These comments are about:

1. using the word 'definition', instead of 'specification';
2. how functionalist reduction avoids problems that beset other kinds of reduction;
3. the state of the literature about functionalism and reduction.

(1) We have so far mostly said 'specified' and 'specification'. We will also say, since the literature often does: 'defined' and 'definition'. Beware: these words of course connote both: (a) arbitrary verbal stipulation (cf. Humpty Dumpty in *Through the Looking Glass*: 'When I use a word, it means just what I choose it to mean—neither more nor less'); and (b) staying faithful to a given meaning of the word (cf. lexicography). But as we will stress: these connotations are often misleading for our cases in philosophy of mind, of science in general, and of physics. The 'definition' of a state or concept by its functional role is very often neither a stipulation, nor faithful to some given meaning.

Agreed, logical behaviourism did aim to define (the words for) mental states and concepts in behavioural terms, faithfully to the words' meanings. Before functionalism (and especially Lewis' work), this endeavour had long been thought to be beset by problems of logical circularity. Again, a standard pedagogic example is that the logical behaviourist wants to define belief that it is raining along the lines of being disposed to pick up an umbrella if going outdoors. But, as we said, this seems to stumble on the apparent need to also assume that the agent desires to stay dry—a mental state. And since the logical behaviourist's natural strategy for defining desire to stay dry is, again, along the lines of being disposed to pick up an umbrella if going outdoors, and this strategy apparently needs to also assume that the agent believes it is raining... clearly, a vicious logical circle of definition looms. It is this sort of logical circle that Lewis' idea of *simultaneous unique definition* avoids at a stroke.

Besides, as we will explain: this proposal for avoiding circularity does not at all depend on taking a definition to be (a) stipulative or (b) faithful to a given meaning. So one can equally well speak of *simultaneous unique specification*.

(2) In Sects. 7.2 and 7.3, we will prepare for our advocacy of functionalist reduction (Sects. 7.4 and 7.5) by discussing the enterprise of reduction: first in general (Sect. 7.2) and then *a la* Nagel (Sect. 7.3). The general idea of reduction will be that a problematic discourse (or theory) is shown to be part of an unproblematic discourse (or theory), by: (i) suitably specifying or defining, in terms of the unproblematic discourse, the words (or concepts or properties) of the problematic discourse; and thereby (ii) recovering the problematic discourse's claims, usually by deduction. Note that this idea is much less specific than the functionalist reduction we advocate. For (i) is less specific than the idea of unique specification by a functional role; and the deduction in (ii) is not underwritten by *mandatory* i.e. inferred bridge laws.

In this enterprise there are three kinds of problem, or at least objection, that a programme of reduction, is liable to face. We spell them out in Sect. 7.2.2. But since in later Sections, and in our other papers, they will be a template for discussing and assessing Nagelian reductions and functionalist reductions, it will help set the stage if we announce them here, as follows.

1. *Faithlessness*: The proposed reduction is faithless to the original problematic discourse. That is, using just the concepts of the unproblematic discourse, one cannot faithfully specify some concept of the problematic discourse.
2. *Plenitude*: The unproblematic discourse provides many specifications of some concept(s) of the problematic discourse: many specifications that are equally good, so that one cannot choose between them in a non-arbitrary way—so that they are also equally bad. In effect, the objection is: 'existence of a specification is all too easy, but uniqueness is unobtainable'.
3. *Scarcity*: This is the opposite of Plenitude: the unproblematic discourse cannot provide even one good specification of some concept(s) of the problematic discourse. In effect, the objection is: 'uniqueness of the specification is all too easy, but existence is unobtainable'.

These labels will recur. In this paper and its companions, we will see both:

(a) examples of these problems, for both philosophical and physical programmes of reduction; and
(b) more positively: how functionalist reduction claims the existence and uniqueness of a concept's specification, thus avoiding the objections of Plenitude and Scarcity.

Besides, as we announced in this Section's preamble: in our other papers' examples of (b), the existence and uniqueness claim is indeed a *theorem*.

Also, beware on two counts. (i): We have for simplicity stated these problems, (1) to (3), as if they were mutually exclusive; but we shall see that in fact they mingle with each other. (ii): We have stated these problems as about *specifications*. And 'specification' is a word that, we admit, tends to be ambiguous between the role, and the realizer i.e. occupant of the role. ('Definition', on the other hand, lacks this ambiguity: one naturally hears it as referring only to the role, or the words that express the role.) But this ambiguity is a convenience, indeed an advantage, for us. For reduction and functionalism will face objections of both sorts: about many, or no, roles; and about many, or no, realizers. So we have here used 'specification' for simplicity, i.e. so as to concisely cover the various cases.

(3) We believe the literature in philosophy of science about functionalism and reduction has tended to forget the way they can be combined, as above, in functionalist reduction. Recall the preamble of this Section, where we lamented what we dubbed a 'second missed opportunity' and the tendency of philosophers of science to ignore the Canberra Plan. The questions arise: (i) whether our contention is right; and (ii) if so, why this has happened.

As to the first question (i): We admit that we have not attempted a systematic survey. But we note that a recent comprehensive survey of 'scientific reduction' (van Riel and van Gulick, 2019)—and more important, the great majority of the literature it cites—works with what we called the *binary contrast* between reduced and reducing: not the *ternary contrast* of functionalist reduction, that we sketched above (and that matches what has been called the *two steps* of the Canberra Plan; cf. Braddon-Mitchell and Nola (2009, 7–9, 185–190, 267–269)).

Much of the literature also sees functionalism as offering a *non*-reductive relation between the higher and lower levels (or theories). The idea here is that the fact that the words (or concepts or properties) at the higher level/theory are functional, i.e. are each specified by a functional role, makes the higher level/theory suitably dependent on, or rooted in, the lower level/theory—but *without* being reduced to it. This is broadly similar to, indeed sometimes combined with, another widespread strategy for securing a *non*-reductive relation between higher and lower levels (or theories): namely to say that the higher level supervenes on, is determined by, the lower.[3]

As we have said: following Lewis, we will reject the claim of non-reduction. Besides, the sense of reduction given by functionalist reduction will be a deduction of the reduced theory. It will be like Nagelian reduction in using bridge laws—but with the difference that the bridge laws are themselves deduced! The details will be in Sect. 7.5.

As to the second question (ii): Of course, one can only speculate. But we surmise that the facts just mentioned—i.e. the binary contrast; and functionalism being, like supervenience, widely thought to articulate non-reductionist relations between levels—have been influential. We suspect also that there is a tendency to think that functionalism is a doctrine *confined to*, or at least plausible only in, the philosophy of mind. A case in point is that van Riel and van Gulick, for all their merits, discuss functionalism only within their Sections on philosophy of mind (viz. their Sections 3.3, 3.4, 4.5). Besides, they discuss only the idea of functional role: not the crucial consequential ideas of unique simultaneous specification, and of—again, the need for a ternary contrast!— derived bridge laws. Indeed, of Lewis' articles on the topic, they cite two, both on the philosophy of mind (viz. his 1969 (a brief discussion of multiple realisability), and 1972). But they do not cite his definitive 1970: which discusses theories in general (it does not mention philosophy of mind), and which presents with full rigour both: (i) how to extract from a theory unique simultaneous specifications of several, even many, terms, and (ii) how to derive bridge laws, and so deduce the reduced theory.[4]

[3] Beware: this strategy stumbles on Beth's theorem. Namely: for first-order languages, supervenience i.e. determination is, surprisingly, equivalent to explicit definability, and so to reduction. In philosophy of science and mind, this was first pointed out by Hellman and Thompson (1975), and later stressed by various authors, e.g. Butterfield (2011a, Section 5.1, pp. 948–951). Cf. Dewar (2019) and footnotes 20 and 47.

[4] Our criticism is not meant to single out van Riel and van Gulick (2019): it has many merits. Our point is just that its emphases are part of a pattern: witness the fact that Lewis (1970), despite being definitive and general, has fewer citations than his (1972).

For both our first and second questions, we also recommend Lewis' 'Reduction of mind' (1994). Its second half (pp. 421f.) is specialist: Lewis rebuts fashionable views, e.g. the 'language of thought' hypothesis, about Brentano's problem of intentionality, i.e. the question what determines the contents of mental states like beliefs and desires. But the first half summarizes—and updates—the position we have reported in Sects. 7.1.1.1 and 7.1.1.2. We especially commend:

So much by way of introducing how functionalist reduction has much more to it than just the idea of functional roles. Of course, for our papers' purposes, we do not need to endorse the functionalist philosophy of mind that served in Sect. 7.1.1.1 as our main illustration. (But as it happens, we do endorse it.) All we really need is that functionalist reduction fits cases, indeed many cases, in spacetime theories; and so it gives a good sense of 'spacetime functionalism'.

But in this paper we also want to urge the more general thesis that in the general enterprise of reduction, it is a powerful and plausible model, well able to address issues such as multiple realizability, meaning variance, reductions being partial, and the three problems of Faithlessness etc. listed in (2) above. We will take up this thesis from Sect. 7.2 onwards. But first, we set the stage by sketching: how functionalist reduction applies to space and time (Sect. 7.1.2), and how this relates to the work of Torretti (Sect. 7.1.3).

7.1.2 Functionalism About Spacetime

As we said in this Section's preamble and Sect. 7.1.1, our over-arching message is that functionalist reduction applies well to space and time. Both its main claims (i.e. (i) to (iii) of this Section's preamble), and the various issues and possible objections that these claims raise (e.g. those we labelled Faithlessness, Plenitude and Scarcity), apply well to several much-studied cases of theories of space and time. But since, unfortunately, the recent literature on spacetime functionalism has not articulated or assessed these applications, we aim do so—in our other papers.

But to complete this announcement of our general aims, we should say here what is the analogue, for spacetime, of the contrast between 'mental' and 'material' that, as we saw, functionalism in the philosophy of mind assumed at the outset, so as to make its claims.

That is; we should say what is the problematic/unproblematic contrast, with spacetime or spatiotemporal concepts allocated to the problematic side. The recent literature on spacetime functionalism has considered two main choices.

Lewis' advocacy of supervenience as reductive, his answer to the objection from *qualia*, his use of two-dimensional semantics to analyse the necessary *a posteriori*, and his answer to the idea that the term 'pain' is a rigid designator (pp. 412–421).

What Lewis says about the last topic (pp. 420–421) is also relevant to our point that, unfortunately, 'functionalism' is often understood as non-reductive. Thus Lewis writes: 'It is unfortunate that this superficial question [in effect, a question about the best nomenclature for the occupant or realizer of a role] has sometimes been taken to mark the boundary of 'functionalism'. Sometimes so and sometimes not—and that's why I have no idea whether I am a functionalist' (p. 421). We will briefly return to this in Sect. 7.4.3.2.

We also suspect, perhaps cheekily, that people have not sufficiently taken up functionalist reduction simply because its exposition in Lewis (1970) occurs only in the last two Sections (p. 441 et seq.). For most of the paper is taken up with a rigorous logical (though beautifully clear!) exposition of simultaneous unique definition within the first, i.e. reduced, theory. In short: we suspect the last two Sections have been ignored.

First, there is the contrast between spacetime and the physics of matter and radiation: (intuitively, the problematic 'void' or 'receptacle', vs. the unproblematic 'stuff'). Using this contrast, spacetime functionalism is evidently related to relationism about space and time; and to what has recently been called the 'dynamical approach' to chrono-geometry. Thus the focus is on how the physics of matter and radiation contributes to determining, or perhaps even determines, or even explains, chrono-geometry. It is *this* contrast that we will be concerned with: our overall point being, again, that the older literature already gave us examples of such a functionalist understanding of chrono-geometry in terms of matter and radiation.[5]

Second, there is the contrast between: (a) spacetime taken to be a continuum (as it is in all our established theories), and (b) some non-continuum underpinning that spacetime is supposed to approximate (or to be a coarse-graining of, or to emerge from)—an underpinning proposed by some speculative framework, usually a programme in quantum gravity. Thus some discussions of spacetime functionalism (especially in the philosophy of quantum gravity literature) invoke this contrast.[6]

But we will abjure this second contrast, and stick to the first one.[7] That is: we will take spacetime functionalism to focus on how the physics of matter and radiation contributes to determining, or perhaps even determines or explains, chrono-geometry; and so as closely related to relationism, and the 'dynamical approach' to chrono-geometry.[8]

---

[5] The 'dynamical approach' is especially associated with Harvey Brown (especially his 2006); indeed, it was a member of his school, Eleanor Knox, who first coined the phrase 'spacetime functionalism' as a label for a position she favoured, and saw as close to Brown's.

Recalling from this Section's preamble, and Sect. 7.1.1.1, that functionalist reduction proceeds in two stages, and uses a ternary contrast, not a binary one, you will ask: what will be our ternary contrast? That is: what will be our 'second theory', accepted independently of (or after) the first one? In short: our answer, in subsequent papers, will vary from case to case; but the text's contrast between spacetime and matter-and-radiation will be the unifying theme.

[6] Agreed: here, our labels 'problematic' and 'unproblematic' become strained. For in the present state of knowledge, it is spacetime, i.e. our established theories positing a spacetime continuum, that should be called 'unproblematic', and the speculative quantum gravity programme that should be called 'problematic'. But labels aside, the intended analogy between the mind and spacetime cases is as clear as for the first contrast. Namely: just as mind is best understood in terms of mental concepts' webs of relations to each other, and to material concepts, so also spacetime is best understood in terms of spatiotemporal concepts' webs of relations to each other, and to the concepts of a postulated 'non-spacetime' theory.

[7] Agreed: it might be worthwhile to assess some emergent-spacetime research programmes using the taxonomy of problems that we will develop for our versions of spacetime functionalism. In particular, some of these programmes' efforts to obtain a spacetime continuum may face the problems we label as Faithlessness, Plenitude and Scarcity (cf Sect. 7.2.2). Thanks to Julius Doboszewski for this point.

[8] We should mention here, since we will be advocating Lewis' account of functionalism and reduction, that he himself did not espouse spacetime functionalism in our sense. In fact, he did not work in detail on philosophy of physical geometry. But his main view was substantivalist: spacetime is an object, the mereological fusion of its regions, that have various spatiotemporal relations to each other. Besides, in his mature metaphysical system, these relations are, in his

7.1.3 Connections with the Work of Torretti

Clearly, the project of this paper and its companions puts us in "the land of Torretti". For much of his scholarly writing has been about the philosophy and history of geometry, especially of physical geometry and thus, after the advent of relativity theory, of chrono-geometry. For example, we will see already at the start of our second paper that two of his main books discuss in detail our first two examples of spacetime functionalism, viz. the work of Lie (building on Helmholtz) on the "problem of space", and the work of Malament (building on Robb), on simultaneity in special relativity.

Nor is it just the philosophy and history of physical geometry that links our project to Torretti's work. For of course Torretti has written a lot about reduction, and inter-theoretic relations in general, not just in geometry but across all of physics; and he has advocated the semantic (also called: structural) conception of scientific theories, against the traditional syntactic conception assumed by writers like Nagel. So by the end of this paper, after we finish our advocacy of functionalist reduction, it will be clear that various projects beckon: for example, comparing functionalist reduction with Torretti's treatments of reduction and related topics.

We will not have space for details about any of these projects. But in Sect. 7.6, we will discuss three: the comparison of our and Torretti's treatments of reduction; and two topics in the philosophy and history of the axiomatic method. Let us also, here at the outset, reassure Torretti that despite our inclining more than he does to 'scientific realism' and 'reductionism'—witness this paper's invoking Lewis and Nagel—we will agree with him about many details. As always, it is a matter of getting beyond the slogans and 'isms'. See also the reassurances at the start of Sect. 7.2.

7.2 The Enterprise of Reduction

In Sect. 7.2.1, we first introduce reduction as a strategy for legitimizing a problematic discourse; namely by showing it to be really a part of an unproblematic discourse. We then take one discourse 'being a part' of another as a matter of the latter giving specifications or definitions of the concepts of the problematic discourse, in such a way that the unproblematic discourse, augmented with these specifications, then implies the claims of the problematic discourse. We illustrate this with some historically influential programmes of reduction, including from the philosophy of mathematics; and we stress the need for deduction, not mere postulation.

In Sect. 7.2.2, we introduce three sorts of problem, which we label *Faithlessness*, *Plenitude* and *Scarcity*, that are liable to beset reduction—whatever one's exact conception of it. These labels are helpful for classifying the various objections

jargon, perfectly natural and external. But we will *not* need these doctrines in this paper, or in our others. We only need Lewis' treatment of functionalism and reduction.

that specific conceptions or examples of reduction face; as we will see in the other papers' examples of chrono-geometry.

Four initial disclaimers: or perhaps better, reassurances. (1): Obviously, reduction and functionalism are large and much-contested topics; and we have not the space to fully defend the traditional accounts of them—in short: Nagel's for reduction, Lewis' for functionalism—that this paper, and its companions, will adopt. Although we will of course indicate our defence, not least by citation: it suffices for us that these accounts fit perfectly our examples in our companion papers. After all, there is no point in fighting over the words4Sec5 'reduction' and 'functionalism'.

2. Note that in this paper and its companions, our favoured word is 'reduction', not 'reductionism'. We will *not* be concerned with either of two 'big-picture' *reductionisms*:

    (a) the 'unified science' picture, that reality, or science, is arranged in a sequence or hierarchy of 'levels' (levels of scale, or of description): with a physical level (roughly: collection of theories), that individually or collectively reduce a chemical level or collection of theories; that individually or collectively reduce a biological level or collection of theories etc.; or
    (b) the 'emergent spacetime' picture, mentioned at the end of Sect. 7.1.2, that spacetime is emergent from a fundamental level that is not spatiotemporal.

    We stress this because: firstly, picture (a) is much discussed (criticised!), both in general philosophy of science, and in our wider culture; and secondly, picture (b) is common in quantum gravity research, and thereby in some recent philosophical literature on 'spacetime functionalism'. But we will have no need to endorse, or even assess, either of these big-picture reductionisms.[9]
3. Similarly, we will not need to endorse, or even formulate, scientific realism (though as it happens, we do endorse some modest formulations). Agreed, we assume that the endeavour of interpreting physical theories, especially as regards how matter and radiation "mesh with" chrono-geometry, makes sense. But this assumption is entirely compatible with, for example, being a constructive empiricist: witness van Fraassen's discussion of what an interpretation of a physical theory is (1991, pp. 8–12). Our discussion of Torretti will also briefly return to this topic (Sect. 7.6.1).
4. Nor will we, or the precursor spacetime functionalists we celebrate in other papers, be committed to strongly realist metaphysical views, such as Lewis'. Yes, we will endorse the objectivity of reference and of truth, as part of endorsing Lewis' functionalism with its unique definitions. And yes, this means we face challenges like Putnam's model-theoretic argument, and its precursors

[9] Agreed: we will start our discussion of reduction, in Sect. 7.2.1.1, by mentioning some historically influential proposed reductions that were ambitious and philosophical, like the 'unified science' picture. But these are just by way of example. In the same vein, note that our main claims will not depend on the labels, 'problematic' and 'unproblematic'. To be sure: in some cases, the discourse to be reduced is unproblematic, but a reduction remains of interest. Cf. footnote 6.

like Newman's objection to Russell's structural realism: challenges urging that by our realist lights (Russell's structural realist lights), reference and truth are all too easy to attain. (Button (2013, Part A) is a thorough presentation.) But there are various cogent replies to these challenges. And some of these are significantly less 'gung-ho' or ardent in their realism than Lewis' own reply, which invokes a strong doctrine of objective similarity.[10] So one does not face a forced choice between, say, just Putnam's views and Lewis': there are intermediate options. We will say more about this in Sect. 7.4.2. But the main point is that in this and our other papers, we will not need to choose between these options. The reason for this flexibility will lie in the specificity of the scientific contexts we are concerned with. Thus consider the property that is centre-stage in our second paper's first example: free mobility, as a property of rigid bodies. This property is sufficiently close to observation, and sufficiently rich in its relations to other properties, that we can be confident that we grasp it—that very property, *pace* Putnam!—according to various of these intermediate replies. And similarly for the other properties, simultaneity, congruence etc., that will figure in our other paper's examples; cf. the end of footnote 2.

7.2.1 The Problematic, and How to Legitimize it

Philosophical problems and projects often begin with a contrast between: some discourse (roughly: a set of concepts, and claims involving them) that is believed to be problematic; and another discourse that is believed to be unproblematic.

Some well-known examples are as follows:

(i) Mind and matter: i.e. mental concepts and claims are problematic, while material (or bodily) concepts and claims are not.
(ii) Ethics and factual description: i.e. ethical (and maybe other evaluative) concepts and claims are problematic, while factual descriptive concepts and claims are not.
(iii) Pure mathematics and the empirical: i.e. pure mathematics is problematic, the main question being, since we seem to have no experience of mathematical objects such as numbers: how do we know mathematical truths? On the other hand, empirical concepts and claims seem unproblematic, since rooted in our experience.

[10] Lewis proposes that the extent to which a property encodes objective similarity—which he dubs: how natural it is—is a feature the property has across all of modal reality, wholly independent of contingencies such as what are the laws of nature. So this is indeed 'limning the true and ultimate structure of reality' (Quine, 1960, p. 202). Among less gung-ho replies, Taylor's proposal, which he calls a 'vegetarian substitute' to Lewis, is to relativize similarity and associated notions to a theory (1993, especially Section IVf., p. 88f.); and van Fraassen answers Newman and Putnam from an empiricist and pragmatic perspective (2008, pp 229–235).

(iv) The unobservable and the observable: concepts and claims about the unobservable are problematic, while observable concepts and claims are not.

Of course, in all examples the concepts at issue in either the problematic or unproblematic discourse are only vaguely delimited; and authors vary about how to be more precise.[11]

Of course, responses to each of these contrasts vary greatly. Some reject the contrast as a mistake. For example, they diagnose the mistake as rooted in a simplistic picture of how some part of our language (or other practices) works. So what seemed problematic is, in fact, not. Some accept the contrast and conclude: 'So much the worse for the problematic: though it may have meritorious uses, it is, as it stands, cognitive rubbish'. Some accept the contrast and yet conclude: 'We cannot dismiss the problematic as rubbish, albeit useful rubbish; so we must revise our previous views about why it is problematic'.[12]

But we will focus on another more irenic response: that the problematic discourse must be legitimized by appeal to the unproblematic discourse. Here of course: (a) it may only be some part of the problematic discourse that gets legitimized; and (b) to win legitimacy, one might also use ingredients (concepts and claims) from discourses other than the unproblematic one originally delineated.

Again, there are various versions of this response. One version is *conceptual analysis*: each problematic concept is to be analysed in terms of unproblematic concepts—and so legitimized. It is this version that is espoused by the Canberra Plan, mentioned in Sect. 7.1; a well-known example being Jackson (1998). Of course, the given problematic concept might be vague, so that its *analysans* will either be correspondingly vague, or be an analysis of a 'precisification' of the concept. Another version is Carnap's notion of *explication* (cf. Stein, 1992, 280–282; Beaney, 2004). This is like conceptual analysis, but more revisionary: it allows that the given concept is defective in some way, and if so, it provides a precise version aiming to rectify the defective aspect.

Another version is Russell's idea of *logical construction*: (which he also called 'logical fiction'). The problematic concept is to be replaced by one that is precisely defined in terms of unproblematic ingredients, in such as way as to mimic its properties; or again, with some allowance of revision: to mimic its desirable or correct properties (1918, p. 122, p. 144; 1924, p. 160). This version, with its word 'construction', introduces another theme we have so far been silent about. Namely: whether the unproblematic discourse is assumed to have tools, such as set theory or

[11] For our use of 'concept' and 'state', recall footnote 2. References for each example, out of countless that could be given, are: (i) Smith and Jones (1986); (ii) Harman (1977, Part I); (iii) Benacerraf (1973); (iv) van Fraassen (1980, Chap. 2). For variety, we have here chosen expository references that are *not* committed to a reduction. Section 7.2.1.1 will discuss reductions for these examples.

[12] Again, there are many examples of these responses. The 'mathematical atheism' of Field (1980) exemplifies the second, eliminativist, response for example (iii) above. The condemnation of the 'naturalistic fallacy' by Moore (1903, Chap. 2) exemplifies the third, '*sui generis*', response for example (ii).

mereology, with which to *construct* entities (whether objects or properties) that it is not initially given, or thought of, as including (as being 'ontologically committed to'). If so, the power of the unproblematic discourse to secure a reduction is, in general, increased. That is: provided one is willing to identify the objects or properties at issue within the problematic discourse with such constructions—a misgiving we will return to, in Sect. 7.2.2 and also later.

We will not need to choose between these options of analysis, explication and construction (including constructing objects and properties). So we need a word to cover them all. As discussed in (1) of Sect. 7.1.1.2, we will say **define, definition** and thus also '**definiens**' and ' **definiendum**'.

This usage is analogous to the practice in logic books of calling a universally quantified bi-conditional with a single predicate F on the left-hand side, $(\forall x)(Fx \equiv \Phi(x))$ where F does not occur in the open sentence $\Phi(x)$ on the right, a **definition** of F in terms of the vocabulary in Φ. This analogy has three aspects:

(i) such a bi-conditional determines the extension of F in terms of the (extensions of) the vocabulary occurring in $\Phi(x)$; but also,
(ii) there is no mandatory implication that the definition must be true, or nearly true, to any pre-existing meaning of F;
(iii) nor is it implied that F and Φ are co-extensive in other "worlds", i.e. in domains other than the given one.

The flexibility stated by (ii) and (iii) will be important for functionalism. It is also an advantage that the words, *definiens* and *definiendum*, are established jargon (unlike, say, *specificans* and *specificandum*). So here, F is the *definiendum*, and Φ is the *definiens*. Note: these Latin tags are often used—and we will also use them—not for the linguistic item, but for what they express or denote. So on this usage, one says, in functionalist jargon: the *definiens* is the role, and the *definiendum* is the realizer. (In general, both role and realizer are properties, states or concepts: cf. footnote 2.)

But we admit: this analogy of usage is also limited, in three ways. First: one naturally hears 'definition' as referring to the *definiens*, not the *definiendum*; or in functionalist jargon: to the role, not the realizer. But we will sometimes want to refer to the realizer; and for this, the word 'specification' is often more natural than 'definition' (as we noted in (ii) at the end of (2) in Sect. 7.1.1.2). Second: the word 'definition' connotes (outside a logic book!) both (a) arbitrary verbal stipulation and (b) faithfulness to a given meaning; and these connotations are stronger than for the alternative word, 'specification'. (Cf. (1) in Sect. 7.1.1.2.) Third: 'definition' does not connote—and in logic books, usually excludes—logical construction of the kinds just mentioned; but we will use 'definition' widely, as also covering constructions.

So far, we have gestured at the various versions of the irenic response, by considering how to treat the problematic *concepts*: analysis vs. explication vs. logical construction. But there is a corresponding variety in the treatment of the problematic discourse's use of its concepts, i.e. its *claims*.

For these claims, there is a pre-eminently natural way to try and achieve the response's basic aim of legitimizing the problematic by appeal to the unproblematic.

Namely: show that the problematic discourse's claims *'are really a part of'* the unproblematic discourse's claims. Here again: (a) one might be, in part, an eliminativist—maybe only a favoured subset of the problematic discourse's claims get legitimized; and (b) one might augment the unproblematic discourse with ingredients from other discourses. But the main point is: since claims are expressed in language, *'are really a part of'* is here naturally read as *'can be deduced from'*.[13]

Thus we arrive at the core notion of *reduction*. That is: *'Legitimize the problematic by defining (in our liberal sense) its concepts in terms of the unproblematic, in such a way that one can then derive the problematic claims (or at least some favoured subset of them) from within the unproblematic realm (perhaps augmented with other ingredients)'*.

We shall make this more exact in Sect. 7.2.2 onwards. But we end this Section by listing some historically influential examples of proposed reductions (Sect. 7.2.1.1).

7.2.1.1 Some Proposed Reductions

There are of course many historically influential cases of reduction, including examples (i) to (iv) above; albeit with varying precision and success. For brevity, we will talk only of discourse, not separately of concepts and claims.

(i) Mind and matter. The obvious case is logical behaviourism's project to reduce mental discourse to discourse about behaviour. This looks wrong to most people, since the behaviourist apparently denies that mental states are real.[14] But in Sect. 7.1.1.1, we saw how functionalism in the philosophy of mind combined the merits of logical behaviourism with accepting the reality of mental states, viz. as neurophysiological states.

(ii) Ethics and factual description. The obvious case is neo-naturalism in meta-ethics: the project to reduce ethical discourse to factual descriptive discourse. The idea is that Moore's allegation of a 'naturalistic fallacy' was wrong: there is a longer and subtler analysis or explication of 'good', 'right' etc. in terms of facts, for example about what satisfies human desires.[15]

(iii) Pure mathematics and the empirical. Here, the obvious historical case is the project to reduce mathematics, not to empirical discourse, but to something allegedly even more 'secure': to logic, or more plausibly, to set-theory. Thus

[13] Saying this shows that we will take a body of claims, a theory, to be a set of sentences or propositions, rather than a set of models, as in the semantic or structural conception of theories. In Sect. 7.3.1, we will briefly defend this: in short, it will not matter to anything we say. Note also that (as mentioned above): among the ingredients that augment the unproblematic so as to enable deduction, there might be tools of construction, such as set theory.

[14] The standard reference is Ryle (1949); but note that this over-simplifies Ryle's views (Tanney, 2015, especially Sections 8, 9).

[15] Cf. Hurley (1989), Jackson (1998), Lewis (1989): all three explicitly invoke functionalism's idea of simultaneous unique definition, to be described in Sect. 7.4. Again, this is part of the Canberra Plan; cf. the Chapters by Colyvan and Robinson in Braddon-Mitchell and Nola (2009).

Frege, Russell and Whitehead claimed to derive from logic, first arithmetic, and then (leaning on previous authors' methods) the rest of pure mathematics, e.g. analysis. Assessing this claim is a large, and still controversial, task—and not for us.[16] But even if one objects, as well one should, that these authors' 'logic' is really set-theory in disguise: it remains an enormous collective achievement by many authors, from ca. 1880 to 1920, to cast pure mathematics as set-theory, and indeed, to derive much of it in an axiomatized set-theory using a logic as sparse as first-order predicate logic.

(iv) Unobservable and observable. Let us begin with Bishop Berkeley's phenomenalism. Famously, this takes such a logically weak, i.e. wide, interpretation of 'unobservable' as to utterly reverse (i)'s view that the mental is problematic. The idea is: discourse about the material world external to my sensory experience is problematic, while my sensory experience is unproblematic; and so the former should be reduced to the latter. 'Speaking with the vulgar', as the eighteenth-century phenomenalist *savant* would put it: material objects like chairs exist in the external world, for example in this room. But such claims as 'there is a chair in this room' are to be analysed as very long conjunctions of conditional statements about mental ideas.

Although nowadays (iv) seems unbelievable, its influence within the analytic tradition, on epistemology and philosophy of science, has been enormous: including for our topic of 'theory' vs. 'observation' in philosophy of science. Empiricists and positivists, suspicious of concepts and hypotheses about the unobservable, have urged that these should be analysed as compendiously summarizing observable concepts and claims. And some took 'observable' in an avowedly mentalistic, i.e. idealistic, way: *not* as ordinary observable properties of material objects, of 'moderate-sized specimens of dry goods', as Austin memorably put it (1962, p. 8). This was often combined with—indeed inspired by—example (iii). Think of Russell's writings (e.g. *Our Knowledge of the External World*, 1914) propounding a phenomenalist metaphysics, and corresponding foundationalist epistemology, for empirical knowledge, both everyday and scientific. And among the logical empiricists, think of Carnap's *Aufbau* of 1928.

This leads to the research programmes which our other papers discuss as spacetime functionalism. For the relationist tradition is that the attribution of geometry to space itself, or of chrono-geometry to spacetime itself, is problematic. It seems to outstrip empirical warrant, since we only experience the material, i.e. ponderable matter and (perhaps) radiation. But it might yet be legitimized by a reduction. Hence research programmes such as the 'causal theory of time' and Machian approaches to dynamics.

[16] Among countless references, we recommend Potter (2020, Chapters 10–13, 31, 37, 42) as a survey of logicism; while Potter (2000) is a monograph focussing on arithmetic, but spanning from Kant to Carnap.

7.2.2 *Problems of Faithlessness, Plenitude and Scarcity*

We already in (2) of Sect. 7.1.1.2 introduced these three words as labels for three problems or objections that a programme of reduction is liable to face. We can now develop and illustrate them, using some of Sect. 7.2.1's examples. Of course, not all problems apply to all versions of the examples. For instance, an objection of Faithlessness ('you are faithless to the meanings of the concept you claim to reduce') will apply much better to a version of reduction whose definitions claim to be conceptual analyses, rather than those whose *definiens* makes no claim to synonymy. (Recall from Sects. 7.1.1.2 and 7.2.1, our liberal i.e. logically weak usage of 'definition', 'define', etc. as like 'specification' 'specify' etc.)

We will also see that the problems mingle. For example, a problem of Plenitude can underpin one of Faithlessness; and Faithlessness can underpin Scarcity. But despite this variety and mingling, we will see in the sequel that the labels are worthwhile. For they help us classify the problems or objections that beset reduction—and also functionalism.

In line with Sect. 7.2.1's adoption of the word 'definition' as our preferred term, we will present these problems as about definitions. But we again note, in the light of the word's connotations, that for us, definitions: (i) do not need to be either arbitrary stipulations or faithful to a pre-existing meaning, (ii) can use construction tools like set theory. We also note again that the problems or objections, 'More than one!' or 'None!', can be alleged—not just for the *definiens*, to which the word 'definition' naturally attaches, but also—for the *definiendum*. To put it in functionalist jargon: there can be a problem of 'More than one realizer!', or 'No realizer!'. But these versions of the problems will become prominent only in Sects. 7.3 and 7.4. So happily, in the rest of this Section, we can construe 'definition' as referring to the *definiens*.

7.2.2.1 Faithlessness

"The proposed definition is faithless to (the meaning of) the *definiendum*. And to make the objection stick: no amendment will work. That is: if you use only the concepts of the unproblematic discourse, you cannot write down a faithful definition of the concept. So you cannot derive the claims of the problematic discourse."

Although this objection is, obviously, more plausible against a reduction whose definitions claim to be conceptual analyses of, or synonymous with, their reduced concepts (their *definienda*), it can also apply to a reduction with a more liberal conception of definition. The objector will say: "Even by your liberal undemanding standard of what it takes to define, you cannot succeed: the *definiendum* has no definition, even in your liberal sense, that uses only your avowedly unproblematic concepts."

This objection also takes another form, often as important as the first just stated. It is one of over-shooting, rather than under-shooting. Namely: any definitions using the concepts of the unproblematic discourse will entail claims that are not part of the problematic discourse, and so are unwelcome.

We see both forms of the objection in a famous example: Benacerraf's (1965) critique of reductions of arithmetic to set-theory, i.e. of set-theoretic definitions of the natural numbers—definitions that form the core of our example (iii) in Sect. 7.2.1.1. This example, being about pure mathematics, has the further advantage of being as sharply defined as one could hope a discussion of meanings to be.

Benacerraf's critique starts from the fact that there are several worked-out reductions of arithmetic to set theory. He takes two examples: the first defines the natural numbers as the Zermelo ordinals, i.e. $0 = \emptyset$, $1 = \{\emptyset\}$, $2 = \{\{\emptyset\}\}$, $3 = \{\{\{\emptyset\}\}\}$ etc.; and the second defines them as the von Neumann ordinals, i.e. $0 = \emptyset$, $1 = \{\emptyset\}$, $2 = \{\emptyset, \{\emptyset\}\}$, $3 = \{\emptyset, \{\emptyset\}, \{\emptyset, \{\emptyset\}\}\}$ etc. That is: the reductions agree that $0 = \emptyset$; but then Zermelo takes each natural number to be the singleton of its successor, while von Neumann takes each natural number to be the set of smaller numbers.

But both reductions make claims about numbers that are utterly alien to arithmetic. They agree on some such claims. For example, they both say that 1 is an element of 2: $1 \in 2$. But they also make mutually contradictory claims that are alien to arithmetic: the first says that 1 is not an element of 3, $1 \notin 3$; whereas the second says that 1 is an element of 3, $1 \in 3$. Thus with both the agreed, and the mutually contradictory, claims: there is a problem of *over-shooting*.

Benacerraf argues that it makes no sense to try to assess such claims. And it hardly matters whether the reductions' claims agree or disagree: even an agreed claim like $1 \in 2$ seems faithless to the meaning of 'natural number'. So he concludes, as Mercutio did in *Romeo and Juliet*: 'a plague on both your houses'. Indeed: he concludes that *all* such set-theoretic reductions are wrong, since any of them will imply such claims alien to arithmetic. And he ends by proposing that numbers are not *objects* at all: thus supporting a structuralist philosophy of mathematics.

We will not need to pursue this example, or assess Benacerraf's structuralism about numbers. But these problems or objections about definitions in the foundations of arithmetic are very similar to those about definitions in the foundations of *geometry* that we will discuss, in Sects. 7.6.2 and 7.6.3 and our other papers.[17] For

[17] Shapiro (2000, Chapter 10) is an introduction to the issues. But note that Benacerraf was re-discovering an old theme. Potter (2000, Sections 3.2–3.5) describes how already in 1888, Dedekind articulated Benacerraf's "over-shooting" critique, and a version of structuralism about numbers, as immune to it. The similarity of the problems shows up in Potter's Section headings, 'Existence' and 'Uniqueness' (of natural numbers). For compare how our problem of Plenitude makes existence/uniqueness of the *definiendum* easy/difficult, respectively; while our problem of Scarcity makes existence/uniqueness of the *definiendum* difficult/easy, respectively.

Besides, there are several other precursors—indeed, luminaries like Frege and Quine; (thanks to Alex Oliver for these cases). (1): Already in the *Grundlagen* (1884: paragraph 69, pp. 80–81) Frege himself puts the over-shooting objection to his own official definition of the number of a concept (*immediately* after propounding the definition!). He asks: 'do we not think of the extensions of concepts as something quite different from numbers?'; and goes on to say that we say 'one extension of a concept is wider than another' (i.e. in modern jargon: is a superset of another),

the moment, we just note that Benacerraf's example of Faithlessness turns on the idea of there being two—indeed many—equally *bad* definitions of numbers as sets: bad because they over-shoot, viz. by entailing (when taken together with set-theory) claims alien to arithmetic. This idea of many definitions (or reductions) leads to the problem of Plenitude.

7.2.2.2 Plenitude

"The unproblematic discourse provides several, even many, definitions of some concept(s) of the problematic discourse. These definitions are equally good, in that they enable deductions of the problematic discourse's claims. So one cannot choose among them in a non-arbitrary way. So they are also equally bad."

Agreed: we have already seen the theme here—'many equally good, and so equally bad': existence of a definition trivialized, and so uniqueness ruled out—in Benacerraf's example. No surprise: as we announced at the start of Sect. 7.2.2, the three problems mingle: a problem of Plenitude can underpin one of Faithlessness. But for us, it is worth articulating Plenitude as a separate problem, for two reasons.

The first is general, and negative. It is about how equipping the unproblematic discourse with construction tools like set theory can engender a problem of Plenitude. This will lead in to our second reason. Although this reason gives no general answer to the problem, it is positive and is worth stating, since it is specific to our papers' projects. The two reasons also differ as regards the realizer-role, or *definiendum-definiens*, contrast. As we noted at the start of Sect. 7.2.2: the problem, 'More than one!' can arise for the realizer, i.e. the *definiendum*, just as much as for the role, i.e. the *definiens*. And indeed: the first reason will concern a Plenitude of definitions in the sense of *definiens*, while the second will concern a Plenitude of the *definiendum*—of realizers.

---

while 'certainly we do not say that one number is wider than another'. But after discussion, he concludes that, although 'it is not usual to speak of a Number as wider or less wide than the extension of a concept, . . . neither is there anything to prevent us speaking this way'. That is, Frege bites the bullet, and allows the definition to mildly revise what we say. To put the point in Carnap's jargon: Frege says it is enough to give an explication. Dummett (1991, p. 177–179) endorses Frege's moves. First, he raises the objection of Faithlessness. But then he urges that since nothing Frege will prove, or argue for, turns on the arbitrarily chosen features of his *definiens* that go beyond the received sense of the *definiendum*, Frege's choice of *definiens* is legitimate. He sums up: 'Benacerraf's problem simply does not arise for Frege'. (2): Quine rehearses the same considerations in *Word and Object*'s discussion of the various ways set-theorists and philosophers define an ordered pair. He admits that each proffered definition overshoots in the way we have discussed, but breezily says that he doesn't care (1960: Section 33, p. 166, Section 53, p. 238). We shall return to this theme—whether to care that one is Faithless—in Sect. 7.6.2.

We also note that a cousin of Benacerraf's structuralism, called 'structural realism', is prominent in recent philosophy of science. It relates to functionalism through, for example, the use of Ramsey sentences. But it does not bear closely on our main claims; so we postpone a proper discussion to another paper.

(1) *The plethora of mock-ups*: The first point is, in short, that Benacerraf's example of two set-theoretic reductions of arithmetic is the proverbial tip of the iceberg. Often, once we see one strategy for defining concept(s) of the problematic discourse that enables deductions of that discourse's claims, we can instantly see that there are many similar strategies (at least, on a liberal notion of definition, as adopted in Sect. 7.2.1). "If you can secure a reduction with this strategy, then here are many similar strategies that work equally well." The unproblematic discourse in Benacerraf's example, i.e. set theory, illustrates this point very well.

Given almost any structure, according to almost any understanding of the word 'structure', we can build a set-theoretic mock-up of it.[18] Namely, we use iterated curly brackets to represent the internal organization of the structure. But once we do this one way, we can instantly see how to do it in many other similar ways, with different conventions e.g. about trivial matters like the order in which items are listed in an n-tuple, or how to define an ordered pair (Quine's example in footnote 17). Nor is this just a matter of some half-dozen easily imagined alternative conventions. There are countless ways to define some ghastly forest of curly brackets in which to cast a reduction: and thereby make it incomprehensible, no matter how lucid it was before being consigned to the forest. In short, set-theoretic mock-ups of the given structure are cheap indeed: two a penny, a dime a dozen.

Besides, we can build such mock-ups from initial objects, properties or relations (which we place on the lowest or first few ranks of our set-theoretic hierarchy) that are utterly alien to the subject-matter of the problematic discourse. For example, we can build a mock-up of a given structure that involves, say, cats, by using sets with only prime numbers as initial objects (or indeed, with only pure sets).

Nor is this threat specific to set theory. Thus logicians and metaphysicians distinguish set theory from its cousin, mereology. But mereology has sufficiently strong constructive resources to engender the threat. Where we intuitively say there is one cat, Tibbles, on the mat, mereology tells us there is a plethora of objects, differing by the"mental subtraction" of a single hair, all of them equally good deservers of the name 'Tibbles': this is *the problem of the many* (cf. Lewis, 1993).

How to respond to this dismaying plethora? The first thing to say is a reassurance. This plethora does *not* trivialize the enterprise of reduction. Agreed, the plethora of (set-theoretic or mereological) objects means that if there is one reduction, i.e. one deduction of claims that are formulated in terms of appropriately defined objects, there is a plethora of other such deductions—almost all of them unthought of, and incomprehensible since cast in some ghastly forest of curly brackets. But for all that, it is still a substantive enterprise to exhibit just one reduction. Two obvious examples, one from mathematics and one from philosophy, are: (i) Bourbaki's informal axiomatisation of pure mathematics in set theory, and (ii) Carnap's attempt

[18] We say 'almost any understanding' so as not to presuppose sets: we mean, roughly, a plurality of objects with properties and relations among them. And we say 'almost any structure', to signal that some vast structures may need proper classes rather than sets: in which case, one could talk of a 'class-theoretic mock-up'.

in the *Aufbau* at a phenomenalist reduction of ordinary talk (example (iv) in Sect. 7.2.1.1). Neither Bourbaki nor Carnap worried about the prospect of these countless complicated alternative reductions. They knew they had enough work to do, to show in detail just one successful reduction; (cf. Carnap, 1963, p. 16).

As we see matters, there are two broad strategies for responding to the plethora. One can propose requirements that reduce it ('pruning'). Or one can argue that the Plenitude is okay ('acceptance'). And one can combine these strategies: first prune, and then accept the remainder.

As examples of the pruning strategy, one can require any or all of the following:

(i) the definition(s) must provide a synonym of the concept(s) of the problematic discourse; and-or that
(ii) the reduction must use—the mock-up must be built from—objects, properties and relations in the subject-matter of the problematic discourse, rather than items in some alien subject-matter; and-or that
(iii) the deduction of claims, and-or the construction of the mock-up, must be suitably short or natural.

These requirements are undoubtedly tenable, although vague (especially 'short' and 'natural'). But one must admit that they lead to wider, and mutually related, philosophical controversies about e.g. synonymy, and the distinction between the conventional and the substantive; and thereby to the second strategy, of acceptance.

Some aspects of acceptance are straightforward. One should often just take in one's stride that there can be different, equally convenient, conventions about e.g. ordering the items in an n-tuple, or how to define ordered pairs. For often, nothing important can turn on one's choice of convention. We expand on this in (2) below; (and again, cf. footnote 17 about not caring about being Faithless). But as we just mentioned, there are deeper controversies hereabouts. Recall Putnam's model-theoretic argument against the objectivity of reference, raised in (4) of this Section's preamble (and echoed above by our mention of cats and prime numbers). At its simplest, Putnam's challenge is: 'how can you be sure to succeed in imposing the requirement (ii), that a reduction use cats not some alien subject-matter like prime numbers?' Or more pointedly: 'my model-theoretic argument shows that you cannot succeed in imposing such a requirement'.

But obviously, and as we said in (4) at the start of Sect. 7.2, there are several tenable replies to Putnam's and similar challenges; and this paper and its sequel will not need to choose among them. It is just that in discussing Plenitude, we are duty-bound to point out this challenging form of it. All the more so, since in Sect. 7.4.2 we will advocate the functionalist idea that the functional role of each problematic concept has a unique realizer; thereby providing simultaneous definitions of each of the problematic concepts (recall Sect. 7.1.1.1). So the challenge is that these uniqueness claims will stand refuted by a dismaying plethora of realizers: gruesome *Doppelgangers* whose existence seems guaranteed by the constructive power of set theory, and-or of mereology. In Sect. 7.4.2 we will return to the question what is the best general response to this challenge. But whatever it is, we will (as we said in (4)) maintain the objectivity of reference, and so stick to our uniqueness claims, in view of the rich scientific context of the properties, like free mobility, with which we are concerned.

(2) *Plenitude is controlled by representation theorems*: But the discussion in (1) also reveals a *positive* aspect of Plenitude: an aspect which will apply to the examples in our other papers. Namely: for physical theories, there is often a precise and unproblematic version of the distinction touched on in (1), between the conventional and the substantive. It is the distinction between: choices of the unit of length, the origin and orientation of spatial coordinate axes etc.; and the coordinate-independent facts (such as the pure-number ratio of two objects' lengths, both expressed in e.g. metres). For formulations of specific physical theories, this distinction can be precise, and so easily treated.

Indeed, there is a considerable tradition of treating it, in foundational studies in geometry and spacetime theories. Namely, in *representation theorems* whose gist can be stated, in the philosophical jargon we have adopted so far, as follows. Given the geometry or spacetime theory, the realizer of each functional role extracted from the theory is *not* unique. But this is as it should be. For the non-uniqueness reflects the theory not being committed to any single convention about the unit of length, the origin and orientation of spatial coordinate axes etc. Thus the representation theorem for a geometry or spacetime theory says:

(i) For each functional role extracted from the theory, every realizer of that role is related to every other such by a transformation T that embodies changing one's choices of the unit of length, the origin of axes etc.
(ii) Besides, this transformation is to be the same for all the different functional roles, in the natural sense. Namely: if you fix on realizer r_i of role R_i for the various i, while for some role R_j I fix on a transformed realizer $T(r_j)$ of R_j: then I must also fix on the corresponding transform $T(r_i)$ (using the same T) as realizer of role R_i, for $i \neq j$.

In mathematical jargon, one says that:

(i) the realizer is 'unique up to an appropriate transformation T of units and coordinates'; and
(ii) the realizers taken collectively e.g. in an n-tuple are unique up to a *common* transformation T.

In short: although the realizer is non-unique—there is Plenitude—we have complete control and understanding of the variety of realizers, as arising from different conventional choices. The Plenitude is welcome, and right. There should *not* be uniqueness: it would dictate to us a single convention. As mentioned, we will see this in our examples.

7.2.2.3 Scarcity

"The unproblematic discourse cannot provide even one definition of some (at least one) concept of the problematic discourse."

Again, the reasons for the objection vary from case to case. As we mentioned, an objection of Faithlessness can prompt one of Scarcity: "you cannot write down a faithful definition". Examples (i) and (ii) in Sect. 7.2.1.1 give well-known cases. For (i) ('mind and matter'), some say that the 'raw feel' of pain cannot be faithfully defined in material vocabulary. For (ii) ('ethics and facts'), Moore says that 'good' cannot be faithfully defined in naturalistic vocabulary (cf. footnote 12). Besides, the objection need not be based on requiring definitions to give conceptual analyses or synonyms. For even on a liberal notion of definition, there can be what we called under-shooting and-or over-shooting.

Sad to say, but here, as in life: scarcity makes for thieving. That is: we connect here with Russell's quip about the advantages of theft over honest toil. As we mentioned at the end of Sect. 7.2.1.1, much of Russell's writing, especially in his phase as a logical atomist, proposed reductions in our sense. He required that the *definiendum* must be—not declared, but—*shown* to have, thanks to the *definiens*, the properties of the original problematic entity (in his jargon, the 'metaphysical entity' (1918) or 'supposed entity' (1924)). He is thus opposed to so-called *implicit definition*, i.e. to thinking it is enough, for justifying a concept or discourse, to give a set of postulates ('axioms') containing the concept and from which one's claims about it can be deduced. How do you know—he might say—that your deduction sets out from safe ground? (In Sect. 7.6.2 we will return to this, in connection with (a) the Frege-Hilbert controversy about implicit definition, and (b) Torretti's work.)

Hence Russell's famous *bon mot* about 'theft over honest toil'. It is in his *Introduction to Mathematical Philosophy* (1919), in his discussion of deducing the truths of arithmetic from logic, or what we today would call 'set-theory' (cf. example (iii) in Sect. 7.2.1.1):

> The method of 'postulating' what we want has many advantages; they are the same as the advantages of theft over honest toil. Let us leave them to others and proceed with our honest toil. (1919, p. 71)

Thus 'theft' is here the dogmatic postulation of entities by implicit definitions, viz. as being those things that obey certain axioms or postulates; and 'toil' is here the work of finding judicious definitions (including constructions) so that the *definienda* can be shown using logical inference alone to satisfy the claims made about them. And as we said: Scarcity makes for theft.[19]

---

[19] Agreed: advocates of conceptual analysis often do not stress that they must capture all or most of the claims about the *analysandum* once it is interpreted in terms of the *analysans*. At least they do not stress this, with 'capture' meaning 'derive'. That is hardly surprising since, as we have seen: in most cases of philosophical interest, such a derivation is a very tall order. Nevertheless: reduction, as we understand the enterprise, is thus obliged. And as we saw in the quotes above, Russell himself accepted this obligation: as did Carnap (1963, p. 16).

Not that we wish to put Russell or his *bon mot* on a pedestal. Oliver and Smiley (2016, 272) call it 'one of the shoddiest slogans in philosophy'; and their reasons echo one of our themes for our spacetime examples, viz. that writing down the right functional role, or analysis, of a concept, *before* any 'construction' begins, can take considerable 'honest toil'. Thus they point out that : (i) Russell originally aimed it, unfairly, at Dedekind's treatment on continuity, and (ii) 'it assumes

7.2.2.4 Answering These Three Problems in the Sequel

So much by way of stating the problems, or objections, of Faithlessness, Plenitude and Scarcity. We will see illustrations of them, and of how they can be answered, as regards reduction (in Sect. 7.3), functionalism (in Sects. 7.4 and 7.5) and spacetime theories (in our other papers). Broadly speaking, the situation will be that:

(1) In general, we will allow some Faithlessness—we do not require a reduction's (nor a functionalist reduction's) definitions to provide synonyms. For recall that in Sects. 7.1.1.2 and 7.2.1, we adopted a liberal i.e. logically weak usage of 'definition' etc. as like 'specification' etc. We will see in Sect. 7.3.1 that this fits with the core idea of Nagelian reduction, viz. what logicians call 'definitional extension'. For the Nagelian, a bridge law—a proposition that enables the deduction of the reduced theory—can be a definition in this logicians' liberal usage of the term. It can even involve constructions: which, in general, logicians' usage of 'definition' excludes.

But here, we must stress the distinctive features of functionalist reduction: especially its ternary, not binary, contrast of vocabularies (cf. the start of Sect. 7.1). For definitions extracted from the initial theory tend to be more Faithful than the definitions given by the later or independent theory (which are therefore more naturally called 'specifications').

Think of Sect. 7.1.1.1's philosophy of mind example, especially its two-premise inference to the derived bridge law, that pain is C-fibre firing. Here, the initial theory is everyday mental and material-behavioural discourse. The later or independent theory is neurophysiology. The functional definition of 'pain' extracted from the former has a much better claim to be Faithful to the meaning of 'pain' than does the "definition"—hence: better called 'specification'—given by neurophysiology. We will see that this pattern is typical. By and large: (i) the functional definitions extracted from the initial theory have a good claim to provide synonyms; while (ii) the specifications of the same term given by the later or independent theory do not. In particular, for the authors and theorems we celebrate in our other papers:

(i) the initial theory's functional definitions of 'freely mobile', 'simultaneous', etc. are certainly explications, and have a strong claim to be synonyms, or conceptual analyses, of the concepts at issue; while
(ii) the specifications of these notions given by the later or independent theory are not analyses or synonyms or explications: they give novel information, for example that free mobility requires a Riemannian metric of constant curvature.

(2) As to Plenitude: In the literature within the philosophy of science on reduction (i.e. disregarding functionalist reduction: as in Sect. 7.3), the problem of a plenitude of definitions, i.e. many *definiens*, is hardly discussed. There are two obvious reasons: one creditable, one less so. The creditable reason is that reduction is 'uphill work'. It is in general hard to find for each *definiendum* in the problematic discourse,

[wrongly] that we already know what we want ... the examples from Dedekind show just how much honest toil it takes to discover—to formulate precisely—just what it is that we want'.

even just one *definiens* that, taken together with other such, secures a deduction of all the problematic discourse's claims. In short: one faces a problem of Scarcity, rather than Plenitude. The less creditable reason is that philosophers of science tend to ignore the logical and metaphysical issues that led, in (1) of Sect. 7.2.2.2, to the plethora of mock-ups: to the fact that if there is one reduction, there are very many. In effect, philosophers of science assume that some combination of (1)'s pruning and acceptance strategies will control the plethora. (Agreed: sometimes, with good reason: for example, we will see that Nagel proposed that each *definiens* should be short and conceptually homogeneous.)

But here we again need to recall the realizer-role, or *definiendum-definiens*, contrast. As we noted at the start of Sect. 7.2.2: the objections, 'More than one!' or 'None!', can be alleged for the realizer, for the *definiendum*, just as much as for the role, for the *definiens*. And the objection 'More than one realizer' goes of course by the name *multiple realizability*. This, we of course admit, *is* much-discussed in the literature on reduction—and we will address it in Sect. 7.3. So in short: multiple realizability amounts to Plenitude of the *definiendum*, but not of the *definiens*. Anyway, for us, the more important point will be, as we noted in (2) of Sect. 7.2.2.2: the authors and theorems we celebrate in our other papers each give a representation theorem that yields a controlled and welcome Plenitude—of realizers, of the *definiendum*.

(3) As to Scarcity:— In Sect. 7.3.5, we will discuss this problem in the sense we introduced: namely lack of a definition, a *definiens*. It *is* emphasised in the literature on reduction: mostly under the heading of *multiple realizability*. For although *multiple* realizability means there are many ways to realize a predicate of the reduced discourse (theory), each way is sufficient but not necessary; and this means that multiple realizability makes for lack of a definition, i.e. for Scarcity. But we shall argue that multiple realizability is not really a problem for reduction, but only for reductionism. Scarcity also arises under the heading of *circularity*. But this problem will be answered by functionalism's idea of simultaneous unique definition, already introduced in Sect. 7.1.1. Thus the problem of circularity will prompt our transition to functionalism, in Sect. 7.4.

7.3 Reduction Based on Definitional Extension

So much by way of generalities about reduction and about the three problems of Faithlessness etc. that a reduction can face. In this Section, we discuss reduction more precisely, in the jargon of philosophy of science—but without considering functionalism. We begin with the formal notion of *definitional extension* (Sect. 7.3.1), and then discuss how it has been modified, especially by Nagel (Sect. 7.3.2). Then we describe how the three problems of Faithlessness etc. play out for this account of reduction. As we have just announced (Sect. 7.2.2.4), they do get illustrated, and in part get answered, albeit under different headings—the best known of which are *multiple realizability* and *circularity*. We shall nevertheless organize our discussion, using our three labels, 'Faithlessness' etc., for three Subsections (Sects. 7.3.3, 7.3.4, and 7.3.5). This discussion prepares us for the next main Section on functionalism (Sect. 7.4).

7.3.1 Definitional Extension

We now recall the notion of reduction articulated by Nagel (especially 1961, pp. 354–358; 1979, pp. 361–373)); and Hempel (1966, especially pp. 75–77) . In this Section, we begin with its formal core, called *definitional extension*. In the next, we will discuss the informal conditions Nagel and Hempel add to it.

We take the *relata* of reduction to be theories, i.e. bodies of claims. We will also take a theory to be a deductively closed set of sentences. To this, an objection will be made immediately, i.e. quite apart from the topic of reduction. Namely: this *syntactic conception* of a theory as a set of sentences is in any case wrong, and should be replaced by the *semantic (or structural) conception* of a theory as a set of models: and this will prompt some other treatment of reduction—presumably as something like subset-hood of sets of models. Indeed, Torretti himself would surely make this objection, since he rejects the syntactic conception and endorses the semantic one, in a version similar to that of Sneed and Stegmüller (Torretti 1990, pp. 109–160; 1999, pp. 407–416). We will return to this objection, and to Torretti, in Sect. 7.6.1. But here we just set it aside, since *pace* Torretti, we believe the recent literature contains convincing replies. Here, we emphasise:

(i) Lutz's detailed arguments for there being no material difference between the conceptions: cf. his (2017a), focussed on a three-cornered debate between Halvorson, Glymour and van Fraassen, and his (2017b: especially Section 5.2), focussed on Newman's objection to "structural realism". (Hudetz (2019a) builds on the former; Halvorson (2019, pp. 107–11, 172–174) introduces the debate.)
(ii) Niebergall's studies of inter-theoretic reduction (2000, 2002), which argue against a semantic or structural understanding of it.

We allow that the language in which a theory is written is natural rather than formal. But we presume that the languages, and so the theories, we are concerned with have rules of deductive inference that make the enterprise of reduction as deduction, announced in Sect. 7.2.1, reasonably precise. Then, taking a theory as a deductively closed set of sentences, the formal core of reduction is that one theory T_t ('t' for 'top', or 'tainted') *is reduced to* another theory T_b ('b' for 'bottom' or 'better') iff:

> by adding to T_b, a set D of definitions, one for each vocabulary item in the language of T_t, one can, within this augmented $T_b + D$ (i.e. using its underlying logic, and any set of its sentences, including the definitions in D), deduce every sentence of T_t.

In such a case: we say that T_t *is a definitional extension* of T_b.

This is a standard idea in formal logic. Here of course, the theories are cast in a formal language, almost invariably a predicate logic with the vocabulary items being predicates, and maybe also functional expressions and singular terms. So for predicates, the formal proposal is that each n-place predicate F of the language of T_t gets as a definition a universally quantified biconditional stating F to be co-extensive with some open formula Φ within the language of T_b that has n free variables. Thus the definition, which is to be an element of D, looks like:

$$(\forall x_1) \ldots (\forall x_n)[F(x_1, \ldots, x_n) \equiv \Phi(x_1, \ldots, x_n)].$$

Recall the start of Sect. 7.2.1 with our liberal, i.e. logically weak usage of 'definition' and cognate words: we do not require that the definition be faithful to a pre-existing meaning, or that F and Φ are co-extensive in domains other than the given one.

But we emphasise that despite this liberality or logical weakness, providing the set of definitions D is not a matter of 'theft': it is 'honest toil'. To obtain a deduction of all of T_t, the definitions in D will have to be *judiciously chosen*; and it is easy to write down simple examples of T_b and T_t for which it is impossible to formulate such definitions.

In philosophy of science, the jargon is of course *bridge law*, rather than 'definition'. But we shall mostly keep to 'definition' (or 'specification'), for two reasons (cf. (1) in Sect. 7.2.2.4):

(i) 'bridge law', like 'correspondence rule', connotes various controversies from mid-twentieth century philosophy of science, about whether they are contingent hypotheses (even laws of nature?) or stipulations (always or sometimes?): controversies which we will be able to avoid;
(ii) our advocacy of functionalist reduction (Sect. 7.5) will be clearer if we reserve 'bridge law' for its *derived bridge laws*.

So much for the form of definitions of T_t's predicates. Similar proposals are made for its functional expressions $f, g, \ldots$ and individual constants $a, b, \ldots$: (though these usually involve admissibility conditions requiring that a function be single-valued and that an individual constant have a bearer (Hodges, 1997, p. 52)). But in philosophical logic (and so quite independently of one's account of reduction), it is standard practice to simplify matters by eliminating these expressions in favour of predicates (i.e. within T_t—before reduction). Thus a functional expression with n arguments is eliminated in favour of a $(n + 1)$-place predicate, and a singular term is eliminated in favour of a 1-place predicate, in the spirit of Russell's theory of descriptions; (e.g. Quine (1960, Sections 37, 38); of course, Russell himself saw that theory as a good example of 'logical construction': cf. Sect. 7.2.1.1).

Of course, this is just the beginning of a large topic in logic. Among the questions to be addressed are:

(i) What about many-sorted languages?
(ii) What about the choice of logic, for example allowing higher-order not just first-order quantification?

(iii) How does the notion of definition just sketched, called *explicit definition* by logicians, relate to what they call *implicit definition*—which is the logicians' precise version of philosophers' notion of *supervenience* or *determination*?

(iv) And when one notices that the sketch just given will not provide new objects—for the quantifiers $(\forall x_1)$ etc. range merely over T_b's given domain of quantification—one naturally asks: What about endowing T_b (or its language) with 'construction tools', so that the reduction can indeed build new objects? The obvious candidates for such tools are of course, set theory and mereology: cf. the mock-ups discussed in Sect. 7.2.2.2.

However, we do not need to develop answers to these questions, or even to taxonomise the possible rival answers and choices. It will be enough for this paper (and its companions) that we here raise the questions; so that we can later see how the various proposals we discuss illustrate various answers and choices.[20]

What we of course must do is assess definitional extension being adopted as the formal core of the conception of reduction. This we undertake in the following Sections: the focus will of course be on the *pros* and *cons* of the informal conditions that might get added to definitional extension.

7.3.2 Nagel's Modification of Definitional Extension

Agreed: it was shown long ago (in the 1960s) that definitional extension was wrong—extensionally wrong, so to speak—as a description of reduction between scientific theories. Examples were given showing that definitional extension was too weak, i.e. some examples of definitional extension are not reductions. And other examples showed that it was too strong, i.e. some examples of reductions are not definitional extensions.

For the most part, these examples were offered as criticisms of Nagel's (1961) account of reduction. Our answer is, in short: 'Unfair to Nagel!' There are two concerns here: (1) about what Nagel actually said about reduction; and (2) about whether he was right. We shall discuss them in turn: though only briefly, since more details—broadly, *pro* Nagel—are elsewhere.[21]

---

[20] For rigorous details about the notion of definitional extension, and about answers and choices for these questions, cf. for example, Boolos and Jeffery (1980, pp. 245–249), Button and Walsh (2018, Sections 5.1–5.5) and Hodges (1997, Sections 2.3, 2.6.2, and 5.5). Note that the idea of 'implicit definition' in (iii) above, originally due to Padoa (in 1900), is *not* the idea usually understood by this phrase, that was advocated by Hilbert, and denounced by Frege. We will return to clarify both ideas in Sect. 7.6.2.1. And in Sect. 7.3.4 we will briefly relate definitional extension to the question when two theories count as equivalent.

[21] Cf. Butterfield (2011a Sections 3.1, 3.2; 2014, Sections 1.2, 4); and the papers by Dizadji-Bhamani et al. and Schaffner cited below.

(1) *Nagel's account*: One must distinguish definitional extension from Nagel's own account. Nagel did not say that reduction is just definitional extension, for two reasons.

First, there is a topic that this Section has so far set aside. Namely: do the vocabularies of the two theories T_t and T_b *overlap*? That is, do they have any (non-logical) vocabulary in common? In some examples, both in general philosophy and in philosophy of science—including examples we have mentioned—the answer is 'No'. Thus elementary arithmetic makes no mention of set theory; the wave theory of light makes no mention of electromagnetism.[22] But in many cases, the answer is 'Yes'. And in the vaguer and more complicated cases of discourses, rather than scientific theories, which we treated in Sect. 7.2.1, it is likely to be very difficult to sift out two sets of claims, each cast in one vocabulary, with a view to deducing the one from the other, once it is augmented with suitable definitions (in our liberal sense) of the one's terms. Thus consider examples (i) and (ii) in Sects. 7.2.1 and 7.2.1.1: i.e. mind and matter; and the ethical and the factual: or within the philosophy of science, theory and observation ((iv) of Sect. 7.2.1.1), and the case of interest to spacetime functionalism—a theory of chrono-geometry and a theory of matter-and-radiation.

In any case, Nagel allows the non-logical vocabularies to overlap, or even be identical; (in which case he speaks of 'homogeneous reduction'; otherwise, of 'heterogeneous reduction'). And whether or not they overlap, he is not committed to the form of the definitions, i.e. the bridge laws, always being as in Sect. 7.3.1, i.e. to their being understood as in a logic book. It is enough that the bridge laws state connections—make assertions using both vocabularies—in such a way that, once they are added to T_b, T_t can be deduced (Dizadji-Bahmani et al. (2010, p.398); Schaffner (2012, p. 538)).

Second, even for cases where the definitions, the bridge laws, take the form in Sect. 7.3.1, Nagel added to definitional extension, further informal conditions—and in advance of much of the 1960s criticism, to boot (1961, pp. 358–363). These conditions were motivated by the idea that the reducing theory T_b should *explain* the reduced theory T_t; and following Hempel, he conceived explanation in deductive-nomological terms (cf. Schaffner, 2006, pp. 380–382; 2012, p. 536). Thus he says, in effect, that T_b reduces T_t iff:

(i) T_t is a definitional extension of T_b; and
(ii) in each of the definitions of T_t's terms, the *definiens* in the language of T_b must *play an explanatory role* in T_b. This is a matter of it being reasonably short, and conceptually unified; so it cannot be, for example, a long and heterogeneous disjunction.

[22] Or rather: that is so, in a suitably historically sensitive sense of 'wave theory of light', e.g. up till 1870; since the success of Maxwell's theory, most expositions of the wave theory of light of course emphasise that light is electromagnetic waves. We will return to the issue of meanings shifting over time.

Agreed, condition (ii) with its phrases, 'playing a role' and 'being reasonably short, and conceptually unified', is vague. And even if one made it precise, many reject Nagel's (and Hempel's) deductive-nomological account of explanation that motivates it, while still wanting reduction to include explanation. So a consensus about the account of reduction will require a consensus about the much-contested concept of explanation. Obviously, we cannot settle such controversies here. Suffice it to say that if there are scientific reductions exemplifying the Nagelian account, that makes a good case that Nagel's added condition (ii) answers the first part of the above criticism, i.e. the allegation that definitional extension is too weak. That is, Nagel can reply: 'Yes it is too weak, but condition (ii) disposes of the objection'.

Besides, Nagel replied to the second part of the above criticism, i.e. the allegation that definitional extension is too strong. The idea of the criticism, is that in many cases where T_b reduces T_t, T_b corrects, rather than implies, T_t. One standard case is Newtonian gravitation theory (T_b) and Galileo's law of free fall (T_t). This T_t says that bodies near the earth fall with constant acceleration. This T_b says that as they fall, their acceleration increases, albeit by a tiny amount. But surely T_b reduces T_t. To which, Nagel's reply is 'I agree': a case in which T_t's laws are a close approximation to what strictly follows from T_b *should* count as reduction. (Nagel called this 'approximative reduction' (1979, pp. 361–363, 371–373); cf. also Hempel (1965, p. 344–346; 1966, pp. 75–77).) Besides, cases where T_b corrects T_t do not always involve merely quantitative approximation, with no change of the concepts involved. Thus, Schaffner's proposed modification of Nagel's account requires only that there be a strong analogy between T_t and its corrected version, i.e. the theory that strictly follows from T_b (1967, p. 144; 1976, p. 618). For a recent study of approximation and analogy, as they apply here, cf. Fletcher (2019).

(2) *Right?* Is Nagel's account of reduction (as modified, e.g. to allow approximative reduction or an analogue of T_t) right? This of course raises controversies, e.g. about how reduction relates to explanation, that we cannot hope to resolve. (Arguably, in the present state of knowledge, no one could.) But our advocacy of functionalist reduction, and the examples in our other papers, do not need a general defence of Nagel's account. After all, what matters scientifically and conceptually is—not the best conceptual analysis or explication of the word 'reduction', but—to understand the various relations between scientific theories. So it will be enough for us that, as modified by functionalism, Nagel's account fits the examples in our other papers. Nevertheless, we submit that over the years, it has stood up well. Here we commend: Dizadji-Bahmani et al. (2010) who defend it, as modified by Schaffner, against a battery of objections (p. 400f.); and Schaffner (2012), who gives a historical survey of its reception including modifications by himself (pp. 539–549), a partial defence and a detailed example from optics (pp. 551–559). Cf. also Niebergall (2000, 2002).

But our advocacy of functionalist reduction will need some details of how the three problems we have labelled Faithlessness etc. play out for Nagel's account of reduction: details which will match some of what these authors say. So to this, we now turn.

7.3.3 *Faithlessness as a Problem for Nagelian Reduction*

Clearly, a reduction based on definitional extension is liable to face a problem, or objection, of Faithlessness, simply because a 'definition', as we (and the logic books) use the term, need not be Faithful to the pre-existing meaning of a term in T_t. Nor is this liability lessened by Nagel's modifications of definitional extension. For as to his informal condition (ii) in Sect. 7.3.2: a short and conceptually unified *definiens* can be Faithless to the *definiendum*'s pre-existing meaning. And allowing approximative reduction and-or analogy does not secure that the *definiens* used is Faithful.

Clearly, how severe this problem is—how convincing the objection is—will vary from case to case. Broadly speaking, it will be more of a problem for reductions as definitional extensions in general philosophy (i.e. metaphysics, philosophy of mind and ethics), where there is a strong tradition of requiring the definitions to be conceptual analyses or explications (cf. Sect. 7.2.1.1), than for scientific reductions as described by philosophy of science, which of course makes no such requirement on the definitions, i.e. bridge laws. (Obviously, this contrast holds good whether one takes reductions just as definitional extensions, *a la* Sect. 7.3.1, or follows Nagel in adding conditions like (ii): it turns just on whether definitions must be Faithful.)

But we should also notice that (as we announced in Sect. 7.2.2.4) the ternary contrast involved in functionalist reduction will make a difference. That is: we will see that the functional definitions extracted from the initial theory can claim to be Faithful to the meaning, even in scientific reductions like our examples of spacetime functionalism; (they will also be short and conceptually unified, as Nagel requires). On the other hand, the definitions—better called 'specifications'—given by the later or independent theory are not, nor aim to be, Faithful to a pre-existing meaning.

In any case, the strategy for replying to the problem must be the same for reductions in (1) general philosophy and in (2) science: viz. as we put it above, 'to allow a little Faithlessness'. We briefly discuss the cases (1) and (2).

1. *In philosophy*: The concepts of T_t, the concepts of the initially problematic discourse, are usually vague. And even setting aside vagueness, criteria for the identity of concepts (properties etc.: cf. footnote 2) are much disputed in general philosophy. Should we judge identity by some notion of logical equivalence, or more finely (i.e. hyper-intensionally), or more coarsely e.g. up to some nomological equivalence? (Cf. e.g. Oliver (1996, pp. 16, 20–25, 44).) So it is hardly surprising that there is rarely a consensus about whether a reduction has been faithful to T_t's concepts. (Or 'faithful enough': recall Sect. 7.2.1's allowance that reduction capture only part of the problematic discourse's concepts and claims.) In philosophy as a whole, one of the most discussed examples is *qualia*, also known as "raw feels"—cf. example (i) in Sects. 7.2.1.1 and 7.2.2.3. Here, the standard example is pain. Whether a reduction proceeds by conceptual analysis or explication (Sect. 7.2.1) or by definitional extension as in Sect. 7.3.1, the reduction will necessarily express the *definiens* of 'pain' *in words* (perhaps technical, as well as everyday). And some people deny that any amount, or any

type, of discursive information, necessarily expressed in words, can capture the 'quale', the 'raw feel', of pain. But of course, this is not the place to address this denial.[23] Here, we just note the obvious strategy for replying to the problem: allow a little Faithlessness.

2. *In science*: In philosophy of science, the problem of Faithlessness is discussed under the heading of *meaning variance*, especially in discussing the nature of scientific progress. Recall that 't' in T_t could stand for 'tainted' as well as 'top'; and 'b' in T_b for 'better' as well as 'bottom'. When one theory is succeeded by another, it is natural, on a cumulative view of the transition, to think the successor theory gives a reduction of the predecessor. But usually, the successor will not imply exactly the predecessor. And as we said in Sect. 7.3.2, this is not always a matter of just quantitative approximation (cf. Newtonian gravitation theory succeeding Galileo's law of free fall). In some cases, the *definiens* in the language of T_b, of a vocabulary item in T_t, is not completely Faithful to the item's meaning. And this point is not just a matter of being content, as a matter of one's philosophical method, to be revisionary, or to admit Carnapian explications. Undoubtedly, transitions from one scientific theory to another often involve conceptual change that is significant enough to make the *definiens* Faithless. But as we said above: in our spacetime examples, the *functional* definitions will not face this problem of Faithlessness.

7.3.4 Plenitude as a Problem for Nagelian Reduction

As we said in (2) in Sect. 7.2.2.4: the problem of Plenitude, of there being too many definitions of a *definiendum*, is hardly discussed in the philosophy of science.[24] The focus is instead on Scarcity; cf. the next Section. But the problem is worth stating, as it affects definitional extension or any notion of reduction built on it, such as Nagel's notion. For although our statement of the problem, in (1) of Sect. 7.2.2.2, made no mention of definitional extension (which we had not yet introduced), it is clear that invoking definitional extension does nothing to avoid the threatened plethora of equally good—and so equally bad—definitions. Rather, invoking definitional extension just makes the threat more precise. Similarly,

[23] As we mentioned at the end of Sect. 7.1.1.2: we in fact reject the denial, and associated views like epiphenomenalism, since we endorse a Lewis-Armstrong-style functionalism about mind. But nothing in this or our other papers will turn on this.

[24] The problem should be distinguished from what has come to be called 'Newman's objection': which was originally made by Newman against Russell's (1927), but is now seen as a problem for various forms of 'structural realism'. For while Plenitude is a matter of there being too many definitions , Newman's objection is a matter of it being *logically guaranteed* that there is a *realizer*, a *definiendum*. The idea then is that this guarantee is a problem since structural realism wants the existence of the *definiens* to be its main substantial assertion. As mentioned in footnote 17, we postpone discussion of structural realism to another paper.

invoking definitional extension does not mitigate how the strategy, 'allow a little Faithlessness', that we endorsed in our liberal, i.e logically weak, construal of 'definition' and in Sect. 7.3.3 aggravates the problem of Plenitude.

Besides, it hardly helps to appeal to reduction having requirements additional to definitional extension, such as Nagel's condition (ii), that a *definiens* be short and unified. For some constructions using set-theory and-or mereology, as described in (1) of Sect. 7.2.2.2, will yield, when applied to a short and unified *definiens*, a rival—since equally short and unified—*definiens*. That is: there could be a plethora of definitions that each satisfy the additional requirements. We made essentially this point in different words, i.e. independently of definitional extensions, at the start of the response 'acceptance' in (1) of Sect. 7.2.2.2—emphasising that one can often take it in one's stride.

But here, we should also note *another kind of plenitude*. For it yields a moral about meanings, that will return in Sect. 7.6.2. It is not a plenitude of ways to make a given T_t a definitional extension of a given T_b; but a plenitude of surprising cases of definitional extension. This plenitude is well recognised by logicians and model-theorists who work on formal relations between theories such as definitional extension (and its cousins, studied in the theory of definability: cf. Button and Walsh (2018, Chapter 5)). That is: they are aware that often, one theory turns out to be a definitional extension of another, even though we usually think of them as being about very different topics, or indeed as contradicting one another about a given topic.

Agreed: on reflection, it is not surprising that this should happen, despite the theories' different intended topics or claims. For the definitions that yield a definitional extension are *not* required to respect any pre-existing meanings of the terms in either theory. So this allows what one might call definitional extension 'thanks to Faithlessness': or 'thanks to typographic accident'. Besides, over the course of time, the mathematical community's knowing of a definitional extension can shift the meanings even of central mathematical words. For example, think of the nineteenth-century rigorization of analysis: at the end of that process, but not at the beginning, mathematicians could understand 'real number' as 'an equivalence class of Cauchy sequences of rational numbers'. This of course echoes the theme of meaning variance, between scientific theories, as discussed in (2) at the end of Sect. 7.3.3.

But this plenitude of definitional extensions is worth emphasising to philosophers of science. It should give them pause in their discussions of formal notions of theoretical equivalence.[25] For in some cases, the surprising relation of definitional extension is symmetric. That is, each of two theories, that we think of as about very different topics or as making contradictory claims, is a definitional extension of the other. This is called being 'definitionally equivalent'. Hudetz (2019b, pp.

[25] Recent discussions include Butterfield (2018, Section 5), De Haro (2020), Dewar (2019), Halvorson (2019), Hudetz (2019a,b) and Weatherall (2018a,b). For general arguments against formal analyses of theoretical equivalence, cf. Sklar (1982) and Coffey (2014).

60–62, especially Prop. 4) gives a telling example. One can formalize Minkowski and Euclidean geometry for $\mathbb{R}^4$—which we usually think of as inequivalent, indeed contradictory—as definitional extensions of each other. For each of them can be formulated as a definitional extension of the theory of the real line; and then each can 'recover' the other, by 'building' from the real line. So each 'contains' the other. Besides, this situation cannot be readily overcome by appealing instead to other model-theoretic notions of equivalence for theories. For definitional equivalence is stronger than several other natural notions of equivalence studied in model theory; (Button and Walsh, 2018: Proposition 5.10, p. 117; and cf. the discussion of Feferman's theorem in ibid. pp. 118–120, and in Niebergall (2000, pp. 44, 52)). The overall moral here—that one cannot expect formal structures to completely capture meanings—will return in Sect. 7.6.2.

So much by way of stating the problem of Plenitude, in relation to definitional extension. Fortunately, as we noted in (2) of Sect. 7.2.2.2: the problem will not affect the authors and theorems we celebrate in our other papers. Their representation theorems will give a controlled and welcome Plenitude.

7.3.5 Scarcity as a Problem for Nagelian Reduction: Multiple Realizability and Circularity

For Nagelian reduction, the problem or objection of Scarcity is: 'there is not even one definition of a certain *definiendum* of T_t, that enables (along with other definitions) the deduction of all T_t's claims'. As we see matters, there are two main ways this problem arises, which we will discuss in turn: (1) multiple realizability and (2) circularity.[26] Or rather, the problem of Scarcity *seems to* arise. For we will argue: (i) following Sober (1999): that multiple realizability is not really a problem for Nagelian reduction; and (ii) as presaged in Sect. 7.1.1: that the problem of circularity is solved by functionalism's idea of *simultaneous unique definitions*.

Note that (1) and (2) relate differently to Sect. 7.1's distinction between binary and ternary contrasts of vocabulary. (1) relates to the usual binary contrast invoked in discussions of reduction. For (1) does not consider functional definitions in T_t: the focus is just on instances of a predicate in T_t's vocabulary satisfying disparate predicates in T_b's vocabulary. But on the other hand, (2) involves the ternary contrast: for it is about functional definitions within T_t.

(1) *Multiple realizability: no worries*: Very often, the instances of a concept (predicate) F in one theory T_t vary greatly in how they are described by another theory T_b. Indeed, they vary greatly even in respects that are candidates to occur in the *definiens* of F in a putative Nagelian reduction of T_t to T_b. This is *multiple realizability*. Recall from the end of Sect. 7.1.1.1, the philosophy of mind's standard

[26] For why multiple realizability threatens Scarcity of definitions, despite being a plenitude of realizations at the level of T_b, cf. (3) at the end of Sect. 7.2.2.4.

example: pain might be one brain state in humans, another in molluscs (Lewis, 1969, p. 25).[27]

Such cases are often pressed as objections to the reductionist picture of a hierarchy of levels (of scale or of description), with reductions between successive levels. As we said in (2) of the preamble to Sect. 7.2: we have no brief to endorse the reductionist picture. And we agree that multiple realizability undoubtedly makes higher-level theories autonomous from lower-level theories; (in another jargon: theories in the special sciences autonomous from theories in the basic sciences). Here, *autonomy* means, so to speak, never having to care: to develop a theory of capital growth, you never need to consult chemical theories; and for a theory of genetics, you never need to consult nuclear physics.

But on the other hand, we do *not* agree that multiple realizability, and it being so widespread, is an objection to a broadly Nagelian conception of reduction. Agreed: it certainly implies that either:

(a) the *definiens* (of T_t's multiply realized concept) that is given by T_b is very disjunctive; or
(b) using a non-disjunctive *definiens*, the reduction is in an obvious sense *local*.

We also agree that either (a) or (b) makes for T_t being autonomous in the above sense. For (a) implies that to do science in terms of the vocabulary (the concepts) of T_t, while avoiding hopeless complexity and confusion, we will not care about the many varied disjuncts in the *definiens*. And similarly, (b) implies that to do science with T_t—to describe and further investigate the patterns of co-occurrence of the vocabulary, the concepts, of T_t—we will not care about (b)'s single local reduction: for we are focussed on the T_t-patterns, that we can see to be realized in many ways other than *via* this local reduction.

But we follow Sober (1999) in maintaining that (a) and-or (b) do *not* make trouble for Nagelian reduction. To explain this, it will be clearest to distinguish the contentions: (i) that multiple realisability provides an argument against reduction; and (ii) that it provides an argument against Nagel's account of reduction. The idea of (i) is that the *definiens* of a multiply realisable concept (predicate) will have to be so disjunctive that it cannot enter into scientific explanations, and-or cannot enter into laws. The idea of (ii) is that, as we reported in Sect. 7.3.2, Nagel himself required that the *definiens* play a role in the reducing theory T_b; and in particular, it cannot be a very heterogeneous disjunction.

[27] Some say that there might even be *infinitely* many ways, according to the taxonomy provided by the vocabulary (the concepts) of T_b, to be an instance of (to realize) T_t's predicate F. This is the idea of *supervenience* or *determination*: that the T_t-facts merely supervene on, are determined by, the T_b-facts, and a *definiens* would have to be an infinite disjunction. Cf (iii) in Sect. 7.3.1. But we doubt there are such truly infinite cases; and if they occur, we doubt their scientific importance (cf. Butterfield (2011a, Sections 4.1, 5.1, pp. 940–944, 948–951; 2011b, Sections 4.2.3, 5.2.3 and 6.3.4, at pp. 1070, 1089, 1100, 1127). Anyway, such cases will not occur in our other papers' examples. So we will only consider finite disjunctions.

Our reply is that (i) is wrong; and that while (ii) is right about Nagel, it hardly matters. (For more details, cf. Butterfield (2011a, Section 3.1.1, (5), p. 933, Section 4.1.1, p. 941): Dizadji-Bahmani et al. (2010, p. 406) and Schaffner (2012, pp. 543–544) are concordant replies.) As to (i), we endorse a persuasive reply by Sober (1999). Sober says: a disjunctive *definiens* in the language of T_b for a concept (predicate) F that occurs in T_t is no bar to a deduction of a law of T_t involving F and other concepts in T_t (perhaps each also with disjunctive *definiens* within T_b). Nor is it a bar to this deduction being an explanation of the law. Sober sums up this reply as a rhetorical question (1999, p. 552): 'Are we really prepared to say that the truth and lawfulness of the higher-level generalization is inexplicable, just because the… derivation is peppered with the word 'or'?' We agree with Sober: of course not!

This reply also provides our reply to (ii). Agreed, and as we reported: Nagel himself vetoed very heterogeneously disjunctive *definiens*. Following Sober, we think this was unnecessary. But agreed: if you veto them, one has *local* reductions. But we need have no quarrel with such a veto. As we said in (2) of Sect. 7.3.2, there is surely no single best sense of 'reduction'. And a stronger sense along these lines, i.e. requiring non-disjunctiveness, will unquestionably make for reductions narrower in scope. What really matters, scientifically and philosophically, is to assess, in any given scientific field, just which such reductions hold good, and how narrow they in fact turn out to be. Cf. also Sober (1999, pp. 558–559).

So much by way of discussing multiple realizability, as a form of Scarcity that threatens reductions based on definitional extensions. Fortunately, as we announced in (3) of Sect. 7.2.2.4: the problem will not confront our examples about space and time. In each example, the *definiendum* predicate in T_t is *not* multiply realizable in the relevant sense. Its instances do *not* vary greatly in respects that are mentioned in the *definiens* provided in the vocabulary of T_b. For example, in the first case-study: although the instances of the *definiendum* predicate '… is a freely mobile rigid body' of course differ from one another in their properties, they do *not* differ from one another in the geometric properties that enter in to the *definiens* for this predicate.

(2) *Circularity avoided*: Finally, we note how Nagelian reduction (more generally: reduction based on definitional extensions) runs up against an apparent problem of logical circularity. This counts as a problem of Scarcity, since if this circularity were vicious, the enterprise would fail—there would be no definitions fit for their purpose. But we will urge that happily, the problem is by and large, *only* apparent—thanks to functionalism.

The problem is simply stated. When we try to formulate definitions for *all* all the non-logical vocabulary (say: predicates) of T_t in terms of that of T_b, it seems that the *definiens* for a predicate (concept) F of T_t may well need to also use another predicate (concept) G of T_t; while also the *definiens* for G needs to use F—a logical circle. We saw this illustrated, already in Sect. 7.1.1.1, for logical behaviourism (without the T_t/T_b distinction): i.e. for the reduction of mental discourse to material and behavioural discourse, with reduction taken as conceptual analysis, so that definitions must be Faithful to the pre-existing meanings of mental terms. The problem seems likely to beset the enterprise of finding definitional extensions, whenever the reduced discourse or theory T_t includes a reasonably large or rich set of propositions mixing its predicates (concepts), F, G etc. And all the more likely, if the definitions are required to be Faithful to the given meanings of F, G etc.

In reply, the first thing to say, of course, is just to admit that 'yes, there can be such obstacles to definitional extension'. After all, the advocate of reduction never claimed that definitions of each of T_t's non-logical vocabulary (predicates), in terms of T_b's non-logical vocabulary, can in all cases be constructed so as to derive T_t from T_b. (Recall Sect. 7.3.1's definition of 'definitional extension'.) On the contrary, the advocate of reduction recognises that the enterprise of reduction is a risky business—it can fail. Indeed, it can fail even if one allows the definitions not to be Faithful; and even if—in line with the allowances in the italic schema of Sect. 7.2.1—one seeks only to derive a certain subset of T_t, and-or one allows an augmented T_b.[28]

Besides, the problem was historically influential: advocates of reduction recognised it. Think of how, in philosophy of mind, the threat of circularity was lodged as an objection to logical behaviourism. And in philosophy of science, the logical empiricists' efforts to reduce theory to observation (example (iv) in Sect. 7.2.1.1) were beset by it. To take one famous example: think of how in his 'Testability and meaning' Carnap lowered his sights from defining the (putatively theoretical) predicate 'is soluble' in terms of observational predicates such as 'is in water' and 'dissolves', and proposed only what he called 'reduction sentences' such as, in logical notation: $(x)[InWater(x) \supset [Soluble(x) \equiv Dissolves(x)]]$ (1936: pp. 439–444) (Cf. also Braithwaite (1953, 66–68, 76–79).)

But as we announced in Sect. 7.1.1: in fact, *all is not lost*. That is: in many cases—scientifically important cases, and philosophically important cases—the threat of circularity is only apparent. This is the distinctive insight of functionalism: each concept of the problematic discourse or theory is vindicated (and so claims involving them can be vindicated) by displaying a *simultaneous definition* of each of them, in terms of the unproblematic concepts.

But beware: this insight needs to be stated carefully, so as to respect the distinction between the binary and ternary contrasts of vocabularies, introduced already in Sect. 7.1; i.e. the distinction between the first and second steps of the Canberra Plan (cf. (3) in Sect. 7.1.1.2). That is: the functionalist idea of simultaneous

28 This admission is like the point often made by mathematicians and logicians about axiom systems that are said to 'implicitly define' the (non-logical) words within them (or the concepts referred to be those words). Namely, that 'implicit definition' is a misnomer, since in general, one cannot extract genuine definitions of each term from the axiom system. As it is often put: the elementary analogy with solving n simultaneous linear equations for n unknowns is misleading. We shall return to this, especially in Sect. 7.6.2.

definitions of each of many terms applies in a single given theory (which we called the 'initial theory'); whose vocabulary is divided into:

(i) terms that are taken as unproblematic (understood); and
(ii) others that are each defined (simultaneously) as the unique occupant of a certain role spelt out using the terms in (i).

This is the binary contrast. On the other hand, there can be *another* theory, accepted later than or independently of the first: a theory that also specifies these occupants—in terms different from both the classes (i) and (ii). In this situation, i.e. if there is such a theory, we have a ternary contrast of vocabularies—and the possibility of derived bridge laws ('definitions' in our weak sense) that imply a reduction.

So the care is needed because:

(a) we philosophers all first learn about functional roles, and simultaneous functional definitions etc., in the context of a single theory, with examples like logical behaviourism and the empiricist attempt to define theoretical terms in terms of observational ones—all a matter of the binary contrast, and the first step of the Canberra Plan: while on the other hand,
(b) the jargon of reduction, and our mnemonic notations T_t and T_b ('t' for 'top', 'b' for 'bottom'), and the usual examples like thermodynamics being reduced to statistical mechanics (which was Nagel's main example), all fit the situation where another theory is added, later or independently—all of which means one is considering a ternary contrast, and the second step of the Canberra Plan.

So much by way of a warning. We now spell out, first, the functionalist insight about simultaneous definitions, with its binary contrast (Sect. 7.4); and then, functionalist reduction, with its ternary contrast (Sect. 7.5).

7.4 Functional Roles and Simultaneous Definitions

In Sect. 7.1.1, we introduced functionalism by briefly explaining, in turn, the ideas of:

(a) a functional role, e.g. of a mental state or concept, and its realizer or occupant;
(b) simultaneous unique definitions of those occupants;
(c) a theoretical identification, e.g. of a mental state with a brain state, being compulsory, i.e. being the conclusion of a valid argument with true premises (premises that describe the unique occupant of a functional role in two ways); rather than just recommended as ontologically parsimonious—so that functionalism gives reductions.

In this Section and the next, we give more details: (a) in Sects. 7.4.1 and 7.4.2, (b) in Sect. 7.4.3, and (c) in Sect. 7.5. So as we warned at the end of Sect. 7.3: (a) and (b), and so all of Sect. 7.4, will be about a binary contrast of vocabularies within a single theory. But Sect. 7.5 will return us to the scenario of two theories, the second accepted later than or independently of the first: the usual scenario of reduction, though now involving a ternary contrast of vocabularies

7.4.1 Functional Roles

Undoubtedly, many concepts are *functional*, in philosophers' sense of that term. That is: *the concept is individuated by—i.e. uniquely specifiable by—its pattern of relations to other concepts*. This pattern is then called the concept's *functional role*; and the concept is the *occupant* or *realizer* of the role. Following our usage since Sect. 7.2.1, we shall often say, instead of 'specification': 'definition'—without connotations of synonymy.

As we have discussed, pain is the standard example from the philosophy of mind. A sketch of the functional role is: to be in pain *is* to be in a state that is (a) typically caused by tissue damage, (b) typically causes aversive behaviour, i.e. avoidance of the damage's cause, and (c) is related in such-and-such ways to other mental states, for example, implying the emotion of distress, and typically causing both belief that one is in pain and an intention to avoid the damage's cause.

Of course, details vary as regards the four main words: (1) 'concept', (2) 'relations', (3) 'definition' ('specification') and (4) '(unique) occupant'; (though the ensuing variety in versions of functionalism need not spell incompatibility). We will treat these in order, with '(unique) occupant' getting a Section of its own (Sect. 7.4.2).

1. *The concept*: The concept at issue can be a property, or what many would rather call a way of thinking, a mode of presentation, of a property; (often called 'a concept of the property'). In some examples (including that of pain), what is specified is more naturally called a 'state', or 'state of affairs'. As we said in footnote 2 (and also touched on later, e.g. (1) of Sect. 7.3.3), this variety is not worrisome. For 'concept , 'property' and 'state' are philosophical terms of art, and we will not need to decide on an exact usage: nor on exact criteria of individuation, nor on an exact range of possibilities across which the concept or what-not is claimed to be uniquely specified. What is important for us, both in this paper and our others, is just that the concept or property specified:

   (i) can be a physical quantity: our other papers will have examples like being freely mobile (a property of bodies), and being simultaneous (a relation between events);
   (ii) can have properties and relations (maybe many such) beyond those by which one specifies it: this of course opens the door to it being specified in a different, independent, way—and thus to functionalist reduction (cf. Sect. 7.5).

   We emphasise that there is nothing problematic about (ii). Just think of an attributive definite description such as 'the tallest Swede now alive'. Whoever that person is, he or she has countless properties not encoded in that description: properties which—if only we knew them—we could use to refer the person. Such examples make vivid how the idea of specifying something by its pattern of relations to other things applies equally well to concrete objects, like people, as to concepts and properties. Think of drama: Ian McKellen is the occupant of the role, Macbeth; and Judi Dench the occupant of the role, Lady Macbeth (both in a famous production in the 1970s). Similarly, Lewis' own exposition sketches a detective story: the detective uniquely specifies the culprit or culprits by their actions and their relations to other people and things (1972, p. 250–251: cf. Sect. 7.4.3.1 below).

2. *The relations*: The relations to other concepts can be a matter of: either

    (a) causation, whether deterministic or probabilistic (as suggested by the 'typically caused' in our sketched functional role for pain); or
    (b) law-like association, i.e. Hume's 'constant conjunction' of properties without the interpretation as causation (as in many quantitative non-causal putative laws, e.g. Ohm's law); or
    (c) logical relations, such as implication (e.g. being in pain implies being in distress—as in the sketch above).

    Of these three types of relation, philosophers focus on (a) and (b); and accordingly, often say 'causal-nomological role' instead of 'functional role'. But we shall keep to the phrase, 'functional role': not least because in the roles figuring in our other papers' spacetime examples, logical relations like implication will be *much* more prominent than causal relations. Note also that when a functional role invokes a relation of any of these types, (a), (b) and (c), it may well be qualified by a 'typically' or 'normally'. Thus our sketch for pain said 'typically caused by tissue damage'. (Besides, these words 'typically' and 'normally', and especially the latter, might mean more than a simple statistical majority: 'normal' connotes 'norm' and thus 'goal' or 'purpose'—for example in biology, the proper functioning of the organism. In this way, the 'function' in 'functionalism' has a slight connotation of purpose. But though this is a theme in philosophy of biology, it will not be an issue for us.)

3. *The definition*: In our logically weak usage (since Sects. 7.1.1.2 and 7.2.1), 'definition' need not be either (a) stipulatively defining a new word, or (b) conceptual analysis: where (a) and (b) deploy some agreed notion of logical or metaphysical equivalence. A definition can merely state a reference (extension) for the word, whether pre-existing or new, that is unique, i.e. unambiguous, for the context and purposes at hand. Thus the context and purposes may require much less than faithfulness to a pre-existing use, and-or much less than a statement of the reference in all "possible worlds", i.e. much less than uniqueness on some agreed notion of logical or metaphysical equivalence. Cf. Sect. 7.3.3 on allowing a little Faithlessness. What matters is only that the definition coheres

with the claims using the word that we hold true in the context and purposes at hand.[29]

This logically weak usage may suggest that such definitions are not much of a topic: small beer, as we say in England. But not so. And not just because of the idea (coming up in Sect. 7.4.3) of simultaneous definitions of several words. For even when aiming—within a limited context, and allowing a little Faithlessness—for just one definition, it can be a very considerable achievement to give a definition of a word using the vocabulary prescribed by the context, that coheres with the claims we hold true.

We already touched on this in footnote 19, when endorsing Oliver and Smiley's point, *a propos* of Dedekind, about 'just how much honest toil it takes to discover—to formulate precisely—just what it is that we want' (2016, p. 272). Their example is Dedekind's definition of natural numbers in terms of the successor relation, by what are now called 'Peano's axioms'. For details, including a comparison of Dedekind, Peano and Frege, and also Dedekind's categoricity theorem (i.e. that all models of the axioms are isomorphic), cf. Potter (2000, pp. 81–89).

Furthermore, our other papers will give examples in geometry and spacetime physics. For example: the definition of 'freely mobile' for rigid bodies, that was articulated by Helmholtz, Lie and their successors, and the definition of 'simultaneity' for special relativity in terms of causation, that was articulated by Malament (following work by Robb and others), are both considerable achievements. Although the context is limited (viz. rigid bodies and special relativity, respectively), and only one word or phrase is defined, it is a considerable achievement to *prove* uniqueness of reference—analogous to Dedekind's categoricity theorem. This leads in to Sect. 7.4.2.

7.4.2 The Unique Occupant

The claim that there is a unique occupant (realizer) of a given functional role can be challenged in various ways: and of course, most of the variety will arise from the specific details of the role we are considering. Much that one might say by way of challenge, and much one might say by way of response, is straightforward, or even common sense: especially when the functional role is of a concrete object like a person.

Challenge: 'There are many players of the role 'Macbeth'. Besides, one might say there is no single role, since interpretations of the part vary'. Response: 'But I only claim uniqueness relative to the context of a specific production, with the role's interpretation agreed by the director and players'. Challenge: 'But what about

[29] Agreed: if we are undertaking a reduction with a definitional extension, we ask for more: the definition, taken together with others and with the claims of the reducing theory, must imply the claims of the reduced theory. Cf. Sect. 7.3.1.

understudies?' Response: 'OK: I only claim uniqueness for a specific production, on a specific night'.

Similarly, about philosophically interesting concepts or properties like pain. Challenge: 'Pain is multiply realized: there are many players of the pain-role. Besides, one might say there is no single pain-role, since there is flexibility and vagueness about what to include in it.[30] Response: 'But I only claim uniqueness relative to a kind, which may well be narrower than a species. As Lewis says: '[pain] might even be one brain state in the case of Putnam, another in the case of Lewis. No mystery: that is just like saying that the winning number is 17 in the case of this week's lottery, 137 in the case of last week's' (1969, p. 25): cf. (1) in Sect. 7.3.5.

There is also the Challenge of no realizers, rather than many. Consider eliminativism about folk-psychological propositional attitudes like belief, or about the theoretical entities (whether properties, relations or objects) of yesteryear's theories. Challenge: 'Nothing fits the role of belief, or desire; or the role of phlogiston, or caloric'. Again, there are two straightforward Responses, each of which is surely right in many cases. First, one just agrees with eliminativism: the role in question is unrealized, and the theory or discourse it comes from, should be abandoned (cf. the 'cognitive rubbish' response in Sect. 7.2.1). Second, one says that the role has near-realizers, and one of the nearest is near enough to deserve the name in question ('belief' or 'phlogiston', as the case may be). And of course, one must admit that there can be ambiguity and vagueness: perhaps none of the near-realizers counts unambiguously as nearer than all the others.

Agreed; both these Responses look like qualifications of functionalism's leading idea, viz. unique realizers of functional roles. But no worries. For of course, no functionalist claims that any role, any pattern of relations, that you can extract from (think of, or write down, within) any theory or discourse, must have a unique realizer. The claim is instead (as we shall see in detail in Sect. 7.4.3) that the advocate of a theory claims that it is sufficiently informative (logically strong) that each of the entities (whether properties, relations or objects) that it newly introduces is the unique realizer of a pattern of properties and relations that can be extracted from the theory. Besides, the pattern can be extracted systematically, in the same way for all the new entities. To put the claim in terms of language, not entities: the functionalist claims that each of the terms the theory newly introduces can be defined by the term's pattern of occurrence in all the assertions of the theory. Besides, these definitions can be extracted systematically from the theory, and presented simultaneously.

For our purposes, especially in our other papers, what matters is that the claim of a unique occupant (realizer) holds good: indeed, *provably so*. To put it in terms of the Challenges, the straightforward Responses apply: for instance, one responds to multiple realization by limiting the context enough to prove the uniqueness.

[30] Recall the traditional objection to logical behaviourism, that it ruled out what seems possible: for example, the conjunct in our sketched pain-role, 'typically causes aversive behaviour', seems to rule out perfect-actor "super-Spartans" who never flinch when in pain.

Let us illustrate with the example of simultaneity, mentioned at the end of Sect. 7.4.1. In the limited context of special relativity, there is a theory about causal connectability of spacetime points that is sufficiently informative that one can prove there to be a unique equivalence relation satisfying certain conditions that, most would agree, are part of the meaning of the term 'simultaneity'. To put it in the jargon of functionalism: one proves that there is a unique occupant of the simultaneity-role; and the term 'simultaneity' can be thereby defined; besides, the definition is Faithful to the pre-existing meaning of the term. And for our papers' other examples, the situation will be similar: one proves unique occupancy of the relevant role. (Or in some cases: uniqueness *modulo* a choice of units and coordinate system—a controlled and welcome Plenitude; cf. (2) in Sect. 7.2.2.2.)

But agreed: not all that one might say hereabouts, by way of challenge to the functionalist claim of unique occupancy, or by way of response, is straightforward. Already in (4) at the start of Sect. 7.2, and at the end of (1) in Sect. 7.2.2.2, we acknowledged that Putnam's model-theoretic argument, and Newman's objection to Russell's structural realism, make a deep and general challenge about reference: roughly, that for realists like us (or Lewis), reference and truth are all too easy to attain. (In (1) of Sect. 7.2.2.2, the specific challenge was a problem of Plenitude: the plethora of mock-ups.[31]) In reply, we emphasised that there are several cogent responses to these challenges, between which we (and our other papers) do not have to choose; (similarly, cf. footnotes 17 and 24 about postponing structural realism). But here, in the more specific context of functionalism, we should give more detail about our views on reference. For in the next Section we will report Lewis' (1970; 1972) expositions, especially of functional definition. And since it assumes that the O-terms used by a functional definition—the *old* or *original* terms: what Sect. 7.2.1 called 'unproblematic terms'—have a reference, we owe some discussion of how *they* get a reference.

In short, our answer is twofold: partly Lewisian and partly not. The Lewisian part is: *causal descriptivism*, so labelled (and endorsed) by him in his reply to Putnam's argument (1984: pp. 226–227). That is: a descriptivist account of reference that:

(i) takes to heart various points engendered by the criticisms launched by the causal theory of reference (1984: p. 223); and
(ii) couches its descriptions in largely causal terms—and thereby, urges Lewis, often accounts for puzzle cases at least as well as a causal theory.

But care is needed. *Global descriptivism* is the view that for *all* the terms of our language, their reference is determined solely by requiring any assignment of candidate referents to render true, *according to that assignment*, whatever we assert. Evidently, this view will make truth all too easy to attain. And indeed: Lewis

[31] As we noted in (3) in Sect. 7.2.2.4 and footnote 26: the labels 'Plenitude' and 'Scarcity' can be confusing in the present context. For they were introduced in Sects. 7.2.2.2 and 7.2.2.3 as about, respectively, having too many, or no, definitions. But in this Section, the focus has been on having too many, or no, occupants (realizers) of a definition: a different topic.

argues that the lesson of Putnam's model-theoretic argument is precisely that global descriptivism is *false* (1984: 224, 226). Again, we agree: we read Putnam the same way. Besides, we agree with Lewis that this conclusion is *not* avoided by making one's global descriptivism also causal, i.e. by also adopting causal descriptivism's (ii) above. That would be of no avail; since for global descriptivism, this tactic amounts to invoking 'just more description, just more theory'. So an advocate of causal descriptivism, such as Lewis or us, must keep their descriptivism *non*-global; and so, in order to answer Putnam, they 'must seek elsewhere [than causation] for the saving constraint' (p. 227).

This is the point at which Lewis turns to his theory of natural properties (expounded in his 1983)—and at which we part company with him. For while we admire the way this theory fulfils several needs in his overall metaphysical system, it uses a notion of similarity this is given once for all, across all of modal reality. In particular, it is not relativized to either a theory or a possible world. This we admit that we cannot believe. And so as we said in (4) of Sect. 7.2, we must seek reference's 'saving constraint' in some other direction.

The upshot is that like Lewis, we endorse causal descriptivism and deny global descriptivism (even using causal descriptions); but as a reply to Putnam's model-theoretic argument, we would opt for a less gung-ho realism than Lewis'.[32] Obviously, this is not the place to develop this position. We have no space; and anyway, our position is not unusual. Causal descriptivism has been defended both in philosophy of language (e.g. Kroon, 1987), and in philosophy of science, specifically as part of defending scientific realism (Psillos, 1999, 2012). (Our endorsement of scientific realism will return in Sect. 7.6.)

7.4.3 Simultaneous Unique Definitions

In Sects. 7.1.1 and 7.3.5, we saw how all the concepts in some relevant set being definable or specifiable by some of their relations to each other and to other concepts faces a threat of logical circularity. (For the choice between 'definition' and 'specification', cf. (1) in Sect. 7.1.1.2, the start of Sect. 7.2.1 and (3) in Sect. 7.4.1.) For if each of two concepts is to be defined, in part, by its relations to the other, we apparently cannot define either of the concepts without a circularity. Similarly for more than two concepts: there could be a sequence of putative definitions of

[32] A wrinkle about terminology. Nowadays, some (e.g. Janssen-Lauret and MacBride (2020)) use 'global descriptivism' for the weaker doctrine that the reference of all terms is settled *en bloc* by total theory. This is weaker than our definition above, since it makes no claim that the only constraint on reference-assignment is making true whatever we assert. So it can be combined with the sort of 'reference magnetism' Lewis espoused; or with some other constraint that is not 'just more theory'. So there is no disagreement here.

concepts, where the last definition invokes the concept defined by the first—a logical circle. Besides, in branches of philosophy where functionalism has been attractive, such logical circles seem all too likely. Recall the example of how a logical behaviourist might try to define belief that it is raining.

But in fact, this threat can be answered. We *can* make perfectly good sense of simultaneously defining, with no logical error, each concept in a set. We only need all the definitions to be implied by a sufficiently rich body of information, which spells out the functional roles of each concept in the set. And this requires simply that the body of information be true only if for each such role, there is a unique occupant or realizer of that role.

As so often, it is in Lewis' work that the idea is stated most clearly. So this Section summarizes his exposition. We begin with a parable from his (1972): Sect. 7.4.3.1. This leads to the Ramsey and Carnap sentences of a theory, and how to modify them: (Sect. 7.4.3.2; also drawing on his (1970)). This stage-setting prepares us for the *denouement*: explicit simultaneous definitions of many terms—partly in terms of each other: Sect. 7.4.3.3.

7.4.3.1 A Parable

Lewis (1972) begins with an example of simultaneous unique specifications of, not concepts, but people—in a country-house detective story.

> We are assembled in the drawing room of the country house; the detective reconstructs the crime. That is, he proposes a theory designed to be the best explanation of phenomena we have observed: the death of Mr. Body, the blood on the wallpaper, the silence of the dog in the night, the clock seventeen minutes fast, and so on. He launches into his story:
>
>> *X*, *Y* and *Z* conspired to murder Mr. Body. Seventeen years ago, in the gold fields of Uganda, *X* was Body's partner ... Last week, *Y* and *Z* conferred in a bar in Reading ... Tuesday night at 11:17, *Y* went to the attic and set a time bomb ... Seventeen minutes later, *X* met *Z* in the billiard room and gave him the lead pipe ...
>
> And so it goes: a long story. Let us pretend that it is a single long conjunctive sentence ...
>
> Suppose that after we have heard the detective's story, we learn that it is true of a certain three people: Plum, Peacock and Mustard [respectively] ...
>
> We will say that Plum, Peacock and Mustard together *realize* (or are a *realization* of) the detective's theory ... We may also find out that the story is not true of any other triple ...
>
> In telling his story, the detective set forth three roles and said that they were occupied by *X*, *Y* and *Z*. He must have specified the meanings of the three terms '*X*', '*Y*' and '*Z*' thereby ... They were introduced by an implicit functional definition, being reserved to name the occupants of the three roles. When we find out who are the occupants of the three roles, we find out who are *X*, *Y* and *Z*. Here is our theoretical identification. (1972, p. 250–251)

The point is crystal-clear, especially from the last paragraph. In short: we interpret the detective's story as truly indicting Plum, Peacock and Mustard iff they as a triple are the unique realization of it; and it is true, i.e. true of some or other trio of culprits, iff some triple of people are the unique realization of it.[33]

7.4.3.2 Ramsey Sentences and Carnap Sentences—Modified

With a little notation, we can compendiously state both Lewis' general idea and the battery of explicit definitions.[34] As in Lewis' parable (and his 1970), we adopt the notation: T-terms and O-terms. But as Lewis says: 'T-term' need not mean 'theoretical term', and 'O-term' need not mean 'observational term'. In the parable, he wrote: 'O does not stand for 'observational'. Not all the O-terms are observational terms, whatever those may be. They are just any old terms' (1972, p. 250). Similarly in his (1970), he writes:

> I do not understand what it is just to be a theoretical term, not of any theory in particular, as opposed to being an observational term (or a logical or mathematical term).[A footnote endorses Putnam's article 'What Theories Are Not'.] I believe I do understand what it is to be a T-term: that is, a theoretical term introduced by a given theory T at a given stage in the history of science. If so, then I also understand what it is to be an O-term: that is, any *other* term, one of our *original* terms, an *old* term we already understood before the new theory T with its new T-terms was proposed. An O-term can have any epistemic origin and priority you please. It can belong to any semantic or syntactic category you please. Any old term can be an O-term, provided we have somehow come to understand it. And by understand I mean "understand"—not "know how to analyze." (1970, p. 428)

So despite the letters 'T' and 'O', Lewis' proposals are not only about the theory-observation distinction. Indeed, these papers' influence in the years since 1970 has led to the framework of functional definition being applied to many of philosophy's contrasts between what Sect. 7.2.1 called 'the problematic' vs. 'the unproblematic': for instance, the ethical vs. factual contrast, as in footnote 15. In short, it has led to the Canberra Plan.

Happily, Button and Walsh (2018, p. 55) suggest a helpful mnemonic to replace the overly restrictive, and misleading, 'theory' and 'observation': namely, 'T' stands

---

[33] Here, 'true' means 'completely true'. This logical strength prompts a clarification. Shortly after the quoted passage, Lewis writes:

> A complication: what if the theorizing detective has made one little mistake? He should have said that Y went to the attic at 11:37, not 11:17. The story as told is unrealized, true of no one. But another story is realized, indeed uniquely realized: the story we get by deleting or correcting the little mistake. We can say that the story as told is *nearly realized*, has a unique *near-realization*... In this case the T-terms ought to name the components of the near-realization... But let us set aside this complication for the sake of simplicity, though we know well that scientific theories are often nearly realized but rarely realized, and that theoretical reduction is usually blended with revision of the reduced theory.

Well said. Indeed in (1) of Sect. 7.3.2, we saw Nagel, Schaffner and Fletcher also say what 'we know well'.

[34] In this Section and the next, we are very indebted to Adam Caulton: whose insightful comparison of Lewis' and Carnap's views we have regretfully suppressed, for the sake of space.

for *troublesome*, and 'O' stands for *okay*. We will from now on use this mnemonic. But the main point is as in the quote: the O-terms are understood, their reference is settled (*modulo* the deep issues set aside at the end of Sect. 7.4.2!); while the T-terms are yet to be understood, either because they are new or because they are problematic.

So we begin by assuming we have a theory T, for which there is some distinction between troublesome and okay terms: T-terms and O-terms. We take T as a long conjunction of claims, which we call the *postulate*. The leading idea will be to use the patterns of relations to the O-terms, that the T-terms enjoy, to fix the meanings of the T-terms. And this will involve commitment to unique realizations, which will yield explicit definitions, one for each T-term.[35]

In the theory T, let us make all the theoretical terms $t_1, \dots t_n$ used in T explicit by writing $T(t_1, \dots t_n)$. Here, we treat all theoretical terms as first-order, i.e. as names. This is as in Lewis, who says that this choice

> 'is of no importance. It is a popular exercise to recast a language so that its non-logical vocabulary consists entirely of predicates; but it is just as easy to recast a language so that its non-logical vocabulary consists entirely of names (provided that the logical vocabulary includes a copula). These names, of course, may purport to name individuals, sets, attributes, species, states, functions, relations, magnitudes, phenomena or what have you; but they are still names. Assume this done, so that we may replace all T-terms by variables of the same sort' (1972, p. 253).[36]

Now form the *realization formula* of T by replacing each T-term in all its occurrences with the same variable, $x_1, \dots x_n$ respectively (where we of course assume that none of $x_1, \dots x_n$ already occur in T, even as bound variables). Call this formula: $T(x_1, \dots x_n)$. Any n-tuple of entities that satisfies the realization formula is a *realization* of T. The entities *realize* T.

The *Ramsey sentence* of T is just the claim that T is realized. That is to say, it is:

$$\exists x_1, \dots, x_n T(x_1, \dots, x_n). \tag{7.1}$$

The Ramsey sentence has exactly the same purely okay consequences as T; i.e. it has as consequences exactly the same sentences containing *only* okay terms as does T. But the Ramsey sentence has no purely troublesome consequences—*a fortiori* no purely troublesome consequences entailed by T—since it contains no

[35] So in our other papers, we will use this notation in examples of defining a spacetime structure as the unique realizer of some role. Broadly speaking: it will be vocabulary about the physics of matter and radiation that are the O-terms, and vocabulary about chrono-geometry that are the T-terms.

[36] This may seem nonchalant: Lewis (1970, p. 429) gives a longer defence of this choice. But agreed: questions remain, both in general (e.g. 'How does this bear on the usual construal of Newman's objection as depending on the Ramsey sentence using second-order quantifiers?') and for our own advocacy of spacetime functionalism (e.g. 'In our examples, can the defined spacetime structures, e.g. a metric, be named?'). We address these questions in our other papers.

troublesome vocabulary.[37] So the Ramsey sentence serves as a completely adequate okay surrogate for the original theory T.

Thus the residual content of T lies in what T says by way of "implicit definitions" of the T-terms. That seems to be encapsulated in what is called the *Carnap sentence* of T. This says that if T is realized, then it is realized by $t_1, \ldots, t_n$. It says: if anything realizes T, then $t_1, \ldots, t_n$ does. Thus the *Carnap sentence* of T is:

$$[(\exists x_1, \ldots, x_n T(x_1, \ldots, x_n)) \supset T(t_1, \ldots, t_n)] . \tag{7.2}$$

Conjoining the Carnap sentence with the Ramsey sentence entails T. But the Carnap sentence entails no purely okay sentences except logical truths (logically valid formulas). So its role is apparently to interpret the troublesome terms. Thus Lewis writes: 'it does seem to do as much toward interpreting them as the postulate itself does. And the Ramsey and Carnap sentences between them do exactly what the postulate does' (1970, p. 431). But Lewis will shortly argue that these sentences need to be *modified.*

First, let us press the question: what do the variables x_i range over? As we know, Lewis is a realist. He sees no ontological distinction between "observable" and "theoretical" *entities*: where 'a theoretical entity is something we believe in only because its existence, occurrence, etc. is posited by some theory—especially some recent, esoteric, not-yet-well-established scientific theory' (1970, p. 428).[38] He admits only a distinction between troublesome (i.e. yet-to-be-understood, e.g. because newly introduced), and okay (i.e. already understood), *terms*. Thus he writes

> I am also not planning to "dispense with theoretical entities." Quite the opposite. The defining of theoretical terms serves the cause of scientific realism. A term correctly defined by means of other terms that admittedly have sense and denotation can scarcely be regarded as a mere bead on a formal abacus.... Theoretical entities are not entities of a special category, but entities we know of (at present) in a special way. (1970, p. 428)

So pinning down newly-introduced entities using a theory is like the detective in the parable identifying a suspect. We should think of the entities corresponding to the T-terms as *out there, somewhere*, just like the people X, Y and Z: we just need sufficient detail in our theory to pick them out.

In this endeavour, Lewis distinguishes three cases: *unique, failed, and multiple, realization* (1970, pp. 431–433). In presenting Lewis' parable, we emphasised the first of these. But the situation will also be clear enough in the other cases.

---

[37] To be precise: its purely troublesome consequences are merely all the logical truths (logically valid formulas) containing only T-terms.

[38] Note that although 'entity' connotes 'object', and Lewis mentions some objects as his examples of theoretical entities, viz. living creatures too small to see and the dark companions of stars, Lewis' taking all T-terms as names means that properties and relations, like physical magnitudes or a spacetime metric, will count as theoretical entities. Cf. the end of footnotes 2 and 36.

1. *T is uniquely realized.* Here, the Ramsey sentence is true; and the Carnap sentence 'clearly gives the right specification' (p. 431), i.e. determines the correct unique referents for $t_1, \ldots, t_n$.
2. *T is not realized.* Then 'the Carnap sentence says nothing about the referents of the T-terms' (p. 432). But here, we should distinguish two sub-cases:

   (a) T is nowhere near being realized. Lewis gives a familiar example: we know what phlogiston is supposed to be, or what it would be were it to exist, even while we also know that it doesn't exist. So in this sub-case, the Carnap sentence's silence seems wrong, i.e. overly modest. Where it is silent, one should be opinionated: one should say that the T-terms do not refer.
   (b) T is *nearly* realized. That is: We want to say that a certain n-tuple comes near enough to realizing T. Then as in the quotation from Lewis in footnote 33, the T-terms surely name the components of the near-realization. In such a case, the Carnap sentence can probably be regarded as correctly specifying the referents, viz. by taking the term-introducing theory to be some theory T' that is similar to T (maybe logically weaker, i.e. implied by T) and that *is* realized by T's near-realization.

3. *T is multiply realized.* Here again, the Carnap sentence seems wrong to Lewis, because unduly instrumentalist. He writes:

   > In this case, the Carnap sentence tells us that the T-terms name the components of some realization or other. But it does not tell us which; and there seems to be no non-arbitrary way to choose one of the realizations. So either the T-terms do not name anything, or they name the components of an arbitrarily chosen one of the realizations of T. Either of these alternatives concedes too much to the instrumentalist view of a theory as a mere formal abacus. Neither does justice to our naive impression that we understand the theoretical terms of a true theory, and without making any arbitrary choice among realizations. We should not accept Carnap's treatment in this case if we can help it. Can we? (1970 p. 432).

In short, Lewis' idea here is that unique realization is preferable to multiple realization, and we ought to say that multiply realized theories have denotation-less T-terms. He also urges two reasons why in scientific theorizing, unique realization is a reasonable, not extravagant, hope (1970 pp. 433):

(i) 'I am not claiming that there is only one way in which a given theory *could* be realized; just that we can reasonably hope that there is only one way in which it *is* realized.' So in the jargon of possible worlds: multiple realization is still allowed *across possible worlds*, just not within a world. (This leads to some subtleties we need not report—since they are not defects, though they have influenced philosophers' usage of the word 'functionalism'; cf. pp. 436–437, Lewis (1994, p. 419–421) and our footnote 4.) In fact, the T-terms are probably *not* rigid designators ('logically determinate names'; cf. also p. 435–436).
(ii) The O-terms, whose interpretations are fixed, are a large and miscellaneous set; specifically, they are not confined to being observational. (This point yields an answer to a theorem of Winnie.)

Here, we would add a third reason to be hopeful. Namely: unique realization does not require T to establish *every* fact about the T-term's referents: just enough facts to secure uniqueness. There can be plenty of room for us to discover more about these referents. This of course leads to the *ternary* contrast of vocabularies, and the *functionalist reduction* that we advertised in Sect. 7.1: which we develop in Sect. 7.5.

We can sum up this discussion—specifically, the proposal that with no or multiple realization, the T-terms do not refer—as proposed modifications of the Ramsey and Carnap sentences. Thus the Ramsey sentence Eq. 7.1 is to be replaced by the *modified Ramsey sentence*, which says that T is *uniquely* realized:

$$\exists y_1, \ldots, y_n \forall x_1, \ldots, x_n \left[T(x_1, \ldots, x_n) \equiv (y_1 = x_1 \;\&\; \ldots \; y_n = x_n)\right]. \tag{7.3}$$

Thus Lewis writes:

> The Ramsey sentence has exactly the same O-content as the postulate of T; any sentence free of T-terms follows logically from one if and only if it follows from the other.[footnote suppressed] The modified Ramsey sentence has slightly more O-content. I claim that this surplus O-content does belong to the theory T—there are more theorems of T than follow logically from the postulate alone. For in presenting the postulate as if the T-terms have been well-defined thereby, the theorist has implicitly asserted that T is uniquely realized. (1972, pp. 253–254)

The Carnap sentence is to be replaced by three postulates, as follows: which together say that T is uniquely realized iff it is realized by $t_1, \ldots, t_n$. (Here, it is important to adopt a system of logic that allows bearerless terms, so that a formula such as $(\exists x)(x = t_1)$ is not a logical truth. Lewis adopts a system by Scott (1970, p. 430).)

1. If T is uniquely realized, then it is uniquely realized by $t_1, \ldots, t_n$:

$$(\exists y_1, \ldots, y_n \forall x_1, \ldots, x_n \left[T(x_1, \ldots, x_n)\right. \\ \left.\equiv (y_1 = x_1 \;\&\; \ldots \; y_n = x_n)\right]) \supset T(t_1, \ldots t_n)). \tag{7.4}$$

   Lewis calls this the *modified Carnap sentence* (1972, p. 254); it is logically implied by the Carnap sentence.
2. If T is not realized at all, then $t_1, \ldots t_n$ don't refer:
   $(\neg\exists x_1, \ldots, x_n T(x_1, \ldots, x_n) \supset (\neg\exists x \; x = t_1 \;\&\; \ldots \; \neg\exists x \; x = t_n))$
3. If T is multiply realized, then $t_1, \ldots t_n$ don't refer:
   $(\exists x_1, \ldots, x_n T(x_1, \ldots, x_n) \;\&$
   $\neg\exists y_1, \ldots y_n \forall x_1, \ldots x_n (T(x_1, \ldots x_n) \equiv (y_1 = x_1 \;\&\ldots\; y_n = x_n))) \;\; \supset \;\; (\neg\exists x \; x = t_1 \;\&\ldots$
   $\neg\exists x \; x = t_n)$.

Lewis (1972) introduces a helpful concise notation. He uses **boldface** names and variables to denote n-tuples. So our original statement of the theory that displayed the T-terms, $T(t_1, \ldots, t_n)$, is now written as $T[\mathbf{t}]$; the realization formula

$T(x_1, \ldots, x_n)$ is written as $T[\mathbf{x}]$; and the Ramsey sentence is written as $\exists \mathbf{x} T[\mathbf{x}]$. So Lewis' modified Ramsey sentence, Eq. 7.3, is now written as:

$$\exists_1 \mathbf{x} T[\mathbf{x}]; \quad \text{that is: } \exists \mathbf{y} \forall \mathbf{x} (T[\mathbf{x}] \equiv \mathbf{y} = \mathbf{x}); \tag{7.5}$$

and similarly, the modified Carnap sentence, Eq. 7.4, is now written as:

$$\exists_1 \mathbf{x} T[\mathbf{x}] \supset T[\mathbf{t}]; \quad \text{that is: } (\exists \mathbf{y} \forall \mathbf{x} (T[\mathbf{x}] \equiv \mathbf{y} = \mathbf{x})) \supset T[\mathbf{t}]. \tag{7.6}$$

Lewis also points out (1972, p. 254) that using this notation, we can state conditions 2 and 3 just above, about failed and multiple realizations (respectively), as a conditional. We just need to adopt the neat device of a necessarily denotationless term, say $*$. (For more reasons why this is neat, cf. Oliver and Smiley (2013).) Thus consider

$$\neg \exists_1 \mathbf{x} T[\mathbf{x}] \ \supset \ (\mathbf{t} = *). \tag{7.7}$$

Here, $\mathbf{t} = *$ means that each t_i is denotationless. For if $*$ is some chosen necessarily denotationless name, then $*$ is $\langle *, \ldots * \rangle$ and $\mathbf{t} = *$ is equivalent to the conjunction of all the identities $t_i = *$.

7.4.3.3 Simultaneous Explicit Definitions

Lewis' proposal in Sect. 7.4.3.2, that the T terms refer iff the realization formula $T(x_1, \ldots x_n)$ (also written $T[\mathbf{x}]$) is uniquely realized, immediately yields *explicit definitions* for each of the T-terms.

For if as usual we write ι for the description symbol, i.e. we read $\iota x F(x)$ as 'the F', then the conjunction of:

(i) the modified Carnap sentence, Eq. 7.4 or equivalently Eq. 7.6, and
(ii) our 'veto' on denotation for the cases of no or multiple realizations, i.e. Eq. 7.7, is logically equivalent to the battery of definitions:

$$t_1 =_{df} \iota y_1 \exists y_2, \ldots, y_n \forall x_1, \ldots, x_n \left(T(x_1, \ldots, x_n)\right.$$
$$\left. \equiv (y_1 = x_1 \ \& \ \ldots \ y_n = x_n)\right),$$

$$\ldots$$

$$t_n =_{df} \iota y_n \exists y_1, \ldots, y_{n-1} \forall x_1, \ldots, x_n \left(T(x_1, \ldots, x_n)\right.$$
$$\left. \equiv (y_1 = x_1 \ \& \ \ldots \ y_n = x_n)\right).$$

Or more compactly, in the boldface notation: the conjunction of (i) and (ii) is logically equivalent to

$$\mathbf{t} = \iota \mathbf{x} T[\mathbf{x}].^{39} \tag{7.8}$$

Immediately after exhibiting these explicit definitions, Lewis sums up.

> This is what I have called functional definition. The T-terms have been defined as the occupants of the causal roles specified by the theory T; as the entities, whatever those may be, that bear certain causal relations to one another and to the referents of the O-terms.
>
> If I am right, T-terms are eliminable—we can always replace them by their definientia. Of course, this is not to say that theories are fictions, or that theories are uninterpreted formal abacuses, or that theoretical entities are unreal. Quite the opposite! Because we understand the O-terms, and we can define the T-terms from them, theories are fully meaningful; we have reason to think a good theory true; and if a theory is true, then whatever exists according to the theory really does exist.
>
> I said that there are more theorems of T than follow logically from the postulate alone. [Cf. the last displayed quote in Sect. 7.4.3.2, just after Eq. 7.3; and cf. (1970, pp. 438–440)] More precisely: the theorems of T are just those sentences which follow from the postulate together with the corresponding functional definition of the T-terms. For that definition, I claim, is given implicitly when the postulate is presented as bestowing meanings on the T-terms introduced in it. (1972, p. 254)

Let us also sum up by returning to the central example of mind and body. So let T be the folk psychology of mental states. It is a catalogue of platitudes about belief, desire, etc., connecting those mental states to public situations and behavioural dispositions, whose descriptions are given using the O-terms; and the T-terms refer to the mental states. T is taken to assert the *existence* of inner states that fulfil the roles of appropriately mediating the connections between situations and behaviour. (Recall the detective-story parable.) The existence of such states is a fact *beyond* the situation-behaviour correlations. T may fall way short of giving necessary and sufficient conditions for them; and it is in general a contingent matter, varying from possible world to possible world, *which* states they may happen to be. Recall comments (i) and (ii) just before Eq. 7.3. This is *functionalism* about mind and body.

7.5 Functionalist Reduction

All the props are now on stage. We have developed a Nagelian account of reduction, taking in to account the problems of Faithlessness, Plenitude and Scarcity (Sects. 7.2 and 7.3); and Lewis' account of functional definition (Sect. 7.4). We now put these together (again following Lewis) to get functionalist reduction: the second step of

[39] A minor clarification. The equivalence requires an appropriate stipulation about the truth-values of identity statements using descriptions that are improper, i.e. not realized or multiply realized. But no worries: as Lewis says (1970, pp. 430, 438; 1972, footnote 11, p. 254), the stipulations in Scott's system are appropriate.

the Canberra Plan. As we announced in Sect. 7.1, it is the *golden oldie* that, sadly, the literature on functionalism and reduction has forgotten.

Lewis' main idea is that thanks to Sect. 7.4's functional definitions of the T-terms, one can show that a Nagelian reduction of T by some later or independent theory, T^* say, has no need to *postulate* bridge laws (linking the vocabularies of T and T^*) as separate empirical hypotheses. For the functional definitions we extracted from T mean that a version of the bridge laws—viz. with each T-term replaced by its functional definition using O-terms—can be a *theorem* of T^*. There is no bar against this, since these versions do not contain any vocabulary alien to T^*. (Lewis calls these versions the *definitionally expanded bridge laws*.) And if they *are* theorems of T^*, then: T^* alone yields the bridge laws (by the transitivity of identity); and so T^* alone reduces T. (In Sect. 7.3.1 we introduced the mnemonic labels T_t and T_b for the reduced and the reducing theories. But in this Section, it will be clearer to follow Lewis' notation of T and T^*.)

Of course, we already reviewed this idea in Sect. 7.1.1.1's example of pain and C-fibre firing. T was the folk psychology of mental states from which functional definitions of mental terms like 'pain' were extracted ; and T^* was a neurophysiological theory, accepted later than or independently of folk psychology, which we took to include the theorem that C-fibre firing is the unique occupant of the pain-role. So T^* asserts that C-fibre firing is typically caused by tissue damage, typically causes aversive behaviour etc. So this theorem is the relevant definitionally expanded bridge law. Then as we said in Sect. 7.1.1.1, it follows by transitivity of identity that pain = C-fibre firing.[40]

So our job now is to state this idea more precisely and generally. This will also make it clear how not only the bridge laws, but also the whole of T is definitionally implied by T^*.

To state the idea precisely, it will be clearest to begin with the simplest scenario, where: (i) the T-terms have retained the interpretations they received when first introduced (i.e. as expounded in Sect. 7.4.3.2) and (ii) 'this is a reduction in which T survives intact; not, what is more common, a simultaneous partial reduction and partial falsification of T by T^*' (1970, p. 441). Afterwards, we can turn to the more common and complicated scenario, which Lewis' account of course admits just as Nagel and Schaffner do: (recall Sects. 7.3.2 and 7.3.3).

First, Lewis stresses that T^* can be miscellaneous; recall our disavowing reductionism, in (2) at the start of Sect. 7.2. But all the terms of T^* are taken as being understood (interpreted), just as the O-terms in Sect. 7.4 were. Besides, one may as well use 'O-term' to include also the 'okay' terms of T^* by which T^*'s newly-introduced, or 'troublesome', terms—the T^*-terms—got their interpretations. Then Lewis proposes to label as an O^*-term, whatever is either an O-term or T^*-term.

[40] 'Transitivity of identity' is usually understood as identity of objects; and this accords with Lewis' taking all T-terms as names not predicates (1972, p. 253, quoted in Sect. 7.4.3.2). But we can also understand it as co-extensiveness of predicates; cf. footnotes 2 and 36.

So here, Lewis is at first registering what we dubbed the *ternary* contrast of vocabularies; but then he introduces the O^* notation, so as to focus our attention on the binary contrast of greatest interest in reduction: the contrast between the T-terms of T, and the 'rest', i.e. the O^*-terms in the sense just stipulated. For instance, 'tissue damage', 'aversive behaviour' are O-terms, but they can enter the functional definition of the T^*-terms. Thus Lewis writes

> The reducing theory T^* need not be what we would naturally call a single theory; it may be a combination of several theories, perhaps belonging to different sciences. Parts of T^* may be miscellaneous unsystematized hypotheses which we accept, and which are not properly called theories at all. Different parts of T^* may have been proposed or accepted at different times, either before or after T itself was proposed... T^*, or parts of T^*, may introduce theoretical terms; if so, let us assume that these T^*-terms have been introduced by means of the same O-vocabulary which was used to introduce the theoretical terms of T. This is possible regardless of the order in which T and T^* were proposed. Any term that is either an O-term or a T^*-term may be called an O^*-term. (1970, p. 441)

Now suppose that the following sentence is a theorem (a claim) of T^*: $T(\rho_1, \ldots \rho_n)$, where $\rho_1, \ldots \rho_n$ are all O^*-terms, including, possibly, definite descriptions constructed using other O^*-terms. Here, $T(t_1, \ldots t_n)$ is of course the original postulate of T, just as at the start of Sect. 7.4.3.2; and $T(\rho_1, \ldots \rho_n)$ simply substitutes ρ_i for each occurrence of t_i therein. Lewis calls $T(\rho_1, \ldots \rho_n)$ the *reduction premise* for T (1970, p. 441), or the *weak reduction premise* for T (1972, p. 255).

Notice that the original postulate $T(t_1, \ldots t_n)$ follows from the reduction premise, taken together with the *bridge laws*, as usually formulated, viz.: $\rho_1 = t_1, \ \ldots \ \rho_n = t_n$. (Also, the bridge laws (thus formulated) follow from the reduction premise together with the postulate of T and the functional definitions of the $t_1, \ \ldots \ t_n$.) *But where do we get the bridge laws?*

The traditional view is that the bridge laws are separate, empirical hypotheses. In that case, says Lewis (p. 442), one may choose whether to posit T^* and the bridge laws, and so derive T, or else to posit T^* alone, in which case we will have to posit T separately.

But then Lewis asks us to consider the case in which T^* has as theorems *definitionally expanded bridge laws*:

$$\rho_1 =_{df} \iota y_1 \exists y_2, \ldots y_n \forall x_1, \ldots x_n \left(T(x_1, \ldots x_n) \equiv (y_1 = x_1 \ \& \ \ldots \ y_n = x_n)\right), \text{etc.}$$

whose right hand sides match the definitions in Sect. 7.4.3.3, cf. Eq. 7.8. Unlike the case for the original bridge laws, the definitionally expanded bridge laws contain only O^*-terms. So there is no problem of *vocabulary* preventing these laws being theorems of T^*.

And if they *are* theorems of T^*, then it follows by sheer logic (Leibniz's Law, the transitivity of identity) that $t_1 = \rho_1$, etc. So there is no choice in the matter: T^* reduces T, without the need for empirical hypotheses. Lewis sums up:

> If T^* yields as theorems a reduction premise for T, and also a suitable set of definitionally expanded bridge laws for T, then T^*—without the aid of any other empirical hypothesis—

> reduces T.... The reduction of T does not need to be justified by considerations of parsimony (or whatever) over and above the considerations of parsimony that led us to accept T^* in the first place. (1970, p. 443; similarly, 1972, p. 255)[41]

So much by way of summarising how bridge laws, and thereby the reduced theory T, might be deduced from the reducing theory T^*—in the simplest scenario, where (i) the T-terms have retained their original interpretations, and (ii) T survives intact, rather than being partially reduced and partially falsified. Lewis ends by briefly discussing the more common and complicated scenario, where 'the original theory is falsified while a corrected version is reduced. If T is thus partially reduced and partially falsified, or revised for any other reason, do the T-terms retain their meanings?' (1970, p. 445). He considers two approaches. The first is prompted by Feyerabend's views. But we shall briefly report only the second: which we favour (as also Lewis seems to) and which leads us back to our previous topics of *causal descriptivism* (endorsed near the end of Sect. 7.4.2) and *near-realizations* (footnote 33).

On this approach, we say that the T-terms name the nearest near-realization of the *original* T. Then so long as the corrections are not too extreme, the T-terms may denote, and we can make sense of the idea that adjustments to T involve learning more about the objects thus denoted. This meshes well with causal descriptivism. Thus Lewis:

> we permit the T-terms to name components of the nearest near-realization of T, even if it is not a realization of T itself. For after T has been corrected, no matter how slightly, we will believe that the original version of T is unrealized. We will want the T-terms to name components of the unique realization (if any) of the corrected version of T. They can do so without change of meaning if a realization of the corrected version is also a near-realization of the original version.
>
> According to this position, we may be unable to discover the meanings of theoretical terms at a given time just by looking into the minds of the most competent language-users at that time.... This situation is surprising, but it has precedent: a parallel doctrine about proper names has recently been defended.[a footnote cites Kaplan's famous 1968 paper 'Quantifying in'] (1970, p. 446)

Indeed, near-realizations and the causal theory of reference will be one of the themes in our discussion of Torretti...

7.6 Glimpsing the Land of Torretti

As we announced in Sect. 7.1.3: now that we have developed our account of functionalist reduction, various projects in relation to Torretti's large and magisterial *oeuvre* beckon us. Even once we postpone to our other papers the projects in the

---

[41] For brevity, we set aside: (i) Lewis' discussion of the *auxiliary reduction premise* (1970, p. 443) and the *strong reduction premise* (1972, p. 255); (ii) Lewis' examples (1970, p. 443–444; 1972, p. 256–258—which is mind and body); and (iii) Lewis' discussion of adding definitionally expanded bridge laws to T^* if they are not theorems already (1970, p. 444–445).

philosophy and history of geometry, there is much to do. For Torretti has written a lot about reduction, and related topics like scientific realism, the syntactic vs. semantic conceptions of theories, and structuralism (in its many senses): not just in geometry, but across all of physics.

Besides, we already noted, from Sect. 7.2 through to 7.5, several important topics we have had to postpone. For example, as regards reduction: defending the syntactic approach to theories; and as regards functionalism: comparing Lewis' and Carnap's treatment of theoretical terms, and assessing Newman's objection to the Ramsey-sentence approach, and to "structural realism". So we owe a discussion of these topics, especially in relation to Torretti's writings.

But for reasons of space, we must postpone these projects. So here, as an appetizer for them and as a courtesy to Torretti, we will only discharge three small obligations to him and his work. First, we will say a little about the obvious project beckoning us: to compare functionalist reduction with Torretti's discussions of reduction (Sect. 7.6.1). Then (Sects. 7.6.2 and 7.6.3) we will exhibit two closely-related links between us and Torretti: both concern the philosophy and history of the axiomatic method.

7.6.1 Reduction: A Peace-Pipe

We admit at the outset a major difference between Torretti's and our discussions of reduction. Like the majority of the literature, he uses a binary contrast between theory and observation, whereas we have advocated a ternary contrast. But when one looks at the details of his discussions, there is a good deal of convergence between him and us.

At first, this may be surprising. For it is clear from the previous Sections that our intellectual outlook is close to figures such as Nagel and Lewis. These figures are more empiricist, more Humean, more scientific realist, and less holist about semantics than, we surmise, Torretti would like. (At least, so it seems to us, since after all, Torretti is also a profound Kant scholar.) But going beyond vague 'isms' and philosophical slogans: we find we agree with several of Torretti's main claims—and we surmise that figures like Nagel and Lewis would too.

Let us briefly smoke this peace-pipe. As tobacco acceptable to both parties, we make the obvious choice: the rejection of a neutral observation language, and the issues to which it leads. We will begin with Torretti's writings, and then concur with him.

Torretti denies that from direct experience, unleavened by conceptualization—that 'blooming, buzzing confusion', as William James put it—one can define worthwhile scientific concepts, let alone derive scientific knowledge. Expressed more positively: Torretti's view is that all observation subsumes its object under a general concept. This view is developed in several places in Torretti's *oeuvre*. For example, one finds it in his essay *Observation* (1986: p. 4, Section 5, and p. 21; labelled as the 'principle of conceptual grasp') . It is announced as a leading theme at the start of *Creative Understanding* (1990: pp. ix, 1, 5–7) whose Chapter 1 is a reworking of 1986, and whose Chapter 2 gives an extended critique of logical empiricists' distinction between theory and observation (especially Section 2.4.3 about Carnap). It also appears in Torretti's *Philosophy of Physics* (1999: pp. 402–404, 421), and of course also in his Kantian scholarship (e.g. 2008, pp. 81–82).

In all these works, Torretti's development of this view is clear, gracious and often witty: as always with his writing. We cannot resist quoting a passage where Torretti goes one better than the famous metaphor invented by a kindred spirit: viz. Quine's web of belief.

> ... if all our knowledge of physical objects is corrigible, it must be self-correcting, for there is no outside authority to which one could turn for help. Quine's famous dictum that "our statements about the external world face the tribunal of sense experience not individually, but only as a corporate body"... is apt to be misleading. For in the trial of empirical knowledge the defendants are at once the prosecution, the witnesses, and the jury, who must find the guilty among themselves with no more evidence than they can all jointly put together. (1990, p. 7; cf. 1986, p. 9)

Of course, this view faces challenges, the obvious one being: without raw experience to adjudicate scientific claims, how can we secure the objectivity of scientific knowledge? In particular, how can we refute the threat of incommensurability, urged by Kuhn and Feyerabend? Torretti answers these challenges, in various places within *Creative Understanding* and *Philosophy of Physics* (1990: pp. x-xi, 44; and 79–81, referring back to C.1-C.3 of pp. 32–33; 1999: 421–430). Broadly speaking, he gives various reasons for objectivity, even continuity, in scientific change, despite our lack of a theory-neutral observation language; among them the many ways that even our most arcane theoretical knowledge is rooted in our everyday world. We cannot resist quoting another passage where, again, he goes one better than the famous metaphor invented by a kindred spirit: in this case, Neurath's ship.

> In Neurath's ship, the neat steel turrets of theory are built on and bridged by the wooden planks of common sense, which may be worn and musty but are indispensible to keep afloat the enterprise of knowledge. Physicists who advocate different theories do not 'practice their trades in different worlds', for there is but one world for them to wake up to, namely, the world they are in together with the persons they love and the goods they yearn for ... (1999, p. 404; cf. also 421f.)

With all this, we concur; (and we daresay Lewis, and to a large extent Nagel, would concur). Recall our previous scepticism about reductions of empirical knowledge to experience (cf. Sect. 7.2.1.1), and more specifically, our joining Lewis in taking the O-terms to be, not 'observational', but 'old' or 'okay' i.e. already-understood (cf. Sect. 7.4).

We also concur with much else in Torretti's discussions. Let us report two examples, from many that could be given. We choose them because they connect to previous Sections' discussions:

1. Torretti's rejection of 'reference without sense', i.e. his rejection of Putnam's proposal in the mid-1970s to answer the challenge of incommensurability by

adopting a causal theory of reference in which Fregean senses and-or Carnapian intensions have no place (1990: pp. 51–70; echoed at 1999, p. 422);

2. Torretti's critique of the treatment by the Sneed-Stegmüller school of structuralist analysis of scientific theories—to which, overall, he is sympathetic (1990, p. xi, 109f.; 1999, p. 412–417, 424)—of its "problem of theoretical terms"; this critique is in (1990: pp. 129–130, 134–137: reviewed pp. 161–162) and endorsed in (1999: p. 414, note 18).

As to (1), our agreement with Torretti is obvious. We recall that we joined Lewis in *causal descriptivism* (Sect. 7.4.2); and that this account of reference has since been taken up in detailed defence of scientific realism (Psillos, 1999, 2012).

As to (2), we shall give a bit more detail, since the proposals being criticised are less well known. The first thing to note (as Torretti does at the start of his discussion: 1990, p. 114) is that this "problem of theoretical terms" is *not* what goes by that name for the logical empiricists—and indeed, for the rest of us: viz. the semantic and epistemological questions, how theoretical terms get their meanings, and how we can be warranted in applying them. Instead, it is (roughly!) that for some terms in a theory, every way of establishing that they apply in a real situation presupposes that the theory 'has an actual model', i.e. applies in that very situation or some other one. Sneed calls such terms 'theoretical'. More precisely, since Sneed et al. adopt the semantic conception of theories: it is—not predicates, but—relations-in-extension in models, especially functions representing physical quantities, that are theoretical for Sneed.

Sneed's problem about these "terms" is that although one might confirm the presupposition by finding an actual model (a successful real application of the theory), doing so will involve applying the term (measuring the function)—and so again presuppose some actual model. Thus a regress, or circularity, looms—prompting the question: 'what empirical claim is being made by someone who holds such a theory?' (1990: p. 115).

In both Sneed and the ensuing literature, classical mechanics (in a traditional point-particle formulation) is a frequent source of examples. Thus an apparent example of this problem is the oft-noted interdependence of Newton's laws and the concept of inertial frame: namely, that the laws are to be true of motions only if they are described in inertial coordinates, while inertial frames get defined as those in which Newton's laws hold—threatening a circle.

Sneed's own answer to this question is that the empirical claim must involve only non-theoretical terms. This answer gets articulated in terms of models that assign extensions only to such terms, the theoretical terms being simply ignored: they are called 'partial potential models' (1990: p. 116, 121).

Torretti describes this answer in detail but—unsurprisingly, given his rejection of non-conceptual observation—rejects it as rooted in 'foundationist scruples'. Indeed, he rejects the problem as a 'pseudoproblem, stemming from a refusal to countenance a genuinely creative understanding of natural phenomena' (ibid. p. xi, p. 134). Similarly, he describes in detail, but also rejects, Gähde's alternative proposal for how to define theoretical terms (pp. 121–128). And he diagnoses the same defect, i.e. a foundationist belief in non-conceptual observation, in Ludwig's account of the interpretation of physical theories (pp. 131–137). For both authors, the illustrations are again from classical mechanics, especially the interdependence of Newton's laws and inertial frames, noted above (pp. 117, 122–123, 136–137).

So far as we can judge, Torretti's critique is right. But we will not pursue details. We are just pleased to note a possible project for anyone sympathetic to the Sneed-Stegmüller school of structuralism. For Sneed's problem of a regress, or circle, of presupposed successful real applications of a theory reminds us of the threat of logical circles of definition, which the functionalist—specifically Lewis—shows can be overcome. As we saw, there is nothing incoherent, or even suspicious, about the idea of simultaneous unique definition (cf. Sect. 7.4.3). So maybe the same idea can help with Sneed's problem.[42]

But that is work for another day. Here, we must curtail our comparison of our and Torretti's general views about reduction and related topics. In the rest of this Section, we will turn to two more specific (and related) links between us and him. Both concern the philosophy and history of the axiomatic method, and what we have called the problem, or objection, of Faithlessness. The first link (Sect. 7.6.2) is general, and relates mostly to the functionalist idea of simultaneous unique definition. It leads to the second link (Sect. 7.6.3): which is specific, and relates mostly to reduction.

7.6.2 Comparing Functionalist Definition with Implicit Definition

In this Section, we will compare the functionalist idea of simultaneous unique definition (or specification) with the idea that a mathematical axiom system defines its terms, or some of them—in some sense of 'define'. This idea is often called *implicit definition*. But as Torretti and many authors point out, the word 'definition' is misleading, since it suggests both uniqueness and Faithfulness to a given meaning—both features that in general fail in this context.

This idea was at the centre of discussions among mathematicians in the second half of the nineteenth century: especially about geometry, as mathematicians

[42] Certainly, it seems to fit the example of the interdependence of Newton's laws and inertial frames. That is: we propose that one can reasonably take classical mechanics (in a point-particle formulation of the sort Sneed et al. consider) as introducing simultaneously terms for time and length (and so: frame), and for mass: terms that are held by the theory to be uniquely realized in such a way as to make the theory true—including Newton's laws, using these terms, being true, i.e. true for motions described in inertial coordinates.

Presumably, the starting point for this project would be Chapter III of Sneed (1979) and Chapter II.3 of Balzer et al. (1987). But the gap in the literature is wide. Sad to say: so far as we know, neither of the two sides—Lewis and other Canberra Planners, and the Sneed-Stegmüller school—refers to the other.

responded to the rise of non-Euclidean geometries. So—fortunately for us—it is discussed in detail by Torretti in his magisterial *Philosophy of Geometry from Riemann to Poincaré* (1978). This book is a treasure-trove, and we have learnt a lot from it: not just about this comparison, but about much else—especially the Helmholtz-Lie *Raumproblem* which will be our second paper's first example of spacetime functionalism. But in this Section, we must stick to the comparison. (Sect. 7.6.3 will focus on geometry and look ahead to our second paper.)

More specifically, the idea was at the centre of the famous controversy between Frege and Hilbert, that was triggered by Frege's response to Hilbert's *Grundlagen der Geometrie* (1899). Thus their central dispute was 'over the nature of axioms, i.e., over the question whether axioms are determinately true claims about a fixed subject-matter or re-interpretable sentences expressing multiply-instantiable conditions', as Blanchette's fine review puts it (2018, Section 3). Frege gave the first answer, Hilbert the second: and Torretti sides firmly with Hilbert.[43] As Hilbert put it to Frege:

> Every theory is naturally only a scaffolding or schema of concepts, together with their necessary mutual relations, and the basic elements (*Grundelemente*) can be conceived in any way you wish. If I conceive my points as any system of things, e.g. the system *love, law, chimney-sweep*, ... and I just assume all my axioms as relations between these things, my theorems, e.g. the theorem of Pythagoras, will also hold of these things. In other words, every theory can always be applied to infinitely many systems of basic elements. It suffices to apply an invertible univocal transformation [i.e., a bijection] and to stipulate that the axioms hold correspondingly for the transformed things. [...] This property is never a shortcoming of a theory and is, in any case, inevitable. (Torretti, 1978, p. 251; also endorsed at Torretti, 1999, p. 409)[44]

By 1900, Hilbert's view of axioms (often called a 'formal' or 'structural' view) was already dominant. Not only did Hilbert's book and other writings deploy it to prove relative consistency and independence results, in unprecedented detail. And not only was this work very influential, in the ensuing years, in moulding the modern conception of model theory and formal semantics. Also, as Torretti's discussions bring out: already in 1900, several contemporaries shared Hilbert's 'formal' or 'structural' view of axioms. For example, Torretti cites Padoa (a member of Peano's school) saying in 1900 that the undefined symbols of a deductive theory are 'entirely devoid of meaning' and that axioms 'far from stating *facts*, i.e. *relations* between the

[43] One might similarly summarise the dispute as about whether it is defect of an axiom system that it can be realized, i.e. made true, by very diverse interpretations ('models' as we nowadays say in logic and model theory)— with Frege saying 'Yes' and Hilbert saying 'No'. (The dispute is often summarised as about whether axioms *implicitly define* their terms (with Frege saying 'No' and Hilbert saying 'Yes'); but as we mentioned, 'implicit definition' is a misleading term.) We recommend Potter's discussions (2000: 87–94; 2020, 124–132).

[44] In fact, Hilbert already took this view, some ten years before: witness the similar remark in 1891, made while waiting for a train in a station waiting-room: 'one must be able to say at all times—instead of points, straight lines and planes—tables, chairs and beer mugs': cf. Kennedy (1972, p. 133).

ideas represented by the undefined symbols, are nothing but *conditions* with which the undefined symbols must comply' (1978, p. 226).[45]

Against this background, we have two main points to make. The first (Sect. 7.6.2.1) will be that some nineteenth-century discussions of axiom systems contain striking precursors of simultaneous definition. Since we are by no means historians, we will here be wholly indebted to Torretti's scholarship. Our second point (Sect. 7.6.2.2) is about how one should understand *logical consequence*. Our view of this will imply that, although Hilbert undoubtedly "beat" Frege as regards historical influence, there was more right on Frege's side than he is usually given credit for. Besides: more right than Torretti gives him credit for. So here, we regret to say, we have a philosophical disagreement with Torretti. At first sight, this may seem bad manners in a *Festschrift*. But we trust that with his liberal and gentlemanly outlook—and his relish for philosophical debate—Roberto will forgive us!

7.6.2.1 Precursors of Functionalist Definition

Both our points, and our historical debt to Torretti, are best introduced by the closing passage (pp. 252–253) of Section 3.2.10 of Torretti (1978), which is entitled 'Axioms and definitions. Frege's criticism of Hilbert' (Section 3.2 is entitled 'Axiomatics'). For the passage itself will serve to establish our first point, about historical precursors of simultaneous definition. And Torretti's views, especially at the end of the passage, will be the springboard for our second point.

First, Torretti criticises the phrase 'implicit definition', on the grounds that it is in general impossible to extract from an axiom system explicit definitions of its primitive terms. He also notes that this is disanalogous from the situation often cited

[45] Below, we will mention another example, Pieri. We also note three historical points. (1): Torretti (1999, p. 408–414) portrays this view as prompting the semantic or structural (as against syntactic) view of theories that, as we saw in Section 6.1, he favours. (2): Gray (2008, Section 4.1, p. 176f.) makes an interesting case that this broad development represented a rise of modernism, in a sense analogous to that in art and literature. (3): The Peano school's endorsement of the Hilbertian structural view of axiom systems also surfaced in the philosophy of arithmetic: where it leads us back to Russell's quip about 'theft over honest toil'. Thus recall from Sect. 7.2.2.1's discussion of Faithlessness, especially footnote 17, that Frege "already played the role of Benacerraf". That is: he objected to his own definition of natural number; and then replied that the Faithlessness (specifically: over-shooting) was a price worth paying for the benefit of an otherwise successful definition. *A propos* of this, Alex Oliver points out to us (p.c.) that Peano also made this objection in 1901; and the reply that the price is right came from Russell, in his *Principles of Mathematics*. Writing with his usual verve (but lack of argument!), he in effect just outfaced Peano: 'To regard a number as a class of classes must appear, at first sight, a wholly indefensible paradox. Thus Peano remarks that "… these objects have different properties". He does not tell us what these properties are, and for my part I am unable to discover them' (1903/2010: Section 111, p.115). Similarly in later writings: Russell had the 'structuralism' of Peano, Dedekind and Hilbert in mind when he condemned the 'method of postulating' as 'theft': cf. Sect. 7.2.2.3, especially footnote 19, and Linsky (2019, Section 3).

as an analogue, viz. extracting the n roots of n simultaneous linear equations.[46] On this second point, Torretti also cites in his support, Frege, who writes as follows; (it will be clearer in Sect. 7.6.2.2 why he does so):

> "If we survey the whole of Mr. Hilbert's definitions and axioms, they will be seen to be comparable to a system of equations with many unknowns; for in each axiom you normally find several of the unknown expressions 'point', 'line', 'plane', 'lie on', 'between', etc., so that only the whole, not particular axioms or groups of axioms, can suffice to determine the unknowns. But does the whole suffice? Who can say that this system is solvable for the unknowns, and that these are unambiguously determined?" (note 148, p. 402).

Then (p. 252), Torretti briefly mentions Pieri (who, like Padoa, was a member of Peano's school). Torretti's previous passages about him (pp. 224–226, 250–251) describe how his papers of 1899 and 1900:

(a) urged the implicit/explicit contrast under the (respective) labels 'real'/'nominal';
(b) were in the mathematical vanguard, in that Pieri wished to liberate geometry from any appeal to spatial intuition, and also sided (implicitly!) with Hilbert's view of axioms, against Frege's; although also Pieri:
(c) preferred to reserve the word 'definition' for explicit/nominal definitions—a practice that logicians nowadays generally follow, and that Torretti endorses.[47]

Then Torretti reports that the phrase 'implicit definition', and the analogy with simultaneous equations (and so for us functionalists: the idea of simultaneous definition), seems to have been first used eighty years earlier, by the geometer Gergonne in his "Essai sur la théorie des définitions" (1818). (Quine (1964, p. 71) also cites Gergonne.) But Torretti ends by emphasising the critique of the idea of implicit definition with which he began on p. 252. Thus he writes

> Gergonne observes that a single sentence which contains an unknown word may suffice to teach us its meaning. Thus, if you know the words *triangle* and *quadrilateral* you will learn the meaning of *diagonal* if you are told that"a quadrilateral has two diagonals each of which divides it into two triangles". [Now Torretti quotes Gergonne ...]
> "Such phrases, which provide an understanding of one of the words which occurs in them by means of the known meaning of the others, might be called *implicit* definitions, in contrast with the ordinary definitions, which we would call *explicit*. There is evidently between the latter and the former the same difference as between solved and unsolved equations. One sees also that, just as two equations with two unknowns simultaneously determine both,

[46] In Sect. 7.4.2 we agreed that this is in general impossible, but argued that this is no objection to Lewisian functionalism.

[47] Two further points, which will be centre-stage in our second paper. (1): Beware: 'Implicit definition', as used *nowadays* by logicians, means something else. The idea is due to Padoa (Torretti, 1978, pp. 226–227). It is essentially equivalent to metaphysicians' notion of supervenience/determination (cf. footnote 20). But it is utterly precise, and logically weaker than explicit definition (in general—though equivalent to it, for first-order languages, by Beth's theorem: cf. Boolos and Jeffery, 1980, p. 245f., Hodges, 1997, p. 149). Thus proving explicit undefinability by showing implicit undefinability is called *Padoa's method* (Hodges, 1997, p. 58). (2): For us, Pieri's own axiomatisation of geometry has the further interest that, following the tradition of the Helmholtz-Lie *Raumproblem*, it is based on the idea of rigid motions.

two sentences which contain two new words, combined with other known words, can often determine their sense. The same can be said of a greater number of words combined with known words in a like number of sentences; but, in this case, one must perform a sort of elimination which becomes more difficult as the number of words in question increases." (1818, p. 23; Gergonne's italics).

[Then Torretti concludes ...] Gergonne has grasped well a familiar linguistic phenomenon and has given it an appropriate name. But his systems of simultaneous implicit definitions are something evidently very different from abstract axiom systems. In these, all designators and predicators behave, if you wish, as unknowns, and no process of elimination can lead to fix their meanings, one by one. We ought not to burden Gergonne with the paternity of the rather unfortunate description of axioms as implicit definitions. (1978, p. 253)

The first thing to say about this passage is that the quotation from Gergonne is indeed a striking precursor of functionalists' idea of simultaneous definition. It is striking for its contrast between known and unknown words (cf. Lewis' O-terms and T-terms), for its idea of simultaneous definition—and for its early date. Thus ends our first point.

7.6.2.2 Logical Consequence Is not Formal: Faithlessness and Frege

As we announced, our second point is not historical, but philosophical; and marks a disagreement with Torretti. We think it worth displaying because it has a deep origin: in different ways of thinking about *logical consequence*. In short: we say that logical consequence is *not* a formal notion. It is *not* fully captured by the textbook definition of subset-hood among sets of models. But Torretti thinks it is thus captured.

Thus in Section 3.2.2 of his (1978), entitled 'Why are axiomatic theories naturally abstract?', he lays out a formal semantics of a fragment of English sufficient to state mathematical propositions (called 'm-English': pp. 192–195). It is essentially the familiar logic-textbook semantics of (maybe higher-order) predicate languages; with the usual allowance of any set of objects as the domain D of quantification, and any assignment of semantic values therein to the non-logical vocabulary (i.e. elements of D to individual constants, subsets of D to monadic predicates, sets of ordered pairs from D to dyadic predicates etc.). Then he defines logical consequence in the usual way as subset-hood among sets of models.

A sentence S is a logical consequence of a set of sentences K if, and only if, every interpretation of $K \cup \{S\}$ which satisfies K also satisfies $\{S\}$. We use the abbreviation $K \models S$ ... This relation does not depend on a particular interpretation of K and S. Indeed, we can replace all interpretable words in K and S by meaningless letters ... and it will still make sense to say that $K \models S$. The mathematician who studies an axiomatic theory need not worry about the referents of its sentences ... The important thing is that, for every conceivable interpretation of the theory, if the axioms are true, then the theorems are also true. (p. 195)

In reply to this, we propose a return to how we all think about logical consequence, before we study the modern textbook in logic class: a return, we daresay, to how everyone thought about it before the twentieth-century advent of formal semantics, and Hilbert's "beating" Frege in the way we reported in this Section's preamble. This means: logical consequence as a relation between a sentence S and another S' (or: a set of interpreted sentences K) that is *Faithful* to how they are understood (intended) in the language concerned. The relation is: if S' (understood as intended) is or were true (or: all the sentences of K are or were true), then S (understood as intended) *must* be true.

Our point is, of course, not about the *word* 'logical consequence'. We are happy to give that to Torretti and the logic books; and to say instead 'entailment', or 'implication'. Our point is that this relation, entailment, does not involve *dis-interpreting* any words. In particular, it does not try to capture, or theorise about, the compulsoriness of the inference from S' to S—what Wittgensteinians used to call the "force of the logical must"—by dividing the language's vocabulary into two sets: non-logical vocabulary that is then to be interpreted (given semantic values) in an arbitrary way, irrespective of what the words in fact mean, and logical vocabulary whose interpretation is fixed.

Obviously, our point is not original. Countless discussions in philosophical logic, for example about the differences between natural and artificial languages, emphasise that when one is faced with the countless entailments (equivalently: valid arguments) in a natural language, one naturally tries to find patterns, every instance of which is an entailment (a valid argument). And one soon finds many such patterns that are:

(i) a matter of where in the entailment (the argument) a small number of words—such as 'and', 'not', 'all', 'some'—occur, and yet:
(ii) wholly independent of what other words ('red', 'tall', ...) occur.

Thus is formal logic born.

Besides, it is natural to make some sort of toy-model semantics: toy-models of how the world could have been different, to represent the phrase above 'or were true'. And once we focus on the patterns just mentioned, it is natural to represent the entailment's (the valid argument's) independence of the other words in (ii), by:

(a) letting the toy-models assign to the words in (ii) arbitrary semantic values; and then
(b) checking that whatever assignment a toy-model makes, S comes out true in the toy-model provided that S' does.

Thus are formal semantics and model theory born, with their textbook definition of logical consequence.

The point now is: these toy-models, these arbitrary assignments, can only be expected to be appropriate for those entailments (valid arguments) that *are* instances of such patterns: i.e. for those that turn solely on the placing of the few words selected in (i). But countless entailments (valid arguments) are *not* such instances. For there are countless 'meaning connections' ('analytic connections') between our words which cannot be plausibly claimed to be due to a discernible 'logical form', lurking just below 'surface grammar'. Agreed, it is plausible that 'all bachelors are unmarried' is necessary, simply because 'bachelor' means 'unmarried male', so that it instantiates the textbook's logically valid formula $(\forall x)((Ux \wedge Mx) \supset Ux)$. But such cases are exceptional. How, for instance, should we 'analyse away' the connection between the meanings of a specific colour predicate, and of '... is coloured'. Thus consider: S' := 'this pencil is blue'; S := 'this pencil is coloured'. S' entails S. But predicate logic will render these sentences along the lines, with 'this pencil' as an individual constant p: $B(p)$ and $C(p)$; and the latter is by no means a logical consequence, in Torretti's sense, of the former: $B(p) \not\models C(p)$. (We take this example from Coffa (1983), whom we discuss shortly.)

As indeed countless discussions say: this point has a long and varied history. Various authors have espoused various different conceptions of logical form, or 'deep structure', into which they hoped to analyse or 'regiment' natural language, so that all inferences are rendered formal—by the lights of their preferred notion of logical form. But there is no consensus about logical form, nor even agreement about the desiderata for it.[48] We do not need details of this variety, let alone an assessment of the various authors' proposals. All we need is to emphasise the gap—wide and poorly mapped—between entailment in natural language and these formal calculi. Let us briefly do so with two historically influential examples: Wittgenstein's troubles with elementary propositions in 6.3751 of the *Tractatus*, and Davidson's analysis of adverbs, and advocacy of events as genuine objects (particulars).

In short: Wittgenstein hoped to completely analyse all necessity as the unproblematic combinatorial necessity of truth-tabular tautologies, and so maintained that the elementary propositions in which analysis would end must be logically independent of each other (i.e. all truth combinations are genuinely possible). So he could never give an example of such a proposition, but only told us that, thanks to the mutual exclusion of 'red and 'green', propositions about colour in a patch of space or one's visual field were not elementary.[49]

And again, in short: Davidson noticed (1967) that adverbs and adverbial phrases modifying descriptions of action give entailments (valid arguments), such as in the sequence: 'John quickly buttered the toast with the knife'; so: 'John quickly buttered the toast'; so: 'John quickly buttered'. With ordinary objects, like John, the toast and the knife, as the referents of the singular terms, these entailments far outstrip the power of predicate logic to make them valid by a formal rule. For with these referents, predicate logic will render these sentences along the lines: 'ButterQuick(John,the toast,the knife)'; and 'ButterQuick(John,the toast)'; and 'ButterQuick(John)'. But despite the typography, these predicates are distinct,

[48] Indeed, the alleged contrast between logical form and 'surface grammar' is itself very questionable: cf. Oliver (1999).

[49] So this causes trouble for combining the doctrines of the *Tractatus* with the sort of phenomenalism Carnap espouses in the *Aufbau*. It is also, of course, an Achilles heel of the *Tractatus*: it prompted Wittgenstein to return to philosophy, and write his 'Remarks on logical form' in 1929; cf. e.g. Potter (2020, pp. 337–338, 409–410).

indeed of different polyadicities; and predicate logic recognizes no compulsory 'meaning connections' between them.[50]

These two examples are enough—more than enough for this paper!—to show that there is much more to entailment in natural language than one sees in formal logic; and that it is very hard to get a satisfactory theory of that 'much more'. And in that endeavour, we tend to endorse traditional ideas of analyticity and synonymy: we side with Carnap, against Quine (cf. Stein, 1992, 275–278, 282–285). So here ends our excursus in to philosophical logic: we can now return to Torretti.

As we have said, Torretti sides with Hilbert against Frege. He does not mince words: 'To demand like Frege that [...] shows a lack of understanding of logical consequence that is indeed astonishing in the founder of modern logic. ... Frege's obtuseness is truly baffling' (1978, p. 251: endorsed at his 1999, p. 408f). But as we said in this Section's preamble; our view is that there was more right on Frege's side than this allows.

And we have been pleased to find that long ago, Coffa gave a similar *apologia* in his (otherwise laudatory) review (1983) of Torretti (1978). We shall quote Coffa liberally, since he makes the points clearly and wittily. And more important, his discussion lays the ground for our return to geometry in Sect. 7.6.3. (Cf. also the discussion in Coffa's posthumous book (1991: 48–57, 381–381).)

Coffa considers the work of Beltrami and Klein (principally in 1868 and 1871, respectively) on the independence of Euclid's fifth, parallels postulate, from the rest of Euclid's axioms: i.e. on providing a model of non-Euclidean geometry—principally what is now called the 'Beltrami-Klein' model (Torretti, 1978, pp. 125–137). Coffa starts with an incontrovertible point of philosophical logic about such theorems. Then he uses this point to make his historico-critical remark—favouring Frege against Hilbert, and Carnap against Quine: thereby lodging his, and our, criticism—courteous and minor!—of Torretti.

The point of logic starts from the fact that an independence theorem, saying that an axiom A is independent of the other axioms, X say, must countenance non-isomorphic models: one (or some) making true both A and all of X, and another (or some others) making true not-A and X. So in accepting such a theorem, we cannot take the meaning of the terms in the sentences of X to be so constraining, so logically rich, that an interpretation making all the sentences of X true—a model of X—must be unique (up to isomorphism, of course). Accepting the theorem as showing that A is not provable from X requires us to be liberal, undemanding, about meanings. We must admit: 'the models making true not-A and X are legitimate: in particular, no worse as models of X than the models making true A and X'.

[50] Of course, Davidson, like Quine, advocates predicate logic as the preferred language for logical forms or 'regimentations'. So he sees this situation as an argument for an ontology of events, in this example a human action. For then one has regimentations like, for our first two sentences: $(\exists x)((Butter(x) \wedge By(x, John) \wedge Quick(x) \wedge Of(x, toast) \wedge With(x, knife))$, and $(\exists x)((Butter(x) \wedge By(x, John) \wedge Quick(x) \wedge Of(x, toast))$. So our first inference becomes formally valid in predicate logic, i.e. an instance of dropping a conjunct in the scope of an existential quantifier.

Of course, this point underlies the rise, in the years 1870 to 1900, of the Hilbertian 'structural' view of axioms. And as we said at the start of Sect. 7.6.2: thereafter this view became well-nigh universal. For example, consider this quotation from Weyl in 1934: 'a science can never determine its subject-matter except up to an isomorphic representation' (1934, p. 95–96). (Of course, nowadays quotations like this tend to remind philosophers of Putnam's model-theoretic argument.)

But, says Coffa, it is also legitmate to *reject* such an independence theorem. An objector—call him 'Gottlob'!—can object that to treat both sorts of model on a par, as equally legitimate, is to be *Faithless* to the intended meanings of one or more terms. Which terms, and in what way faithless, will of course vary from case to case: but Gottlob's strongest objections of Faithlessness will probably be directed at the interpretations given to some terms in the axiom A, by models making true not-A and X.

Coffa's point here is not to urge that the objector is *right*; but only to urge that he is not *silly*. He agrees that it is nowadays well-nigh universal practice to understand the enterprise of interpreting an axiomatic theory in the sense 'famously advanced by Hilbert in his *Grundlagen* (in effect, the doctrine that axiomatics ignores the meanings of all non-logical signs)' (1983, p. 687). But he goes on:

> ...[we should not ...] conflate axiomatics with formalism; the former aims to identify within a class of claims C a manageable subclass that contains all of the information conveyed by C, the latter invites us to ignore the meanings of some words and to infer only in conformity with the remaining content.
> The point of this distinction may be clarified by recalling a doctrine of the celebrated Prof. Schmall, who argues that someone who accepts the claim
>
> (*) this pencil is blue
>
> is still free to choose whether to assert or deny the claim
>
> (**) this pencil is colored,
>
> for, he says, (**) is no part (does not follow from, is not a consequence of) what (*) says. His reasoning is this: first we substitute 'long' for 'blue' in (*) and 'short' for 'colored' in (**); then we notice that the pencil in question is, in fact, long. QED.
>
> Prof. Schmall's routine response to the routinely astonished faces of his listeners is that his reasoning in no way differs from Klein's widely acclaimed proof of the independence of Euclid's parallel postulate, so that his own claim stands or falls with Klein's.
> On the face of it, Prof. Schmall seems to have a point. The centuries-old question concerning the parallel postulate was not, one may safely assume, whether once we ignore in Euclid's axioms the meanings of 'point, 'line' and all other geometric words, the logical skeleton of information surviving that semantic slaughter still contains one redundant claim—the surviving content of the parallel postulate. The question was, of course, whether what the parallel postulate says (about points, lines, etc.) is part of the information conveyed by the remaining axioms.
> The standard move at this point is to draw a distinction between formal and material consequence. What Klein and Schmall deal with, we are told, is *formal* consequence; the other sort is not their concern. Perhaps the reason why Schmall's audience gets angry is because *they* are thinking of material consequence. End of solution.
> The formal-material distinction is what one might call an anaesthetic distinction. There is nothing technically wrong with it, but its normal tendency is to desensitize its users to philosophical problems. Material consequence is a fancy name for plain old consequence; B is a material consequence of A iff what B says is part of what A says. Formal consequence is

> what Klein and Schmall (should) have in mind. Far from solving the problem, the distinction only allows us to pose it in new terms: how come that a formal answer is being offered to what is clearly a material question?
> In Schmall's case the answer is easy: he is silly. Was Klein silly? Was the entire geometric community silly? Frege, of course, thought so: Profs. Schmall and Hilbert were, in his opinion, guilty of precisely the same blunder. And one is bound to miss the force of his point if one jumps into Hilbert's arms as quickly as, say, Quine and Torretti. (p. 687–688).

Thus Coffa argues that the mathematical community's well-nigh universal acceptance of Hilbert's viewpoint was a *decision*, not a rational necessity: albeit one made with good reasons. He goes on:

> As I see it, the formalism that Hilbert eventually forced into the field of logic was not a discovery of the essence of axiomatics but an invention, a clever expedient inspired by a number of circumstances, prominent among them, the developments in nineteenth century geometry, which were hard to accommodate except within a formalist framework. In the course of that development it became clear that nothing was clear about geometric primitives except what the appropriate axioms implicitly stated about them, so that any extra-axiomatic considerations involving the meanings of these terms came to be seen as not only irrelevant but positively obtrusive. That Klein's proof was recognized as a proof of independence by the geometric community ... is no evidence of the fact that geometers had seen the essence of axiomatics but, rather, of the fact that they had tacitly reached a decision that Hilbert would make explicit three decades later: that as far as geometry is concerned, the meanings of geometric terms really does not matter, except to the extent that it is determined by the logical words in the axioms involved. The overall story of the episode leading to this decision, and its philosophical echoes, is told by Torretti much better than by anyone else I know; but I wish he had not put formalism at the beginning of his account, as the truth underlying the whole process, but at the end, as the result of a decision called for by the contingent course of geometric history. (p. 688–689)

To conclude: although we admit to being *ingenus* about the history of geometry, we concur with Coffa's distinction between axiomatics and formalism, and his suggestion that Frege's position was legitimate.

7.6.3 Beltrami's Model as an Example of Reduction—And an Analogy

We turn to another link between Torretti's work and our views in this paper. Namely, between Beltrami's Euclidean model for hyperbolic geometry (in 1868) as described by Torretti (1978), and our account of reduction, and the possible objection against reductions that we labelled *Faithlessness*.

So this will be a specific historical example of the issues about definition, and the axiomatic method, raised in Sect. 7.6.2.2. Indeed, we want to develop a striking analogy mentioned by Coffa in his review of Torretti (1978); cf. also Coffa (1991, pp. 48–57). It is an analogy between:

(i) Beltrami's conception of his project in 1868; and
(ii) Frege's and Russell's logicism, i.e. their effort to reduce arithmetic to logic (in modern terms: to set-theory); which we discussed in Sect. 7.2.1.1.

It is this analogy that will give us the vivid illustration of the *Faithlessness* objection.[51]

Thus Coffa writes (1983, p. 684):

> "An Essay of Interpretation of non-Euclidean Geometry"' [i.e. (Beltrami 1868)] contains what is now regarded as the first effort to produce a model of hyperbolic geometry. Although it is sometimes said that Beltrami's discovery was a death-blow to Kantianism, it emerges from Torretti's account that there is nothing in the"Essay" that could have caused anything but glee to a Kantian.
>
> To begin with, Beltrami believes that the only way to legitimize a geometric doctrine is to show that it can somehow be reduced to Euclidean elements. Indeed, *reduction* rather than *interpretation* is the decisive notion in Beltrami's work; for what he does to hyperbolic geometry is precisely what Frege and Russell aimed to do to arithmetic a few years later: to reduce a certain obscure and dubious doctrine to another one, which we understand and find well grounded.
>
> The main conclusion of Beltrami's"Essay" was that 2-dimensional hyperbolic geometry is no more than a fragment of Euclidean geometry in disguise; for it is, in fact, the geometry of a perfectly Euclidean surface of constant negative curvature named by Beltrami the 'pseudosphere'. Moreover, since 3-dimensional hyperbolic geometry is, as far as Beltrami can tell, not reducible to Euclidean geometry, he concluded that no"real substratum" underlies it. Therefore, even though, as Kant had implied, 3-dimensional hyperbolic geometry can be developed analytically (i.e., purely conceptually) without inconsistency, no real-intuitional geometric substratum for it could be offered. Beltrami's tacit premise, one may safely assume, is the Kantian doctrine that our intuitions necessarily conform to Euclidean laws.

As we emphasised: we are not historians. But we concur with Coffa's description of the case: which, as he says, matches Torretti's. And we of course endorse the end of Coffa's second paragraph, which mirrors our Sect. 7.2.1's discussion of reducing the problematic to the unproblematic. So we will take it that Beltrami sees himself as showing a reduction of hyperbolic geometry to Euclidean geometry, by giving as a model of two-dimensional hyperbolic geometry a 'perfectly Euclidean surface of constant negative curvature': which he calls a 'pseudosphere'. And here, 'reduction to Euclidean geometry' connotes 'legitimation from the viewpoint of Euclidean geometry'.

So according to this analogy:

(a) the advocate of hyperbolic geometry is like an advocate of arithmetic; and
(b) the advocate of the reduction to Euclidean geometry—-according to Torretti and Coffa: Beltrami himself, the displayer of the Euclidean model—is like the set-theorist reducer of arithmetic.

[51] Cf. Sects. 7.2.2.1 and 7.3.3. But our rationale for expounding this example is not just to illustrate this paper's themes. It also sets up another analogy, that we explore in our second paper: between Beltrami's construal of hyperbolic geometry, and how a modern relativist who advocates a conformally flat spacetime would construe that paper's second example of spacetime functionalism: viz. Robb's axiom system for causal connectability. There is also another analogy hereabouts, which is both recent and mathematically deep. Beltrami's embedding of 2-dimensional hyperbolic geometry in Euclidean space is a 'baby version' of the famous Nash embedding theorem, that any compact Riemannian manifold can be isometrically embedded in Euclidean space; cf. e.g. Tao (2016).

Let us compare the two cases, using our earlier discussion of reduction, especially the objection of Faithlessness (cf. Sects. 7.2.2.1 and 7.3.3). For convenience, we will call these advocates 'H' and 'E', respectively. So beware: Beltrami is E, not H. Rather, H is some gung-ho advocate of hyperbolic geometry as being well able to "stand on its own two feet". So H is a person who becomes historically possible only after 1868. Indeed: at a pinch, we can take 'H' to stand for Helmholtz![52] And E is not *any* Euclidean geometer, or philosopher of geometry (e.g. Kant): E is precisely the advocate of Beltrami's reduction (to Euclidean geometry).

The first point to make is the obvious one: that H and E are in danger of misunderstandings, i.e. of speaking at cross-purposes. For they are liable to mean different things by the same words such as 'geodesic' or 'straight'. For a hyperbolic straight line, i.e. geodesic in Beltrami's model (on the pseudosphere), is of course *not* a geodesic of the embedding Euclidean geometry. And a hyperbolic triangle is not a Euclidean triangle since its angles do not sum to 2π. And so on.

So let us, more specifically, compare the cases in terms of *Faithlessness*. Thus we can envisage: H will accuse E of Faithlessness to the meanings of his, H's, words:

> I, H, do not mean by 'geodesic' a curve in Euclidean space that is not straight according to Euclidean geometry!

And H might well go on to accuse E of what Sect. 7.2.2.1 called 'over-shooting': just as Benacerraf accused a set-theoretic reduction of arithmetic of over-shooting—yielding claims in the reduced theory (here: hyperbolic geometry) that are alien to it.[53]

There is also a notable contrast between the two analogues: about the chronological order, and one might say 'order of understanding', of the reduced and reducing theories. And this contrast will prompt another objection of Faithlessness—in the opposite direction, from E to H.

Thus: In the reduction of arithmetic to set theory, arithmetic was used successfully for millennia before set theory was even formulated. (And that successful use surely means it was in some good sense 'understood': albeit not analysed, and maybe also, problematic in ways that rightly prompted the logicists' enterprise of reduction.) In short: the reduced theory, arithmetic, 'was there first'.

But in the reduction of hyperbolic geometry to Euclidean geometry, it is the *reducing* theory that was used successfully for millennia (and again: surely in some

---

[52] We take up Helmholtz in our second paper. Agreed, Gauss in his unpublished work envisaged that a non-Euclidean geometry could be the real geometry of space: so, cheekily, 'H' could also stand for Gauss.

[53] This is not to say the objection is right: recall footnotes 17 and 45. Note that H might also accuse E of what Sect. 7.2.2.2 called 'Plenitude'. Thus the variety of ways to identify numbers with sets, which Benacerraf emphasised—and we called 'Plenitude'—corresponds to e.g. different placements of the Beltrami pseudosphere within $\mathbb{R}^3$. So H might accuse E of having many equally good, and therefore equally bad, placements of the hyperbolic space, within $\mathbb{R}^3$. (We write $\mathbb{R}^3$, but the affine Euclidean space would be more accurate: no matter—nothing turns on this.)

good sense 'understood', albeit not analysed, and maybe also, problematic), before the reduced theory, hyperbolic geometry, was even formulated.

And this of course makes an objection of Faithlessness in the opposite direction, i.e. from E to H, also tenable. For the long history, the entrenchment, of Euclidean geometry makes E's use of words such as 'geodesic' or 'straight' much more natural for us. This is of course reminiscent of Frege's complaint against Hilbert: cf. the discussion in Sect. 7.6.2.2. Thus E might say:

> H's use of 'geodesic' etc is faithless to my/Euclidean/the proper meaning. Agreed: Beltrami shows us how to define H's words 'geodesic' etc.—which for clarity we should really write as 'geodesic(hyp)' etc.—viz. as what I call an appropriately *curved* line in $\mathbb{R}^3$, in such a way that all H's claims are truths of my/Euclidean/the proper geometry. In short, hyperbolic geometry is a part of Euclidean geometry. Namely: a small part—about a surface of constant negative curvature. But sadly, H presents his theory with misleading words. For example, when H writes 'geodesic', he really should write e.g. 'geodesic(hyp)'.

7.7 Conclusion

We will not take the space to summarize *post facto* the main ideas—like simultaneous unique definition, functionalism as reduction and the Canberra Plan—that this paper has argued for. Those ideas are clear enough from Sect. 7.1.1's announcement of them. Instead, let us consider them in relation to spacetime, and so look ahead to our other papers.

Our complaint has been that these ideas—though standard undergraduate fare even forty years ago—are neglected in the recent literature on spacetime functionalism. Agreed, that might signal merely different concerns from the bulk of philosophers now writing about functionalism (and related topics like levels of description); and so it might merely signal a different use of the term, 'functionalism'. Though the homonymy might be regrettably misleading, no real harm would be done.

But we think some harm is done. For there are examples of spacetime functionalism, *stricto sensu*, to be had. Besides, these examples are "hiding in plain sight", in either the philosophical literature (especially about relationism) or in the physics literature about relational or Machian traditions of dynamics: these literatures including some precise—indeed, impressive—results.

Thus we think that someone, aware of the functionalist tradition we have reviewed, who sought for whatever reason to find examples of 'spacetime functionalism', would in short order think of these examples. And since the examples are impressive, the recent literature's detaching the phrase 'spacetime functionalism' from the tradition's concern with simultaneous unique definition and reduction, is regrettable. Hence our aim, in our other papers, to celebrate these examples of spacetime functionalism.

Acknowledgments We are grateful to: audiences at talks in Cambridge UK, Harvard (Black Hole Initiative), MIT, Munich, New York (MAPS), Oxford, and the 'Quantum information structure of spacetime' Network. For conversations and comments on previous versions, we are very grateful: to David Chalmers, Grace Field, Sam Fletcher, Eleanor Knox, Dennis Lehmkuhl, James Read, Alex Roberts and Bobby Vos; especially to Erik Curiel, Sebastian De Haro, Josh Hunt, Alex Oliver and Bryan Roberts; and above all, to Adam Caulton for—as ever—such insight and generosity. We are also very grateful to Cristián Soto, not least for his patience.

References

Austin, J. (1962). *Sense and sensibilia*. G.J. Warnock (Ed.). Oxford: Oxford University Press.
Balzer, W., Moulines, C., & Sneed, J. (1987). *An architectonic for science*. Dordrecht: Reidel.
Beaney, M. (2004). Carnap's conception of explication: From frege to husserl. In *Carnap brought home; the view from jena* (pp.117–150). Chicago: Open Court.
Beltrami, E. (1868). Saggio di interpretazione della geometria non-euclidea. *Giomale di matematiche, 6*, 284–312. (Reprinted in his collected works, *Opere matematiche* (1902, Milan) Volume I, 374–405.)
Benacerraf, P. (1965). What numbers could not be. *The Philosophical Review, 74*, 47–73.
Benacerraf, P. (1973). Mathematical truth. *Journal of Philosophy, 70*, 661–679.
Blanchette, P. (2018). The Frege-Hilbert controversy. *Stanford Encyclopedia of Philosophy*. https://plato.stanford.edu/entries/frege-hilbert/
Boolos, G., & Jeffery, R. (1980). *Computability and logic*, 2nd edn. Cambridge: Cambridge University Press.
Braddon-Mitchell, D., & Nola, R. (2009). *Conceptual analysis and philosophical naturalism*. MIT Press: Bradford Books.
Braithwaite, R. (1953). *Scientific explanation*. Cambridge: Cambridge University Press.
Brown, H. (2006), *Physical relativity*. Oxford: Oxford University Press.
Butterfield, J. (2011a). Emergence, reduction and supervenience: A varied landscape. *Foundations of Physics, 41*, 920–959.
Butterfield, J. (2011b). Less is different: Emergence and reduction reconciled. *Foundations of Physics, 41*, 1065–1135.
Butterfield, J. (2014). Reduction, emergence and renormalization. *Journal of Philosophy, 111*, 5–49.
Butterfield, J. (2018). On Dualities and equivalences between physical theories. In N. Huggett, B. Le Bihan, & C. Wüthrich (Eds.), *Philosophy beyond spacetime*. Oxford: Oxford University Press. http://philsci-archive.pitt.edu/14736/. Forthcoming (abridged).
Button, T. (2013). *The limits of realism*. Oxford: Oxford University Press.
Button, T., & Walsh S. (2018). *Philosophy and model theory*. Oxford: Oxford University Press.
Carnap, R. (1936). Testability and meaning. *Philosophy of Science, 3*, 419–471.
Carnap, R. (1963). Intellectual autobiography. In P. A. Schilpp (Ed.), *The philosophy of rudolf carnap: The library of living philosophers* (vol. XI). Open Court.
Coffa, J. A. (1983). Review of Torretti (1978). *Nous, 17*, 683–689.
Coffa, J. (1991). *The semantic tradition from kant to carnap: To the Vienna station*. L. Wessels (Ed.) Cambridge: Cambridge University Press.
Coffey, K. (2014). Theoretical equivalence as interpretative equivalence. *British Journal of Philosophy of Science, 65*, 821–844.
Davidson, D. (1967). The logical form of action sentences. In N. Rescher (Ed.), *The logic of decision and action*. Pittsburgh: University of Pittsburgh Press.
De Haro, S. (2020). Theoretical equivalence and duality. *Synthese, 198*, 5139–5177. https://doi.org/10.1007/s11229-019-02394-4
Dewar, N. (2019). Supervenience, reduction and translation. *Philosophy of Science, 86*, 942–954.

Dizadji-Bahmani, F., Frigg R., & Hartmann S. (2010). Who's afraid of Nagelian reduction? *Erkenntnis, 73*, 393–412.
Dummett, M. (1991). *Frege: Philosophy of mathematics*. London: Duckworth.
Field, H. (1980). *Science without numbers*. Hoboken: Blackwell.
Fletcher, S. (2019). Counterfactual reasoning within physical theories. *Synthese, 198*, pp. 3877–3898. https://doi.org/10.1007/s11229-019-02085-0
Frege, G. (1884). *The foundations of arithmetic* (Translated and edited by J. Austin 1974). Hoboken: Blackwell.
Gray, J. (2008). *Plato's ghost*. Princeton: Princeton University Press.
Halvorson, H. (2019). *Logic in the philosophy of science*. Cambridge: Cambridge University Press.
Harman, G. (1977). *The nature of morality: An introduction to ethics*. Oxford: Oxford University Press.
Hellman, G., & Thompson, F. (1975). Physicalism: Ontology, determination and reduction. *Journal of Philosophy, 72*, 551–564.
Hempel, C. (1965). *Aspects of scientific explanation*. New York: Free Press.
Hempel, C. (1966). *Philosophy of natural science*. New York: Prentice-Hall.
Hodges, W. (1997). *A Shorter Model Theory*. Cambridge: Cambridge University Press.
Hudetz, L. (2019a). The semantic view of theories and higher-order languages. *Synthese, 196*, 1131–1149.
Hudetz, L. (2019b). Definable categorical equivalence. *Philosophy of Science, 86*, 47–75.
Hurley, S. (1989). *Natural reasons: Personality and polity*. Oxford: Oxford University Press.
Jackson, F. (1998). *From metaphysics to ethics: A defence of conceptual analysis*. Oxford: Oxford University Press.
Janssen-Lauret, F., & MacBride, F. (2020). Lewis's global descriptivism and reference magnetism. *Australasian Journal of Philosophy, 98*, 192–198. https://doi.org/10.1080/00048402.2019.1619792
Kennedy, H. (1972). The origins of modern axiomatics: Pasch to peano. *The American Mathematical Monthly, 79*, 133–136
Kroon, F. (1987). Causal descriptivism. *Australasian Journal of Philosophy, 65*, 1–17.
Lewis, D. (1966). An argument for the identity theory. *Journal of Philosophy, 63*, 17–25.
Lewis, D. (1969). Review of 'art, mind, and religion'. *Journal of Philosophy, 66*, 23–25.
Lewis, D. (1970). How to define theoretical terms. *Journal of Philosophy, 67*, 427–446.
Lewis, D. (1972). Psychophysical and theoretical identifications. *Australasian Journal of Philosophy, 50*(3), 249–258.
Lewis, D. (1983). New work for a theory of universals. *Australasian Journal of Philosophy, 61*, 343–77.
Lewis D. (1984). Putnam's paradox. *Australasian Journal of Philosophy, 62*, 221–236.
Lewis, D. (1989). Dispositional theories of value. *Proceedings of the Aristotelian Society, LXIII*, 113–137.
Lewis, D. (1993). Many, but almost one. In K. Campbell, J. Bacon, & L. Reinhardt (Eds.), *Ontology, causality and mind: Essays on the philosophy of D. M. Armstrong* (pp. 23–38). Cambridge: Cambridge University Press.
Lewis, D. (1994). Reduction of mind. In S. Guttenplan (Ed.), *A companion to the philosophy of mind* (pp. 412–431). Hoboken: Blackwell.
Linsky, B. (2019). Logical construction. Stanford Encyclopedia of Philosophy. https://plato.stanford.edu/entries/logical-construction
Lutz, S. (2017a). What was the syntax-semantics debate in the philosophy of science about? *Philosophy and Phenomenological Research, 95*, 319–352.
Lutz, S. (2017b). Newman's objection is dead. Long live Newman's objection! https://philsci-archive.pitt.edu/13018/
Moore, G. (1903). *Principia ethica*. Cambridge: Cambridge University Press.
Nagel, E. (1961). *The structure of science: Problems in the logic of scientific explanation*. San Diego: Harcourt.

Nagel, E. (1979). Issues in the logic of reductive explanations. In *Teleology revisited and other essays in the philosophy and history of science*. New York: Columbia University Press; reprinted in Bedau and Humphreys (2008); page reference to the reprint.
Niebergall, K.-G. (2000). On the logic of reducibility: Axioms and examples. *Erkenntnis, 53*, 27–61.
Niebergall, K.-G. (2002). Structuralism, model theory and reduction. *Synthese, 130*, 135–162.
Oliver, A. (1996). The metaphysics of properties. *Mind, 105*, 1–80.
Oliver, A. (1999). A few more remarks on logical form. *Proceedings of the Aristotelian Society, 99*, 247–272.
Oliver, A., & Smiley, T. (2013). Zilch. *Analysis, 73*, 601–613.
Oliver, A., & Smiley, T. (2016). *Plural logic*. Oxford: Oxford University Press.
Potter M. (2000). *Reason's nearest kin*. Oxford: Oxford University Press.
Potter, M. (2020). *The rise of analytic philosophy, 1879–1930*. Oxfordshire: Routledge.
Psillos, S. (1999). *Scientific realism: How science tracks truth*. Oxfordshire: Routledge.
Psillos, S. (2012). Causal descriptivism and the reference of theoretical terms. In A. Raftopoulos & P. Machamer (Eds.) *Perception, realism, and the problem of reference*. Cambridge: Cambridge University Press.
Quine, W. (1960). *Word and object*. Cambridge: MIT Press (new edition: 2013).
Quine, W. (1964). Implicit definition sustained. *Journal of Philosophy, 61*, 71–74.
Russell, B.(1903/2010). *The principles of mathematics*. Crows Nest: Allen and Unwin (2010 reprint by Routledge).
Russell, B. (1918). The philosophy of logical atomism. In *The monist, 28*, 495–527; **29** (Jan., April, July 1919): 32–63, 190–222, 345–80. Page references to *The Philosophy of Logical Atomism*, D.F. Pears (ed.), La Salle: Open Court, 1985, 35–155.
Russell, B. (1924). Logical atomism. In D. F. Pears (Ed.), *The philosophy of logical atomism*. La Salle: Open Court. 1985, 157–181: also in *The Collected Papers of Bertrand Russell: vol. 9, Essays on Language, Mind and Matter: 1919–1926*, J.G. Slater (ed.), pp. 160–179; 2001: London and New York.
Russell, B. (1927). *The analysis of matter*, London: Kegan Paul.
Russell, B. (1919). *Introduction to mathematical philosophy*. New York and London.
Ryle, G. (1949). *The concept of mind*. London: Hutchinson.
Schaffner, K. (1967). Approaches to reduction. *Philosophy of Science, 34*, 137–147.
Schaffner, K. (1976). Reductionism in biology: Prospects and problems. In R. Cohen, et al. (Eds.), *PSA 1974* (pp. 613–632). Dordrecht: Reidel.
Schaffner, K. (2006). Reduction: The Cheshire cat problem and a return to roots. *Synthese, 151*, 377–402.
Schaffner, K. (2012). Ernest Nagel and reduction. *Journal of Philosophy, 109*, 534–565.
Shapiro, S. (2000). *Thinking about mathematics*. Oxford: Oxford University Press.
Sklar, L. (1982). Saving the noumena. *Philosophical Topics, 13*, 89–110.
Smith, P., & Jones, O. (1986). *The philosophy of mind*. Cambridge: Cambridge University Press.
Sneed, J. (1979). *The logical structure of mathematical physics*. Dordrecht: Reidel, Pallas Paperbacks.
Sober, E. (1999). The multiple realizability argument against reductionism. *Philosophy of Science, 66*, 542–564.
Stein, H. (1992). Was Carnap entirely wrong, after all? *Synthese, 93*, 275–295.
Tanney, J. (2015). Gilbert ryle. Stanford Encyclopedia of Philosophy. https://plato.stanford.edu/entries/ryle/
Tao, T. (2016). Notes on the Nash embedding theorem. https://terrytao.wordpress.com/2016/05/11/notes-on-the-nash-embedding-theorem/
Taylor, B. (1993). On natural properties in metaphysics. *Mind, 102*, 81–100.
Torretti, R. (1978). *Philosophy of geometry from riemann to poincaré*. Kufstein: Reidel.
Torretti, R. (1986). Observation. *British Journal for the Philosophy of Science, 37*, 1–23.
Torretti, R. (1990). *Creative understanding: Philosophical reflections on physics*. Chicago: University of Chicago Press.

Torretti, R. (1999). *Philosophy of physics*. Cambridge: Cambridge University Press.
Torretti, R. (2008). Objectivity; a Kantian perspective. *Royal Institute of Philosophy: Supplement, 63*, 81–94.
van Fraassen, B. (1980). *The scientific image*. Oxford: Oxford University Press.
van Fraassen, B. (1991). *Quantum mechanics: An empiricist view*. Oxford: Oxford University Press.
van Fraassen, B. (2008). *Scientific representation*. Oxford: Oxford University Press.
van Riel, R., & van Gulick, R. (2019). Scientific reduction. Stanford Encyclopedia of Philosophy. https://plato.stanford.edu/entries/scientific-reduction/
Weatherall, J. (2018a). Why not categorical equivalence? https://arxiv.org/abs/1812.00943
Weatherall, J. (2018b). Theoretical equivalence in physics. *Philosophy Compass, 14*, e12592–e12591. https://arxiv.org/abs/1810.08192.
Weyl, H. (1934). Mind and nature. In *Mind and nature: Selected writings in philosophy, mathematics and physics*, P. Pesic (Ed.) (2009). Princeton: Princeton University Press.

Chapter 8
Intertheoretic Reduction in Physics Beyond the Nagelian Model

Patricia Palacios

Abstract In this chapter, I defend a pluralistic approach to intertheoretic reduction, in which reduction is not understood in terms of a single philosophical "generalized model", but rather as a family of models that can help achieve certain epistemic and ontological goals. I will argue then that the reductive model (or combination of models) that best suits to a particular case study depends on the specific goals that motivate the reduction in the intended case study.

8.1 Introduction

Intertheoretic reductions, whereby a theory is said to reduce to another, play an important role in modern physics. But under what conditions a theory reduces to another, and what is achieved by reduction? Nagel (1961) famously attempted to offer a general structure of scientific reduction, in which this relation was understood in terms of the logical deduction of the reduced theory from the union of the reducing theory and bridge laws. However, this approach to reduction proved inadequate to describe even the most paradigmatic cases of reduction in physics. Despite its limitations, the Nagelian model—and revised versions of it (e.g. Dizadji-Bahmani et al., 2010; Van Riel, 2011; Sarkar, 2015)—continues nowadays being regarded as the standard philosophical model of reduction (Batterman, 2001) and it has been considered by some as the background philosophy of reductive enterprises in physics such as the reduction of thermodynamics to statistical mechanics (Frigg, 2008; Butterfield, 2011b; Dizadji-Bahmani et al., 2010).

I will argue in this contribution that although revised versions of the Nagelian model can actually explain some cases of reduction in physics, they do not suffice to explain the most important examples of reduction in physical sciences, including the alleged reduction of thermodynamics to statistical mechanics. Thus, I will contend

P. Palacios (✉)
Department of Philosophy, University of Salzburg, Salzburg, Austria
e-mail: patricia.palacios@sbg.ac.at

C. Soto (ed.), *Current Debates in Philosophy of Science*, Synthese Library 477,
https://doi.org/10.1007/978-3-031-32375-1_8

that in order to have a better understanding of reduction one needs to consider alternative models of reduction that focus on the role of limits and approximations as well as on the structural connection between the theories to be compared.

In Creative Understanding (1990), Roberto Torretti already distinguishes between five alternative models of intertheoretic reduction, pointing out the importance of each of them. Here I will follow this pluralistic approach to reduction stressing the different epistemic and ontological goals that these models of reduction fulfill. I will argue then that the reductive model (or combination of models) that best suits to a particular case-study depends on the specific goals that motivate the reduction in the intended case study.

This chapter is organized as follows. In Sect. 8.2, I will discuss different epistemic and ontological functions that reduction plays in physics. In Sect. 8.3, I will describe the main features of the Nagelian model and revised versions of it, pointing out its virtues and limitations. In Sect. 8.4, I will address Kemeny and Oppenheim's model of reduction and I will contend that the model can account for some cases of (eliminative) reduction of a theory by another. However, I will argue that the model is very limited in scope, since it depends on a sharp distinction between theoretical and observational terms and does not impose any structural relationship between the reduced and the reducing theory. In Sect. 8.5, I will analyse Schaffner's model of reduction, arguing that this model can successfully account for cases in which the reduced theory is modified by the reducing theory. In Sect. 8.6, I will analyze Nickles' distinction between $reduction_1$ and $reduction_2$. I will argue that although his proposal constitutes a step forward towards a pluralistic approach to reduction, the distinction between $reduction_1$ and $reduction_2$ is less sharp than what he suggests, since many cases of reduction in physics combine both models of reduction. In Sect. 8.8, I will analyse the structuralist model of reduction, emphasizing the epistemic roles that this model fulfills.

8.2 The Goals of Intertheoretic Reduction

Intertheoretic reductions play an outstanding role in physics, but what is achieved by the reduction of a theory to another is not always clear. Traditionally, perhaps due to the influence of Nagel (1949, 1961, 1970), the reduction of a theory T_2 to another T_1 has been taken to amount to the *explanation* of the reduced theory T_2 by the reducing theory T_1. This idea is also present in some reductive enterprises in physics. For instance, Rudolf Clausius, one of the first physicists to formulate the Second Law of thermodynamics, believed that the atomic theory might be helpful in explaining why entropy never decreases for isolated systems. Similarly, Ludwig Boltzmann, the father of statistical mechanics, justified his own attempts to connect the Second Law of Thermodynamics with the atomic theory as a way of giving a statistical mechanically based explanation of the Second Law of Thermodynamics. As he puts it in an 1866 paper: "It is the purpose of this article to give a purely analytical, completely general proof of the second law of thermodynamics." (Klein, 1973, op., cit., 57).

However, believing that explanation is the only epistemic function of intertheoretic reduction would be an understatement. In fact, intertheoretic reductions can often serve to *justify* the success of the reduced theory. This is particularly important in cases in which the reduction of a theory to another does not seek to eliminate the reduced theory but rather to retain it as a useful device for making predictions. For instance, the reduction of Newtonian gravitation to general relativity may be understood as giving a justification for why Newtonian gravitation was so successful in the past, at the same time, this reductive relation may be used to justify the further use of Newtonian theory as a convenient tool for making predictions in cases in which the velocity of the body is small compared to the speed of light. Torretti (1990) points out that in order to legitimize the use of the reduced theory, there needs to be a structural relationship between the reduced and reducing theories, which assures that the reduced theory will be successful in all relevant cases. As it will be seen in the following sections, this requirement imposes an important restriction upon the models of reduction that will be suitable to describe successful reductions in physics.

In general, one is interested in reducing theories that have been well established in the scientific community and that have an existence prior to the construction of the reducing theory. As Nickles (1973, p. 200) points out, the success of the reduced theory generally imposes constraints on the physical variables and their mathematical relations in potential successor theories. This means that the reduced theory can work as a *heuristic guide* for the development of the new theory. The attempt of reducing the Second Law to the kinetic theory can be seen as playing a heuristic role in the construction of statistical mechanics. Indeed, it was the attempt of reducing the Second Law, which led Boltzmann to consider molecular distribution functions rather than the complete set of variables and to establish a close connection between entropy and probability.[1] The reduction of the thermodynamic theory of phase transitions to statistical mechanics can also be seen as playing a heuristic role in the development of a statistical mechanical treatment of the phenomena, which followed the lead of thermodynamics by defining phase transitions in terms of discontinuities in the derivatives of the free energy.

Apart from helping to the construction of the reducing theory, the reduction of a theory to another can also play a role in the *acceptance* or *consolidation* of the new theory. As it has been said above, usually one is interested in reducing theories that have made successful predictions. The ability of the reducing theory to recover those predictions and to account for the same (or approximately the same) range of phenomena that were successfully explained by the reduced theory can lead then to the consolidation of the reducing theory. A good example of this is the reduction of optical theory of light to Maxwell's theory of electromagnetism. According to the

[1] See Brush (2006) and Blackmore (1995) for a historical analysis on the heuristic role played by attempt of reducing the Second Law in the development of statistical mechanics.

optical theory of the eighteenth century, light consisted of material particles. This theory, which was later replaced by the electromagnetic theory, could successfully account for a wide range of phenomena such as simple reflection, refraction and prismatic dispersion. The recovery of these predictions by the new electromagnetic theory of light, according to which light consists of a series of wave-like changes in a disembodied electromagnetic field, led finally to the acceptance of Maxwell's theory.[2]

Since the recovery of the predictions of the reduced theory by the reducing theory can play a role in the acceptance of the reducing theory, the incapacity of recovering those predictions can also frustrate inter-theory reductions. For example, by the time Boltzmann was attempting to derive the Second Law from a microscopic theory, the atomic hypothesis was not well established in physics, since many physicists, under the influence of positivist philosophy, were reluctant to accept it, in that it postulated the existence of entities that were too small to be observed (Blackmore, 1995). The reduction of the Second Law to statistical mechanics was then regarded as a strategy to increase the confidence in the latter theory. Unfortunately, different objections pointing out manifest contradictions between the Second Law and Boltzmann's theory delayed the acceptance of the latter. One of these objections was proposed by Loschmidt (1876), who showed that there was a contradiction between one of the basic premises of Boltzmann's theory, i.e, the reversibility of all purely mechanical motions, and the Second Law, which implies an irreversible increase of entropy. Another objection was due to Poncaré (1889) and Zermelo (1896), who proved that any mechanical system constrained to move in a finite volume and with fixed total energy will eventually return to the specified initial conditions. This means that the system can in principle evolve towards a state of lower entropy, which is incompatible with the Second Law. Since the Second Law was widely accepted in the physics community, the apparent incompatibility between this law and statistical mechanics led most physicists to think that the latter was false (Blackmore, 1995; Brush, 2006). In fact, despite Boltzmann's attempts to undermine these objections by arguing that the mechanical viewpoint did not have any consequences in disagreement with experience, the acceptance of the atomic theory and statistical mechanics had to await until the 20th century. Only after statistical mechanics was accepted within the physics community, the attempt to reduce the Second Law of thermodynamics led to the development of several approaches that attempted to *correct* the reduced theory. For example, approaches in which the original strict version of the Second Law of thermodynamics was modified so as to allow for entropy fluctuations in equilibrium (e.g. Einstein, 1910; Greene and Callen, 1951; Dauxois et al., 2002). The main goal of these reductive

[2] There were some predictions of electromagnetic theory which were incompatible with the predictions of traditional physical optics, such as the exponentially decaying penetration of electromagnetic waves into the surface of a reflecting opaque object. However, these inconsistencies were settled quickly in favor of a modification of the reduced theory. The modified version of traditional optics was considered to be totally reducible to the electromagnetic theory and this reduction led finally to the acceptance of the latter theory (Worrall, 1989, p. 148).

approaches was no longer to consolidate the reducing theory but rather to correct the reduced theory according to the reducing theory.

Apart from the epistemic roles mentioned above, reduction also aims to support the relative *fundamentality* of the reducing theory with respect to the reduced theory. Fundamentality is a crucial function of reduction, since it can account for its asymmetric character. Loosely speaking, fundamentality can be understood in the sense that the reducing theory has some "advantage" over the reduced theory or it is "better" than the reduced theory. In the philosophical literature, one can distinguish at least four different ways in which a theory T_1 is said to be more fundamental than the other T_2 . (1) Empirical power: T_1 accounts for every observation within the scope of T_2 and for new observations outside T_2's domain, (2) Aesthetic virtues: T_1 is "simpler", "more elegant" or more "systematized" than T_2, (3) Empirical correctness: T_1 is more empirically correct than T_2, iv) Ontological supremacy: T_1 has the right ontology or at least an ontology that is closer to the truth than the ontology of the T_2. Let me stress that, although all reductions are asymmetric with respect to fundamentality, not all reductions aim to demonstrate fundamentality in the same sense. This will become clear in the analysis of the different models of intertheoretic reduction developed in the rest of the chapter.

So far, by focusing on a few examples, we have distinguished between six different goals of reduction: *explanation*, *justification* and *correction* of the reduced theory, as well as *development*, *consolidation* and *fundamentality* of the reducing theory. Yet the more one looks into different case studies regarded (sometimes contentiously) as successful reductions in physics, the more epistemic and ontological goals one finds. It is not the aim of this chapter to give an exhaustive account of all the roles that reductions can play in physics, but rather to focus on a subset of them that are particularly important for the examples that will be considered.

8.3 The Nagelian Model

Now that we have mentioned some important epistemic and ontological goals of intertheoretic reduction, different questions of philosophical interest arise. For instance, what are the most important features of successful reductions in physics? What is the (logical) structure that characterizes intertheoretic reductions in physics? What are the necessary and sufficient conditions for successful reductions in physics?

In 1949, Nagel famously argued that all these questions have a unique answer associated with the formal structure of intertheoretic reductions. These ideas were then developed in his 1961 book "The Structure of Science". For Nagel, there is a general way of characterizing the main components and the logical structure of the different types of reduction in science. According to this view, every reduction can be constructed as a series of statements, in which one of them, i.e. the reduced theory, is the conclusion while the others, i.e. the reducing theory and auxiliary assumptions, are the premises. In the case of *homogeneous reductions*, in which all the specific matter terms in the reduced theory T_2 are present in the reducing theory T_1, the formal structure is straightforwardly that of a deductive argument. In the case of *inhomogeneous reductions*, in which at least one descriptive term in the reduced theory does not occur in the reducing theory, the reducing theory needs to be supplemented by "rules of correspondence" or "bridge laws" (BL), which establish connections between the distinctive terms of the reducing theory and the terms of the reduced theory. Once the reducing theory has been suitably supplemented by bridge laws, inhomogeneous reduction, like homogeneous reductions, embodies the pattern of deductive arguments.[3]

Thus, according to this view, there are two main properties that serve to give a general characterization of reductions in science:

- Derivability: The laws of the reduced theory can be logically derived from the laws of the (augmented) reducing theory plus auxiliary assumptions.
- Connectability: In the case of inhomogeneous reductions, there are bridge laws, which connect the vocabulary of the reducing theory with the vocabulary of the reduced theory.

Nagel (1961) also pointed out that if the reduction is not trivial, the bridge laws should constitute scientific hypotheses, which should be susceptible of empirical confirmation or disconfirmation.[4]

This general characterization of intertheoretic reductions in terms of logical deduction has important virtues if we think of the goals of intertheoretic reduction mentioned in the previous section. In fact, the greatest advantage of understanding reduction in terms of logical deduction, is that logical deduction is truth-preserving. This means that if T_1 and BL as well as the auxiliary assumptions are all true, then the reduced theory must also be true. In other words, if we accept the validity of the new theory T_1, then on the basis of the logical relation between $T_1 \cup BL$ and T_2, we are forced to accept the validity of T_2 under certain auxiliary assumptions. Straightforwardly, this leads to the justification of the use of T_2 as a useful device for making predictions and can also lead to explain the success of T_2 from the perspective of T_1. Furthermore, given that in most cases, the reduced theory T_2 has proven to be empirically successful, showing that T_2 is a logical consequence of the augmented theory $T_1 \cup BL$, implies that in the augmented theory, one can recover all the successful predictions of T_2. One can then use this structural relation between T_2 and the augmented theory $T_1 \cup BL$ as an argument for the consolidation or acceptance of the new theory T_1, which in general replaces the old theory T_2. Furthermore, reduction understood in terms of logical deduction can play a heuristic role in the development of the reducing theory. Indeed, knowing

[3] This idea can also be put in terms of *definitional extension* of a theory. The core idea of Nagelian reduction is that a theory T_1 reduces another T_2, if and only if T_2 can be defined as a definitional extension of T_1, which means that T_2 can be shown to be a sub-theory of the augmented theory $T_1 \cup BL$ (Butterfield, 2011a).

[4] See Nagel (1970), Dizadji-Bahmani et al. (2010), and Schaffner (2012) for a detailed discussion on the status of bridge laws.

that the goal is to derive the theory T_2 from the new theory T_1 can lead to the construction of suitable bridge laws and auxiliary assumptions that will help deriving the original theory T_2. Finally, strict Nagelian reductions can also serve to support the relative fundamentality of T_1 over T_2. Indeed, if T_2 is a logical consequence of the augmented theory $T_1 \cup BL$, and the converse does not hold, then $T_1 \cup BL$ can be thought of as being more general than T_2.

In spite of its advantages, the Nagelian model of reduction has been challenged by a number of philosophers. One of the most important criticisms is that even in the most paradigmatic cases of reduction in science, the reduced theory does not strictly follow from the reducing theory plus bridge laws (Feyerabend 1962, Sklar, 1967, Schaffner 1967, 2012).[5] Consider the example of the reduction of the Galilean law of a freely falling body to the Newtonian theory regarded by Nagel (1961) as a paradigmatic case of homogeneous reduction. It has been often pointed out (e.g. Schaffner, 1967; Torretti, 1990) that even in this case, the reduced theory cannot be derived from the reducing theory, since the two theories are strictly speaking "inconsistent". In fact, while Galileo's law asserts that the acceleration of a freely falling body near the earth's surface is constant, Newtonian theory implies that acceleration varies with the distance of the falling body to the earth's center of mass. This means that, at most, what can be derived from the reducing theory is an approximation to the reduced theory and not the reduced theory itself.

In 1970, Nagel explicitly recognized the use of approximations in intertheoretic reductions. However, he argued that approximations were not incompatible with his model, since they can take part in the auxiliary assumptions needed to derive the reduced theory:

> More generally, though no statistical data are available to support the claim, there are relatively few deductions from the mathematically formulated theories of modern physics in which analogous approximations are not made, so that many if not all the laws commonly said by scientists to be deducible from some theory are not strictly entailed by it. It would nevertheless be an exaggeration to assert that in consequence scientists are fundamentally mistaken in claiming to have made such deductions. It is obviously important to note the assumptions, including those concerning approximations, under which the deduction of a law is made. (p. 363)

It is important to note, however, that if the reduced theory T_2 can only be approximatively deduced from the reducing theory T_1, then the reduced theory T_2 can no longer be considered to be embedded in T_1, since T_1 and T_2 differ. This has important consequences for the epistemic roles that approximative Nagelian reductions may fulfill. In fact, if T_2 cannot be derived from T_1, one cannot justify the empirical success of T_2 only on the basis of its logical relationship with T_1. *Prima*

[5] Feyerabend (1962) strongly criticized the model by pointing out contentious issues associated not only with the derivability condition, but also with the connectability condition. His criticism to the connectability condition came from his "incommensurability thesis", according to which, all scientific vocabulary, including observational terms, are globally infected by the theory in which they functioned. Nagel (1970) replied to these objections with an incisive criticism to the incommensurability thesis.

facie, this is not so problematic since the Nagelian model gives a prominent role to the auxiliary assumptions. In fact, according to this model, T_2 is not deduced from T_1 alone but from T_1 and auxiliary assumptions, and in the case of inhomogeneous reductions, from T_1, BL and auxiliary assumptions. If the auxiliary assumptions are true, and the reducing theory is correct, then we can infer the (approximate) correctness of the reduced theory under certain conditions. For instance, in the reduction of Galileo's law of free fall to Newtonian mechanics, if one assumes that the distance traveled by the falling body from its starting position to the surface of the earth is very small compared to the radius of the earth, then we can assume, by a good approximation, that the acceleration takes a constant value. Under these assumptions we can effectively derive Galileo's law of free fall and justify its success when the distance traveled by the free falling body is small compared to the radius of the earth.

There are however important differences between approximative reductions and strict reductions. First, approximative reductions, in contrast to strict reductions, rely strongly on auxiliary assumptions involving approximations. If these auxiliary assumptions are false or if they fail to constitute good approximations, one can neither justify nor warrant the success of the reduced theory. Second, approximative reductions only imply the approximative correctness of the secondary theory. This means that there is always a risk that the secondary theory may lead to inaccurate results.

Given these differences, one may be tempted to consider strict Nagelian reductions and approximative Nagelian reductions as two different models of reduction. Sklar (1967) even considers the possibility of restricting the term "reduction" to cases of strict reduction. However, as Sklar himself recognizes, there are important cases of reduction in physics, which constitute approximative reductions, and so restricting the term "reduction" to cases of strict reduction would not fit well with the scientific practice. Nagel's (1961; 1970) motivation for considering the two cases to fit into the same model of reduction is that both types of reduction have the same structure. This means both approximative Nagelian reductions and strict reductions can be constructed as a deductive argument in which the reduced theory follows from the reducing theory plus bridge laws and auxiliary assumptions. The main difference between strict reductions and approximative reductions is, according to Nagel, that the latter requires auxiliary assumptions involving approximations. Although I believe that it is convenient to distinguish between these two types of reduction, I will follow Nagel in assuming that since they have approximately the same structure, they should be considered to fit into the same model of reduction. The aspect in which I disagree with Nagel is in that all approximative reductions are of the same kind. In fact, there are cases of approximative reductions that can be constructed as deductive arguments in which the reduced theory follows from the premises, such as the reduction of Galileo's law of free falling bodies to the laws of Newtonian mechanics discussed above. However, there are many cases of approximative reductions in physics, including the reduction of the Second Law, in which the reduced theory does not follow from the reducing theory and auxiliary assumptions. For instance, there are cases in which the theory that can be derived from the reducing theory and auxiliary assumptions is not the original reduced theory but an analogous version of the reduced theory. I will argue in Sect. 8.5 that these cases are better captured by the Schaffner model. Yet there are other cases of approximative reductions that cannot be conveniently put in terms of deductive arguments. I will discuss these cases in Sects. 8.7 and 8.8.

8.4 Kemeny and Oppenheim's Model

In order to offer a solution to the issues related with the derivability condition in Nagel's model, Kemeny and Oppenheim (1956) suggested an alternative general model of reduction, in which they relinquish the idea of linking the structures of the reduced and the reducing theory. In this model, intertheoretic reduction is defined indirectly relative to a set of observational data. The model proposes that a theory T_2 is reduced by means of T_1 relative to observational data O, iff (1) the vocabulary of T_2 contains terms that are not part of the vocabulary of T_1, (2) any part of O explainable by means of T_2 is explainable by T_1, and (3) T_1 has a greater systematic power than T_2, which means that T_1 has the ability to predict as wide a range of phenomena as possible from as little data as possible.

The main problem of this model is that it seems too weak to account for many cases of reduction in science. The weakness of this model is due to the fact that it does not impose any structural relationship between T_1 and T_2. Indeed, it only requires that the two theories to make the same observational predictions within the range of phenomena covered by T_2. The problem is that without positing a structural link between T_1 and T_2, the reduction does not allow one to justify the success of T_2 on the basis of T_1. This means that one cannot use the reduction of T_2 to T_1 to justify the use of T_2 under certain conditions. Furthermore, without a structural relationship between T_1 and T_2, one cannot explain T_2 on the basis of T_1 and one cannot argue for the ontological fundamentality of T_1 over T_2. It seems therefore that the only epistemic goals of the Kemeny and Oppenheim's model is to help consolidate T_1, by showing that this new theory makes the same (or similar) predictions as the older theory T_2, and to support the relative fundamentality of T_1 over T_2 in the epistemic sense that T_1 has greater systematic power than T_2. Cases of "reduction" that only lead to the consolidation of the reducing theory are cases in which the reduced theory is in fact replaced by the reducing theory and the reduced theory is overthrown. An example is the reduction of the phlogiston theory by the oxygen theory of combustion or the caloric theory by the energetic theory of heat.

From what has been said above it is clear that, if the Kemeny and Oppenheim model is to account for some reductions in physics, they have to be cases of total replacement of a theory by another. Sklar (1967) suggests, however, that the Kemeny and Oppenheim model is not even capable to account for such cases, since it requires a sharp distinction between observational and theoretical terms, which has been shunned by most philosophers of science. Furthermore, the model requires that the two theories predict the same observational results within the scope of T_2, which is hardly ever satisfied. Leaving aside the problem of the theoretical/observational distinction, I believe that a revised version of the Kemeny and Oppenheim model that only requires the two theories to make similar predictions for a relevant set of observables can be weak enough to account for historical cases of *eliminative reductions*, which have the only purpose of consolidating the new theory by supporting a relative epistemic fundamentality of the reducing theory over the reduced theory. The replacement of Ptolemy's theory of planetary motion by Copernicus' theory can be considered as a paradigm example of this kind of reduction. Now, one may object that cases of total replacement, in which the previous theory is overthrown, do not deserve to be called "reductions" at all (Sklar, 1967). Although I do not have a strong opinion on this issue, I believe that it is convenient to call them *eliminative reductions*, because they fulfill at least two of the most important goals of intertheoretic reduction, namely, the consolidation of the new theory and the support of a relative fundamentality of one theory over another in the sense of systematic power.

8.5 The Schaffner Model

Schaffner (1967) took a different route from Kemeny and Oppenheim and proposed a model of direct reduction that was supposed to give a more general characterization of intertheoretic reductions than the Nagelian model. This model says that reduction occurs *if and only if*:

1. All the primitive terms $q_1, \ldots, q_n$ appearing in the *corrected* reduced theory T_2^* appear in the reducing theory T_1 (in the case of homogeneous reductions) or are associated with one or more of T_1's terms with the help of reduction functions, i.e. bridge laws.
2. T_2^* is derivable from T_1, when T_1 is conjoined with bridge laws.
3. T_2^* corrects T_2 in the sense that it provides more accurate experimentally verifiable predictions than T_2 in almost all cases, and should also indicate why T_2 was incorrect (e.g. crucial variables are ignored), and why it worked as well as it did.
4. T_2 should be explicable by T_1 in the sense that T_1 yields T_2^* as a deductive consequence.
5. The relations between T_2 and T_2^* should be one of *strong analogy*, which means that T_2^* bears a *close similarity* to T_2 or produces numerical predictions which are "very close" to T_2's.

Schaffner's contention was that this model should be thought of as a "general reduction paradigm" of intertheoretic reductions in science and he tries to show that other approaches to reduction such as the Nagelian model and the Kemeny and Oppenheim's model are just special cases of this generalized model.[6] A similar view has been recently defended by Dizadji-Bahmani et al. (2010), who argue that a slightly revised version of this model, which they called "the Generalized Nagel-Schaffner Model" (GNS), gives us the right analysis of intertheoretic reductions. Or, in other words, "tells us the right story about how synchronic intertheoretic reduction works" (p. 393).[7] An important difference between the GNS model and Schaffner's original proposal is that the GNS model (like Nagel's model) emphasizes the role of auxiliary hypotheses, which do not seem to play an important role in the original formulation of Schaffner's model. According to the GNS model, reduction is the deductive subsumption of a corrected version of T_2, i.e. T_2^*, under T_1, where the deduction involves (1) first deriving a restricted version, T_1^*, of the reducing theory T_1 by introducing boundary conditions and auxiliary assumptions and then (2) using bridge laws to obtain T_2^* (Dizadji-Bahmani et al., 2010, p. 398).

Like the Nagelian model, the Schaffner model has been target of a number of criticisms. One of these criticisms is that it relies on a syntactic view of theories (Suppe, 1974). Another is that the status of bridge laws is not clear (Sklar, 1995). A third problem is that bridge laws are incompatible with multiple realizability (Kitcher, 1984). The fourth issue is that the fact that there are no restrictions for the auxiliary assumptions makes this model too liberal. Finally, it has been pointed out that the meaning of strong analogy is too vague to allow us give a precise characterization of reduction. Dizadji-Bahmani et al. (2009) addressed all these problems and offered a compelling reply to each of them. I will not discuss these problems here, but instead I will focus on a different problem, which was not addressed by them, namely, the generality of the model.

As I see it, the biggest advantage of the Schaffner model, or the GNS model, is that it can correctly account for those cases of reduction in which the theory that is deduced from the reducing theory T_1 plus auxiliary assumptions and bridge laws is not the original secondary theory T_2, but a *corrected* version of the theory, i.e. T_2^*. Sommerfeld's modification of physical optics according to Maxwell's electromagnetic theory can be considered as a paradigmatic case of this type of reduction (Schaffner, 1967, 2012). On the basis of Maxwell's equations, Sommerfeld modified Fresnel's famous sine and tangent laws for the ratio of the relative amplitudes of incident, reflected, and refracted polarized light. This

[6] In 1977 and 2012 he reiterates this idea by proposing an even more general model, which he baptized as "The general reduction replacement model". This model was supposed to be general enough to have "the reduction paradigm" as a limiting case, which for its part yields the Nagel's model as limiting case.

[7] Dizadji-Bahmani et al. (2010) restrict their analysis to the so-called synchronic intertheoretic reductions, which they define as "the reductive relation between pairs of theories which have the same (or largely overlapping) domains of application and which are simultaneously valid to various extends." (p. 394) It is important to point out, however, that Schaffner (1977, 2012) did not restrict his analysis to this kind of reduction. In fact, the paradigmatic case of reduction that he presents is the reduction of physical optics to the electromagnetic theory, which should be better regarded as a "diachronic reduction", in which a theory historically replaces the other.

modified version of Fresnel's laws is an approximation or close analogy of the original laws of physical optics, in the sense that it produces predictions that are very close to the predictions of physical optics.[8]

One can see then that Schaffner's reduction leads crucially to the correction of T_2 by T_1. One can infer from this that T_1 is more fundamental than T_2 in the sense that T_1 is more empirically correct than T_2. It is important to note, however, that this kind of reduction does not generally lead to the acceptance of T_1. In fact, in general, reductions that aim to modify the reduced theory according to the reducing theory are based on reducing theories that have been previously accepted in the physics community.

Although Schaffner (1967) believes that his model converges towards the Nagelian model when T_2 and T_2^* are identical, I believe that it is important to keep these two models apart, not only because they have different structures but also because they fulfill different epistemic roles. The structural difference between the Nagelian model and the Schaffner model is significant: in the former case, the theory that can be derived from auxiliary assumptions and bridge laws is the original theory T_2, whereas in the latter case one derives a different theory, namely T_2^*, which is only strongly analogous to the original. Interestingly, in some examples of reduction in physics, physicists have the option of "strengthen" the auxiliary assumptions in order to derive the original theory T_2 or to "weaken" these assumptions to derive a theory T_2^* that modifies or corrects T_2. An example of this is the Second Law of thermodynamics, which states that the entropy of an isolated system cannot decrease. Without invoking the thermodynamic limit in which the number of particles goes to infinity, the best one can do in statistical mechanics is to entail an analogous theory that allows for the possibility of statistical fluctuations, which may cause temporary decreases of entropy. There are important approaches that attempt to develop a generalized form of thermodynamics that incorporates fluctuations of thermodynamic parameters around equilibrium. In these approaches, the Second Law of thermodynamics is modified accordingly to include fluctuations. In the refined formulation of the Second Law, i.e. T_2^*, it is only the entropy averaged over the fluctuations that monotonically increases and reaches a maximum at equilibrium (e.g. Einstein, 1910; Tisza and Quay, 1963; Callen and Kestin, 1960; Mishin, 2015; Valente, 2021). The aim of such approaches is precisely to expand the scope of thermodynamics by introducing statistical elements, which can be applied to cases in which fluctuations cannot be neglected, such as nanometer-scale systems. Since in these approaches, there exists a theory T_2^* that modifies T_2 (original formilation of the Second Law), they have been regarded as a successful cases of Schaffner reduction (Dizadji-Bahmani et al., 2009, Callender, 2011).[9]

[8] See Schaffner (2012) and Worrall (1989) for a detailed analysis of this reduction.

[9] Whether this is indeed a case of Schaffner reduction is still unclear. In fact, in order to prove that this is in fact a case of Schaffner reduction, one would need to show that the theory T_2^* can be actually derived from statistical mechanics T_1. This is not easy to prove since at least some of these approaches (Tisza and Quay, 1963; Valente, 2021) seem to constitute an amalgamation of statistical

It is important to note, however, that if one takes the thermodynamic limit, then statistical fluctuations become negligible. This is because the relative fluctuations of a macroscopic observable A in pure phases of an extensive system vanish in the large N limit (Gross, 2001a):

$$\frac{\langle A^2\rangle - \langle A\rangle^2}{\langle A\rangle^2} \propto \frac{1}{N}. \tag{8.1}$$

This means that, if one takes the thermodynamic limit, one can in principle derive the exact Second Law under certain assumptions. Furthermore, in the thermodynamic limit, Zermelo's objection becomes blunted as the recurrence time is much longer than any observation time (Gross, 2001a; Rau, 2017).

My contention is that the choice of the auxiliary assumptions determines the suitable model of reduction, and this in turn depends on the specific goals of the intended reduction and case study under investigation. For instance, if the main goal is to recover the exact predictions of the secondary theory, for instance in order to consolidate the reducing theory, which has not been well-established in the scientific community, then one may "strengthen" the assumptions in order to derive the original theory T_2. Boltzmann's use of the thermodynamic limit and other idealizations such as the so-called Boltzmann-Grad may have had this purpose (e.g. Boltzmann, 1877, 1885). However, if the reducing theory has been consolidated and the main goal of the intended reduction is to correct the reduced theory according to the reducing theory, for instance, in order to derive a theory that makes more accurate predictions, then one will tend to "weaken" the auxiliary assumptions and try to derive a theory T_2^* that is only analogous to the original T_2. Apart from the finite approaches mentioned above, there are interesting approaches to statistical mechanics that aim to account for non-extensive systems without taking the thermodynamic limit. The motivation for such approaches is that in non-extensive systems such as small systems and long-range interacting systems, fluctuations must be taken seriously (Gross, 2002).

Another example that illustrates that the choice of auxiliary assumptions may lead to pursue different models of reduction is the case of phase transitions that I will discuss in more detail in the next section. In thermodynamics, first order phase transitions are defined in terms of discontinuities in the first derivatives of the free energy. In order to recover this theory of phase transitions from statistical mechanics, and also to recover the same predictions of thermodynamics, the most important approaches invoke the thermodynamic limit by which the number of particles and the volume of the system go to infinity. This reduction has led to the consolidation of the statistical mechanical theory of phase transitions rather than to a modification of the thermodynamic treatment. There are, however, various attempts to develop a finite theory of phase transitions without invoking the thermodynamic

mechanics and thermodynamics, rather than a derivation of an alternative an thermodynamic from statistical mechanics.

limit, which have been mainly motivated by the fact that, in some particular cases, the thermodynamic limit makes no sense or is bad approximation. This is for example the case of "small" or non-extensive systems, where the linear dimension is of the characteristic range of the interaction between the particles. Gross (2001b) and Casetti and Kastner (2006) have developed a topological theory of phase transitions without using the thermodynamic limit, in which phase transitions are entirely determined by topological peculiarities. In an important sense, this theory can be regarded as correcting the thermodynamic theory of phase transitions by offering an alternative definition of the phase transitions that can be applied to non-extensive systems. Nonetheless, it remains controversial whether the statistical mechanical theory of phase transitions that invokes the thermodynamic limit should be entirely replaced by a finite theory of phase transitions. One of the arguments against this replacement is that in most cases the thermodynamic limit makes calculations more treatable. Another argument is that in many cases the thermodynamic limit allows to remove irrelevant details, such as finitary effects, and it allows us to give a rigorous definition of phase transitions. Finally, there is the empirical observation that in most cases the thermodynamic theory of phase transitions works just fine (Palacios, 2019; Mainwood, 2006).

The point that I want to make here is that none of the discussed models of reduction explains better than the other the "essence" or "the general structure" of reduction, nor can one be formulated in terms of the other. These are two different models of reduction and whether there is reduction à la Nagel, reduction à la Schaffner, or other kind of reduction depends, as we have seen, on the intended case study and the goals of that particular reduction. I also want to stress that there seems to be a trade-off between the goal of "consolidating" the reducing theory and "correcting" the reduced theory. In fact, if the reducing theory has not been well established in the scientific community yet, deriving a theory that differs from the original secondary theory T_2 with respect to the phenomena that are well-described by the latter theory, may lead to the rejection of the new theory. Therefore, at this stage, scientists may tend to strengthen the assumptions in order to derive the original secondary theory T_2. The attempt to reduce the Second Law of thermodynamics by taking the thermodynamic limit can be regarded as an outstanding example of this. However, once the theory T_1 has been consolidated, the main purpose of reduction can be to correct the theory T_2 according to T_1, for example, in order to construct a theory T_2^* that makes more accurate predictions. At this stage scientists may try to achieve a reduction à la Schaffner.

Finally, it is also important to point out that Schaffner's reduction requires the existence of a theory T_2^* that is analogous to T_2 and that is derived from T_1. Sometimes, especially in cases of theory replacement, this theory does not exist, or, if it exists, it cannot be derived from T_1 plus bridge laws and auxiliary assumptions. If the modified theory does not exist, but we can derive the original theory under specific assumptions that restrict the range of application of the reduced theory T_2, we have a case of Nagelian or approximative Nagelian reduction. An example of this kind of reduction is the reduction of Galilean laws to Newtonian laws discussed above. Now, if the modified theory does exist but cannot be derived (deduced) from the reducing theory, for instance, because the two theories are inconsistent, one may have a reduction that neither corresponds to the Nagelian nor to Schaffner's model of reduction. As we will discuss it in Sect. 8.6, the partial reduction of Newtonian mechanics to the general theory of relativity and the partial reduction of Newtonian mechanics to quantum mechanics in the classical limit $\hbar \to 0$ are good candidates for the latter.

8.6 Nickles' Approach

In contrast to Nagel and Schaffner, Nickles (1973) rejects the idea that there is a single model of reduction that is capable of capturing the logical structure of reductions in science.

> In this paper I reject the widespread view that reductions of scientific theories are all of one basic type. I agree with those who hold that intertheoretic reduction involves the relation of a theory to its special case, but wish to emphasize in how many different ways one theory may constitute a special case of another." (Nickles, 1973, p. 181).

In his analysis, Nickles distinguishes between two different models of reduction attached to different scientific functions or purposes: $reduction_1$, which corresponds to the Nagelian model and $reduction_2$, which consists in the recovery of one theory from another by applying a set of limiting operations or other appropriate transformations. According to his view, whereas $reduction_1$ amounts to the *explanation* of a theory by another and to *ontological economy*, $reduction_2$ has a *heuristic* and *justificatory* role. Heuristic in the sense that it helps in the construction of the new theory and justificatory in the sense that it helps explain why the previous theory was successful. For him, the different models are also associated to different types of reduction, whereas $reduction_1$ serves to account for "domain-combining reductions", in which there are two different theories describing the same rage of phenomena at different levels of description, $reduction_2$ serves to account for "domain-preserving" reductions, in which one theory is a successor of the other.[10]

Although I agree with Nickles' contention that there are different models of reduction that can be associated with different epistemic or ontological goals, I disagree with various aspects of his approach. First of all, I disagree with his generalization of $reduction_1$ in which the Schaffner model is presented as a revised version of the Nagelian model, and therefore as a special case of $reduction_1$. As we have seen above, the Nagelian model and Schaffner's model have different epistemic goals and therefore shall be considered as two different models of reduction. Second, I disagree with his idea that $reduction_1$ and $reduction_2$ serve to account for different types of reduction associated with specific goals. As I will argue

[10] Domain-combining and domain-preserving have been sometimes understood in terms of "synchronic" and "diachronic reductions" respectively (Dizadji-Bahmani et al., 2009).

below, many cases of reduction in physics combine these two models of reduction, which suggests that the distinction between $reduction_1$ and $reduction_2$ and their association to specific functions is less sharp than what Nickles suggested.

Since I have discussed the Nagelian model in the previous sections, I will now focus on Nickles' notion of $reduction_2$. Nickles (1973, p. 197) characterizes $reduction_2$ as follows:

> $Reduction_2$ Let O_i be a set of intertheoretic operations, then a theory T_2 $reduces_2$ to another T_1 iff $O_i(T_1) \rightarrow T_2$, where the arrow represents "mathematical derivation" understood in a broad sense including not only logical deduction but also limiting operations and approximations of many kinds.

Now, as Palacios (2019) points out, one should note that mathematical operations such as limits and other approximations are performed not on the theory itself but on functions (or equations) representing physical quantities. Therefore a more precise schema needs to be formulated in terms of the relevant quantities and not directly on the theories to be compared. Furthermore, as Nickles himself recognizes, in order for $reduction_2$ to hold, it is necessary that the mathematical operations performed on T_1 make physical sense. Although he is not explicit about what he means by "physical sense", one can interpret this constraint as signifying that after applying a set of mathematical operations on T_1, the resulting (limit) theory, let us call it T_1^*, can still describe realistic behavior. Taking the limit of a constant of nature to zero, for example, may result in a (limit) theory that does not account for realistic behavior unless this limit is adequately explained. Similarly taking the limit of a parameter such as temperature or the number of particles to infinity may also be illegitimate if these limits are not adequately justified. Thus, a more precise characterization of $reduction_2$ suggested by Palacios (2019) is the following:

> $Reduction_2^*$ Given a set of intertheoretic operations O_i, a quantity Q_1 of T_1 $reduces_2^*$ another quantity Q_1^* of T_1^* iff (i) $O_i(Q_1) = Q_1^*$ and (ii) the mathematical operations O_i make physical sense.

A special case of $reduction_2$ is limiting reduction, which refers to cases in which the transformations correspond to mathematical limits. Following Palacios (2019), we can express limiting reduction as follows:

> *Limiting reduction* Let Q^1 denote a relevant quantity of T_1, Q_1^* a relevant quantity of T_2, then a quantity Q_1^* of T_1^* *limiting reduces* to a corresponding quantity Q^1 of T_1 iff (i) $\lim_{x\rightarrow y} Q_x^1 = Q_1^*$ (where x represents a parameter appearing in T_1 and (ii) the limiting operation makes physical sense.

If we understand $reduction_2^*$ as a reduction between functions or equations representing physical quantities, then it is clear that $reduction_2^*$ can be combined with other models of reduction such as the Nagelian or the Schaffner model in order to achieve the (partial) reduction of a theory to another. Interestingly, Nickles recognizes this and explicitly says that in some cases $reduction_2$ may help clarify the analogous relationship between T_2 and T_2^* in the Schaffner model:

> There is no reason to think that $reductions_2$ can in all cases take over the work of Schaffner's analogy relation, but when it can we shall have T_1 $reducing_1$ T_2 approximatively by $reducing_1$ T_2^*, which in turn $reduces_2$ to T_2." (p. 195)

I will now push this idea forward and argue that in many cases, $reduction_1$ and $reduction_2$ may be combined to achieve the same epistemic goals and that $reduction_2$ may play a crucial role not only in the explanation of the analogous relationship between T_2 and T_2^*, but also in the choice of the auxiliary assumptions that help achieving the (Nagelian) reduction of a theory to another.

A good example that illustrates that $reduction_1$ and $reduction_2$ may be combined to achieve the same epistemic roles is the case of phase transitions mentioned in the previous section (see also Palacios, 2022). First order phase transitions are defined as discontinuities in the first derivatives of the free energy. In statistical mechanics, the free energy is given by the following expression:

$$F(K_n) = -k_B T \ln Z, \tag{8.2}$$

where K_n is the set of coupling constants, k_B is the Boltzmannian constant, T is the temperature, and Z is the canonical partition function, defined as the sum over all possible configurations:

$$Z = \Sigma_i e^{H_i} \tag{8.3}$$

Since the Hamiltonian H is usually a non-singular function of the degrees of freedom, it follows that the partition function, which depends on the Hamiltonian, is a sum of analytic functions. This means that neither the free energy, defined as the logarithm of the partition function, nor its derivatives can have the discontinuities that characterize first-order phase transitions in thermodynamics. Taking the thermodynamic limit $N \rightarrow \infty$, $V \rightarrow \infty$ allows one to recover the discontinuities in the derivatives of the free energy and therefore allows to derive the thermodynamic treatment of phase transitions. In sum, what the most important approaches to phase transitions do is to take the limit of a given quantity Q, i.e. the free energy, in order to reproduce the values of the corresponding quantities in the limit theory T_1^*, in which $N = \infty$. In other words, one proves that the values of the derivatives of the free energy in finite statistical mechanics SM converge towards the values of the corresponding quantities evaluated in infinite statistical mechanics SM_∞, such that:

$$\lim_{N \to \infty} Q_N^{SM} = Q^{SM_\infty}.$$

Since the singularities in the derivatives of the free energy are successfully obtained in the limit theory SM_∞, one can construct bridge laws that relate the thermodynamic derivatives of the free energy with the corresponding quantities in "infinite statistical mechanics" and *deduce* the thermodynamic treatment of phase transitions (Butterfield, 2011b; Palacios, 2019, 2022). In summary, the reduction of first order phase transitions to statistical mechanics involves the following basic steps: (1) a *limiting reduction* between the relevant quantities of T_1 (i.e. SM) and T_1^* (i.e. SM_∞) by the use of the thermodynamic limit. (2) The use of bridge laws that relate the derivatives of the thermodynamic free energy with the corresponding quantities in infinite statistical mechanics. (3) A Nagelian reduction between T_1^* and T_2 (the thermodynamic treatment of phase transitions) with the help of bridge laws and auxiliary assumptions (e.g. lattice structure, a particular kind of degrees of freedom, ranges of values of the degrees of freedom, etc.).[11] In a similar way, I believe that the attempt to reduce the Second Law to statistical mechanics by using the thermodynamic limit can also be regarded as a case that combines both limiting reduction and Nagelian reduction. Crowther (2019) offers an interesting analysis of the attempts of reducing both general relativity and quantum field theory to quantum gravity and suggests that this is a potential case that combines Nagelian reduction and Nickles' concept of $reduction_2$.

It is important to note that in the previous examples, *limiting reduction*, which for Nickles would correspond to $reduction_2$, is playing a role in what he calls "domain-combining" reductions, by which two different theories describe the same range of phenomena at different levels of description. This means—against what Nickles suggests—that $reduction_2$ is not restricted to the so-called "domain-preserving" reductions nor has a merely heuristic and justificatory role. In fact, the reduction of the thermodynamic treatment of phase transitions to statistical mechanics can be interpreted as having an ontological goal, in the sense that it can suggest that phase transitions are nothing over and above the result of atomic behavior. This behavior can also be interpreted as having an explanatory role, in the sense that it has the potential of explaining the thermodynamic behavior of phase transitions from the perspective of statistical mechanics.

One should also note that in the case of phase transitions, the thermodynamic limit, i.e. limiting reduction, helps deriving the exact treatment of phase transitions. This means that this reduction does not lead to the modification of the theory of phase transitions and, therefore, cannot be covered by the Schaffner model. Interestingly, finite approaches to phase transitions that do not invoke the thermodynamic limit (Gross, 2001a) allow one to derive an alternative theory of phase transitions for specific case-studies (non-extensive systems). This reduction appears to fit well with the Schaffner model. In other words, in the case of phase transitions, scientists seem to have the option of deriving the original thermodynamic treatment of phase transitions by combining limiting reduction and Nagelian reduction, or deriving an alternative theory of phase transitions that corrects the thermodynamic treatment

[11] The reduction of phase transitions is still a controversial issue in the philosophical literature. Batterman (2001), for example, has famously argued against the reduction of phase transitions pointing out the "singular" nature of the thermodynamic limit. Butterfield (2011b), Norton (2011), and Palacios (2019), among others, have replied to these arguments suggesting that the "singular nature" of the thermodynamic limit is not incompatible with the reduction of phase transitions to statistical mechanics.

without invoking the thermodynamic limit, i.e. without using limiting reduction. This latter case is a good candidate for Schaffner reduction.

A question that arises is whether $reduction_2$ should always be combined with Nagelian reduction, in other words, whether $reduction_2$ always plays the role of straighten the assumptions that allow one to derive the original theory T_2. Although in the cases examined above this seems to be the role of limiting reduction, there is no reason to think that the role of limiting reduction is restricted to straighten the assumptions that allow for Nagelian reduction. Indeed, as Nickles pointed out, in some cases, limiting reduction, or more generally, $reduction_2$ may be useful in spelling out the notion of strong analogy. For example, $reduction_2$ can indirectly help understanding the relationship between T_2 and T_2^* by offering a precise definition of "closeness". In fact, a quantitative comparison between the values of the quantities Q_1 evaluated in the finite theory T_1 and the values of the corresponding quantities Q_1^* evaluated in the limit theory T_1^* can in some cases give us information about how close are the values of the relevant quantities in the theory that can be derived from T_1, i.e. T_2^*, to the values of the corresponding quantities in the original theory T_2. For instance, in the case of phase transitions, if the values of the quantities evaluated in finite statistical mechanics are close to the values of the corresponding quantities in infinite statistical mechanics $Q_{N_0}^{SM} \approx Q^{SM_\infty}$, we can infer that the values of the quantities Q^{TD^*} in the modified thermodynamic theory TD^* that can be derived from SM are also close to the values of the quantities evaluated in original thermodynamic theory of phase transitions TD. In this particular sense of numerical approximation, $reduction_2$ can be useful to understand the strongly analogous relationship between T_2 and T_2^*.

But one should also allow for the possibility of obtaining $reduction_2$ even in cases in which the Nagelian and Schaffner reduction fail. As Nickles points out, only rarely will all the equations of a theory T_2 be reduced to the equations of T_1 and only rarely a whole theory will reduced to another. In most cases, what is achieved is a *partial* reduction, in which T_1 and T_2 stand for theory parts. I believe that $reduction_2$ can be especially useful in cases of partial reductions. A good candidate for $reduction_2$ is the partial reduction of classical mechanics to the general theory of relativity. Fletcher (2019) has offered a detailed analysis of this reduction and has suggested that the partial reduction of classical mechanics to general relativity is not the result of a logical derivation of the general theory of relativity from classical mechanics, but rather of giving a specific interpretation to the limit operation $c \to \infty$ that meets the explanatory demand of intertheoretic reductions. Another example (although I admit a contentious example) is the *partial* reduction of classical mechanics to quantum mechanics in the classical $\hbar \to 0$ limit. Feinzeig (2020) has argued that giving a certain interpretation of the $\hbar \to 0$ limit allows one to show that the predictions given by quantum mechanics for expectation values are in many cases "close" to the corresponding predictions in classical physics. In this case there is no logical deduction of the corrected theory from the reducing theory, but one can still use this limiting relationship to explain why the predictions of classical mechanics are nearly accurate in many systems.

An advantage of $reduction_2$ is that it does not require the corrected theory T_2^* to be derivable (i.e. logically deducible) from T_1 and in this way serves to account for many cases of reduction in which logical deduction does not seem to play an important role.

To sum up, in agreement with Nickles (1973), I have argued in this section that $reduction_2$ shall be distinguished from other models of reduction such as the Nagelian model or the Schaffner model. However, contra Nickles, I have pointed out that there are no specific epistemic goals or functions associated with $reduction_2$. Indeed, in some cases, $reduction_2$ may be used to straighten the assumptions that allow for (approximative) Nagelian reduction and in this way it may help obtain a reduction that aims for the consolidation of the reducing theory, or the explanation of the reduced theory. In other cases, it may help spelling out the analogical relationship between T_2 and T_2^* in Schaffner-type reductions. Yet in other cases, may help accounting for partial reductions that aim at the justification of the success of the reduced theory T_2, even in cases in which the Nagelian and the Schaffner reduction fail.

8.7 The Structuralistic Model of Reduction

Torretti (1990) discusses another model of intertheoretic reduction, which for some reason has been largely ignored in the Anglo-American discussion. This is the model of Balzer et al. (1987), which is part of the so-called structuralist program and is characterized by the use of informally stated set-theoretical predicates ("informal" in the sense that there is no formal language employed). This model has the advantage of giving a precise reconstruction of intertheoretic relations and of offering a precise notion of intertheoretic approximation. Without going into so many details, the structuralist notion of reduction can be characterized as follows (Moulines, 1984, p. 53): Let T_1 be a theory consisting of a class M_p of potential models (e.g. one potential model contains a set of particles, a set of springs together with their spring constants, the masses of the particles, as well as their positions and mutual forces as functions of time), a subclass M of actual p models (e.g. the subclass of potential models satisfying the system's equation of motion), and a set I of intended applications ("pieces of the world" to be explained, predicted or technologically manipulated). Correspondingly, let T_2 consist of M'_p, M' and I'. T_2 is reducible to T_1 by means of F iff there are M'_r, I'_o so that:

1. $M'_r \subseteq \mathrm{M}'_p$ and $M'_r \cap \mathrm{M'} \neq \emptyset$
2. $\emptyset \neq I'_o \subseteq I' \cap M'_r$
3. $F : M'_r \longmapsto M_p$ and F is many-one.
4. $\overline{\mathrm{F}}(M'_r \cap M') \subseteq M$
5. $\overline{\mathrm{F}}(I'_o) \subseteq I$

It is clear from this definition that this model does not require semantic predicate-by-predicate connections (in the sense of Nagel's bridge laws), nor the deducibility of statements, although it is compatible with them. In this sense, this model is weaker than the Nagelian model of reduction. As stated above, an important advantage of this model of reduction is that if one introduces appropriate topological uniformities defined on the classes M_p and M'_p, this model of reduction can give a precise characterization of approximative reductions that could serve to account not only for cases of limiting reduction but also for other approximation relations. More precisely, by introducing a uniformity, which imposes a topology on an otherwise unstructured set, they define a class of admissible "blurs" (degrees of approximation admitted between different models) as elements of a uniformity, such that if a pair $\langle x, y\rangle$ is an element of "blur" u (where $x \in M_p$, $y \in M'_p$), then x and y approximate each other at least to the degree given by u. For real numbers (or any other standard metric, where the absolute value of the different of elements is meaningful), each blur is determined by a particular ϵ:

$$u_\epsilon = \{\langle x, y\rangle : |a - b| < \epsilon\}.$$

Interestingly, they define an *upper bound* of admissibility, meaning that beyond a degree of inaccuracy at or above the upper bound, the approximation becomes unacceptable (Balzer et al., 1987, p. 330).

One can see then that the structuralistic account of reduction opens the possibility of discussing matters of strong analogy and approximative reduction on a less informal level than other models of reduction, and in this sense it has the potential of improving the approximative Nagelian model and Nickles' model of $reduction_2$. The problem is then shifted to the task of showing that some of the interesting cases of reduction fit into this account. Although the model has led to detailed reconstructions of particular examples of reductions in physics, such as the reduction of rigid body mechanics to Newtonian particle mechanics (Sneed, 1971) and Kepler's planetary theory to Newtonian particle mechanics (Moulines, 1980), it is not at all clear whether all relevant reductions in physics can be reconstructed in this way.

Another problem of this account is that reconstructions in this sense can only be made *a posteriori*, which means that the structuralist account of reduction cannot play a heuristic role in the construction of reducing theories. However, this does not mean that this approach is useless, since a proper reconstruction of reduction in this sense has the potential to help specify the approximation relations between the reduced and the reducing theory and in this way to give a solid justification for the approximate success of the reduced theory. In fact, this model has inspired recent reductive analysis of specific cases of reduction in physics, in which the approximation relations between the reduced and the reducing theory have been topologically specified (e.g. Scheibe, 1997; Fletcher, 2015; Feinzeig, 2020). Whether this approach can help stating the fundamentality of the reducing theory with respect to the reduced theory is controversial. Indeed, a big limitation of this model is that it does not specify the sort of transformation F that relates the domains in the two theories. Therefore, it allows for cases that are not generally considered to be successful (ontological) reductions. For example, it could account for cases in which there is an *ad hoc* mathematical relationship between the domains of two theories that are completely alien to each other. A possible solution to this problem comes from specifying the sort of relationship admissible in successful reductions. For instance, Moulines (1984) has argued that in order to have a real reduction at the ontological level, the sets of physical individuals (domain of a theory) need to be in a biunivoque correspondence, in other words, the ontological reductive link needs to be a one-to-one function. However, the requirement of biunivoque correspondence may appear too strong to account for reductions in the most relevant cases. One could try to weaken this requirement by requiring just a functional correspondence in one direction, which would be close to functionalist approaches to reduction (Kim, 1998). Whether a functional correspondence between the two theories is enough to establish the fundamentality of a theory with respect to another is still matter of controversy in the philosophical literature (e.g. Batterman, 2001, Ch. 5).

In sum, we have seen that the structuralist approach to reduction has the potential to make precise the meaning of approximative relations between theories. In this way, it has the potential to improve the approximative Nagelian model and Nickles' concept of $reduction_2$. The problem is that reductions in this sense are reconstructed *a posteriori*, which means that they cannot play an heuristic role in the construction of reducing theories. The extent to which this approach to reduction can help establishing the fundamentality of a theory over another depends on a specification of the sort of transformation F that relates the domains in the two theories.

8.8 Conclusion: A Pluralistic Approach to Reduction

In this chapter, I have defended a pluralistic approach to reduction, in which reduction is not understood in terms a single philosophical "generalized model", but rather as a family of models of intertheoretic reduction that can help achieve certain epistemic and ontological goals. By analyzing historical cases of reduction in physics, I have distinguished between six different functions of reduction: *explanation*, *justification* and *correction* of the reduced theory, as well as *development*, *consolidation* and relative *fundamentality* of the reducing theory. I have also argued that not all cases of reduction in physics aim to achieve the same goals. In fact, I have suggested that there are interesting trade-offs between some of the goals typically associated with reduction. More specifically, I have suggested that there is a trade-off between the correction of the reduced theory and the consolidation of the reducing theory. Thus, I have suggested that the model of reduction that best suits to a particular case study depends on the specific goals that underlying the intended reduction. For instance, cases in which the main goal is to consolidate the reducing theory instead of modifying the reduced theory may fit better with the

Nagelian model. On the other hand, cases that aim to correct the secondary theory according to the reducing theory may fit better with Schaffner's model of reduction. I have also argued that there are cases, in which in order to achieve the goals of the intended reduction, it is necessary to combine the Nagelian or Schaffner's model with other models of reduction that give a more precise notion of approximation, such as Nickles' model of $reduction_2$ and the structuralist approach.

Fortunately, philosophers of science present us with a number of notable models of reduction (by no means restricted by the models that I have discussed here), which help us understand the nature of the most important cases of reduction in physics. Defining reduction according to a single model of reduction, such as the Nagelian model, makes the term "reduction" too restrictive and incapable to account for cases that satisfy the most important epistemic and ontological goals of scientific reductive programs. A consequence of this is that some successful cases of reduction may be wrongly classified as failures of reduction because they do not meet the conditions of the standard model. For instance, phase transitions have been regarded sometimes as a failure of reduction because (among other reasons) the thermodynamic treatment cannot be *deduced* from finite statistical mechanics (e.g. Batterman, 2001; Anderson, 1972). Having a better understanding of reduction that allows us to account for successful reductive programs in physics is important not for nomenclature issues, but because it can give us a better understanding of the relationship between different theories.

References

Anderson, P. W. (1972). More is different. *Science, 177*(4047), 393–396.

Balzer, W., Moulines, C. U., & Sneed, J. D. (1987). *An architectonic for science: The structuralist program* (Vol. 186). Springer Science & Business Media.

Batterman, R. W. (2001). *The devil in the details: Asymptotic reasoning in explanation, reduction, and emergence.* Oxford University Press.

Blackmore, J. T. (1995). *Ludwig Boltzmann: His later life and philosophy, 1900–1906: Book two: The philosopher* (Vol. 174). Springer Science & Business Media.

Boltzmann, L. (1877). *Über die Beziehung zwischen dem zweiten Hauptsatze des mechanischen Wärmetheorie und der Wahrscheinlichkeitsrechnung, respective den Sätzen über das Wärmegleichgewicht.* Sitzungsbericht der Akadamie der Wissenschaften, Wien (Vol. II, pp. 67–73).

Boltzmann, L. (1885). Über die möglichkeit der begründung einer kinetischen gastheorie auf anziehende kräfte allein. *Annalen der Physik, 260*(1), 37–44.

Brush, S. G. (2006). Ludwig boltzmann and the foundations of natural science. In *Ludwig Boltzmann (1844–1906)* (pp. 65–80). Springer.

Butterfield, J. (2011a). Emergence, reduction and supervenience: A varied landscape. *Foundations of Physics, 41*(6), 920–959.

Butterfield, J. (2011b). Less is different: emergence and reduction reconciled. *Foundations of Physics, 41*(6), 1065–1135.

Callen, H. B., & Kestin, J. (1960). An introduction to the physical theories of equilibrium thermostatics and irreversible thermodynamics.

Casetti, L., & Kastner, M. (2006). Nonanalyticities of entropy functions of finite and infinite systems. *Physical Review Letters, 97*(10), 100602.

Dauxois, T., Latora, V., Rapisarda, A., Ruffo, S., & Torcini, A. (2002). The hamiltonian mean field model: from dynamics to statistical mechanics and back. In *Dynamics and thermodynamics of systems with long-range interactions* (pp. 458–487). Springer.

Dizadji-Bahmani, F., Frigg, R., & Hartmann, S. (2010). Who's afraid of nagelian reduction? *Erkenntnis*, *73*(3), 393–412.

Einstein, A. (1910). Theorie der opaleszenz von homogenen flüssigkeiten und flüssigkeitsgemischen in der nähe des kritischen zustandes. *Annalen der Physik*, *338*(16), 1275–1298.

Feinzeig, B. (2020). The classical limit as an approximation. *Philosophy of Science* (forthcoming).

Feyerabend, P. K. (1962). Explanation, reduction, and empiricism.

Fletcher, S. C. (2015). Similarity, topology and physical significance in relativity theory. *The British Journal for the Philosophy of Science*, *67*, 365–389.

Fletcher, S. C. (2019). On the reduction of general relativity to newtonian gravitation. *Studies in History and Philosophy of Science Part B: Studies in History and Philosophy of Modern Physics*, *68*, 1–15.

Frigg, R. (2008). A field guide to recent work on the foundations of statistical mechanics. In *The ashgate companion to contemporary philosophy of physics*. Ashgate.

Greene, R. F., & Callen, H. B. (1951). On the formalism of thermodynamic fluctuation theory. *Physical Review*, *83*(6), 1231.

Gross, D. (2001a). Second law of thermodynamics, macroscopic observables within boltzmann's principle but without thermodynamic limit. Preprint, arXiv:cond-mat/0101281.

Gross, D. H. (2001b). *Microcanonical thermodynamics: Phase transitions in "small" systems*. World Scientific.

Gross, D. H. (2002). Thermo-statistics or topology of the microcanonical entropy surface. In *Dynamics and thermodynamics of systems with long-range interactions* (pp. 23–44). Springer.

Kemeny, J. G., & Oppenheim, P. (1956). On reduction. *Philosophical Studies: An International Journal for Philosophy in the Analytic Tradition*, *7*(1/2), 6–19.

Kim, J. (1998). *Mind in a physical world: An essay on the mind-body problem and mental causation*. MIT Press.

Kitcher, P. (1984). 1953 and all that: A tale of two sciences. *The Philosophical Review*, *93*(3), 335–373.

Klein, M. J. (1973). The development of boltzmann's statistical ideas. In *The Boltzmann equation* (pp. 53–106). Springer.

Loschmidt, J. (1876). Ueber den zustand des wärmegleichgewichtes eines system von körpern. *Akademie der Wissenschaften, Wien. Mathematisch-Naturwissenschaftliche Klasse, Sitzungsberichte*, *73*, 128–135.

Mainwood, P. (2006). *Phase transitions in finite systems*. Ph.D. Thesis, University of Oxford.

Mishin, Y. (2015). Thermodynamic theory of equilibrium fluctuations. *Annals of Physics*, *363*, 48–97.

Moulines, C. U. (1980). Intertheoretic approximation: the kepler-newton case. *Synthese*, 387–412.

Moulines, C. U. (1984). Ontological reduction in the natural sciences (1). In *Reduction in science* (pp. 51–70). Springer.

Nagel, E. (1949). The meaning of reduction in the natural sciences. In R. Stauffer (Ed.), *Science and Civilization*. University of Wisconsin Press.

Nagel, E. (1961). *The structure of science: Problems in the logic of scientific explanation*. Hackett.

Nagel, E. (1970). Issues in the logic of reductive explanations. In H. K. K. Munitz (Ed.), *Mind, science and history* (pp. 117–137). SUNY Press.

Nickles, T. (1973). Two concepts of intertheoretic reduction. *The Journal of Philosophy*, *70*(7), 181–201.

Palacios, P. (2019). Phase transitions: A challenge for intertheoretic reduction? *Philosophy of Science*, *86*(4), 612–640.

Palacios, P. (2022). *Emergence and reduction in physics*. Cambridge University Press.

Poncaré, H. (1889). Sur les tentatives d'explication m'ecanique des principes de la thermodynamique. *Comptes Rendus de l'Academie des Sciences*, *108*, 550–553.

Rau, J. (2017). *Statistical physics and thermodynamics: An introduction to key concepts*. Oxford University Press.

Sarkar, S. (2015). Nagel on reduction. *Studies in History and Philosophy of Science Part A*, *53*, 43–56.

Schaffner, K. F. (1967). Approaches to reduction. *Philosophy of Science*, *34*(2), 137–147.

Schaffner, K. F. (1977). Reduction, reductionism, values, and progress in the biomedical sciences. *Logic, Laws, and Life*, *6*, 143–171.

Schaffner, K. F. (2012). Ernest nagel and reduction. *The Journal of Philosophy*, *109*(8/9), 534–565.

Scheibe, E. (1997). *Die Reduktion physikalischer Theorien: Ein Beitrag zur Einheit der Physik*. Springer-Verlag.

Sklar, L. (1995). *Philosophical issues in the foundations of statistical mechanics*. Cambridge University Press.

Sneed, J. D. (1971). *The logical structure of mathematical physics* (Vol. 35). Dordrecht.

Suppe, F. (1974). *The structure of scientific theories*. University of Illinois Press.

Tisza, L., & Quay, P. M. (1963). The statistical thermodynamics of equilibrium. *Annals of Physics*, *25*(1), 48–90.

Torretti, R. (1990). *Creative understanding*. University of Chicago Press.

Valente, G. (2021). Taking up statistical thermodynamics: Equilibrium fluctuations and irreversibility. *Studies in History and Philosophy of Science Part A*, *85*, 176–184.

Van Riel, R. (2011). Nagelian reduction beyond the Nagel model. *Philosophy of Science*, *78*(3), 353–375.

Worrall, J. (1989). Structural realism: The best of both worlds? *Dialectica*, *43*(1–2), 99–124.

Zermelo, E. (1896). Ueber mechanische Erklärungen irreversibler Vorgänge. eine antwort auf Hrn. Boltzmann's „Entgegnung" . *Annalen der Physik*, *295*(12), 793–801.

Chapter 9
Inductive Inferences on Galactic Redshift, Understood Materially

John D. Norton

This paper is dedicated with gratitude to Professor Roberto Torretti, whose work and personal example exercised a formative influence on me. He set a standard of clarity and precision in philosophical writing that I have long sought to emulate. He also showed me that precision in writing and a wicked sense of humor can cohabit. The exhilarating time we spent together in 1983 as Fellows in the Center for Philosophy of Science at the University of Pittsburgh remains as vivid to me as if it happened yesterday.

Abstract A two-fold challenge faces any account of inductive inference. It must provide means to discern which are the good inductive inferences or which relations capture correctly the strength of inductive support. It must show us that those means are the right ones. Formal theories of inductive inference provide the means through universally applicable formal schema. They have failed, I argue, to meet either part of the challenge. In their place, I urge that background facts in each domain determine which are the good inductive inferences; and we can see that they are good in virtue of the meaning of the pertinent background facts. This material theory of induction is used to assess the competing inductive inferences in the debate in 1972 between John N. Bahcall and Halton Arp over the import of the redshift of light from the galaxies.

I thank Siska De Baerdemaeker for helpful comments on an earlier draft.

J. D. Norton (✉)
Department of History and Philosophy of Science, University of Pittsburgh, Pittsburgh, PA, USA
e-mail: jdnorton@pitt.edu

C. Soto (ed.), *Current Debates in Philosophy of Science*, Synthese Library 477,
https://doi.org/10.1007/978-3-031-32375-1_9

9.1 Introduction

Good science is distinguished from other imaginative narratives by the fact that we have good evidence for its extraordinary accounts. If we cannot recover this most basic fact about science, our efforts to develop a cogent philosophy of science have come to nothing. Yet we, as philosophers of science, are faring poorly at the task. Our accounts of inductive inference have so far been a poor fit with the actual practice of science. The failure has stung most for me when I turn to the history of science. For there we see how the scientists found and weighed the evidence advanced in support of our science. In exploring this history, we should be able to use our accounts of inductive inference to identify when the efforts to provide good inductive support for our science have succeeded; and when they have failed.

However, in recounting various episodes in the history of science, I found it hard to use existing accounts of inductive inference to assess the cogency of the inferential claims. In specific cases, I would commonly find some account in the scattered repertoire that would fit. But the fit would be Procrustean. An inference may look initially like a strong inference to the best explanation. But the strength of the inference depended on an assessment of the quality of the explanation. We have no good account of that quality. Take the wonderfully crafted, intricate optical instrument that is the eye. Just why is Darwin's complicated and incomplete explanation of its evolution better than the simplicity of the creationist's single supernatural hypothesis?

In other cases, it was not so hard to fit Bayes' theorem to some judgment of evidential support. The nagging worry was that the scientists themselves were not using Bayes' theorem or even formulating probabilities. They have their own explicitly non-probabilistic ways of proceeding. Worse, the real inductive work was rarely done by the theorem. It was done in the selection of likelihoods and other probabilities. Once the informal reasoning that led to that selection was exposed and appreciated, actually computing Bayes' theorem became a superfluous afterthought. Indeed, shrewd selection of the probabilities seemed able to push Bayes' theorem to give almost any desired result.

Some reflection on the nature of inductive inference was needed. Present accounts of inductive inference are almost exclusively formal and universal. In this regard, they are modeled on accounts of deductive inference. They provide formal schemas or templates for good inductive inferences or for appropriate relations of support; and the schemas are applied by substituting factual content into the slots of the schemas. The schemas are universal in the sense that they can be applied equally in any domain. One merely needs to substitute content from any domain into them.

While each of the accounts in the present repertoire can boast of successes in some specific cases, none succeeded universally. At best the repertoire provides us with a patchwork. One scheme works here; a different one works there. It soon became clear that this patchwork character is actually the essence of the matter. It is central to the material theory of induction, first formulated in Norton (2003). According to it, there are no universally applicable schema for inductive inference.

Rather the licit inductive inferences in each domain are warranted by the true, contingent facts of that domain. This account, elaborated briefly in Sects. 9.2 and 9.3 below, meets the dual requirements for a successful account of inductive inference. First, the account must tell us which items of evidence provide inductive support for which propositions. The background facts of the domain do that. Second, it must provide good reasons for why the particular selections of the account are properly cases of inductive support. Those reasons derive directly from the meaning of those true background facts.

The bulk and the remaining part of the paper, Sects. 9.4–9.12, illustrates the utility the material theory of induction in a particular piece of history of science. It is the debate in 1972 between John N. Bahcall and Halton Arp over the import of the redshift of light from the galaxies. Does this redshift reveal an expansion of the universe, as then standard cosmology asserted? Or does the redshift derive from other processes, as yet not fully understood? We shall see how the material theory of induction allows us to delineate and appraise the inductive inferences of each side of the debate. We shall see that central to the debate was the establishing by each proponent of the background facts needed to support their inductive inferences. They also put considerable effort into impugning the corresponding background facts of their antagonists. We also see how the inductive reach of the two sides depended on whether they could access background facts that would support further inductive inferences.

9.2 The Material Theory of Induction

The central thesis of the material theory of induction is that inductive inferences are not warranted by their conformity with universally applicable schema, but by true background facts.[1] These background facts distinguish those inductive inferences that are well-warranted; and the cogency of the warrant is assured by the meaning of the background facts. Since there is no universally applicable, factual principle of induction, there are no universal warranting facts. Hence each warranting fact obtains only in some limited domain. In such a domain, a warranting fact may be rich enough to induce some formally expressible system or logic of inductive inference. However, that logic will be warranted only within that specific domain and may not apply elsewhere. In this sense, all inductive inference is local.

The simplest argument for the material theory proceeds from the ampliative nature of inductive inference. In such an inference, the conclusion always asserts more factually than the premises. Hence any such inference can fail if attempted in a domain that is inhospitable to the inference. The requirement that the domain be hospitable to that inference is, in its most generic form, the background fact of the domain that warrants the inference.

[1] For considerable elaboration on this sketch of the material theory, see Norton (2021, Ch. 1, 2)

9.3 Material Successes

The present literature on logics of induction is enormous and offers us many competing formal systems designed to be universally applicable. Each system works somewhere but turns out to be at best a contrived fit elsewhere. The material theory of induction explains why each such system works where it does: the background facts are hospitable to it. And it explains why they fail when they do: the background facts are inhospitable.

Since there are so many competing accounts of inductive inference, it is practical here only to provide a few illustrations of how these accounts are accommodated by the material theory of induction. Following the categorization of accounts of inductive inference of Norton (2005), we may group accounts of induction into three broad families. Those in the family of inductive generalization proceed under the principle that an instance confirms the generalization. Its simplest form is enumerative induction: from the evidence that some *A*s are *B*, we infer that all *A*s are *B*. The difficulty with this universal schema is that is almost always fails. It only works if the *A*s and *B*s are very carefully chosen. In spite of millennia of efforts, no general rule has been found for specifying which *A*s go with which *B*s.

The material theory of induction entails that no general rule has been found because there is no general rule to be found. The commonality of instances of enumerative induction is superficial. Each or each grouping of them is warranted by background facts peculiar to the pertinent domain. Marie Curie in 1903 found that a mere tenth of a gram of Radium Chloride was crystallographically like Barium Chloride. She had no hesitation in generalizing from that tenth of a gram to all samples of Radium Chloride. This inductive inference was not warranted by a general schema such as enumerative induction. She did not infer to many other conclusions authorized by this schema: that all samples of Radium Chloride must be less than a tenth of a gram, or all must be in Paris, or all must be prepared by Curie. Rather the specific inference she did make was warranted by a hard-won fact of nineteenth century mineralogy known as "Haüy's Principle": all crystalline substances fall into a small set of families, distinguished by the configuration of the axes characteristic of the shape.[2] Other members of the family include Hempel's satisfaction criterion, Glymour's bootstrap, Mill's methods and arguments from analogy.[3] Each can be accommodated within the material theory of induction by comparable analysis.

A second family of accounts of inductive inference, hypothetical induction, is based on the principle that that ability of some hypothesis to entail the evidence is a mark of its truth. This principle, by itself, assigns the mark too indiscriminately to be viable. Many accounts add further conditions in order to restrict this assignment to cases to which it properly belongs. Each account has superficial plausibility,

[2] For further details of this example, see Norton (2021, Ch. 1)

[3] For the case of analogy, see Norton (2021, Ch. 4).

but none succeed generally. Rather they succeed where they do because of facts obtaining locally in the pertinent domain.

As an added condition, we might require that the successful hypothesis must also be simple. Notoriously, there is no general account of what makes an hypothesis simple. Rather, simplicity is merely a compactly expressible surrogate for background facts that vary from domain to domain.[4] Alternatively, we might add a severe testing requirement: that had the hypothesis failed to entail the evidence, then it would most likely be false. There is no universal way to implement this severe testing condition. It is realized only by determining factually in each domain what is most likely. Finally, in inference to the best explanation, we require that the hypothesis not just entail the evidence, but that it also explains it. This scheme has proven hardest to explicate in material terms simply because there is no precise formal specification of inference to the best explanation. Notably, there is no general, formal characterization of explanation such that its addition to mere deductive entailment of the evidence would boost the inductive strength associated with that entailment.[5]

In the third family of accounts, relations of inductive support are characterized by some explicit calculus. The dominant example is the probability calculus, as employed in Bayesian confirmation theory. There are many cases in which this sort of probabilistic analysis captures inductive support relations well. However, these are cases in which the particular background facts in turn authorize probabilistic relations of support, such as when we reason over samples drawn randomly from a population of known composition. Might the Bayesian system aspire to cover all relations of inductive support? Might it be that all relations of inductive support are just probabilistic facts; and all generalities about inductive support are theorems of the probability calculus?

These aspirations fail. For example, invariance arguments can show that, as the evidence becomes so weak as to be completely neutral, it enters into relations of inductive support that are inherently non-additive, in contradiction with the additivity axiom of the probability calculus. The Bayesian literature has laudably taken on the burden of proving that all relations of inductive support or, alternatively, all distributions of belief, must conform with the probability calculus. These proofs fail for a simple reason of logic. They seek to prove a contingent matter, that all these relations are probabilistic, by deductive means. Hence, they must begin with assumptions that are logically at least as strong as the conclusion sought. These assumptions must then covertly already presume the very thing we seek to prove. Once one knows to look for it, all the proofs are undone merely by demonstrating their circularity.[6]

---

[4] Or so it is argued in Norton (2021, Ch. 6 and 7).

[5] Or so it is argued in Norton (2021, Ch. 8 and 9), where examples of inferences to the best explanation are accounted for materially.

[6] See Norton (2021, Ch. 10–16) for an extended development of this remark and the other criticisms of Bayesianism's universal aspirations.

Finally, the now dominant subjective Bayesianism has compounded these problems by replacing objective relations of inductive support as the primitive notion with a subjective relation of belief. Objective support relations are now somehow to be wrestled from the subjective relations. This immerses the problem of understanding inductive inference in a larger problem of finding the formal relations that govern both inductive support and belief. However once one entangles mere, ungrounded opinions with objective relations of inductive support, it has proven too difficult to disentangle them. The much-vaunted convergence theorems cannot "wash out the priors" in the simple and common case in which two hypotheses entail the same body of evidence. The ratio of the posterior probabilities always remains equal to the ratio of the prior probabilities. The subjective Bayesian's probabilities are then an inseparable amalgam of arbitrary opinion and inductive support.

9.4 Cosmological Redshifts and the Recession of the Galaxies

The remainder of this paper reviews an historical debate in cosmology over the import of the redshift of light from galaxies. It came to a head in 1972 with a lively confrontation between the astronomers John N. Bahcall and Halton Arp. In it we shall see, as the material theory of induction predicts, that the debate hinges on the background warranting facts required to warrant the inferences on each side, even though they are sometimes only present tacitly. They are the assumption *Typicality*, whose truth is required to warrant Bahcall's inferences; and *Proximity*, whose truth is required by Arp's inferences. Much of debate involves efforts on each side to support their own assumption and to impugn the key assumption needed by the other.

A further consequence of the material view is that inductive inferences are possible only in so far as warranting background facts can be secured. We shall see in the debate below that Bahcall's side had considerable reach in its inductive inferences since it could call upon sufficient background facts to warrant them. Arp's side, however, had limited inductive reach precisely because of the dearth of suitable background facts.

The context of this debate was the single most important astronomical finding underpinning modern cosmological theory. According to the thesis of the recession of the galaxies, on average, the galaxies are receding from us with a velocity that increases in direct linear proportion to their distance from us. The fabled origin of the finding is a paper by Edwin Hubble (1929) with the transparent title: "A Relation between Distance and Radial Velocity among Extra-Galactic Nebulae." Establishing the relation required determinations of both velocities of recession and distances to the galaxies (Hubble's "extra-galactic nebulae"). The determination of the distances proved most troublesome and required some ingenious analysis on Hubble's part. Even so, his estimates of distances were almost an order of magnitude too small.

The determination of the velocities, however, seemed straightforward. Hubble had access to Slipher's measurements of frequency shifts in the spectra of galaxies towards the red. These redshifts were interpreted as Doppler shifts, in accord with an effect widely recognized in more local physics: the frequencies of waves emitted by a receding body are diminished by the recession of the source. The pitch of sound is lowered and the color of light is reddened. The Doppler shift had been established for stars within our galaxies. It reveals the motions of binary stars orbiting about each other, for example.[7] If we extend the effect to the galaxies, the magnitude of the redshift provides a convenient observational proxy for the velocity of recession of the galaxies.

From the outset, the connection between redshifts and velocities of recession seems scarcely to have been challenged. In an early paper reporting some of his first redshift measurements, Slipher (1912) remarked that the size of a velocity inferred "...raises the question whether the velocity-like displacement might not be due to some other cause, but I believe we have at the present no other interpretation for it." Hubble (1929) wrote as if the redshift was an uncontroversial proxy for velocity. More careful reading of his writings, however, reveals a prudent caution. By the time of his semi-popular *Realm of the Nebulae*, his view was asserted as (1936, p. 34): "Although no other plausible explanation of red-shifts has been found, the interpretation as velocity-shifts may be considered as a theory still to be tested by actual observations." In his history of cosmology, Kragh (2007, Ch. 3) recounts Hubble's persistent hesitations throughout his life to accept redshifts as betokening velocities of recession. Kragh also recounts the ideas of a number of dissident cosmologists who explored alternatives to velocities of recession as the origin of the galactic redshifts.

These hesitations had little effect on mainstream astronomy. Weinberg (1972, pp. 417–18) summarized the established view as:

> The announcement by Edwin Hubble[8] in 1929 of a "roughly linear relationship between velocities and distances" established in most astronomer's minds the interpretation of the red shift as a cosmological Döppler effect, and this interpretation has survived through the decades until the present.

A brief survey other cosmology texts written around this time and more recently indicates this view as widespread. There is scant if any mention at all of the possibility of another interpretation of the redshift of the galaxies.

[7] Aitken (1918) is a synoptic survey of double stars and gravitationally bound binary star systems. The interpretation of spectral shifts in the light from binary starts as velocity-derived Doppler shifts was then already a standard, heavily exploited method of analysis. Its origins derive from Pickering's August 1889 observation of double lines in a stellar spectrogram (p. 27).

[8] Weinberg's footnote is to Hubble (1929), misdated in an apparent typographical error to 1927.

9.5 The Redshift Controversy

The acceptance of the relationship of redshift and velocity of recession was not untroubled. The trouble was Halton Arp, an observational astronomer who became a strident dissenter. He harbored mounting suspicions that the standard account of the formation of spiral galaxies did not fit well with their observed configurations.[9] These suspicions compounded into a general sense that there were many galaxies whose observed character was similarly "peculiar." The resulting *Altas of Peculiar Galaxies* (1966a) was able to organize these peculiar galaxies into roughly similar groups. The hope expressed in the preface was that the *Atlas* could "not only clarify the working of the galaxies themselves, but reveal physical processes and how they operate in galaxies, and ultimately furnish a better understanding of the workings of the universe as a whole." The hope, we can see, is for science practiced in the Baconian vein. We collect our observations and let the science flow inductively from them.

Arp soon fixated on a quite specific anomaly. As he reported (Arp 1966b) to *Science*, he noted cases of objects that appeared to be physically connected in ways incompatible with the standard interpretation of galactic redshifts as velocities of recession. Most notable were cases in which one object seemed to be ejected from another, so they must be close to each other on cosmic distance scales. Yet these same objects might have greatly differing redshifts. This immediately cast doubt on a simple relationship between distances and redshifts; and thus also on the corresponding relationship of distances and velocities of recession. These "discordant redshifts," as he soon came to call them, formed the basis and the near entirety of Arp's subsequent efforts to impugn the standard view of galactic redshifts.

Arp's concerns could not be ignored. He was well credentialed as an astronomer. He was a Harvard graduate, earned a PhD from Caltech and was a staff member of the Palomar Observatory. His complaints were collecting sympathizers. George B. Field, chairman of the Astronomy Section of the American Association for the Advancement of Science for 1972, noted few opportunities for what he labeled "direct confrontation."[10] He arranged for a debate at the AAAS meeting of December 30, 1972, in Washington, D.C. Halton Arp was to defend his claims of discordant redshifts and John Bahcall, an astronomer at the Institute for Advanced Study in Princeton, would defend the standard view. After the event, papers derived from their presentations, as well as background papers nominated by each debater, were published in the volume, *The Redshift Controversy*.

Field reported (p. 12) that the volume specifically "address[ed] the observational evidence on discordant redshifts." That means that this volume provides us with something tailor-made to the present concerns on evidence and inductive inference:

[9] As Arp reported in the Preface to Arp (1966a).

[10] As he reported in his introduction to Arp and Bahcall (1973, p. 4).

it is a self-contained snapshot at the time of the debate of the evidence for and against the standard interpretation of the galactic redshifts. Here are the leading experts on both sides laying out their strongest cases. Hence the analysis below will be confined largely to the cases laid out in the *Redshift Controversy* volume. This volume does not mark the end of the debate. Arp continued to press his case for discordant redshifts. However, it does seem to mark the end of the fleeting interest of mainstream astronomers. Most commonly, Arp's concerns are not even mentioned in cosmology texts after the time of the debate.

9.6 For Redshifts as Distance Indicators

John Bahcall's contribution to the volume was devoted to defending the standard view that the galaxies are receding from us in a unified motion whose magnitude increases linearly with distance. The inference to this standard view has two components: first, there is a roughly linear relationship between galactic distance and redshift; and second, a galactic redshift is linearly related to a velocity of recession. Bahcall's arguments focused just on the first of these two components, for that was the component directly under threat from Arp's arguments. As a result, Bahcall's contribution was explicitly making the case for "redshifts as distance indicators," these words being the title of his contribution. Bahcall's case is in two parts. The first, to be reviewed here, made the positive case. The second, to be reviewed later, sought to overturn Arp's case for discordant redshifts. To be precise, the goal of Bahcall's positive case is to support this claim:

> *Redshift-Distance Relation.* The redshift of light from the galaxies increases linearly with the distance to the galaxy, confounded but not obscured by small deviations due to local "particular" motions of the galaxies.

Bahcall describes (e.g. pp. 65, 77) the positive case as resting on the passing by the standard view of six tests. That is, the view makes predictions that, if unsuccessful, would undermine the view. The tests are passed, he decides.

While this is a popular way of characterizing the evidential import of the evidence, it does not capture the strength of Bahcall's case. For passing a test can fail to provide the extent of support needed. An hypothesis of cosmic contraction predicts a blue shift in light from the galaxies. That prediction provides a test of the contraction hypothesis. Hubble's (1929) paper reported five galaxies whose blueshifts indicated velocities of approach. They are NGC 6822, 598, 221, 224 (Andromeda) and 3031. Had we known only these data, we would have found the test to be passed. However, we might not find the passing of the test a strong support for the contraction hypothesis, since these are mostly nearby galaxies. Their motions might be local, particular motions, not a manifestation of a larger cosmic motion. Worse, I selected these galaxies from Hubble's data precisely because they manifest a blueshift.

An effective regime of testing must somehow probe more fully the hypotheses under test, so that passing the tests is evidentially more potent that merely parrying a threat. That is the case with Bahcall's tests. They are set up and passed in such a way as to give extensive inductive support to the standard view. To see this, we need to move beyond the simple logic of tests. Viewed materially, this stronger relation of inductive support derives from a background assumption that, if true, can warrant the support relations. That assumption is:

> *Typicality.* The relations among redshifts, distances and absolute magnitudes reported for the galaxies surveyed in the evidence are typical of all galaxies in our vicinity.

The assumption assures us that we have not ended up selecting galactic data that will incorrectly bias our inductive inferences. Our Milky Way galaxy precludes easy extragalactic observations in parts of the sky. *Typicality* assures us that galactic behaviors there conform with those found in other, more accessible parts of the sky. The assumption may at first appear benign. However, it will prove to be at the center of the debate. The assumption does not hold for the fictitious case just proposed of a cosmic contraction. The five galaxies in the data set were selected specifically because of their blueshift. They are not typical.

Bahcall's inferences to the redshift-distance relation are authorized by this assumption, in so far as it is true. We shall see that, built into Bahcall's tests, is evidence directly for the assumption.

9.7 Bahcall's Positive Case

Bahcall concluded his positive case (p. 79) with a summary of the six tests passed by the standard view. They are quoted in turn below, with my commentary interleaved.

> *First*, the original relation between redshift and apparent brightness was tested and found valid for the brightest galaxies over a redshift range that is more than one hundred times larger than the range originally available to Hubble.
>
> *Second*, the average apparent brightness of all galaxies irrespective of type decreases[11] like $(\text{redshift})^{-2}$, with a scatter about the mean relation that can be understood in terms of the range in intrinsic luminosities.

These two tests refer to the most direct approach: one checks that the redshift and distance of many galaxies fall under the same linear relationship. If the assumption of *Typicality* is correct, then that assumption authorizes an inductive inference to the linearity of the relation universally.

The complication is that the determinations of distances to galaxies is difficult. The principal technique relies on the fact that the brightness of a galaxy diminishes with the inverse square of distance. However, to use this inverse square diminution

[11] Brightness decreases as $(\text{distance})^{-2}$. Thus, a linear relationship between distance and redshift is equivalent to brightness decreasing as $(\text{redshift})^{-2}$.

of brightness to determine distance requires knowing the absolute brightness of the galaxy. Two galaxies of same apparent brightness might be placed at very different distances from us. That could happen if the nearer galaxy has a smaller absolute magnitude and the more distant galaxy a greater absolute magnitude. Hubble's (1929) original work included galaxies close enough to us so that individual stars in them could be resolved. This enabled distance determinations based on the characteristic behaviors of stars like the Cepheid variables. By the 1970s, however, the investigations included galaxies well beyond the distances at which their individual stars could be resolved.

Bahcall reports two strategies to solve the problem. Galaxies commonly collect into clusters of several hundred. If the range of galactic types in each cluster is comparable, then the brightest galaxies in each cluster might have similar absolute magnitudes.[12] Differences in their apparent magnitudes could then serve as a surrogate for differences of distances. Hence, in the first test, Bahcall reports the most recent results of Sandage (1972): the redshift-distance relation holds for redshift and apparent magnitudes of the brightest galaxies in each cluster measured.

The second of Bahcall's test uses a different, less satisfactory way of compensating for differences in the absolute magnitudes of the galaxies. The absolute brightness of galaxies varies considerably. However, if we plot redshift against apparent brightness for many galaxies, we would expect the redshift-distance relationship to manifest as an average amongst scattered data, where the scatter is due to the differences of absolute brightness. Bahcall reports that just such a confounded relationship had been reported in Humason et al. (1956) for many galaxies.

The inductive inferences in these first two tests are warranted by the assumption *Typicality*. This assumption can only supply the warrant if it is true. This second test provides a means for Bahcall to give support to the assumption. He stresses (p. 71, his emphasis) that Humason and et al.'s analysis gives the result "for *all types* of galaxies"; and that:

> The most important inference to be drawn from Fig. V [of Humason and et al.] is that galaxy redshifts seem to be good distance indicators for average galaxies, not just for the brightest ellipticals.

The first test gave more accurate results. However, its data was restricted to the brightest galaxies. Might this selection of special galaxies compromise the typicality of the relation found? That a similar, albeit confounded, relation is found for all types of galaxies is inductive support for typicality.

This last inductive inference requires a background warranting fact. It is the assumption that a failure of typicality would most likely arise as different redshift properties for different types of galaxies; and it is otherwise unlikely. Using this fact we proceed from the agreement of redshift-distance relations among the brightest galaxies and all galaxies to infer that a failure of typicality is unlikely.

[12] Sandage (1972, p. 1), a paper cited by Bahcall, concludes just this similarity.

Bahcall's third and fourth tests addresses the confounding of the linear relationship due to local "particular" motions of the galaxies. The confounding must be small, else the linear relation is lost:

> *Third*, the mean redshift of double galaxies was found to be much larger than the difference in redshifts between the two members of each such pair.
>
> *Fourth*, the mean redshift of galaxies in a rich cluster was found to be much larger than the dispersion in redshifts of individual galaxies in the cluster.

These results affirm small confounding in the two cases considered: pairs of close galaxies and galaxies in rich clusters. The assumption of *Typicality* allows us to extend this smallness to all galaxies.

Once again, the inductive inference of the extension depends on *Typicality*. Bahcall notes that (his emphasis, p. 73):

> It is important to note that Page's results [on pairs of close galaxies] refer to double galaxies of *all types*: spirals, ellipticals, and irregulars.

As before, the constancy of the results over all types of galaxies provides inductive support for *Typicality*.

> Fifth, the apparent angular diameter of the brightest galaxies in rich clusters decreases as $(\text{redshift})^{-1}$.

This fifth test relied on a different, geometric method of determining distances. For any fixed, distant object, the distance is inversely proportional to its angular size. Bahcall reported Hubble's demonstration that, for each fixed class of galaxies, angular sizes on average decrease as $(\text{distance})^{-1}$. We also have that the angular diameter of the brightest galaxies decrease as $(\text{redshift})^{-1}$. Combining them we have on average a linear relationship between distance and redshift, for those galaxies mentioned. *Typicality* allows us to extend the relationship to all galaxies.

The final test is:

> Sixth, redshifts that are determined by radio and optical techniques agree to high accuracy.

This agreement is relevant since a velocity-derived Doppler shift in frequency must be the same across the spectrum. Its evidential significance is that it narrows the physical mechanism that might be responsible for the redshift. Through *Typicality*, it precludes mechanisms that produce redshifts by affecting frequencies differentially. However, it does not preclude a gravitational redshift such as general relativity associates with intense gravitational field.

Bahcall concludes, however, that (his emphasis, p. 77): "Note also that the 130 galaxies included in Fig. VIII include galaxies of *all types*." As before, this independence from type provides support for *Typicality*.

9.8 Arp's Discordant Redshifts

Arp's contribution to the *Controversy* volume had an essentially negative goal: to impugn the relationship between redshift and distance of the standard view. The case depended on finding discordant redshifts, that is, galactic bodies at similar distances from us but with significantly differing redshifts.

In the early pages of his contribution, his goal appeared to be a deductive refutation of an exceptionless redshift-distance relation. He prefaced his remarks by recalling how astronomers had become so confident of the redshift-distance relation that they routinely used redshifts to determine distances. He wrote (p. 20):

> ... if we can produce just one example of a redshift difference that cannot be explained as a velocity difference, then we have broken the assumption on which the redshift-distance relation is always applied to derive distances.

Arp, however, overstated the import of just one anomalous case. He continued

> In this eventuality it would then become necessary to reexamine each category of the different kinds of galaxies in order to see whether current distance assignments would need to be revised.

A single counterexample would demonstrate only that the relation had an exception. That left the possibility that the counterexamples were so rare that redshift-distance relation was scarcely compromised. It would be reliable but not infallible.

It soon became clear that Arp did have a stronger result in mind. He mentioned in passing on p. 24 that "... evidence to be produced later in this paper [will show] that nonvelocity redshifts are a quite general phenomenon ..." Establishing that generality would be sufficient to overturn Bahcall's inductive inferences supporting the standard view. For, if the generality is established broadly enough, it would show that *Typicality* fails. The connections between redshift and distance for the galaxies reported in support of the standard view would simply be true of those galaxies explicitly within Bahcall's data and not a broader generality.

To establish this generality, Arp produced with a catalog of instances. The first concerned quasars (pp. 20–31). They are "*quas*i-stell*ar* objects," that is, objects that look like unresolved stars, but manifest high redshift. The redshifts are so high that quasars must be at very remote distances, if the redshift-distance relation applies to them. It then follows that they must also be of extraordinarily great brightness.[13] Arp urged an alternative reading that escaped the need to posit extraordinary brightness. He urged that quasars are distributed statistically such that they are near galaxies in clusters or groups that are closer to us. Arp (p. 29) also reported what he boasted to be "the 'experimentum crucis'." He had found a line of four quasars that he interpreted as so aligned because they were ejected from a nearby galaxy.

[13] This was the standard view at the time of Arp's writing and the one now also accepted. Quasars are now believed to be extremely bright galactic nuclei.

The catalog continued in this vein. Arp turned to galaxies with discordant redshifts. He proceeded thorough various cases. He found that fainter galaxies in clusters tended to have greater redshifts than others in the clusters, for example. He continued with interacting double and multiple galaxies with discordant redshifts; chains of galaxies with discordant redshifts; galaxies connected by filaments with discordant redshifts; and tight groupings of galaxies with discordant redshifts.

Arp's analysis contained an inductive inference that would be targeted for extended criticism by Bahcall. Arp had used proximity in the starfield and other related features as supporting a physical connection between objects so that they are equally distant from us. This assumption was essential to the inference that the discordant redshifts violated the redshift-distance relation. We might express the assumption as:

> *Proximity*. An indicator of equality of distance is proximity of galactic objects in the star field, along with further features such as connecting filaments and alignments compatible with ejection.

The cogency of Arp's case depends upon this assumption. There is an initial plausibility to it. If we see filaments looking as if they are connecting objects in the star field, if we see alignments looking as if one object is ejected by another, perhaps they really are physically connected and thus at the same distance from us. Arp also used his data to support the assumption in a way similar to the way that Bahcall had sought to support *Typicality*. Arp concluded (pp. 55–56):

> It cannot be stressed too strongly, however, that these discordant redshifts are not discovered in just one or two isolated cases that have no relation to each other. But in every case we can test–large clusters, groups, companions to nearby galaxies, companions to middle-distance galaxies, companions linked by luminous filaments, galaxies interacting gravitationally, chains of galaxies–in every conceivable case, we come out with the same answer: the same discordant redshifts for the same general class of younger, fainter galaxies. This evidence, taken together with the same kind of evidence for the quasars–which are a kind of extremely young and, if this evidence is correct, intrinsically faint companion–forms a coherent picture of the kind of galaxies that have excess intrinsic redshifts.

This is evidence for Arp's overall case and, *a fortiori*, his assumption *Proximity*. It is a new inductive inference that is based on a further assumption: Were the indicators of *Proximity* unreliable, we would not recover consistent results such as reported here over many cases; but we likely would recover them if the indicators are reliable.

9.9 Bahcall's Rejoinder

The negative part of Bahcall's contribution to the *Controversy* volume constituted a direct rebuttal to Arp's case. Much of it is a detailed analysis of many cases of discordant redshifts reported by Arp. In each case, Bahcall sought to cast doubt on the discordance. For example, he argued (pp. 83–88) in specific cases that the filaments Arp reported may not be there at all, or may be an artifact of the photographic process, or may just be a chance superposition in our view of bodies widely separated in space. On pp. 107–13, Bahcall summarized the then present case for the standard view of quasars as very bright, very distant objects; and then on pp. 113–15 why Arp's case for discordant redshifts among quasars depended on tendentious statistical analysis.

In relation to general matters of inductive inference, the most interesting of Bahcall's complaints was a general criticism of the method employed by Arp to find discordant redshifts. It is a direct attack on Arp's assumption of *Proximity*. All the indicators of sameness of distance Arp reports could arise purely by chance. Bahcall wrote (p. 82):

> The skies when photographed with large telescopes reveal so many individual objects on any photographic plate that one can find almost any configuration one wants if one just hunts: even stars arranged as four-leaf clovers.

The criticism was raised repeatedly. Bahcall recounted the sustained efforts by Arp to establish a physical connection between the compact luminous galaxy M205 and the normal spiral galaxy NGC 4319, with M205 appearing to lie in an outer spiral of NGC 4319. He matched these efforts with the repeated failure of critics to find the connection. He concluded acerbically (p. 89, his emphasis):

> The moral of the story of the apparent connection between NGC 4319 and M 205 is clear: *Seek and ye shall find, but beware of what you find if you have to work very hard to see something you wanted to find.*

Bahcall then laid out the complaint most systematically as (p. 88, his emphasis):

> REMARKS. The way Arp carries out his observational programs, *searching for peculiarities that are not clearly specified before the observations*, actually prevents one from using the argument that a particular observed configuration is too unlikely to be due just to chance. The reader can see easily that such arguments when applied *a posteriori* may be misleading. Suppose you take a detailed large-scale photograph of Times Square that shows at one time in two dimensions all the people in the Square. The *a priori* probability that all the people have by chance the observed angular separations and apparent configurations seen on the photograph is "negligible"; the number of alternative possibilities is very large. However, the *a posteriori* probability for the observed configuration is unity, just because it is the observed configuration!

The admissibility of *Proximity* then depended on whether Bahcall's complaint could be sustained. The decision lay in a duel of conflicting statistical analyses.

Bahcall proceeded to calculate the number of superposition that would arise accidentally given plausible assumptions about galaxy distributions and found them (p. 92) to match near enough the number observed. Bahcall's complaint was supported in turn by three papers he selected for reproduction in the volume, each finding the arrangements Arp judged significant as consistent with purely chance alignments.

For his part, Arp claimed repeatedly that mere chance could not explain the arrangements. For example, he reported (p. 25) a calculation of a chance of less than one in five hundred of a certain significant alignment of seven quasars with relatively nearby peculiar galaxies. Arp could provide his own repertoire of statistical studies showing that mere chance does not suffice. At least five of the papers he had selected for inclusion in the volume were of such studies. Arp concluded his rejoinder to Bahcall with a plea (p. 129)

> I should like to note that Dr. Bahcall says he has estimated what the probabilities are that the associations I have discussed could be due to chance. Obviously there is always some finite chance that any one association could be accidental. If each association is considered separately, each could be dismissed on these grounds. But what is the chance that two or three or a half dozen could be accidental? Since these cases are independent, their improbabilities multiply, yielding in the end an extraordinarily low figure for the probability of chance occurrence. What value of probability would Dr. Bahcall accept as a demonstration establishing the case? Finally, I would like to ask him, seriously, if discordant redshifts do exist, what he would consider as an observation or a set of observations that he would accept as proof.

Each of these dueling studies proceeds from its own set of background assumptions, most importantly on the chance distribution of various astronomical objects. These assumptions, if true, serve to warrant the inductive inferences underpinning the studies. Evaluating the merits of these dueling studies goes well beyond what can be done here and what needs to be done here. As far as the nature of the inductive inferences are concerned, the duel illustrates the centrality of the warranting assumption *Proximity* in Arp's inferences and that their cogency is to be decided by a determination of this assumption's admissibility.

9.10 Differences in Inductive Reach

According to the material theory of induction, background facts warrant inductive inferences. Thus, the reach of inductive inferences depends upon the richness of the pertinent background facts available. Here Bahcall's standard view and Arp's dissident view differed markedly.

Bahcall's standard view conformed with then current cosmological theorizing. This was a fact to which he drew attention when he commenced his summary of the tests passed by the redshift-distance relation (p. 77):

> Also of great importance is the fact that the specific form of Hubble's law, redshift $\propto$ distance, has a very simple theoretical interpretation. Hubble's law is predicted by all cosmologies that assume the universe is expanding and is (at least locally) homogeneous and isotropic.

Bahcall does not spell out why this is important.[14] Mere conformity with further science is important in itself. However, it has another, narrower importance. The conformity enables an inference to a subset of possible cosmological models. Background cosmological models by themselves could not pick among universes

[14] On pp. 81–82, he reports that large redshifts are evidence for the expansion of the universe; and that the overall results conform with the laws of physics found terrestrially.

that were contracting, expanding or even static (if the cosmological constant were appropriately tuned). Among expanding universes, a linear relationship is recoverable only for nearby galaxies. The observed redshift-distance relation then is the evidence that enables an inference to those cosmological models that are expanding and with a region of linearity corresponding to the distances of galaxies observable. There were indications in the most recent studies reported by Bahcall that this was the limit of the region of linearity. Sandage (1972, p. 1) was already able to report a non-zero deceleration parameter, which is a parameter that gauges deviations from linearity.

The inferences to this set of cosmological models were warranted by the general cosmological theory. Its truth seemed fairly secure. The theory required only that general relativity, the theory of gravity then accepted locally, also apply cosmically; and that the universe is on the largest scale roughly homogeneous and isotropic.

In comparison, Arp's analysis had limited inductive reach, contrary to his Baconian expectations. At best it could only establish a negative: the failure of the redshift-distance relation. Arp could call upon nothing further in existing theory to assist him in inferring more from his results. Indeed, he had to report the awkward fact that his conclusions did not conform with then present science. He wrote (p. 18):

> The explanation for any such noncosmological redshifts is not readily available from current physics. Because the discordant redshifts are overwhelmingly redshifts and not blue shifts, peculiar Doppler velocities cannot be invoked. This is because in the case of high random velocities as many approach (blue shift) anomalies as recession anomalies should be observed. Gravitational redshifts require too much mass and too abrupt local gradients of field strength to be reconciled with observations of diffuse galaxies, even if complicated models could be made to work for the more compact quasars.

Bahcall's report on this difficulty was more forthright (p. 82): "If discordant redshifts truly exist, then the known laws of physics do not apply to some galaxies."

It is presumably for this reason that Arp sought out new physics that would be compatible with his discordant redshifts. The papers he selected for inclusion in the volume included those by Fred Hoyle that proposed non-standard physics, such as a joint paper by Fred Hoyle and Jayant Narlikar that non-velocity redshifts might derive from matter with electrons of low mass.

9.11 Who Won?

The *Redshift Controversy* volume has no one serving as an umpire to declare who won the debate. The two sides made their best cases and the decision was left to readers in the community. In a simple, sociological sense, it is clear that Bahcall and the standard view won. For Arp's work and his notion of discordant redshifts has all but completely disappeared from the literature after the debate. In this sense, the astronomical and cosmological community has decided.

However, it is also clear that the weight of evidence strongly favors this standard view. There was no communal irrationality in the decision. For, as we saw in the last section, the redshift-distance relation conformed with the standard view in cosmology. Its place became even more secure as new results in cosmology were built around it. In contrast, Arp's dissident view could not be developed without in turn developing alternative cosmology and astrophysics.

Arp recognized this difficulty. His later publications continued to make fitful references to such physics. In his 1987 development of his claims on discordant redshifts, speculation on their explanation in non-standard physics, such as "tired light," is isolated in a few concluding pages (pp. 178–84) at the end of the volume. Arp's (1998) publication was entitled with autobiographical candor, *Seeing Red*... It is a boisterous mixture of new observational results favoring discordant redshifts, anecdotes largely of Arp's perceived mistreatment by mainstream astronomy and fragments of alternative science. The last included Hoyle and Narlikar's Machian-based theory of gravity that leads to lower mass electrons (p. 108); Hoyle, Burbidge and Narlikar's 1993 "quasi-steady state" cosmology (p. 238); and Arp's own non-expanding cosmology (pp. 251–52).

While Arp continued to publish his evidence for discordant redshifts, he was unable to secure its acceptance. Reports continued to contradict his claim. Iovino and Hickson (1997), for example, concluded in their abstract: "Our results confirm that projection effects alone can account for the high incidence of discordant redshifts in compact groups."

Arp also required novel astrophysics. His view required quasars not to be distant, extraordinarily bright objects, but dimmer, closer objects, possibly ejected from nearby galaxies. The standard view of quasars grew from strength to strength. They came to be widely accepted as the brilliant nuclei of active galaxies. The nuclei contain, most likely, a supermassive black hole and massive amounts of radiation are emitted as matter falls into the hole. A volume published in 2012 celebrated 50 years of quasar research. The idea of quasars' great brightness and distance had become so well established that Peterson, writing in the preface, needed to remind readers with astonishment that this idea was ever doubted (D'Onofrio et al. 2012, p. vii):

> But over at least the first two decades of quasar research, the question that dominated discussion was whether quasars were indeed at the cosmological distances implied by their redshifts!

Meanwhile, the sort of alternative accounts of quasars required for Arp's view fared poorly. Tang and Zang (2005) found no evidence in astronomical survey data for certain models of how supposedly nearby quasars could have significantly greater redshifts than their parent galaxies.

Chapter 2 of the fifty-year retrospective volume (D'Onofrio et al. 2012) included reminiscences prompted by interview questions from astronomers involved in the history of quasars. Halton Arp and Jayant Narlikar were included in the section

"2.11 Challenging the Standard Paradigm." A narrator first needed to remind readers that once there had been other views (p. 61):

> We now move to an entirely different view of quasars, and in many ways to a different universe. Not everyone back in the 1960s accepted at once that quasars were "fast and far." Several astronomers suggested that quasars were not at the distances implied by their redshift...

Arp (pp. 61–72) reviewed at length the basis of his past and continuing disagreement with the standard view. The weakness of Arp's position became clear when the interview question was "Do you think that some alternative fundamental physics could help explain the quasar phenomenon?" The response was brief and hesitant, mentioning Hoyle and Narlikar's views and suggesting exploration of Bose-Einstein quantum states. The interview concluded with a polite dismissal:

> Thank you, Halton for your considerations. Since the above discussion involves unconventional theories, interested readers may look at the more extended presentation in Halton Arp's book (Arp, 1998).

His views had become so remote from the mainstream that no rebuttal was deemed necessary.

In sum, the standard view thrived, moving from evidential success to evidential success. Arp's view languished. Its basic results remained under challenge. If it was to match the strength of evidential support of standard view, it needed to come up with the alternative science that could make sense of its non-expanding cosmology and its novel astrophysical processes. This was a massive evidential debt that remained undischarged. The longer it remained so, the worse it was for Arp's view.

9.12 Conclusion

The recounting above of the inductive relations connection evidence and theory in the Arp-Bahcall debate may well seem to labor the obvious, at times to the point of tedium. Indeed, I hope that the reader has developed this sense. For the narrative displays the inductive relations of support, while drawing nowhere on general theories of confirmation. There are no enumerative inductions, analogies, inferences to the best explanations or displays of Bayes' theorem. They are not needed to display and assess the cogency of the inductive inferences in the debate. All inductive relations are traced back to their warrant in background facts. We see that a critical goal of each side was to provide further support for their own warranting fact, while at the same time impugning that of the other side. The example shows that, when we begin to explore the details of evidential cases laid out in science in at least this one case, the material approach fits closely with the practice of the science while also giving the means to assess the validity the inferences.

References

Aitken, R. G. (1918). *The binary stars.* (No publisher listed).

Arp, H. (1966a). *Atlas of peculiar galaxies.* California Institute of Technology.

Arp, H. (1966b). Peculiar galaxies and radio sources. *Science, 151*(3715), 1214–1216.

Arp, H. (1987). *Quasars, redshifts, and controversies.* Interstellar Media.

Arp, H. (1998). *Seeing red: Redshifts, cosmology and academic science.* Apeiron.

Arp, H., & Bahcall, J. N. (1973). *The redshift controversy.* W. A. Benjamin, Inc.

D'Onofrio, M., Marziani, P., & Sulentic, J. W. (Eds.). (2012). *Fifty years of Quasars: From early observations and ideas to future research.* Springer.

Hubble, E. (1929). A Relation between distance and radial velocity among Extra-Galactic Nebulae. *Proceedings of the National Academy of Sciences, 15,* 168–173.

Hubble, E. (1936). *The realm of the nebulae.* Oxford University Press/London:Humphrey Milford.

Humason, M. L., Mayall, N. U., & Sandage, A. R. (1956). Redshifts and magnitudes of extragalactice nebulae. *Astrophysical Journal., 61,* 97–162.

Iovino, A., & Hickson, P. (1997). Discordant redshifts in compact groups. *Monthly Notices of the Royal Astronomical Society, 287,* 21–25.

Kragh, H. S. (2007). *Conceptions of cosmos: From myths to the accelerating universe: A history of cosmology.* Oxford University Press.

Norton, J. D. (2003). A material theory of induction. *Philosophy of Science, 70,* 647–670.

Norton, J. D. (2005). A little survey of induction. In P. Achinstein (Ed.), *Scientific evidence: Philosophical theories and applications* (pp. 9–34). Johns Hopkins University Press.

Norton, J. D. (2021). *The material theory of induction.* BSPSOpen/University of Calgary Press.

Sandage, A. (1972). The redshift-distance relation. II. The hubble diagram and its scatter for first-ranked cluster galaxies. *Astrophysical Journal, 178,* 1–24.

Slipher, V. M. (1912). The radial velocity of the andromeda nebula. *Lowell Observatory Bulletin, No. 58, II*(8), 56–57.

Tang, S. M., & Zhang, S. N. (2005). Critical Examinations of QSO Redshift Periodicities and Associations with Galaxies in Sloan Digital Sky Survey Data. *The Astrophysical Journal, 633,* 41–51.

Weinberg, S. (1972). *Gravitation and cosmology: Principles and applications of the general theory of relativity.* Wiley.

Chapter 10
When Does a Boltzmannian Equilibrium Exist?

Charlotte Werndl and Roman Frigg

Abstract We present a definition of equilibrium for Boltzmannian statistical mechanics based on the long-run fraction of time a system spends in a state. We then formulate and prove an existence theorem which provides general criteria for the existence of an equilibrium state. We illustrate how the theorem works with toy example. After a look at the ergodic programme, we discuss equilibria in a number of different gas systems: the ideal gas, the dilute gas, the Kac gas, the stadium gas, the mushroom gas and the multi-mushroom gas.

10.1 Introduction

The received wisdom in statistical mechanics (SM) is that isolated systems, when left to themselves, approach equilibrium. But under what circumstances does an equilibrium state exist and an approach to equilibrium take place? In this paper we address these questions from the vantage point of the long-run fraction of time definition of Boltzmannian equilibrium that we developed in our two papers Werndl and Frigg (2015a,b) (see also Frigg and Werndl 2019; Werndl and Frigg 2017b, 2020). After a short summary of Boltzmannian statistical mechanics (BSM) and our definition of equilibrium (Sect. 10.2), we state an existence theorem which provides general criteria for the existence of an equilibrium state (Sect. 10.3). We first illustrate how the theorem works with a toy example (Sect. 10.4), which allows us to explain the various elements of the theorem in a simple setting. After looking at the ergodic programme (Sect. 10.5), we discuss equilibria in a number of different gas systems: the ideal gas, the dilute gas, the Kac gas, the stadium gas,

C. Werndl (✉)
Department of Philosophy, University of Salzburg, Salzburg, Austria
e-mail: charlotte.werndl@plus.ac.at.

R. Frigg
Department of Philosophy, Logic and Scientic Method, and Centre for Philosophy of Natural and Social Science, London School of Economics and Political Science, London, UK
e-mail: r.p.frigg@lse.ac.uk.

C. Soto (ed.), *Current Debates in Philosophy of Science*, Synthese Library 477,
https://doi.org/10.1007/978-3-031-32375-1_10

the mushroom gas and the multi-mushroom gas (Sect. 10.6). In the conclusion we briefly summarise the main points and highlight open questions (Sect. 10.7).

10.2 Boltzmannian Equilibrium

Our focus are systems which, at the micro level, are measure-preserving deterministic dynamical systems $(X, \Sigma_X, \mu_X, T_t)$.[1] The set X represents all possible micro-states; Σ_X is a σ-algebra of subsets of X; the evolution function $T_t : X \to X$, $t \in \mathbb{R}$ (continuous time) or $\mathbb{Z}$ (discrete time) is a measurable function in (t, x) such that $T_{t_1+t_2}(x) = T_{t_2}(T_{t_1}(x))$ for all $x \in X$ and all $t_1, t_2 \in \mathbb{R}$ or $\mathbb{Z}$; μ_X is a measure on Σ_X, which is invariant under the dynamics. That is, $\mu_X(T_t(A)) = \mu_X(A)$ for all $A \in \Sigma_X$ and all t.[2] The function $s_x : \mathbb{R} \to X$ or $s_x : \mathbb{Z} \to X$, $s_x(t) = T_t(x)$ is called the solution through the point $x \in X$.

A set of macro-variables $\{v_1, \ldots, v_l\}$ $(l \in \mathbb{N})$ characterises the system at the macro-level. The fundamental posit of BSM is that macro-states supervene on micro-states, implying that a system's micro-state uniquely determines its macro-state. Thus the macro-variables are measurable functions $v_i : X \to \mathbb{V}_i$, associating a value with a point $x \in X$. Capital letters V_i will be used to denote the values of v_i. A particular set of values $\{V_1, \ldots, V_l\}$ defines a *macro-state* $M_{V_1,\ldots,V_l}$. If the specific values V_i do not matter, we only write 'M' rather than '$M_{V_1,\ldots,V_l}$'. For now all we need is the general definition of macro-variables. We will discuss them in more detail in Sects. 10.3 and 10.4, where we will see that the choice of macro-variables is a subtle and important matter and that the nature as well as the existence of an equilibrium state crucially depends on it.

The determination relation between micro-states and macro-states will nearly always be many-to-one. Therefore, every macro-state M is associated with a macro-region consisting of all micro-states for which the system is in M. A neglected but important issue is on what space macro-regions are defined. The obvious option would be X, but in many cases this is not what happens. In fact, often macro-regions are defined on a subspace $Z \subset X$. Intuitively speaking, Z is a subset whose states evolve into the same equilibrium macro-state. To give an example: for a dilute gas with N particles X is the $6N$-dimensional space of all position and momenta, but Z is the $6N - 1$ dimensional energy hypersurface. X will be called the *full state space* and Z the *effective state space* of the system. The macro-region Z_M corresponding to macro-state M relative to Z is then defined as the set of all $x \in Z$ for which M supervenes on x. A set of macro-states is complete relative to Z iff (if and only if) it contains all states of Z. The members of a complete set of macro-regions Z_M do not overlap and jointly cover Z, i.e. they form a partition of Z.

[1] This section is based on Werndl and Frigg (2015a,b). For a discussion of stochastic systems see Werndl and Frigg (2017a).

[2] At this point the measure need not be normalised.

Z has to be determined on a case-by-case basis, because the particulars of the system under consideration determine the correct choice of Z. We return to this point in Sect. 10.3. However, there is one general constraint on such a choice that needs to be mentioned now. Since a system can never leave the partition of macro-regions, it is clear that Z must be mapped onto itself under T_t. If such a Z is found, the σ-algebra on X can be restricted to Z and one can consider a measure on Z which is invariant under the dynamics and normalized (i.e. $\mu_Z(Z) = 1$). In this way the measure-preserving dynamical system $(Z, \Sigma_Z, \mu_Z, T_t)$ with a normalized measure μ_Z is obtained.[3] We call $(Z, \Sigma_Z, \mu_Z, T_t)$ the *effective system* (as opposed to the *full system* $(X, \Sigma_X, \mu_Z, T_t)$).

M_{eq} is the equilibrium macrostate and the corresponding macro-region is $Z_{M_{eq}}$. An important aspect of the standard presentation of BSM is that $Z_{M_{eq}}$ is the largest macro-region. The notion of the 'largest macro-region' can be interpreted in two ways. First, 'largest' can mean that the equilibrium macro-region takes up a large part of Z. More specifically, $Z_{M_{eq}}$ is said to be *β-dominant* iff $\mu_Z(Z_{M_{eq}}) \geq \beta$ for a particular $\beta \in (\frac{1}{2}, 1]$. If $Z_{M_{eq}}$ is β-dominant, it is clear that it is also β'-dominant for all β' in $(1/2, \beta)$. Second, 'largest' can mean 'larger than any other macro-region'. We say that $Z_{M_{eq}}$ is *δ-prevalent* iff $\min_{M \neq M_{eq}}[\mu_Z(Z_{M_{eq}}) - \mu_Z(Z_M)] \geq \delta$ for a particular $\delta > 0$, $\delta \in \mathbb{R}$. It follows that if a $Z_{M_{eq}}$ is δ-prevalent, then it is also δ'-prevalent for all δ' in $(0, \delta)$. We do not adjudicate between these different definitions; either meaning of 'large' can be used to define equilibrium. However, we would like to point out that they are not equivalent: if an equilibrium macro-region is β-dominant, there is a range of values for δ so that the macro-region is also δ-prevalent for these values. However the converse fails.

Now the question is: why is the equilibrium state β-dominant or δ-prevalent? A justification ought to be as close as possible to the thermodynamics (TD) notion of equilibrium. In TD a system is in equilibrium just in case change has come to a halt and all thermodynamic variables assume constant values (*cf.* Reiss 1996, 3). This would suggest a definition of equilibrium according to which every initial condition lies on trajectory for which $\{v_1, \ldots, v_k\}$ eventually assume constant values. Yet this is unattainable for two reasons. First, because of Poincaré recurrence, the values of the v_i will never reach a constant value and keep fluctuating. Second, in dynamical systems we cannot expect *all* initial conditions to lie on trajectories that approach equilibrium (see, e.g., Callender 2001).

To do justice to these facts about dynamical systems we revise the TD definition slightly and define equilibrium as the macro-state in which trajectories starting in most initial conditions spend most of their time. This is not a feeble compromise. Experimental results show that physical systems exhibit fluctuations away from equilibrium (Wang et al., 2002). Hence, strict TD equilibrium is actually unphysical and a definition of equilibrium that makes room for fluctuations is empirically more adequate.

[3] The dynamics is given by the evolution equations restricted to Z. We follow the literature by denoting it again by T_t.

To make this idea precise we introduce the long-run fraction of time a system spends in a region $A \in \Sigma_Z$ when the system starts in micro-state x at time $t = 0$:

$$LF_A(x) = \lim_{t\to\infty} \frac{1}{t} \int_0^t 1_A(T_\tau(x))d\tau \text{ for continuous time, i.e. } t \in \mathbb{R}, \quad (10.1)$$

$$LF_A(x) = \lim_{t\to\infty} \frac{1}{t} \sum_{\tau=0}^{t-1} 1_A(T_\tau(x)) \text{ for discrete time, i.e. } t \in \mathbb{Z},$$

where $1_A(x)$ is the characteristic function of A, i.e. $1_A(x) = 1$ for $x \in A$ and 0 otherwise. Note that a measure-preserving dynamical system $(Z, \Sigma_Z, \mu_Z, T_t)$ with the normalized measure μ_Z is *ergodic* iff for any $A \in \Sigma_Z$:

$$LF_A(x) = \mu_Z(A), \quad (10.2)$$

for all $x \in Z$ except for a set W with $\mu_Z(W) = 0$.

The locution 'most of their time' is beset with the same ambiguity as the 'largest macro-state'. On the first reading 'most of the time' means more than half of the total time. This leads to the following formal definition of equilibrium:

BSM α-ε-Equilibrium. Consider an isolated system S whose macro-states are specified in terms of the macro-variables $\{v_1, \ldots, v_k\}$ and which, at the micro level, is a measure-preserving deterministic dynamical system $(Z, \Sigma_Z, \mu_Z, T_t)$. Let α be a real number in $(0.5, 1]$, and let $1 \gg \varepsilon \geq 0$ be a very small real number. If there is a macrostate $M_{V_1^*,\ldots,V_k^*}$ satisfying the following condition, then it is the α-ε-equilibrium state of S: There exists a set $Y \subseteq Z$ such that $\mu_Z(Y) \geq 1 - \varepsilon$, and all initial states $x \in Y$ satisfy

$$LF_{Z_{M_{V_1^*,\ldots,V_l^*}}}(x) \geq \alpha. \quad (10.3)$$

We then write $M_{\alpha\text{-}\varepsilon\text{-}eq} := M_{V_1^*,\ldots,V_k^*}$.

An obvious question concerns the value of α. Often the assumption seems to be that α is close to one. This is reasonable but not the only possible choice. For our purposes nothing hangs on a the value of α and so we leave it open what the best choice would be.

On the second reading 'most of the time' means that the system spends more time in the equilibrium macro-state than in any other macro-state. This idea can be rendered precise as follows:

BSM γ-ε-Equilibrium. Consider an isolated system S whose macro-states are specified in terms of the macro-variables $\{v_1, \ldots, v_k\}$ and which, at the micro level, is a measure-preserving deterministic dynamical system $(Z, \Sigma_Z, \mu_Z, T_t)$. Let γ be a real number in $(0, 1]$ and let $1 \gg \varepsilon \geq 0$ be a very small real number. If there is a macro-state $M_{V_1^*,\ldots,V_l^*}$ satisfying the following condition, then it is the γ-ε equilibrium state of S: There exists a set $Y \subseteq Z$ such that $\mu_Z(Y) \geq 1 - \varepsilon$ and for all initial conditions $x \in Y$:

$$LF_{Z_{M_{V_1^*,\ldots,V_l^*}}}(x) \geq LF_{Z_M}(x) + \gamma \quad (10.4)$$

for all macro-states $M \neq M_{V_1^*,\ldots,V_l^*}$. We then write $M_{\gamma\text{-}\varepsilon\text{-}eq} := M_{V_1^*,\ldots,V_k^*}$.

As above, nothing in what we say about equilibrium depends on the particular value of the parameter γ and so we leave it open what the best choice would be.

We contend that these two definitions provide the relevant notion of equilibrium in BSM. It is important to emphasise that they remain silent about the size of equilibrium macro-regions, and do not in any obvious sense imply anything about seize. Indeed, equilibrium marco-regions being extremely small would be entirely compatible with the definitions. That these macro-regions have the right size is a result established in the following two theorems:

Dominance Theorem: If $M_{\alpha\text{-}\varepsilon\text{-}eq}$ is an α-ε-equilibrium of system S, then $\mu_Z(Z_{M_{\alpha\text{-}\varepsilon\text{-}eq}}) \geq \beta$ for $\beta = \alpha(1-\varepsilon)$.[4]

Prevalence Theorem: If $M_{\gamma\text{-}\varepsilon\text{-}eq}$ is a γ-ε-equilibrium of system S, then $\mu_Z(Z_{M_{\gamma\text{-}\varepsilon\text{-}eq}}) \geq \mu_Z(Z_M) + \delta$ for $\delta = \gamma - \varepsilon$ for all macro-states $M \neq M_{\gamma\text{-}\varepsilon\text{-}eq}$.[5]

Both theorems are completely general in that no dynamical assumptions are made (in particular it is not assumed that systems are ergodic—*cf.* Eq. (10.2)), and hence the theorems also apply to strongly interacting systems.

An important aspect of the above definitions of equilibrium is that the presence of an approach to equilibrium is built into the notion of an equilibrium state. If a state is not such that the system spends most of the time in that state (in one of the two senses specified), then that state simply is not an equilibrium state. In other words, if the system does not approach equilibrium, then there is no equilibrium. Having an equilibrium state and there being an approach to equilibrium are two sides of the same coin.

The theorems make the conditional claim that *if* an equilibrium exits, *then* it is large in the relevant sense. Some systems do not have equilibria. If, for instance, the dynamics is given by the identity function, then no approach to equilibrium takes place, and the antecedent of the conditional is wrong. As with all conditionals, the crucial question is whether, and under what conditions, the antecedent holds. We turn to this issue now. As we have just seen, the question whether there is an equilibrium state is tantamount to the question whether the approach to equilibrium takes place, and so the issue of existence is not merely an inconsequential subtlety in mathematical physics—it concerns one of the core questions in SM.

10.3 The Existence of an Equilibrium Macro-State

We now turn to the core question of this paper: under what circumstances does a Boltzmannian equilibrium macro-state exist? The main message is that for an equilibrium to exist three factors need to cooperate: the choice of macro-variables,

---

[4] We assume that ε is small enough so that $\alpha(1-\varepsilon) > \frac{1}{2}$.

[5] We assume that $\varepsilon < \gamma$.

the dynamics of the system, and the choice of the effective state space Z. The cooperation between these factors can take different forms and there is more than one constellation that can lead to the existence of an equilibrium state. The important point is that the answer to the question of existence is holistic: it not only depends on three factors rather than one, but also on the interplay between these factors. For these reasons we call these three factors the *holist trinity*.

A number of previous proposals fail to appreciate this point. The problem of the approach to equilibrium has often been framed as the challenge to identify *one* crucial property and show that the relevant systems possess this property. We first introduce the trinity in an informal way and illustrate it with examples, showing what requisite collaborations look like and what can go wrong. This informal presentation is followed by a rigorous mathematical theorem providing necessary and sufficient conditions for the existence of an equilibrium state.

10.3.1 The Holist Trinity

Macro-Variables The first condition is that the macro-variables must be the 'right' ones: the same system can have an equilibrium with respect to one set of macro-variables and fail to have an equilibrium with respect to another set of macro-variables. The existence of an equilibrium depends as much on the choice of macro-variables as it depends on the system's dynamical properties. Different choices are possible, and these choices lead to different conclusions about the equilibrium behaviour of the system. This will be illustrated below in Sect. 10.4 with the example of the simple pendulum.[6]

This also implies that if no macro-variables are introduced, considerations of equilibrium make no sense at all. Obvious as this may seem, some confusion has resulted from ignoring this simple truism. Sklar (1973, 209) mounts an argument against the ergodic approach by pointing out that a system of two hard spheres in a box has the right dynamical property (namely ergodicity) and yet fails to show an approach to equilibrium. It hardly comes as a surprise, though, that there is no approach to equilibrium if the system has no macro-variables associated with it in terms of which equilibrium could even be defined.

Dynamics The existence of an equilibrium depends as much on the dynamics of the system as it depends on the choice of macro-variables. Whatever the macro-variables, if the dynamics does not 'collaborate', then there is no approach to equilibrium. For this reason the converses of the Dominance and Prevalence Theorems fail: it is *not* the case that if there is a β-dominant/δ-prevalent macro-region, then this macro-region corresponds to a α-ε-equilibrium/γ-ε-equilibrium. If, for instance, the dynamics is the identity function, then there can be no approach

[6] For further examples see Werndl and Frigg (2015a).

to equilibrium because states in a small macro-region will always stay in this region. Or assume that there is a system whose dynamics is such that micro-states that are initially in the largest macro-region always remain in the largest macro-region and states initially in smaller macro-regions only evolve into states in these smaller macro-regions. Then there is no approach to equilibrium because non-equilibrium states will not evolve into equilibrium. This point will also be illustrated with the example of the simple pendulum in Sect. 10.4.

Identifying Z A number of considerations in connection with equilibrium depend on the choice of the effective state space Z, which is the set relative to which macro-regions are defined. Indeed, the existence of an equilibrium state depends on the correct choice of Z. There can be situations where a system has an equilibrium with respect to one choice of Z but not with respect another choice of Z. One can choose Z too small, and, as a consequence, it will not be true that most initial states approach equilibrium and hence there will be no equilibrium (recall that on our definition the system has no equilibrium if it does't approach equilibrium). One can, however, make the opposite mistake and choose Z too large. If there is an equilibrium relative to some set Z, it need not be the case that an equilibrium exists also on a superset of this set. So Z can be chosen too large as well as too small. That Z can be chosen too large will be illustrated with the example of the simple pendulum in Sect. 10.4.

There is no algorithmic procedure to determine Z, but one can pinpoint a number of relevant factors. The most obvious factors are constraints and boundary conditions imposed on the system. If a system cannot access certain parts of X, then these parts are not in Z. In all examples below we see parts of X being 'cut off' when constructing Z because of mechanical restrictions preventing the system from entering certain regions. Another important factor in determining Z are conserved quantities. Their role, however, is less clear-cut than one might have hoped for. It is not universally true that Z has to lie within a hyper-surface of conserved quantities. Whether Z is so constrained depends on the macro-variables. Consider the example of energy. In some cases (the dilute gas in Sect. 10.6, for instance), equilibrium values depend on the energy of the system (equilibrium states are different for different energies) and hence Z must lie within an energy hyper-surface. In other cases (the oscillator in Sect. 10.4, for instance) equilibrium is insensitive toward changes in the system's energy (the equilibrium state is the same for all energy values) and therefore Z is not confined to an energy hyper-surface. This brings home again the holist character of the issue: Z not only depends on mechanical invariants and constraints, but also on the macro-variables.

The interplay between these factors is illustrated with a simple toy model in Sect. 10.4. Due to its simplicity, it is tangible how the three factors mutually constrain each other and it becomes clear how sensitively the existence of an equilibrium depends on the careful balance of these factors. In Sect. 10.6 we discuss how these considerations play out in different gas systems.

10.3.2 The Existence Theorem

In this subsection we present the Equilibrium Existence Theorem, a theorem providing necessary and sufficient conditions for the existence of an equilibrium state (either of the α-ε or the γ-ε type).[7] Before stating the theorem we have to introduce another theorem, the Ergodic Decomposition Theorem (*cf.* Petersen 1983, 81). An ergodic decomposition of a system is a partition of the state space into cells so that the cells are invariant under the dynamics (i.e., are mapped onto themselves) and that the dynamics within each cell is ergodic (*cf.* Eq. (10.2) for the definition of ergodicity).[8]

The Ergodic Decomposition Theorem says that such a decomposition exists for *every* measure-preserving dynamical system with a normalised measure, and that the decomposition is unique. In other words, the dynamics of a system can be as complex as we like and the interactions between the constituents of the system can be as strong and intricate as we like, and yet there exists a unique ergodic decomposition of the state space of the system. A simple example of the theorem is the harmonic oscillator: the ellipses around the coordinate origin are the cells of the partition and the motion on the ellipses is ergodic.

For what follows it is helpful to have a more formal rendering of an ergodic decomposition. Consider the system $(Z, \Sigma_Z, \mu_Z, T_t)$. Let Ω be an index set (which can but need not be countable), which comes equipped with a probability measure ν. Let Z_ω, $\omega \in \Omega$, be the cells into which the system's state space can be decomposed, and let Σ_ω and μ_ω, respectively, be the sigma algebra and measure defined on Z_ω. These can be gathered together in 'components' $C_\omega = (Z_\omega, \Sigma_\omega, \mu_\omega, T_t)$. The Ergodic Decomposition Theorem says that for every system $(Z, \Sigma_Z, \mu_Z, T_t)$ there exists a unique set of ergodic C_ω so that the system itself amounts to the collection of all the C_ω. How the ergodic decomposition theorem works will be illustrated with the example in the next section.

We are now in a position to state our core result:

Equilibrium Existence Theorem: Consider a measure-preserving system $(Z, \Sigma_Z, \mu_Z, T_t)$ with macro-regions $Z_{M_{V_1,\ldots,V_l}}$ and let $C_\omega = (Z_\omega, \Sigma_\omega, \mu_\omega, T_t)$, $\omega \in \Omega$, be its ergodic decomposition. Then the following two biconditionals are true:

α-ε-equilibrium: There exists an α-ε-equilibrium iff there is a macro-state $\hat{M}$ such that for every C_ω:

$$\mu_\omega(Z_\omega \cap Z_{\hat{M}}) \geq \alpha, \quad (10.5)$$

except for components C_ω with $\omega \in \Omega'$, $\mu_Z(\cup_{\omega\in\Omega'} Z_\omega) \leq \varepsilon$. $\hat{M}$ is then the α-ε-equilibrium state.

[7] The proof is given in Werndl and Frigg (2015a).

[8] It is allowed that the cells are of measure zero and that there are uncountably many of them.

γ-ε-equilibrium: There exists a γ-ε-equilibrium iff there is a macro-state $\hat{M}$ such that for every C_ω and any $M \neq \hat{M}$

$$\mu_\omega(Z_\omega \cap Z_{\hat{M}}) \geq \mu_\omega(Z_\omega \cap Z_M) + \gamma, \tag{10.6}$$

except for components C_ω with $\omega \in \Omega'$, $\mu_Z(\cup_{\omega \in \Omega'} Z_\omega) \leq \varepsilon$. $\hat{M}$ is then the γ-ε-equilibrium state.

Like the theorems we have seen earlier, the Equilibrium Existence Theorem is fully general in that it makes no assumptions about the system's dynamics other than that it be measure-preserving. Intuitively the theorems say that there is an α-ε-equilibrium (γ-ε-equilibrium) iff if the system's state space is split up into invariant regions on which the motion is ergodic and the equilibrium macro-state takes up at least α of each region (the equilibrium region is at least γ larger than any other macro-region), except, possibly, for regions of total measure ε. If we have found a space that meets these conditions, then it plays the role of the effective state space Z.

It is important to note that there may be many different macro-state/dynamics/Z triplets that make the Existence Theorem true. The Theorem gives the foundation for a research programme aiming to find and classify such triplets. But before discussing a number of interesting cases, we want to illustrate the theorem in the simplest possible setting. This is our task in the next section.

10.4 Toy Example: The Ideal Pendulum

Consider an ideal pendulum: a small mass m hanging on a 1 meter long massless string from the ceiling. The mass moves without friction. When displaced, the mass will oscillate around its midpoint. We displace the pendulum only in one spatial direction and so the motion takes place in plane perpendicular to the ceiling. The weight of the bob mg, where g is the gravitational constant, has components parallel and perpendicular to the rod. The component perpendicular to the rod is $-mg\sin(x)$, where x is the angular displacement. This component accelerates the bob, and hence we can apply Newton's second law:

$$m\frac{d^2x}{dt^2} = -mg\sin(x) \tag{10.7}$$

For the simple pendulum the further assumption is made that the angular displacement is small (of absolute value less than 15°). Then $\sin(x) \approx x$, and the equation reduces to:

$$\frac{d^2x}{dt^2} = -gx. \tag{10.8}$$

This equation describes simple harmonic motion.

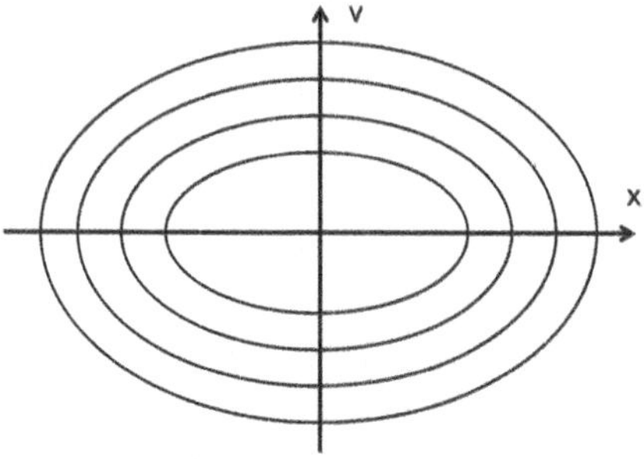

Fig. 10.1 The ergodic decomposition of the harmonic oscillator

That is, the full phase space X is given by the possible angular displacement and angular velocity coordinates (x, v), where the angular displacement is assumed to be less than 15°; and thus the displacement as well as the velocity is bounded from above). Solving the differential Eq. (10.8) above gives

$$x(t) = A\cos(\lambda t - \phi)$$
$$v(t) = \frac{dx}{dt} = -A\lambda \sin(\lambda t - \phi), \tag{10.9}$$

where $\lambda = \sqrt{g}$, A is the amplitude (the maximum displacement from the midpoint), and ϕ is the phase (the shift of the cosine and sinus functions along the time axis). A and ϕ are determined by the initial angular displacement and initial angular velocity.

From these equations we see that the solutions $T_t(x, v)$ are ellipses and the full phase space X is composed of these ellipses. This is illustrated in Fig. 10.1. Σ_X is the Borel σ-algebra on X, and the measure μ_X on the phase space X is the normalized measure that arises by demanding that the system sweeps out equal measures during equal time intervals. Taking these elements together yields the measure-preserving dynamical system $(X, \Sigma_X, \mu_X, T_t)$.

The effective phase space, i.e. the phase space relative to which equilibrium is defined, is in this case identical with X, and thus $(Z, \Sigma_Z, \mu_Z, T_t) = (X, \Sigma_X, \mu_X, T_t)$. We now illustrate the roles the macro-variables, the dynamics and the effective state space play in securing the existence of an equilibrium by discussing different choices and showing how they affect the existence of an equilibrium.

10.4.1 The Role of Macro-Variables

Consider $(Z, \Sigma_Z, \mu_T, T_t)$ with the colour macro-variable v_c, a light bulb that can emit red and white light. So $\mathbb{V}_c = \{r, w\}$, were 'r' stands for red and 'w' for white. The mapping is as follows: if the pendulum is on the right hand side of the midpoint *and* on its way back to the midpoint, then the light is red; the light is white otherwise.

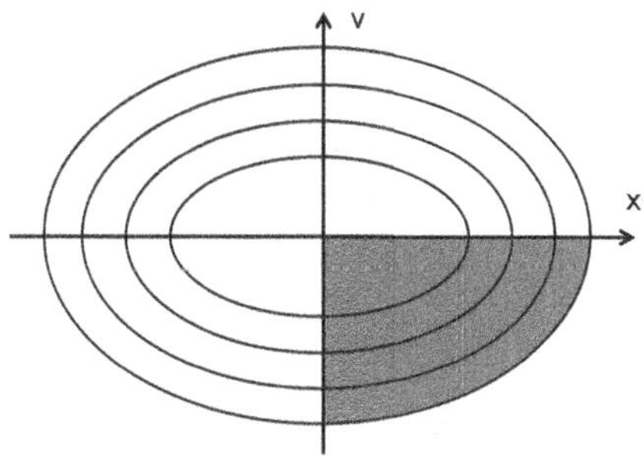

Fig. 10.2 The colour macro-variable v_c: if system's state is in the grey area the light is red; if it is in the white area the light is white

This defines two macro-states M_r and M_w. The macro-region Z_{M_r} is the grey area in Fig. 10.2 and Z_{M_w} is the white area.

Since the ideal pendulum oscillates with a constant frequency, M_w is a 0.75-0-equilibrium of the α-ε type: on each trajectory the light is white for three-quarters of the time and red for one quarter of the time. Thus, by the Dominance Theorem, $\mu(Z_{M_w}) \geq 0.75$ (and we have $\mu(Z_{M_r}) = 0.25$). M_w is in fact also a 0.5-0-equilibrium of the γ-ε type because the systems spends 0.5 more time in M_w than in M_r. Thus by the Prevalence Theorem: $\mu(Z_{M_w}) \geq \mu(Z_{M_r}) + 0.5$ for all Z_{M_r}.

Let us now discuss how the situation presents itself in terms of the Existence Theorem. To this end we first have a look at the ergodic decomposition theorem. The theorem says that Z can be decomposed into components $C_\omega = (Z_\omega, \Sigma_\omega, \mu_\omega, T_t)$. In the case of the harmonic oscillator, the ergodic decomposition is the (uncountable) family of ellipses given by Eq. (10.9) and shown in Fig. 10.1. Each Z_ω is a two-dimensional ellipse determined by the initial energy (the energy is determined by the initial displacement and velocity coordinates (x, v)). The Z_ω are the ellipses themselves; Σ_ω and μ_ω are the standard Borel sets on a line and the normalised measure on a line that arises by demanding that the system sweeps out equal lengths during equal times intervals; and T_t is the time evolution given by Eq. (10.9) restricted to the ellipses. It is easy to see that the motion on each ellipse is ergodic. The decomposition is parameterised by ω, which in this case has a physical interpretation: it is the energy of the system. Ω is the (uncountable) set of energy values between zero and the energy corresponding to a 15° angular displacement, and the measure ν on Ω is the standard Lebesgue measure.

Equation (10.5) holds true for every component $C_\omega = (Z_\omega, \Sigma_\omega, \mu_\omega, T_t)$ because on each ellipse three-quarters of the states correspond to a white light and one quarter to a red light, and hence $\mu_\omega(Z_\omega \cap Z_{M_w}) \geq 0.75$. Hence M_w satisfies the condition for an α-ε-equilibrium with $\alpha = 0.75$ and $\varepsilon = 0$. Likewise, Eq. (10.6) holds true for every component $C_\omega = (Z_\omega, \Sigma_\omega, \mu_\omega, T_t)$ because on each ellipse three-quarters of the states correspond to a white light and one quarter to a red light, and hence $\mu_\omega(Z_\omega \cap Z_{M_w}) \geq \mu_\omega(Z_\omega \cap Z_M) + 0.5$ for all $Z_M \neq Z_{M_w}$. Hence M_w satisfies the condition for an γ-ε-equilibrium with with $\gamma = 0.5$ and $\varepsilon = 0$.

Now consider a different macro-variable v_c'. It is defined like v_c but with one crucial difference: the light is red when the pendulum is on the right side irrespective of whether it is moving towards or away from the midpoint. The light is white

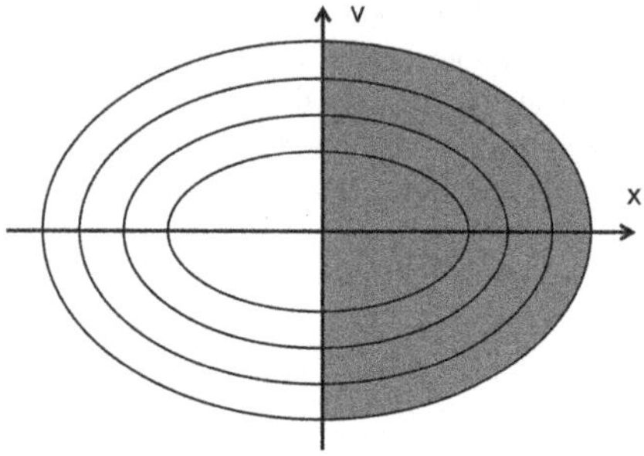

Fig. 10.3 The light bulb macro-variable v'_c: if system's state is in the grey area the light is red; if it is in the white area the light is white

when the pendulum is on the left side or exactly in the middle. This is illustrated in Fig. 10.3, where Z_{M_r} is the grey and Z_{M_w} is the white area. With respect to v'_c the system has no equilibrium. For all solutions the red and the white light are each on half of the time, and both macrostates have equal measure 0.5.

From the vantage point of the Existence theorem, the situation presents itself as follows. Equations (10.5) and (10.6) cannot hold true any more because for every component $C_\omega = (Z_\omega, \Sigma_\omega, \mu_\omega, T_t)$ half of the states correspond to a white light and a half of the states correspond to a red light. Hence the conditions of the Existence Theorem are not satisfied. This example illustrates that a small change in the macro-variable is enough to take us from a situation in which an equilibrium exists to one in which there is no equilibrium.

10.4.2 The Role of the Dynamics

As we have just seen, there exist equilibria of both types for the simple pendulum $(Z, \Sigma_Z, \mu_Z, T_t)$ with the macro-variable v_c. We now change the dynamics: place a wall of negligible width exactly at the midpoint (perpendicular to the plane of motion) and assume that the pendulum bounces elastically off the wall. Denote this dynamics by T'_t. If the pendulum starts on the right hand side, it will always stay on the right hand side. On that side the white and the red light are on half of the time each and so the system has no equilibrium for initial conditions on the right hand side. This violates the condition (in both definitions of equilibrium) that there is at most a small set of initial conditions (of measure $< \varepsilon$) for which the system does not satisfy the relevant equations (Eqs. (10.3) and (10.4) respectively). Hence, the system $(Z, \Sigma_Z, \mu_Z, T'_t)$ with the macro-variable v_c has no equilibrium.

Let us look at the situation through the lens of the Existence Theorem. The ergodic decomposition is now more complicated than above. There are again uncountably many components C'_ω. Yet because of the different dynamics, they are half-ellipses rather than ellipses. More specifically, the index set is $\Omega' = \Omega_1 \cup \Omega_2$, where Ω_1 consist of the uncountably many different values of the energy for systems that start out on the right hand side, and Ω_2 consist of the uncountably many different values of the energy for systems that do not start on the right hand side.

Each Z'_ω is a two dimensional half-ellipse determined by the initial energy and whether the system starts on the right or the left. The sigma-algebra Σ'_ω is the usual Borel σ-algebra on Z'_ω and the measure μ'_ω is the normalized measure that arises if one restricts the measure we considered in the previous subsection to each half-ellipse Z'_ω and divides it by 2. The dynamics on the ellipses is again given by the restriction of T'_t to the half-ellipses Z'_ω. Taking these elements together gives us the components $Cv_\omega = (Z'_\omega, \Sigma'_\omega, \mu'_\omega, T'_t)$, and it is clear that the motion on each component is ergodic.

Now consider the components C_ω that correspond to the case where the pendulum starts on the right hand side. Note that measure μ_Z of all these components taken together is 1/2. Yet half of any of these components is made up of states corresponding to the light being white and the remaining half is made up of states for which the light is red. Consequently, for these components Eqs. (10.5) and (10.6) cannot hold true. Thus the Existence Theorem is not satisfied because the condition is violated that there is at most a small set of initial conditions for which the system does not satisfy the relevant equations.

10.4.3 The Role of the Effective Phase Space

So far we discussed a simple pendulum with a one-dimension position coordinate. Let us now consider the different setup where the pendulum's position coordinate is not one-dimensional but two-dimensional (and, again, we impose the constraint that the maximum displacement in any spatial direction is $\leq 15°$), allowing the pendulum to oscillate in two directions, x and y. We now impose the constraint that the pendulum oscillates along a line going through the coordinate origin, and the time evolution T''_t along this line is in fact the same as above, but now described with two-dimensional angular displacement coordinates. The full state space of the system X'' is thus a three-dimensional ellipsoid: the first two coordinates are the displacement coordinates x and y, and the third coordinate gives the velocity along the line cutting through the origin. $\Sigma_{X''}$ is again the Borel σ-algebra on X'', and $\mu_{X''}$ is the measure that arises when in equal times an equal area is swept out (for the dynamics as explained below). Then $(X'', \Sigma_{X''}, \mu_{X''}, T''_t)$ is a measure-preserving dynamical system.

Now consider the two-dimensional colour macro variable v''_c, which can take three values: red, white and blue. So $\mathbb{V}''_C = \{r, w, b\}$. Because the displacement coordinates are constrained to a line, the displacement coordinate of a solution either oscillates between the first and the third quadrant or between the second and the fourth quadrant. Suppose that if the pendulum is in the first quadrant, the light is red; if the pendulum is in the second quadrant, then the light is blue if the pendulum is on its way to back to the midpoint and white if it moves away from the midpoint or is exactly at the midpoint. If the pendulum is in the third quadrant, then the light is red if the pendulum is on its way back to the midpoint and white if it moves away from the midpoint or is exactly at the midpoint. If the pendulum is in the fourth

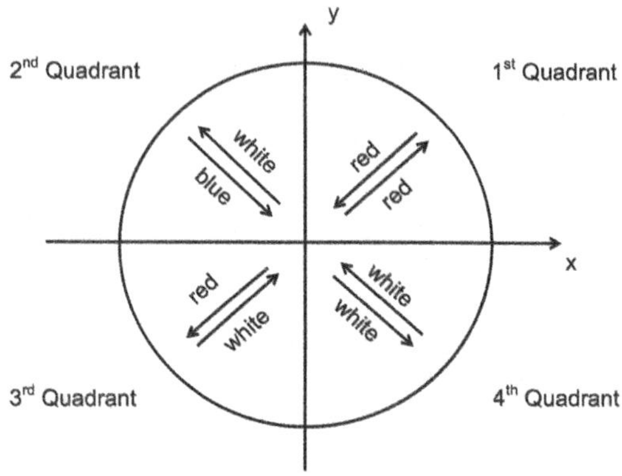

Fig. 10.4 The colour macro-variable v_c''

quadrant the light is white. This is illustrated in Fig. 10.4. It is then easy to see that $\mu_{X''}(X''_{M_w}) = 1/2$, $\mu_{X''}(X''_{M_r}) = 3/8$ and $\mu_{X''}(X''_{M_b}) = 1/8$.

Since the motion of the pendulum lies on a straight line through the midpoint, it always oscillates either between the first and the third quadrant, or between the second and the fourth quadrant. Therefore, for all trajectories with initial conditions either in the first or the third quadrant, the light is red 75% of the time and white 25% of the time; for trajectories with initial conditions in the second and the fourth quadrant the light is white 75% of the time and blue 25% of the time. But neither white nor red is an equilibrium because half of all initial conditions lie on trajectories that only spend 25% of the time in the white state, and the other half of initial conditions lie on trajectories that spend no time at all in the red state. This violates the requirement that initial conditions that don't spend most of the time in equilibrium form a set that has at most measure $\varepsilon \ll 1$.[9]

However, this seems to be the wrong conclusion because intuitively there are equilibria: for initial states with displacement coordinates in the first or the third quadrant the light is red 75% of the time and hence red seems an equilibrium *for states in those quadrants*, and likewise for initial states in the second or forth quadrant for which the light is white 75% of the time. The root of the rift between mathematical criteria and intuition is that we tacitly took the entire state space X'' to be the effective state space Z, and with respect to X'' the conditions for the existence of an equilibrium are not satisfied. But nothing forces us to set $Z = X''$. In fact an alternative choice of Z restores existence. Let Z^1 be the union of the first and the third quadrants. One can then easily construct the effective dynamical system $(Z^1, \Sigma_{Z^1}, \mu_{Z^1}, T'')$, where Σ_{Z^1} is the Borel σ-algebra on Z^1, μ_{Z^1} is the measure $\mu_{X''}$ restricted to Z^1 and T'' is the dynamics restricted to Z^1. It is obvious that for that system the light being red is a an 0.75-0-equilibrium of the α-ε type and a 0.5-0-equilibrium of the γ-ε type. And the same moves are available for the

[9] This example also shows that the largest macro-state need not be the equilibrium state: X''_{M_w} takes up $1/2$ of X'' and yet M_w is not the equilibrium state.

other two quadrants. Let Z^2 be the union of the second or the fourth quadrant. The corresponding effective dynamical system is $(Z^2, \Sigma_{Z^2}, \mu_{Z^2}, T'')$, where Σ_{Z^2} is the Borel σ-algebra on Z^2, μ_{Z^2} is the measure $\mu_{Z''}$ restricted to Z^2 and T'' is the dynamics restricted to Z^2. It is then obvious that the light being white is a an 0.75-0-equilibrium of the α-ε type and a 0.5-0-equilibrium of the γ-ε type.

This example illustrates that the choice of the effective phase space Z is crucial for the existence of an equilibrium. With the wrong choice of Z—the full three-dimensional state space—no equilibrium exists. But if we choose either Z^1 or Z^2 as the effective state space, then there are equilibria.

Let us explain why the Existence Theorem is satisfied for these effective dynamical systems. We first focus on $(Z^1, \Sigma_{Z^1}, \mu_{Z^1}, T'')$. The index set is $\Omega'' = \Omega_3 \times \Omega_4$, where Ω_3 consist of the possible energies of the system and each $\omega_4 \in \Omega_4$, $\Omega_4 = (0, \pi/2]$, denotes an angle and thus a line cutting through the coordinate origin in the first (and therefore also third) quadrant. The measure on Ω'' arises from the product measure $\mu_{\Omega_3} \times \mu_{\Omega_4}$, where μ_{Ω_3} is the uniform measure on the energy values and μ_{Ω_4} is the uniform measure on $(0, \pi/2]$. Each Z''_ω is a two-dimensional ellipse determined by the initial energy and displacement coordinates; the sigma-algebra Σ''_ω is the usual Borel σ-algebra, and the measure μ''_ω is the normalised measure as above on the ellipse Z''_ω. The dynamics on the ellipses is again given by the restriction of T''_t to the ellipses Z''_ω. This gives us the components $C''_\omega = (Z''_\omega, \Sigma_{\omega''}, \mu''_\omega, T''_t)$. Again, it is clear that the motion on each component is ergodic. Now the Existence Theorem is satisfied for the same reason it is satisfied for the pendulum with a one-dimensional position coordinate, namely: Eq. (10.5) holds true for every arbitrary component C''_ω because on each ellipse three-quarters of the states correspond to a red light and one quarter to a white light, and hence $\mu''_\omega(Z''_\omega \cap Z^1_{M_r}) \geq 0.75$. Similarly, Eq. (10.5) holds true for every arbitrary component C''_ω because on each ellipse three-quarters of the states correspond to a white light and one quarter to a red light, and hence $\mu''_\omega(Z''_\omega \cap Z^1_{M_r}) \geq \mu''_\omega(Z''_\omega \cap Z^1_M) + 0.5$ for all $Z^1_M \neq Z^1_{M_r}$. Analogue reasoning for Z^2 shows that also for $(Z^2, \Sigma_{Z^2}, \mu_{Z^2}, T'')$ the Eqs. (10.5) and (10.6) of the Existence Theorem are satisfied.

10.5 A Fresh Look at the Ergodic Programme

The canonical explanation of equilibrium behaviour is given within the ergodic approach. Before looking at further examples, it is helpful to revisit this approach from the point of view of the Existence Theorem. We show that the standard ergodic approach in fact provides a triplet that satisfies the above conditions.

Many explanations of the approach to equilibrium rely on the dynamical conditions of ergodicity or epsilon-ergodicity (see Frigg, 2008) and references therein). The definition of ergodicity was given above (Eq. 10.2). A system $(Z, \Sigma_Z, \mu_Z, T_t)$ is epsilon-ergodic iff it is ergodic on a set $\hat{Z} \subseteq Z$ of measure $1 - \varepsilon$ where ε is a very small real number.[10] The results of this paper clarify these claims. As pointed out in the previous subsection, if the macro-variables are not the right ones, then neither ergodicity nor epsilon-ergodicity imply that the approach to equilibrium takes place. However proponents of the ergodic approach often assume that there is a macro region which is either β-dominant or δ-prevalent (e.g. Frigg and Werndl 2011, 2012). Then this leads to particularly simple instance the Existence Theorem, which then implies that the macro-region corresponds to an α-ε-equilibrium or a γ-ε-equilibrium. More specifically, the following two corollaries hold (for proofs seee Werndl and Frigg 2015a):

Ergodicity-Corollary: Suppose that the measure-preserving system $(Z, \Sigma_Z, \mu_Z, T_t)$ is ergodic. Then the following are true: (a) If the system has a macro-region $Z_{\hat{M}}$ that is β-dominant, $\hat{M}$ is an α-ε-equilibrium for $\alpha = \beta$ and $\varepsilon = 0$. (b) If the system has a macro-region $Z_{\hat{M}}$ that is δ-prevalent, $\hat{M}$ is a γ-ε-equilibrium for $\gamma = \delta$ and $\varepsilon = 0$.

Epsilon-Ergodicity-Corollary: Suppose that the measure-preserving system $(Z, \Sigma_Z, \mu_Z, T_t)$ is epsilon-ergodic. Then the following are true: (a) If the system has a macro-region $Z_{\hat{M}}$ that is β-dominant for $\beta - \varepsilon > \frac{1}{2}$, $Z_{\hat{M}}$ is a α-ε-equilibrium for $\alpha = \beta - \varepsilon$. (b) If the system has a macro-region $Z_{\hat{M}}$ that is δ-prevalent for $\delta - \varepsilon > 0$, $Z_{\hat{M}}$ is a γ-ε-equilibrium for $\gamma = \delta - \varepsilon$.

It is important to keep in mind, however, that ergodicity and epsilon-ergodicity are just examples of dynamical conditions for which an equilibrium exists. As shown by the Existence Theorem, the dynamics need not be ergodic or epsilon-ergodic for there to be an equilibrium.

10.6 Gases

We now discuss gas systems that illustrate the core theorems of this paper. We start with well-known examples—the dilute gas, the ideal gas and the Kac gas—and then turn to lesser-known systems that illustrate the role of the ergodic decomposition and the ε-set of initial conditions that can be excluded. We first discuss a simple example where the dynamics is ergodic, namely a gas of noninteracting particles in a stadium-shaped box. Then we turn to an example of a system with an ε-set that is excluded because the system is epsilon-ergodic, namely a gas of noninteracting particles in a mushroom-shaped box. Finally, we examine a more complicated gas system where there are several ergodic components and an ε-set that is excluded, namely a gas of noninteracting particles in a multi-mushroom box.

[10] In detail: $(Z, \Sigma_Z, \mu_Z, T_t)$ is *ε-ergodic*, $\varepsilon \in \mathbb{R}$, $0 \leq \varepsilon < 1$, iff there is a set $\hat{Z} \subset Z$, $\mu_Z(\hat{Z}) = 1 - \varepsilon$, with $T_t(\hat{Z}) \subseteq \hat{Z}$ for all t, such that the system $(\hat{Z}, \Sigma_{\hat{Z}}, \mu_{\hat{Z}}, T_t)$ is ergodic, where $\Sigma_{\hat{Z}}$ and $\mu_{\hat{Z}}$ is the σ-algebra Σ_Z and the measure μ_Z restricted to $\hat{Z}$. A system $(Z, \Sigma_Z, \mu_Z, T_t)$ is *epsilon-ergodic* iff there exists a very small ε for which the system is ε-ergodic.

10.6.1 The Dilute Gas

A dilute gas is a system a system of N particles in a finite container isolated from the environment. Unlike the particles of the ideal gas (which we consider in the next subsection), the particles of the dilute gas do interact with each other, which will be important later on. We first briefly review the standard derivation of the Maxwell-Boltzmann distribution with the combinatorial argument and then explain how the argument is used in our framework.

A point $x = (q, p)$ in the $6N$-dimensional set of possible position and momentum coordinates X specifies a *micro-state* of the system. The classical Hamiltonian $H(x)$ determines the dynamics of the system. Since the energy is preserved, the motion is confined to the $6N - 1$ dimensional energy hyper-surface X_E defined by $H(x) = E$, where E is the energy of the system. X is endowed with the Lebesgue measure μ, which is preserved under T_t. With help of μ a measure μ_E on X_E can be defined which is preserved as well and is normalised, i.e. $\mu_E(X_E) = 1$ (*cf.* Frigg, 2008, 104).

To derive the Maxwell-Boltzmann distribution we consider the 6-dimensional state space X_1 of one particle. The state of the entire gas is given by N points in X_1. Because the system has constant energy E and is confined to a finite container, only a finite part of X_1 is accessible. This accessible part of X_1 is partitioned into cells of equal size δ^{dg} whose dividing lines run parallel to the position and momentum axes. This results in a finite partition $\Omega_{dg} := \{\omega_1^{dg}, \ldots, \omega_l^{dg}\}$, $l \in \mathbb{N}$ ('dg' stands for 'dilute gas'). The cell in which a particle's state lies is its *coarse-grained micro-state*. An *arrangement* is a specification of coarse-grained micro-state of each particle. Let $N_i$be the number of particles whose state is in cell ω_i^{dg}. A *distribution* $D = (N_1, N_2, \ldots, N_l)$ is a specification of the number of particles in each cell. Several arrangements are compatible with each distribution, and the number $G(D)$ of arrangements compatible with a given distribution D is $G(D) = N! / N_1!N_2! \ldots, N_l!$. Boltzmann (1877) assumed that the energy e_i of particle i depends only on the cell in which it is located (and not on interactions with other particles), which allows him to express the total energy of the system as a sum of single particle energies: $E = \sum_{i=1}^{l} N_i e_i$. Assuming that the number of cells in Ω_{dg} is small compared to the number of particles, Boltzmann was able to show that $\mu_E(Z_{D_{dg}})$ is maximal if

$$N_i = Be^{\Delta e_i}, \tag{10.10}$$

where B and Δ are parameters which depend on N and E. This is the *discrete Maxwell-Boltzmann distribution*, which we refer to as D_{MB}

Textbooks wisdom has it that the Maxwell-Boltzmann distribution defines the equilibrium state of the gas. While not wrong, this is only part of a longer story. We have to introduce macro variables and define Z before we can say what the system's macro-regions are, and only once these are defined we can check whether the dynamics is such that one of those macro-regions qualifies as the equilibrium region.

Let us begin with macro-variables. The macro-properties of a gas depend only on the distribution D. Let W be a physical variable on the one-particle phase space. For simplicity we assume that this variable assumes constant values w_j in cell ω_j^{dg} for all $j = 1, \ldots, l$. Physical observables can then written as averages of the form $\sum_{j=1}^{N} w_j N_j$ (for details see Tolman 1938/1979, Ch. 4). It is obvious that every point $x \in X$ is associated with exactly one distribution D, which we call $D(x)$. Given $D(x)$ one can calculate $\sum_{j=1}^{N} w_j N_j$ at point x, which assigns every point x a unique value. Hence a physical variable W and a distribution $D(x)$ induce a mapping from X to a set of values. Let us call this mapping v, and so we can write: $v : X \to \mathbb{V}$, where $\mathbb{V}$ is the range of certain physical variable. Choosing different W (with different w_j) will lead to different a different v. These are the macro-variables of the kind introduced in Sect. 10.2. A set of values of these variables defines a *macro-state*. For the sake of simplicity we now assume that this set of values would be different for every distribution so that there is a one-to-one correspondence between distributions and macro-states.

The Maxwell-Boltzmann distribution depends on the total energy of the system: different energies lead to different equilibrium distributions. This tells us that equilibrium has to be defined with respect to the energy hyper-surface X_E.[11] States of different energy can never evolve into the same equilibrium and therefore no equilibrium state exists with respect to the full state space X. Now the assumption that the particles of the dilute gas interact becomes crucial. If the particles did not interact, there could be constants of motion other than the total energy and this might have the consequence equilibrium would have to be defined on a subsets of X_E (we discuss such a case in the next subsection). It is usually assumed that this is not the case. The effective state space Z then is X_E, and $(X_E, \Sigma_E, \mu_E, T_t)$ is the effective measure-preserving dynamical system of the dilute gas, where Σ_E is the the Borel σ-algebra of X_E and T_t is the flow of the system restricted to X_E.

We can now construct the macro-regions Z_M. Above we assumed that there is a one to one correspondence between distributions and macro-states. So let M_D be the macro state corresponding to distribution D. The macro-region Z_{M_D} is then just the set of all $x \in X_E$ that are associated with D: $Z_{M_D} = \{x \in X_E \ : \ D(x) = D\}$. A fortiori this also provides a definition of the macro-state $M_{D_{MB}}$ associated with the Maxwell-Boltzmann distribution D_{MB}. Let us call the macro-region associated with that macro-state Z_{MB}. It is generally assumed that Z_{MB} is the largest of all macro-states (relative to μ_E), and we follow this assumption here.[12]

---

[11] Note that this is one of crucial differences between the dilute gas and the oscillator with a colour macro-variable of Sect. 10.4: the colour equilibrium does not depend on the system's energy.

[12] The issue is the following Eq. (10.10) gives the distribution of largest size relative to the Lebesgue measure on the $6N$-dimensional shell-like domain X_{ES} specified by the condition that $E = \sum_{i=1}^{l} N_i e_i$. It does *not* give us the distribution with the largest measure μ_E on the $6N - 1$ dimensional Z_E. Strictly speaking nothing about the size of Z_{MB} (with respect to μ_E) follows from the combinatorial considerations leading to Eq. (10.10). Yet it is generally assumed that the

Even if we grant that Z_{MB} is the largest macro-region (in one of the senses of 'large'), it is not yet clear that $M_{D_{MB}}$ is the equilibrium macro-state (in one of the senses of 'equilibrium'). It could be that the dynamics is such that initial conditions that lie outside Z_{MB} avoid Z_{MB}, or that a significant portion of initial conditions lie on trajectories that spend only a short time in Z_{MB}. To rule out such possibilities one has to look at the dynamics of T_t. Unfortunately the dynamics of dilute gases is mathematically not well understood, and there is no rigorous proof that the dynamics is 'benign' (meaning that it does not have any of the features just mentioned). However, there are plausibility arguments for the conclusion that T_t is epsilon-ergodic (Frigg and Werndl, 2011). If these arguments are correct, then the dilute gas falls under the Epsilon-Ergodicity-Corollary and Z_{MB} is an equilibrium either of the α-ε or the γ-ε type, depending on whether Z_{MB} is β-dominant or δ-prevalent. Moreover, even if the dynamics turned out not to be epsilon-ergodic, it is a plausible assumption that the dynamics is such that the conditions of the Existence Theorem is fulfilled.

Hence, the Maxwell-Boltzmann distribution corresponds to equilibrium as expected. However, the above discussion shows that this does not come for free: we have to accept that Z_{MB} is large and that T_t is epsilon-ergodic, and making these assumptions plausible the choice of the right effective state space Z is crucial. In fact, *relative to X no equilibrium exists* because there are different equilibria for different total energies of the system (as reflected by the Maxwell-Boltzmann distribution, which depends on the total energy E). This shows that the triplet of macro-variables, dynamics, and effective state space has to be well-adjusted for an equilibrium to exists, and that even small changes in one component can destroy this balance.

10.6.2 The Ideal Gas

Now consider an ideal gas, a system consisting of N particles with mass m and *no interaction at all*. We consider the same partitioning of the phase space as above and hence can consider the same distributions and the same macro-variables. One might then think that the ideal gas is sufficiently similar to the dilute gas to regard Z_{MB} as the equilibrium state and lay the case to rest.

This is a mistake. To see why we need to say more about the dynamics of the system. An common way to describe the gas mathematically is to assume that the particles move on a three three-dimensional torus with constant momenta in each

proportion of the areas corresponding to different distributions are the same on X and on X_E (or at least that the relative ordering is the same). Under that assumption Z_E is indeed the largest macro-region. We agree with Ehrenfest and Ehrenfest (1959, 30) that this assumption is in need of further justification, but grant it for the sake of the argument.

direction.[13] This implies that *all* one-particle particle momenta p_i (and hence all one-particle energies $e_i = p_i^2/2m$) are conserved quantities. As a consequence, if an ideal gas starts in a micro-state in which the momenta of the particles are not distributed according to the Maxwell-Boltzmann distribution, they will never reach that distribution. In fact the initial distribution is preserved no matter what that distribution is. For this reason Z_{MB} is not the equilibrium state and the Maxwell-Boltzmann distribution does *not* characterise the equilibrium state. So the combinatorial argument does not provide the correct equilibrium state for an ideal gas, and Z_E is not the effective state space.

This, however, does not imply that the ideal gas has no equilibrium at all. Intuitively speaking, there is a γ-ε-equilibrium, namely the one where all particles are uniformly distributed. To make this more explicit let us separate the distribution D into the position distribution D_x and the momentum D_p (which is a trivial decomposition which can always be done): $D = (D_x, D_p)$. Under the dynamics of the ideal gas D_p will not change over time and hence remain in whatever initial distribution the gas is prepared. By contrast, the position distribution D_x will approach an even distribution D_e as time goes on. So we can say that the equilibrium distribution of the system is $D_{eq} = (D_e, D_p)$, where D_p is the gas' initial distribution. The relevant space with respect to which an equilibrium exists is the hyper-surface Z_p, i.e. the hyper-surfaface defined by the condition that the moment are distributed according to D_p. The relevant dynamical system then is $(Z_p, \Sigma_{Z_p}, \mu_p, T_t)$, where Σ_{Z_p} is the Borel-σ-algebra, μ_p is the uniform measure on Z_p and the dynamics T_t is simply the dynamics of the ideal gas restricted to Z_p.

It is easy to see that with respect to Z_p the region corresponding to D_e is the largest macro-region. The motion on Z_p is ergodic for almost all momentum coordinates.[14] Thus, by the Ergodicity-Corollary, the largest macro-region, i.e. the macro-region corresponding to the uniform distribution, is a γ-ε-equilibrium. There will be some very special momentum coordinates where no equilibrium exists relative to Z_p because the motion of the particles is periodic. However, these special momentum coordinates are of measure zero and for all other momentum coordinates the uniform distribution will correspond to the equilibrium macro-region. Thus this example illustrates again the importance of choosing the correct effective phase space: the ideal gas has no equilibrium relative to Γ_E but an equilibrium exists relative to Z_p.[15]

[13] One also think of the particles as bouncing back and forth in box. In this case the modulo of the momenta is preserved and a similar argument applies.

[14] In terms of the uniform measure on the momentum coordinates.

[15] Another possible treatment of the ideal gas is to consider the different macro-state structure given only by the coarse-grained position coordinates (i.e. the momentum coordinates are not considered). Then the effective dynamical system would coincide with the full dynamical system $(\Gamma, \Sigma_\Gamma, \mu_\Gamma, T_t)$. Relative to this dynamical system there would be an γ-0-equilibrium (namely the uniform distribution). That is, almost all initial conditions would spend most of the time in the macro-state that corresponds to the uniform distribution of the position coordinates.

10.6.3 The Kac Gas

The Kac-ring model consists of an even number N of sites distributed equidistantly around a circle. On each site there is either a white or black ball. Let us assume that $N/2$ of the points (forming a set S) between the sites of the balls are marked. A specific combination of white and black balls for all sites together with the set S is a *micro-state* k of the system, and the state space K consists of all combinations of white and black balls and selection of $N/2$ points between the sites and Σ_K is the power set of K. The dynamics κ of the system is given as follows: during one time step each ball moves counterclockwise to the next site; if the ball crosses an interval in S, it changes colour and if it does not cross an interval in S, then it does not change colour (the set S stays the same at all times). The probability measure is the uniform measure μ_K on K. $(K, \Sigma_K, \mu_K, \kappa_t)$, where κ_t is the t-th iterate of κ is a measure-preserving deterministic system describing the behaviour of the balls (and K is both the full state space X as well as the effective state space Z of the system).

The Kac-ring can be interpreted in several ways. As presented here, the intended interpretation is that of a gas: the balls are described by their positions and their colour is seen as representing their (discrete) velocity. Whenever a ball passes a marked site its colour changes, which is analogous to a change in velocity of a molecule that results from collision with another molecule. The equations of motion are given by the counterclockwise motion together with the changing of the colours (Bricmont 1995; Kac 1959; Thompson 1972). The macro-states usually considered are defined by the total number of black and white balls. So the relevant macro-varible v is a mapping $K \to \mathbb{V}$, where $\mathbb{V} = \{0, \ldots, N\}$. Each value in $\mathbb{V}$ defines a different macro-state. Traditionally these states are labelled M_i^K, where i denotes the total number of white balls, $0 \leq i \leq N$. As above, the *macro-regions* K_i are defined as the set of micro-states on which M_i^K supervenes. It can be shown that the macro-state whose macro-region is of largest size is $M_{N/2}^K$, i.e. the state in which half of the spins are up and half down.

This example is interesting because it illustrates the case where an equilibrium exists even though the phase space is broken up into a finite number of ergodic components. More specifically, the motion of the Kac-ring is periodic. Suppose that $N/2$ is even: then at most after N steps all balls have returned to their original colour because each interval has been crossed once and $N/2$ is even. If $N/2$ is odd, then it takes at most $2N$ steps for the balls to return to their original colour (because after $2N$ steps each interval has been crossed twice). So the phase space of the KAC-ring is decomposed into periodic cycles (together with a specification of S). These cycles are the components of the ergodic decomposition that we encounter in the Ergodic Decomposition Theorem. The Existence Theorem is satisfied and hence a γ-ε-equilibrium exists because on each of these ergodic components, except for components of measures ε, the equilibrium macro-state $M_{N/2}^K$ takes up the largest measure, i.e. Eq. (10.6) is satisfied. Note that there are initial states that do not show equilibrium-like behaviour (that is, the set of initial conditions that do not show an approach to equilibrium is of positive measure ε). For instance, start with all balls being white and let every interval belong to S. Then, clearly, after one step the balls are all black, then after one step they are all white, and so on there is no approach equilibrium (Bricmont 2001; Kac 1959; Thompson 1972). The Existence Theorem is satisfied and hence a γ-ε-equilibrium exists because on each of these ergodic components, except for components of measures ε, the equilibrium macrostate $M^K_{N/2}$

10.6.4 Gas of Noninteracting Particles in a Stadium-Box

Let us now turn to lesser-known examples of gas systems that illustrate the various cases of the Existence Theorem. The first example illustrates the easiest way to satisfy the existence theorem, namely having an ergodic dynamics and a macroregion of largest measure. Consider a stadium-shaped box S (i.e. a rectangle capped by semicircles). Suppose that N particles are moving with uniform speed[16] inside the stadium-shaped box, where the collisions with the walls are assumed to be elastic and it is further assumed that the particles do not interact. The set of all possible states of the system consists of the points $Y = (y_1, w_1, y_2, w_2 \ldots, y_N, w_N)$ satisfying the constraints $y_i \in S$ and $||w_i|| = 1$, where y_i and w_i are the position and velocity coordinates of the particles respectively ($1 \leq i \leq N$). Σ_Y is the Borel σ-algebra of Y. The dynamics R_t of the system is the motion resulting from particles bouncing off the wall elastically (whithout interacting with each other). The uniform measure ν is the invariant measure of the system. (Y, Σ_Y, ν, R_t) is a measure-preserving dynamical system and it can be proven that the system is ergodic (*cf.* Bunimovich 1979).[17] Y is both the full state space X and the effective state space Z of the system.

Now divide the stadium-shaped box into cells $\omega_1^S, \omega_2^S, \ldots, \omega_l^S$ of equal measure δ^S ($l \in \mathbb{N}$). As in the case of the dilute gas, consider distributions $D = (N_1, \ldots, N_l)$ and associate macro-states with these distributions. Macro-variables are also defined as above. It is then obvious that the macrostate $(N/l, N/l, \ldots, N/l)$ corresponds to the macro-region of largest measure.[18] Since the dynamics is ergodic, it follows from the Ergodicity-Corollary that the system has a γ-ε-equilibrium (where $\varepsilon = 0$). More specifically, except for a set of measure zero, for all initial states of the N billiard balls the system will approach equilibrium and stay there most of the time.

---

[16] Speed, unlike velocity, is not directional and does not change when particle bounces off the wall.

[17] Bunimovich's (1979) results are about *one* particle moving in a stadium-shaped box, but they immediately imply the results stated here about n non-interacting particles.

[18] It is assumed here that N is a multiple of l.

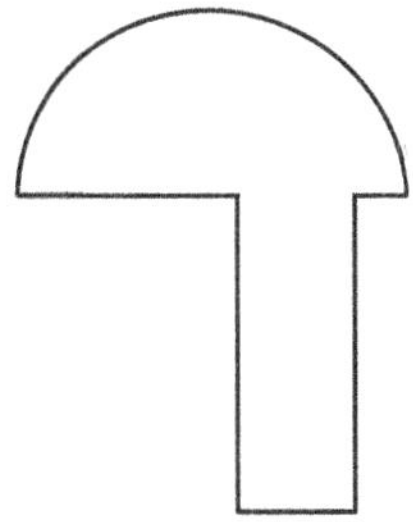

Fig. 10.5 The mushroom-shaped box

10.6.5 Gas of Noninteracting Particles in a Mushroom-Box

The next example illustrates the role of the ε-set of initial conditions that are not required to show equilibrium-like behaviour in the Definition of a γ-ε-equilibrium. For most conservative systems the phase space is expected to consist of regions of chaotic or ergodic behaviour next to regions of regular and integrable behaviour. These mixed systems are notoriously difficult to study analytically as well as numerically (Porter and Lansel, 2006). So it was a considerable breakthrough when Bunimovich (2002) introduced a class of billiard systems that can easily be shown to have mixed behaviour.

Consider a mushroom-shaped box (the domain M obtained by placing an ellipse on top of a rectangle as shown in Fig. 10.5), consisting of the stem St and the cap Ca. Suppose that N gas particles are moving with uniform speed inside the mushroom-shaped box. The collisions on the wall are again assumed to be elastic and, for sake of simplicity, we assume that the particles do not interact. Then the set of all possible states consists of the points $D = (d_1, v_1, d_2, v_2 \dots, d_n, v_n)$, where $d_i \in M$ and $||v_i|| = 1$ are the position and velocity coordinates of the particles respectively ($1 \leq i \leq N$). Σ_D is the Borel σ-algebra of D. The dynamics U_t of the system is the motion of the particles generated by elastic collisions with the boundaries of the mushroom. The phase volume u is preserved under the dynamics. (M, Σ_M, U_t, u) is a measure-preserving dynamical system (and M is both the full state space X and the effective state space Z of the system).

It can be proven that the phase space consist of two regions: an ergodic region and a region with regular or mixed behaviour (i.e. integrable parts are intertwined with chaotic parts). As the stem is shifted to the left, the volume of phase space occupied by the ergodic motion continually increases and finally reaching measure 1 when the stem reaches the edge of the cap.[19] Assume now that the stem be so far to the left that the measure of the ergodic region is $1 - \varepsilon$, in which case the system is ε-ergodic (*cf.* Sect. 10.5). Suppose that the macro-states of interest are distributions $D = (N_{St}, N_{Ca})$, where N_{St} and N_{Ca} are the particle numbers in the stem and

[19] The results in Bunimovich (2002) are all about *one* particle moving inside a mushroom-shaped box, but they immediately imply the results about a system of n non-interacting particles stated here.

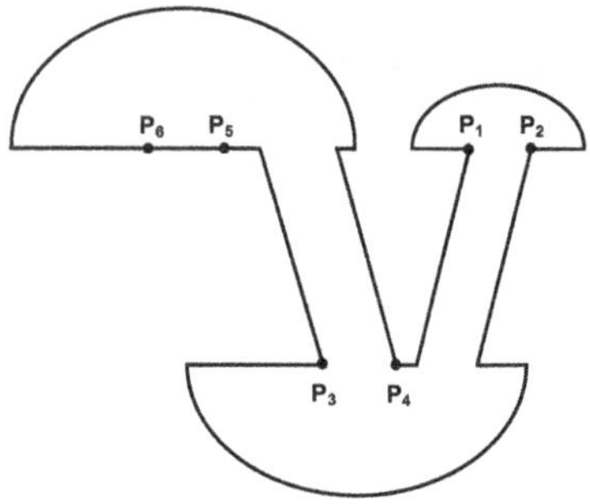

Fig. 10.6 The multi-mushroom-shaped box

cap respectively. We now assume that that the measure of the stem is the same as the measure of the cap. It then follows from the Epsilon-Ergodicity Corollary that the system has a γ-ε equilibrium, namely $(N/2, N/2)$ (*cf.* Bunimovich 2002; Porter and Lansel 2006). This example is of special interest because it is proven that there is a set of initial states of the billiard balls of positive measure that do not show equilibrium-like behaviour. That is, for these initial states the system does not evolve in such a way that most of the time half of the particles are in the stem and half of the ball are in the cap (as is allowed by the definition of an γ-ε-equilibrium).

10.6.6 Gas of Noninteracting Particles in a Multi-Mushroom-Box

In our next and last example the Existence Theorem is satisfied because there are *a finite number of ergodic components* on each of which the equilibrium macro-state takes up the largest measure. Consider a box created by several mushrooms such as the one shown in Fig. 10.6 (the domain TM is constructed from three elliptic mushrooms, where the semi-ellipses have foci P_1 and P_2, P_3 and P_4, P_5 and P_6).

Suppose again that N gas particles are moving with uniform speed inside the box MM, that the collisions on the wall are elastic and that the particles do not interact. The set of all possible states consists of the points $W = (w_1, u_1, w_2, u_2 \ldots, w_n, u_n)$, where $w_i \in MM$ and $||u_i|| = 1$ are the position and velocity coordinates of the particles respectively $(1 \leq i \leq N)$. Σ_W is the Borel σ-algebra on W. The dynamics V_t of the system is given by the motion of the noninteracting particles inside the box, and the phase volume v is preserved under the dynamics. (W, Σ_W, v, V_t) is a measure-preserving dynamical system (and W is both the full state space X and the effective state space Z of the system).

Bunimovich (2002) proved that the phase space consist of 2^N larger regions on each of which the motion is ergodic and, finally, one region of negligible measure ε of regular or mixed behaviour (Bunimovich, 2002).[20] The 2^N ergodic components arise in the following way: each single particle space has two large ergodic regions: one region consisting of those orbits of the particle that move back and forth between the semi-ellipses with the foci P_1, P_2 and P_3, P_4 (while never visiting the semi-ellipse with the foci P_5, P_6). The second ergodic region consists of the orbits that travel back and forth between the semi-ellipses with the foci P_3, P_4 and P_5, P_6 (while never visiting the semi-ellipse with the foci P_1, P_2). Given that the phase space of the entire system is just the cross-product of the phase space of the N single particle spaces, it follows that there are 2^N ergodic components.[21]

Suppose that the macrostates of interest are the distributions $D = (MS, MC)$, where MS is the number of balls in the two stems and MC the number of balls in the three caps of the mushroom, where we assume that the measure of the two stems taken together is the same as the measure of the three caps taken together.[22] Then the system has a γ-ε equilibrium, corresponding to the case where $N/2$ of the particles are in the two stems and $N/2$ of the particles are in the three caps of the mushrooms (cf. Bunimovich 2002; Porter and Lansel 2006). This example is of special interest because it illustrates the case where an equilibrium exists even though the phase space is broken up into a finite number of ergodic components. More specifically, in this case we encounter 2^N ergodic components on each of which the equilibrium macro-state takes up the largest measure and hence Eq. (10.6) is satisfied. Since these ergodic components taken together have total measure $1 - \varepsilon$ and the definition of a γ-ε-equilibrium allows that there is an ε set of initial conditions that do not show equilibrium-like behaviour, it follows that the Existence Theorem is satisfied.

10.7 Conclusion

In this paper we introduced a new definition of Boltzmannian equilibrium and presented an existence theorem that characterises the circumstances under which a Boltzmannian equilibrium exists. The definition and the theorem are completely general in that they make no assumption about the nature of interactions and so they provide a characterisation of equilibrium also in the case of strongly interacting systems. The approach also ties in smoothly with the Generalised Nagel-Schaffner model of reduction (Dizadji-Bahmani et al., 2010) and hence serves as a

---

[20] How small ε is depends on the exact shape of the box of the three elliptic mushrooms (it can be made arbitrarily small).

[21] Again, Bunimovich's (2002) results are all about *one* particle moving inside a mushroom-shaped box, but they immediately imply the results about a system of n non-interacting particles stated here.

[22] This can always be arranged in this way—see Bunimovich (2002).

starting point for discussions about the reduction of thermodynamics to statistical mechanics.

The framework raises a number of questions for future research. First, our discussion is couched in terms of deterministic dynamical systems. In a recent paper (Werndl and Frigg, 2017a) we generalise the definition of equilibrium to stochastic systems. To date there is, however, no such generalisation of the Existence Theorem. The reason for this is that this theorem is based on the ergodic decomposition theorem, which has no straightforward stochastic analogue. So it remains an open question how circumstances under which an equilibrium exists in stochastic systems should be characterised.

Second, macro-variables raise a number of interesting issues. An important question is: how exactly do the macro-variables look like for the variety of physical systems discussed in statistical mechanics? It has been pointed out to us[23] that for intensive variables the exact definition is complicated and can only be done by referring to extensive quantities. A further issue is that many quantities of interest are *local quantities*, at least as long as the system is not in equilibrium (pressure and temperature are cases in point). Such quantities have to be described as fields, which requires an extension of the definition of a macro-state in Sect. 10.2. Rather than associating equilibrium only with a certain value (or a range of values), one now also has to take field properties such as homogeneity into account. We address some of these questions in our other contribution to this book (Chap. 11), but the issue deserves further attention.

Finally, there is a question about how to extend our notion of equilibrium to quantum systems. Noting in our definition of equilibrium depends on the underlying dynamics being classical or the variables being defined on a classical space phase space rather than a Hilbert space, and so we think that there are no in-principle obstacles to carrying over our definition of equilibrium to quantum mechanics. But the proof of the pudding is in the eating and so the challenge is to give an explicit quantum mechanical formulation of Boltzmannian equilibrium.

References

Bricmont, J. (1995). Science of chaos or chaos in science? In P. R. Gross, N. Levitt, & M. W. Lewis (Eds.), *The flight from science and reason. Annals of the New York Academy of Sciences* (Vol. 775, pp. 131–175). New York: The New York Academy of Sciences.

Bricmont, J. (2001). Bayes, Boltzmann and Bohm: Probabilities in physics. In J. Bricmont, D. Dürr, M. C. Galavotti, G. C. Ghirardi, F. Pettrucione, & N. Zanghi (Eds.), *Chance in physics: Foundations and perspectives* (pp. 3–21). Springer.

Bunimovich, L. A. (1979). On the Ergodic properties of nowhere dispersing billiards. *Communications in Mathematical Physics, 65*, 295–312.

Bunimovich, L. A. (2002). Mushrooms and Other Billiards With Divided Phase Space. *Chaos* 11 (4), 802–808.

[23] By David Lavis and Reimer Kühn in private conversation.

Callender, C. (2001). Taking thermodynamics too seriously. *Studies in History and Philosophy of Modern Physics, 32*, 539–553.

Dizadji-Bahmani, F., Frigg, R., & Hartmann, S. (2010). Who's afraid of Nagelian reduction? *Erkenntnis, 73*, 393–412.

Ehrenfest, P., & Ehrenfest, T. (1959). *The conceptual foundations of the statistical approach in mechanics*. Cornell University Press.

Frigg, R. (2008). A field guide to recent work on the foundations of statistical mechanics. In D. Rickles (Ed.), *The Ashgate Companion to Contemporary Philosophy of Physics* (pp. 99–196). London, Ashgate.

Frigg, R., & Werndl, C. (2011). Explaining thermodynamic-like behaviour in terms of epsilon-ergodicity. *Philosophy of Science, 78*, 628–652.

Frigg, R., & Werndl, C. (2012). A new approach to the approach to equilibrium. In Y. Ben-Menahem, & M. Hemmo (Eds.), *Probability in physics* (pp. 99–114). The Frontiers Collection. Springer.

Frigg, R., & Werndl, C. (2019). Statistical mechanics: A tale of two theories? Monist, *102*, 424–438.

Kac, M. (1959). *Probability and related topics in the physical sciences*. Interscience Publishing.

Petersen, K. (1983). *Ergodic theory*. Cambridge University Press.

Porter, M.A., & Lansel, S. (2006). Mushroom billiards. *Notices of the American Mathematical Society, 53*(3), 334–337.

Reiss, H. (1996). *Methods of thermodynamics*. Dover.

Sklar, L. (1973). *Philosophical issues in the foundations of statistical mechanics*. Cambridge University Press.

Thompson, C. J. (1972). *Mathematical statistical mechanics*. Princeton University Press.

Tolman, R. C. (1938/1979). *The principles of statistical mechanics*. Dover.

Wang, G. M., Sevick, E. G., Mittag, E., Debra, J., Searles, D. J., & Evans, D. J. (2002). Experimental demonstration of violations of the second law of thermodynamics for small systems and short time scales. *Physical Review Letters, 89*, 050601.

Werndl, C., & Frigg, R. (2015a). Reconceptualising equilibrium in Boltzmannian statistical mechanics and characterising its existence. *Studies in History and Philosophy of Modern Physics, 49*(1), 19–31.

Werndl, C., & Frigg, R. (2015b). Rethinking Boltzmannian equilibrium. *Philosophy of Science, 82*(5), 1224–1235.

Werndl, C., & Frigg, R. (2017a). Boltzmannian equilibrium in stochastic systems. In M. Massimi & J.-W. Romeijn (Eds.), *Proceedings of the European Philosophy of Science Association* (pp. 243–254). Springer.

Werndl, C., & Frigg, R. (2017b). Mind the gap: Boltzmannian versus Gibbsian equilibrium. *Philosophy of Science, 84*(5), 1289–1302.

Werndl, C., & Frigg, R. (2020). When do Gibbsian phase averages and Boltzmannian equilibrium values agree? *Studies in History and Philosophy of Modern Physics, 72*, 46–69.

Chapter 11
Boltzmannian Non-Equilibrium and Local Variables

Roman Frigg and Charlotte Werndl

Abstract Boltzmannian statistical mechanics (BSM) partitions a system's space of micro-states into cells and refers to these cells as 'macro-states'. One of these cells is singled out as the equilibrium macro-state while the others are non-equilibrium macro-states. It remains unclear, however, how these states are characterised at the macro-level as long as only real-valued macro-variables are available. We argue that physical quantities like pressure and temperature should be treated as field-variables and show how field variables fit into the framework of our own version of BSM, the long-run residence time account of BSM. The introduction of field variables into the theory makes it possible to give a full macroscopic characterisation of the approach to equilibrium.

11.1 Introduction

The central posit of Boltzmannian statistical mechanics (BSM) is that macro-states supervene on micro-states.[1] This leads to a partitioning of the state space of a system into regions of micro-states that are macroscopically indistinguishable, so-called macro-regions. How are macro-states defined and how are the corresponding macro-regions constructed? The standard answer, which goes back to Boltzmann's seminal (1877), is that macro-regions are constructed with what is now called

[1] This paper discusses BSM. For discussion of Gibbsian statistical mechanics see Frigg and Werndl (2021), and for a discussion of the relation between BSM and Gibbsian statistical mechanics see Werndl and Frigg (2017b) and Frigg and Werndl (2019).

R. Frigg (✉)
Department of Philosophy, Logic and Scientific Method, and CPNSS, LSE, London, UK
e-mail: r.p.frigg@lse.ac.uk

C. Werndl
Department of Philosophy, University of Salzburg, Salzburg, Austria
e-mail: charlotte.werndl@plus.ac.at

C. Soto (ed.), *Current Debates in Philosophy of Science*, Synthese Library 477,
https://doi.org/10.1007/978-3-031-32375-1_11

the combinatorial argument, and the largest macro-region is singled out as the equilibrium macro-region.[2]

This argument had considerable successes, most notably that it can establish the Maxwell-Boltzmann distribution as the equilibrium distribution of a dilute gas. At the same time it faces technical limitations and conceptual problems.[3] An important formal limitation is that the argument only applies to systems consisting of non-interacting particles. Such systems are only a small subset of the systems BSM is interested in because the constituents of most systems *do* interact. Even if this technical limitation could be overcome somehow, one would be left with the conceptual quandary of why equilibrium is the state with the largest macro-region. The connection is not conceptual: there is nothing in the concept of equilibrium tying equilibrium to the largest macro-region. But if the connection is not conceptual, what justifies this association?

To solve these problems, we pursued the project of rethinking Boltzmannian equilibrium and proposed an alternative version of BSM (Werndl and Frigg, 2015a,b). While previous approaches often operated 'bottom up' in that they sought to define macro-states and equilibrium in terms of micro-mechanical properties, our approach works 'top down' in that it defines the macro-states and equilibrium in macroscopic terms. For reasons that will become clear soon, we call this the *long-run residence time account of BSM* (LBSM). Discussion of this account have hitherto focussed on real-valued macro-variables, i.e. macro-variables that assign a real number to every micro-state. This, however, leaves open how we should think about situations, where the system's macro-state cannot be adequately described by a finite number of real-valued macro-variables. In this paper we take the next step and explicate how our account deals with such situations by using local variables.

The limitations of real-valued macro-variables become palpable in non-equilibrium situations. As an illustration, consider the expanding gas, an example that has become standard in foundational discussions of BSM. Imagine a gas confined to the left half of a container by a partition wall. When the partition wall is removed, the gas starts spreading and eventually fills the entire container uniformly. Once the gas fills the container uniformly, it is in equilibrium and its macro-state is specified by specific values of the gas' pressure p, temperature T and volume V. The relation between these variables is then given by the Boyle-Charles law, which says that $pV = cT$, where c is a constant. This is possible because these variables assume the same values *everywhere* in the system: in equilibrium, the pressure in the top left corner of the container is the same as the pressure in the bottom right corner, and, indeed, the same as the pressure in every other place in the container. This allows us to assign the gas a single value for pressure and regard this value the pressure of the gas. This is not possible during the system's *approach* to equilibrium. A split second after the partition wall has been removed, there is no

[2] Contemporary discussions of this argument can be found in Albert (2000, Ch. 3), Frigg (2008, Sec. 2), and Uffink (2007, Sec. 4).

[3] For a discussion of these see Uffink's (2007, Sec. 4) and Werndl and Frigg's (2015a; 2015b).

such thing 'the' pressure of the gas. The value of the pressure in the leftmost parts of the container is still almost the same as before the removal of the wall; the value of the pressure in the rightmost pars of the container is still almost zero; and the value of pressure in the middle of the container is somewhere in-between the two. Real-valued macro-variables cannot capture this situation.[4] Pressure is now a *field* that takes a value at every point in the container, and to describe the gas' macro-state a split second after the removal of the wall, one has to specify the pressure field throughout the container.

Rather than describing situations like the expanding gas using fields, thermodynamics deems quantities like pressure undefined in such situations. This purism comes at a high price: the restriction of the scope of thermodynamics to systems in equilibrium. There are three reasons why BSM neither can nor should afford such rigour. First, it has always been one of the main aims of BSM to understand not only equilibrium, but also the *approach* to equilibrium. This requires that the relevant physical quantities are defined also in non-equilibrium situations so that we can trace their time evolution as the system approaches equilibrium. This is impossible if the relevant quantities are simply undefined outside equilibrium.

Second, BSM, as standardly presented, often starts by introducing a partition on the state space, calls the cells of the partition 'macro-states', and then singles out one of the cells (usually the largest) as the equilibrium macro-state.[5] Since equilibrium is unique, it follows that all other macro-states are non-equilibrium macro-states. So BSM in fact always introduces non-equilibrium macro-states together with the equilibrium state. However, calling cells of a partition 'macro-states' lacks motivation (if not legitimacy) as long as no characterisation of these states in terms of macro-variables is available. If cells of a partition are to be macro-stated in more than just name, BSM will have to give a proper macro characterisation of them.

Third, thermodynamics' rigour is out of sync with experimental practice, where quantities like pressure and temperature are measured, and assigned values, even when the system is not in equilibrium. If pressure measurements were performed on a system in a non-equilibrium state like the one that the system is in a short moment after the removal of the partition wall, one would find that the pressure varies considerably between different locations in the container with values in the leftmost parts still being almost the same as before the removal of the wall and values in the rightmost parts still being almost zero. But, on pain of incoherence, one cannot both make such measures and maintain that the quantities measured are undefined. The message we ought to take away from this example is not that pressure is undefined in non-equilibrium situations (and that talk of pressure is meaningless). The message is that the formalism needs to be developed so that it accommodates variables like local pressure.

To account for situations like the expanding gas we need field variables. The project for this paper is to show that LBSM, as formulated in our 2015 papers,

[4] At least if one thinks that local pressure is defined at every point in space.

[5] See, for instance, Goldstein (2001).

can accommodate such variables and to spell out the details. The result of this will be a fully general definition of macro-states that covers both equilibrium and non-equilibrium situations. To this end we first introduce LBSM, state some of its core results, and highlight the limitations of the current formulation (Sect. 11.2). We then distinguish between global and local physical quantities and explicate how local quantities can be accommodated in LBSM trough scalar fields (Sect. 11.3). Formal definitions are only useful if they can be applied to relevant cases. We therefore discuss typical local variables like pressure and show explicitly how they can be defined in our framework (Sect. 11.4). We end with a conclusion (Sect. 11.5).

11.2 The Long-Run Residence Time Account of BSM

Statistical mechanics studies physical systems like a gas in a container, a magnet on a laboratory table and a liquid in a jar. Described mathematically, these systems have the structure of a *measure-preserving dynamical system*, i.e. a quadruple $(X, \Sigma_X, \phi_t, \mu)$.[6] X is the state space of the system, i.e. a set containing all possible micro-states the system can be in. For a gas with n molecules X has $6n$ dimensions: three dimensions for the position of each particle and three dimensions for the momentum of each particle. Σ_X is a σ-algebra on X and μ is a measure on (X, Σ_X), which is required to be invariant under the dynamics: $\mu_X(T_t(A)) = \mu_X(A)$ for all $A \in \Sigma_X$ and all t. The dynamics of the model is given by an *evolution function* $\phi_t : X \rightarrow X$, where $t \in \mathbb{R}$ if time is continuous and $t \in \mathbb{Z}$ if time is discrete. ϕ_t is assumed to be measurable in (t, x) and to satisfy the requirement $T_{t_1+t_2}(x) = T_{t_2}(T_{t_1}(x))$ for all $x \in X$ and all $t_1, t_2 \in \mathbb{R}$ or $\mathbb{Z}$. If, at a certain point of time t_0, the system is in micro-state x_0, then it will be in state $\phi_t(x_0)$ at a later time t. For systems that are governed by an equation of motion such as Newton's equation, ϕ_t corresponds to the solutions of this equation. The *trajectory* through a point x in X is the function $s_x : \mathbb{R} \rightarrow X$, $s_x(t) = T_t(x)$ (and *mutatatis mutandis* for discrete time).

At the macro-level the system is characterised by a set of l *macro-variables* (for some $l \in \mathbb{N}$). From a mathematical point of view, macro-variables are measurable functions from X into another space $\mathbb{V}$. That is $v_i : X \rightarrow \mathbb{V}_i, x \rightarrow v_i(x)$, $i = 1, ..., l$. For example, if v_1 is the magnetisation of the system and the system is in micro-state x, then $v_1(x)$ is the magnetisation of the system when it is in micro-state x. It is important to note that the $\mathbb{V}_i$ can be different for different i. For many of the standard macro-variables the space $\mathbb{V}_i$ is $\mathbb{R}$ because the variables take values in the real numbers. Examples for such macro-variables are internal energy and total magnetisation. For ease of presentation, we restrict our discussion to such variables

[6] The presentation of LBSM follows Werndl and Frigg (2015b). This paper focuses on deterministic systems. The generalisation to stochastic classical systems is spelled out in Werndl and Frigg (2017a), where statements of the relevant definitions and results can be found.

in the remainder of this section. It is important, however, to be clear that there is no assumption that *all* $\mathbb{V}_i$ have to be $\mathbb{R}$. In fact, $\mathbb{V}_i$ can be any space. And this is more than just a mathematical possibility. In the next section we will exploit this flexibility and consider cases in which the $\mathbb{V}_i$ are function spaces, which is precisely what we need to accommodate field variables.

A macro-state is defined by the *values* of a set of macro-variables $\{v_1, \ldots, v_l\}$. We use capital letters V_i to denote the values of v_i and write $v_i(x) = V_i$ to express that variable v_i assumes value V_i when the system is in micro-state x. A macro-state is then defined by a particular set of values $\{V_1, \ldots, V_l\}$. That is, the system is in macro-state $M_{V_1,\ldots,V_l}$ iff $v_1 = V_1, \ldots, v_l = V_l$. In cases where values are real numbers, exact values can sometimes be unsuitable to define macro-states. In such case one can also define macro-states by the macro-variables taking values in a certain interval. One can then say that the model is in macro-state $M_{[A_1,B_1],\ldots,[A_l,B_l]}$ iff $V_1 \in [A_1, B_1], \ldots, V_l \in [A_l, B_l]$ for suitably chosen intervals. Such a move can be useful, for instance, if one wants to take the finite measurement precision of the available laboratory equipment into account. If the macro-states were defined by exact values, but an experiment would not be able to ever give an exact value, we would have to conclude that it is impossible to determine experimentally in which macro-state the system is. This would be unfortunate because macro-states were initially designed to allow physicists to give a description of the system at the macro-level. This problem can be circumvented by defining macro-states through intervals that are chosen in way that takes the precision of the available equipment into account.

Since macro-states *supervene* on micro-states, a system's micro-state uniquely determines its macro-state. This determination relation is normally many-to-one. Therefore, every macro-state M is associated with a *macro-region* X_M consisting of all micro-states for which the system is in M. For a complete set of macro-states the macro-regions form a partition of X (i.e. the different X_M do not overlap and jointly cover X). A set of macro-states is complete if it contains all macro-states that the system can possibly be in. The set can be 'too large' in the sense that it can contain states that the system never assumes; there can, however, be no states the system can be in that is not contained in the set.

One of these macro-states is the equilibrium macro-state of the system. Intuitively speaking, a system is in equilibrium when its properties do not change. This intuition is built into thermodynamics, where a system is said to be in equilibrium when all change has come to a halt and the thermodynamic properties of the system remain constant over time (Fermi, 2000, 4).[7] However, such a definition of equilibrium cannot be implemented in BSM because measure-preserving dynamical systems exhibit Poincaré recurrence and time reversal invariance. As a consequence,

[7] Being in thermodynamic equilibrium is an intrinsic property of the system, which offers a notion of 'internal equilibrium' (Guggenheim, 1967, 7). It contrasts with 'mutual equilibrium' (*ibid.*, 8), which is the relational property of *being in equilibrium with each other* that two systems eventually reach after being put into thermal contact with each other. When defining equilibrium in BSM it is the internal equilibrium that we are interested in.

when the time evolution of a system unfolds without outside influence, the system will eventually return arbitrarily close to the micro-state in which it started. Hence a system starting outside equilibrium (for instance, when the gas was confined to one half of the container) will eventually return to that macro-state. So in BSM no system will remain in any state forever.

This precludes a definition of equilibrium as the state which the system never leaves once it has reached it. The long-run residence time account of BSM aims to stay as close to the thermodynamic definition of equilibrium as the mathematical constraints imposed by measure-preserving dynamical systems permit, and, intuitively, defines equilibrium as the macro-state in which the system spends most of its time in the long run. To give a formal definition, we first have to introduce the concept of the long-run fraction of time $LF_A(x)$ that a system, which is in initial state x at time $t = 0$, spends in a subset A of X:[8]

$$LF_A(x) = \lim_{t \to \infty} \frac{1}{t} \int_0^t 1_A(T_\tau(x))d\tau, \tag{11.1}$$

where $1_A(x)$ is the characteristic function of A: $1_A(x) = 1$ for $x \in A$ and 0 otherwise. Note that long-run fractions depend on the initial condition.

The notion of 'most of its time' can be read in two different ways, giving rise to two different notions of equilibrium. The first introduces a lower bound of $1/2$ for the fraction of time and stipulates that whenever a system spends more than half of the time in a particular macro-state, this state is the equilibrium state. Mathematically, let α be a real number in $(\frac{1}{2}, 1]$, and let ε be a very small positive real number. If there is a macro-state $M_{V_1^*,\ldots,V_l^*}$ satisfying the following condition, then that state is the system's *α-ε-equilibrium state*:

> There exists a set $Y \subseteq X$ such that $\mu_X(Y) \geq 1 - \varepsilon$, and all initial states $x \in Y$ satisfy $LF_{X_{M_{V_1^*,\ldots,V_l^*}}}(x) \geq \alpha$. A system is in equilibrium at time t iff its micro-state at t, x_t, is in $X_{M_{V_1^*,\ldots,V_l^*}}$.

According to the second reading, 'most of its time' refers to the fact that the system spends more time in the equilibrium state than in any other state (and this can be less than 50% of its time). Mathematically, let γ be a real number in $(0, 1]$ and let ε be a very small positive real number. If there is a macro-state $M_{V_1^*,\ldots,V_l^*}$ satisfying the following condition, then that state is the system's *γ-ε-equilibrium state*:

> There exists a set $Y \subseteq X$ such that $\mu_X(Y) \geq 1 - \varepsilon$ and for all initial conditions $x \in Y$: $LF_{X_{M_{V_1^*,\ldots,V_l^*}}}(x) \geq LF_{X_M}(x) + \gamma$ for all macro-states $M \neq M_{V_1^*,\ldots,V_l^*}$. Again, a system is in equilibrium at time t iff its micro-state at t, x_t, is in $X_{M_{V_1^*,\ldots,V_l^*}}$.

[8] We state the definitions for continuous time. The corresponding definitions for discrete time are obtained by replacing the integrals by sums.

It should come as no surprise that these two notions are not equivalent. More specifically, an α-ε-equilibrium is strictly stronger than a γ-ε-equilibrium in the sense that the existence of the former implies the existence of the latter but not vice versa.

These definitions are about the *time* a model spends in the equilibrium state. In contrast to the traditional version of BSM, LBSM does *not* define equilibrium in terms of a macro-region's size. For this reason it is not immediately clear what the two definitions of equilibrium imply about the *size* of the relevant equilibrium macro-regions. It is therefore a result of some importance that the equilibrium regions of α-ε-equilibrium states and of γ-ε-equilibrium states eventually turn out to be the largest macro-regions. So LBSM recovers the standard approach's dictum that equilibrium macro-states are ones with the largest macro-region, but it does not use this as a *definition* of equilibrium (thus avoiding the problem that such a definition is unmotivated), and it does not have to impose any restrictions on the interactions in the system because it eschews appeal to combinatorial considerations (thus avoiding the unwelcome consequence that BSM can only deal with non-interacting systems).

The relevant technical results are as follows. We call a macro-region *β-dominant* if its measure is greater or equal to β for a particular $\beta \in (\frac{1}{2}, 1]$, and we call a macro-region *δ-prevalent* if its measure is larger than the measure of any other macro-region by a margin of at least $\delta > 0$. One can then prove the following theorems (Werndl and Frigg, 2015b):

Dominance Theorem: If $M_{\alpha\text{-}\varepsilon\text{-}eq}$ is an α-ε-equilibrium, then the following holds for $\beta = \alpha(1-\varepsilon)$: $\mu_X(X_{M_{\alpha\text{-}\varepsilon\text{-}eq}}) \geq \beta$.[9]

Prevalence Theorem: If $M_{\gamma\text{-}\varepsilon\text{-}eq}$ is a γ-ε-equilibrium, then the following holds for $\delta = \gamma - \varepsilon$: $\mu_X(X_{M_{\gamma\text{-}\varepsilon\text{-}eq}}) \geq \mu_X(X_M) + \delta$.[10]

It is a consequence of these definitions of equilibrium that a system is not always in equilibrium and that it can fluctuate away from equilibrium. This is a radical departure from thermodynamics. It is therefore worth pointing out that this is not merely a concession to the demands of measure-preserving dynamical systems. Having no fluctuations at all is also physically undesirable. There are experimental results that show that equilibrium is not the immutable state that classical thermodynamics presents us with because systems exhibit fluctuations away from equilibrium (MacDonald, 1962; Wang et al., 2002). Hence adopting a notion of equilibrium that allows for fluctuations increases the empirical adequacy of the theory.

[9] We assume that ε is small enough so that $\alpha(1-\varepsilon) > \frac{1}{2}$.

[10] We assume that $\varepsilon < \gamma$.

11.3 Local Quantities and Field Variables

The examples we gave in the previous section for relevant macro-variables were internal energy and total magnetisation. These are *global variables*. They assign one value to the entire system, rather than a value to each point in space. Global variables contrast with *local variables*, which are variables like pressure, temperature, and local magnetisation density. These variables assign a value to each point in space. Mathematically speaking, global variables are real valued functions in that they assign a real number to every micro-state in X, and that number is the value of the variable for the micro-state. By contrast, local variables assign a value to each point in space, and if the system is not in equilibrium the values at each point in space will typically vary across the system. For this reason these variables have to be treated as *fields*. We have seen an example of this in Sect. 11.1 when we discussed the pressure of a gas shortly after the removal of the partition wall. In this situation there is no such thing as 'the' pressure of the gas and the physical situation is described by a pressure field. In this section we spell out what this means and how fields fit into the framework of LBSM.

Let ordinary physical space be represented by $\mathbb{R}^3$. A *scalar field* on $\mathbb{R}^3$ is a measurable function $f : \mathbb{R}^3 \to \mathbb{R}, \vec{r} \to f(\vec{r})$; i.e. it is a measurable function that assigns each point in space $\vec{r}$ a real number $f(\vec{r})$. Trivially, this definition can be restricted to a subset $S \subseteq \mathbb{R}^3$ and we can say that a scalar field on S is a measurable function $f : S \to \mathbb{R}$. From a formal point of view, saying that quantities like pressure are 'local' means that they are scalar fields. Indeed, pressure and temperature are standard examples of scalar fields. If one wants to restrict the definition of the variable to a particular physical system—for instance to the inside of the container in which the gas is located—then one can say that the variable is scalar field on S, where S is chosen to be the spatial extension of the system.

Let us now consider the set of all scalar fields on $\mathbb{R}^3$ (or S). It is obvious that this set has the structure of a vector space because the linear combination of any two scalar fields is again a scalar field (assuming the standard definition of the multiplication of function with a number and the addition of two functions). Let us denote this space by $\mathcal{F}$. The space could be restricted in all kind of ways, for instance by only allowing continuous or differentiable fields. Whether such restrictions are desirable, or even necessary, depends on the physical situation at hand. At the general level no further restrictions are needed. The space can also be endowed with further structures such as norms, inner products and metrics. Again, whether it is advisable, or even necessary, to introduce such additional structures will depend on the physical quantity and the problem at hand; nothing in what we say about scalar fields at the general level depends on having such additional structures in place.

$\mathcal{F}$ contains all scalar fields on $\mathbb{R}^3$ (or S); that is, all measurable functions $f : \mathbb{R}^3 \to \mathbb{R}$ (or $f : S \to \mathbb{R}$). For example, it contains $f(\vec{r}) = 3$, the function that assigns to every point in space the value 3. It also contains $f(\vec{r}) = |\vec{r}|$, the function that assigns to every point in space the value of it distance from the origin. A particular assignment of values to each point of space is also called a *field configuration*. For example, $f(\vec{r}) = 3$ and $f(\vec{r}) = |\vec{r}|$ are field configurations. So we can say that $\mathcal{F}$ is the space of field configurations.

To see how all this bears on macro-variables in LBSM, recall our observation in the previous section that there is no assumption that *all* $\mathbb{V}_i$ in the definition of a macro-variable have to be $\mathbb{R}$ and that $\mathbb{V}_i$ can in fact be any space. So we are free to take $\mathbb{V}_i$ to be the space $\mathcal{F}$, and doing so is the key to understanding local variables in LBSM. Indeed, local variables are macro-variables for which $\mathbb{V}_i$ is a space of scalar fields; that is, they are macro-variables that assign to every point in the system's state space a scalar field. For this reason the 'values' of local variable v_i is a field configuration. From a mathematical point of view, we can say that local variables are field-valued variables. Hence, we can say that global variables have the mathematical form of real-valued variables and local variables have the mathematical form of field-valued variables.

It is one of the core posits of LBSM that macro-states are defined by the values of a set of macro-variables $\{v_1, \ldots, v_l\}$. Using capital letters V_i to denote the values of v_i, we said that a macro-state was defined by a particular set of values $\{V_1, \ldots, V_l\}$: the system is in macro-state $M_{V_1,\ldots,V_l}$ iff $v_1 = V_1, \ldots, v_l = V_l$. This definition remains valid also after the introduction of fields, but we now have to bear in mind that if a macro-variable is a local variable, then its value is field configuration. One can make this explicit as follows. Assume than that for some $k < l$ all $v_1, \ldots v_k$ are real-valued macro-variables and all $v_{k+1}, \ldots v_l$ are field-valued variables. A macro state is then defined though the set of values $\{R_1, \ldots, R_k, F_{k+1}, \ldots F_l\}$, where we write '$R$' for real numbers and '$F$' for field configurations. This allows us to define the macro-state of the gas we considered in the Introduction through the triple $\{R_V, F_T, F_p\}$, where R_V is the value of the volume of the gas, F_T is temperature field configuration and F_p is pressure field configuration.

LBSM individuates macro-states through values of macro-variables: two macro-states are identical iff all variables assume the same values. This approach presupposes the notion sameness of values. If the values are real numbers, the notion is trivial: the values are the same if the two real numbers are identical. If the values are functions in a function space, macro-states are individuated by equivalent functions. So in the case of local variables we say two values are the same iff the two functions are equivalent. Different choices to spell out 'equivalent functions' are possible and what notion is adopted will depend on the context. The strictest requirement is to say that two functions f and g from $\mathbb{R}^3$ or S to $\mathbb{R}$ are equivalent just in case $f(x) = g(x)$ for all x. In measure-theoretic settings it is also natural to say that f and g are equivalent iff they agree for all x except, perhaps, on a set of measure zero.

The introduction of field-valued variables does not change the fact that macro-states supervene on micro-states. The only thing that has changed is that macro-states are now individuated by field configurations rather than values. So it still is the case that micro-states uniquely determine macro-states and that this determination relation is normally many-to-one. It therefore also still is the case that every macro-state M, now defined in terms of field variables, is associated with a *macro-region* X_M consisting of all micro-states for which the system is in M and that for a complete set of macro-states these macro-regions form a partition of X. For this reasons all other elements of LBSM, in particular the definition of equilibrium, remain unchanged.

As we have previously seen, in some cases it is advisable to define macro-states though intervals rather than exact values. This is straightforward when the values are real numbers, which can easily be ordered in intervals. There is a question how this is best done in the case of field-valued variables. Such a coarse graining could rely, for instance, on a metric on $\mathcal{F}$ and all field configurations that are less than a certain distance away from certain reference field configurations could be seen as belonging to the same macro state. The concrete construction of such a coarse graining depends on the particulars of the situation and there is little one can say at the general level. The point to note here is merely that such coarse grainings can be constructed for field-valued variables just as well as for real-valued variables.

11.4 Physical Realisations

How can relevant physical variables like pressure be defined based on the general formal framework we have outlined in the previous section? A standard way to define local quantities appeals to the so-called Local Equilibrium hypothesis (LEH). In the words of Jou, Vázquez and Lebon the core of LEH is the assumption 'that the system under study can be mentally split into a series of cells sufficiently large to allow them to be treated as macroscopic thermodynamic subsystems, but sufficiently small that equilibrium is very close to being realized in each cell' (2010, 14).[11] As Öttinger (2005, 39), points out, this can be done, for instance, by imagining the system split up into cubes with a side length of 1mm: such cubes contain a large number of molecules (for air at room temperature the number of particles is of the order of 10^{16}), while at the same time being large with respect to the average mean free path of molecules (which is of the order of 10^{-6} m). Such cubes are at once small enough for relevant quantities to be approximately constant and large enough for thermodynamic concepts to apply. The core of LEH then is that the cubes are systems in thermodynamic equilibrium and that therefore thermodynamic concepts can meaningfully be applied to the cubes. Specifically, quantities like temperature, pressure, entropy are rigorously and unambiguously defined in each cube. The values of these quantities remain constant within a single cube while they can vary across different cubes.

This allows us to define a field: the value of $f(\vec{r})$ is the value that the quantity represented by f assumes in the cube in which $\vec{r}$ lies if the system is in micro-state x. This definition covers the relevant cases. Pressure is defined as force per unit area on the surface of the container when the system is in equilibrium, which implies that it assumes the same value all over the surface. This definition does not, as we

[11] For statements and discussions of LEH see Giberti et al. (2019), Jou et al. (2010, 14–15), Spohn (1991, 14), and Öttinger (2005, Ch. 2).

have seen, apply to the entire container as a whole while the gas is spreading; but it applies to a small cube which, by LEH, is in equilibrium. So we can define the pressure in a cube in the same way in which we have previously defined the pressure in the entire vessel. The pressure at point $\vec{r}$ then is simply the pressure in the cube in which the point lies. This unambiguously defines the pressure field across the system.

This definition of the relevant field quantities has the consequence that the field configurations will typically change discontinuously at the boundary between cubes, due to they way in which the cubes are used to define macro-values. This is a drawback because such discontinuous changes are unphysical. This problem can be avoided by changing the definition of local quantities slightly. Rather than first slicing up the system into cubes and then defining the relevant quantities in each cube, one can think of a small cube being placed around each point $\vec{r}$ so that $\vec{r}$ is at the centre of the cube.[12] The size of a the cube will be the same as as above. This allows us to apply LEH to the cube and say that the cube is in equilibrium, which allows us to define quantities like pressure in the cube. The pressure $\vec{r}$ then is the pressure in the cube around $\vec{r}$. This way of introducing local quantities has the advantage that it does not introduce discontinuities in the resulting fields through its use of LEH.

The same moves can be made for other variables. Consider the example of temperature. In the context of statistical mechanics temperature T is usually assumed to to be proportional to the mean kinetic energy of the system's molecules: $T = 2/3k < E_{kin} >$, where $< E_{kin} >$ is the particles' mean kinetic energy and k is a constant. We can now make the same moves as with pressure and define the temperature field. The case of densities like the local magnetisation density is even easier. One simply takes the total magnetisation in a cube and divides it by the volume of the cube. In this way fields for all local quantities can be defined.

11.5 Conclusion

In this paper we have shown how local variables such as the pressure of a gas fit into the framework of LBSM. This is a crucial step forward because a unified treatment of equilibrium and non-equilibrium situations can be given only if such variables are available.

Before drawing the discussion to a close we would like to comment on the relation between the well-known thermodynamic distinction between intensive and extensive variables and our distinction between local and global variables. A variable is *extensive* iff is additive for subsystems. Assume we have two systems S_1 and S_2 and consider a variable v. The values of the variable in the two systems are V_1 and V_2, respectively. Now we merge the two systems to form a new system S.

[12] Nothing depends on this being a cube. The same construction can be made with a sphere.

The variable v is additive iff the value of v in S is $V_1 + V_2$ (Callen, 1985, 10). Assume now that v has the same value in both systems $V_1 = V_2$. If the value of v in S is also V_1, then the variable is intensive (*ibid.*, 38). The most obvious example of an extensive variable is volume, because combining two systems will result in a new system whose volume is the sum of the volume of the systems we started with. Other important examples of extensive variables are internal energy and entropy. Examples of intensive variables are temperature and pressure: combining two systems with pressure p and temperature T will result in a larger system which has again pressure p and temperature T.

The point to note is that the two distinctions do not coincide. There is a certain association between them in that important examples of extensive variables are also global (for instance, volume and internal energy) and important examples of intensive variables are also local (for instance, temperature and pressure). But the association is not perfect. The average magnetisation per site in a lattice system is intensive but not local; and the field that assigns to each point the pressure multiplied by the number of molecules in the system is extensive but not global. So the intensive/extensive and local/global distinctions are logically independent.

Acknowledgments We would like to thank Cristián Soto for inviting us to participate in this project. We also would like to thank Sean Gryb, David Lavis, and Lamberto Rondoni for helpful discussions on the subject matter of the paper.

References

Albert, D. (2000). *Time and chance*. Harvard University Press.

Boltzmann, L. (1877). Über die Beziehung zwischen dem zweiten Hauptsatze der mechanischen Wäermetheorie und der Warhscheinlichkeitsrechnung resp. den Sätzen über des Wärmegleichgewicht. *Wiener Berichte*, *76*, 373–435.

Callen, H. B. (1985). *Thermodynamics: An introduction to thermostatistics* (2nd ed.). John Wiley and Sons.

Fermi, E. (2000). *Thermodynamics*. Dover.

Frigg, R. (2008.) A field guide to recent work on the foundations of statistical mechanics. In D. Rickles (Ed.), *The ashgate companion to contemporary philosophy of physics* (pp. 99–196). Ashgate.

Frigg, R & Werndl, C. (2019). Statistical mechanics: A tale of two theories. *The Monist*, *102*, 424–438.

Frigg, R & Werndl, C. (2021). Can somebody please say what Gibbsian statistical mechanics says? *The British Journal for Philosophy of Science* *72*(1), 105–129.

Giberti, C., Lamberto, R., & Cecilia, V. (2019). O(N) fluctuations and lattice distortions in 1-dimensional systems. *Frontiers in Physics*, *7*, 1–10.

Goldstein, S. (2001). Boltzmann's approach to statistical mechanics. In J. Bricmont, D. Dürr, M. C. Galavotti, G. C. Ghirardi, F. Petruccione, & N. Zanghì (Eds.), *Chance in physics: Foundations and perspectives* (pp. 39–54). Springer.

Guggenheim, E. A. (1967). *Thermodynamics: An advance treatment for chemists and physicists*. North-Holland.

Jou, D., Casas-Vázquez, J., & Lebon, G. (2010). *Extended irreversible thermodynamics*. Springer.

MacDonald, K. C. (1962). An introduction. *Noise and fluctuations*. Wiley.

Öttinger, H. C. (2005). *Beyond equilibrium thermodynamics*. Wiley-Interscience.
Spohn, H. (1991). *Large Scale Dynamics of Interacting Particles*. Berlin. Springer.
Uffink, J. (2007). Compendium of the foundations of classical statistical physics. In J. Butterfield, & J. Earman (Eds.), *Philosophy of physics* (pp. 923–1047). North Holland.
Wang, G. M., Sevick, E. M, Mittag, E., Searles, D. J., & Evans, D. J. (2002). Experimental demonstration of violations of the second law of thermodynamics for small systems and short time scales. *Physical Review Letters*, *89*, 050601.
Werndl, C., & Frigg, R. (2015a). Reconceptualising equilibrium in Boltzmannian statistical mechanics. *Studies in History and Philosophy of Modern Physics*, *49*, 19–31.
Werndl, C., & Frigg, R. (2015b). Rethinking Boltzmannian equilibrium. *Philosophy of Science*, *82*, 1224–1235.
Werndl, C., & Frigg, R. (2017a). Boltzmannian equilibrium in stochastic systems. In M. Massimi, & J.-W. Romeijn (Eds.), *Proceedings of the EPSA15 conference* (pp. 243–254). Springer.
Werndl, C., & Frigg, R. (2017b). Mind the gap: Boltzmannian versus Gibbsian equilibrium. *Philosophy of Science*, *84*(5), 1289–1302.

Chapter 12
Scientific Understanding in Astronomical Models from Eudoxus to Kepler

Pablo Acuña

Abstract In the following essay I present a narrative of the development of astronomical models from Eudoxus to Kepler, as a case-study that vindicates an insightful and influential recent account of the concept of scientific understanding. Since this episode in the history of science and the concept of understanding are subjects to which Professor Roberto Torretti has dedicated two wonderful books—*De Eudoxo a Newton: modelos matemáticos en la filosofía natural* (2007), and *Creative Understanding: philosophical reflections on physics* (1990), respectively—this essay is my contribution to celebrate his outstanding work and career in this volume. I dedicate this piece to Roberto, dear friend and mentor, in gratitude for all his inspirational work and personal support, which has greatly helped me, and many others, to better understand that human wonder we call scientific knowledge.

12.1 Scientific Understanding

The concept of understanding was neglected by contemporary philosophy of science for a long time. Carl Hempel's (1965, 425–433) is a paradigmatic example of the stance adopted by philosophers influenced by logical positivistic principles. In the context of his deductive-nomological model of explanation, he described understanding as a psychological *a-ha!* experience that accompanies scientific explanations. If we consider it as a psychological byproduct of explanations, whether a subject S obtains understanding from an explanation or not, and the type of understanding that S gets from the explanation, depends crucially on subjective

Sadly, during the preparation for publication of this volume, Professor Torretti passed away. May this essay be my contribution to honor and cherish his memory as an outstanding scholar, generous mentor, and dear friend.

P. Acuña (✉)
Institute of Philosophy, Pontificia Universidad Católica, Santiago, Chile
e-mail: ptacuna@uc.cl

C. Soto (ed.), *Current Debates in Philosophy of Science*, Synthese Library 477,
https://doi.org/10.1007/978-3-031-32375-1_12

and context-dependent factors, such as S′ expectations, motivations, background knowledge, etc.

Despite its subjectivity, Hempel did acknowledge an epistemic dimension to understanding. He stated that a phenomenon is understood if its occurrence is *expected*, given the corresponding laws of nature and the initial conditions. Thus, the epistemic dimension of understanding reduces to the notion of explanation. Consequently, given its subjective nature and the subsumption of its epistemic import under the concept of explanation, and following the logical empiricist stance that only the logical aspects of the connection between evidence and scientific explanations are philosophically relevant, Hempel affirmed that understanding is not a concept worth of philosophical inquiry.

Despite the demise of logical empiricism, the proscription of understanding from philosophical consideration continued. At most, related issues like the type of intelligibility of nature that scientific theories convey were treated during the 60s and 70s in the context of discussions about different models of explanation. In the 80s, with the introduction of approaches that made clear the importance of pragmatic issues for explanation, the environment became less hostile for a serious philosophical consideration of understanding.[1] However, although van Fraassen's (1980) pluralistic and pragmatic account recognizes that contextual and subject-based factors are essential for scientific explanation—opening the door for a consideration of understanding *vis à vis* explanation—this approach states that explanation (and consequently understanding) is a pragmatic, but not an epistemic, goal of science.

During the last two decades, though, proposals have been introduced that consider understanding as an essential epistemic goal of science, and as crucial factor in its practice, recognizing its intrinsically pragmatic nature.[2] In his celebrated *Understanding Scientific Understanding*, Henk de Regt (2017) proposes an illuminating account in which understanding is a basic epistemic goal of science, is pragmatic, and is not reduced to explanation. He starts out from a distinction between three senses of the concept of understanding in science. First, we get understanding *of a phenomenon* (UP) when we have an adequate scientific explanation of it. Now, UP is usually accompanied by a subjective psychological experience, which de Regt calls the *phenomenology* of understanding (PU). UP and PU can be recognized in Hempel's received view, that is, understanding is a subjective experience (PU), and it is conveyed by scientific explanations (UP). However, a third sense, understanding *a theory* (UT), must also be acknowledged. By UT, de Regt refers to *the ability to use* a theory, and he evaluates it as a necessary condition for UP.

[1] For a conceptual overview of philosophical stances on scientific explanation, see Woodward (2019).

[2] Torretti (1990) is an early approximation to understanding as a central goal of science. For an overview of the different contemporary stances on understanding, see de Regt and Baumberger (2020). See also the articles in de Regt et al. (2009), and in Grimm et al. (2017).

De Regt claims that scientific explanations are typically given by models (in a very wide sense of the term) built out of theories. Explanatory models represent the phenomena, obeying the constraints imposed by the corresponding theory: models mediate between theory and target phenomena. Following Morgan and Morrison (1999) and Cartwright (1983), de Regt points out that the construction of models is not an algorithmic or merely deductive exercise—scientific theories do not come with a recipe for model construction. This process usually involves idealizations and abstractions, and it requires skills, good sense, and judgment on the scientist's part. Thus, the construction of explanatory models requires the ability to use the corresponding theory.

De Regt connects this ability to the notion of an *intelligible theory*. A theory T is intelligible for a scientist S if S is able to build explanatory models out of T. In de Regt's own words, intelligibility is "the value that scientists attribute to the cluster of qualities of a theory [. . .] that facilitate the use of the theory" (2017, 40). Intelligibility is then a *relational* property, it depends both on the skills and background knowledge of S, and on the qualities of T that fit S′ skills.

Hence, the intelligibility of a theory is in a sense subjective: although independent of PU, the intelligibility of T essentially refers to features of S. However, this subjective dimension does not affect the objectivity of science. Toolkits of intelligibility are acquired and developed by scientists within a community. Thus, there exist public criteria and standards that establish if the lines of reasoning followed by individual scientists in the construction of explanatory models conform to objectivity conditions. Besides, intelligibility, understood as a value, must also conform to the basic values of empirical adequacy and internal consistency—following Longino (1990) and Douglas (2009), de Regt supports a conception of scientific objectivity in which values, including intelligibility, play an essential and constitutive role.[3]

Now, this framework of objectivity leaves plenty of room for the variation of intelligibility standards, both synchronically among scientific (sub)communities, and diachronically across the history of science. That is, certain skills, and the theories that conform to such skills, are rendered as tools for intelligibility and as intelligible, respectively, *relative to a specific context*. Different skills and different types of theories are valued as intelligible by scientists of different communities and/or different times. In other words, intelligibility is a pragmatic concept.

Given the definitions presented, UT and intelligibility are necessary conditions for UP. Explanatory models can be generated from a theory insofar as the corresponding theory is understood and intelligible (UT). Thus, scientific understanding in the sense of UT is an essential epistemic aim of science, and given the characterization offered by de Regt, it is at the same time pragmatic. Furthermore,

[3] In this conception of scientific objectivity, empirical adequacy and internal consistency are also values. That is, although both are basic values that transcend revolutionary changes in the history of science, there are always contexts in which they can be traded for other values. As we will see below, scientific understanding can be obtained from false theories. For a treatment of inconsistency in scientific theories, see Vickers (2013) and Frisch (2014).

there is a virtuously circular interconnection between UP and UT. Scientists use intelligible theories to build explanations, which may turn out successful or not. Success does not depend only on the intelligibility of the theory, but also, and crucially, on the basic values of empirical adequacy and consistency.[4] Now, if the explanations are indeed successful, the skills and forms of intelligibility associated to the theory get vindicated also as providing understanding *of the phenomena*. As a result, those skills and the corresponding qualities of theories can get canonized as paradigms of UT and UP, and applied in subsequent scientific inquiry.

The essential link between understanding and explanation, incarnated in the interconnection between UT and UP, is captured by de Regt's definition of a criterion for understanding a phenomenon (CUP): "a phenomenon P is understood scientifically if and only if there is an explanation of P that is based on an intelligible theory T and conforms to the basic epistemic values of empirical adequacy and internal consistency" (2017, 92). Now, as we said, a theory is intelligible if scientists are able to use it, and this ability depends on skills and background knowledge possessed by the scientist, and also on the qualities of the theory that adapt to such skills. Given the pragmatic nature of UT, the skills and the qualities of theories that are considered as paradigms and/or norms of intelligibility and understanding vary across communities and historical periods. De Regt formulates a criterion for intelligibility of theories (CIT) that openly acknowledges its context-dependency, and that makes room for diachronic and synchronic variation: "a scientific theory [. . .] is intelligible for scientists (in context C) if they can recognize qualitatively characteristic consequences of T without performing exact calculations" (2017, 102).

De Regt offers several case-studies in the history of physics that support his views. Examinations of episodes in the development of Newtonian gravitation, electrodynamics, statistical mechanics and quantum mechanics provide evidence for the diachronic and synchronic variability of standards of intelligibility, supporting in turn de Regt's pragmatic account of scientific understanding. The variation is of course multi-factorially determined, but the success and failure of (types of) theories that conform to different criteria of intelligibility are always crucial. Features like visualization, causality, mechanisms, and mathematical abstraction, come and go as canons of intelligibility of theories (UT), and of the understanding of phenomena (UP), depending on the success or failure of the theories that incarnate those features.

We will now add another case-study that also vindicates de Regt's proposal: the development of astronomical models from Eudoxus to Kepler. As we will see, this crucial and long episode in the history of physics shows with special clarity that UT is a condition for UP, and also the pragmatic nature of scientific understanding.

[4] As we will see below, false models can be successful, and therefore convey UP. However, even in these cases some degree of empirical adequacy is involved.

12.2 Astronomical Models from Eudoxus to Kepler[5]

The development of astronomical geometric models from Eudoxus to Kepler is a crucial stage in the constitution of modern physics. This episode in the history of science spans about two millennia. Eudoxus of Cnidus, in the IV century BC, was the first astronomer to invent a comprehensive geometric model that accounts for naked-eye celestial phenomena. Eudoxus' model strictly follows an aesthetic-metaphysical principle of circular uniform motion for celestial objects. About 500 years later, Claudius Ptolemy, elaborating on ideas introduced earlier by Apollonius and Hipparchus, invented a much more refined and empirically adequate model that would dominate astronomy for about thirteen centuries. Although Ptolemy's astronomy was also based on circular motion, it abandoned uniformity. A serious rival to Ptolemy's model would only be formulated by Nicolaus Copernicus in 1543. Although the introduction of a heliocentric model was indeed a major innovation, Copernicus' model was actually rather conservative: one of its main motivations was the reintroduction of the ancient principle of circular uniform motion. Four decades after the work of Copernicus, Tycho Brahe proposed a model that captured the advantages of Copernicus' over Ptolemy's, but that reinstalled the Earth as the center of the universe. Finally, in 1609, Johannes Kepler, following a truly revolutionary insight, amended Copernicus' model. The resulting system of circular (but not uniform) motion, however, was doomed to fail in the representation of the orbit of Mars by an unacceptable observational margin. Since Kepler also showed that Copernicus', Brahe's and Ptolemy's models were geometrically inter-translatable, the failure of his improved version of Copernicus model also showed the failure of its two rivals. Thus, Kepler's work signed the ruin of the project of circular motion astronomy, and opened a new path in the development of physics with his three laws of planetary motion. As we will see in Sect. 12.3, the following narrative of the rise and fall of circular motion astronomy constitutes a historical case that clearly illustrates the pragmatic character of scientific understanding, and its relevance and necessity for scientific explanations.

12.2.1 Naked Eye Astronomy

The astronomical models we will review were designed to account for celestial phenomena that can be observed with the naked eye. The telescope, in the hands of Galileo, only enters the stage of science at the beginning of the seventeenth century. Such phenomena were basically four: the motion of the stars, the Sun,

[5] For the history of astronomy from Eudoxus to Kepler, see Barbour (2001), Crowe (2001), Dijksterhuis (1986), Dreyer (1953), Evans (1998), Jacobsen (1999), Koyré (1973), Kuhn (1995), Linton (2004), Neugebauer (1975, 1986), and Torretti (2007). Most of the figures below are based on Linton (2004).

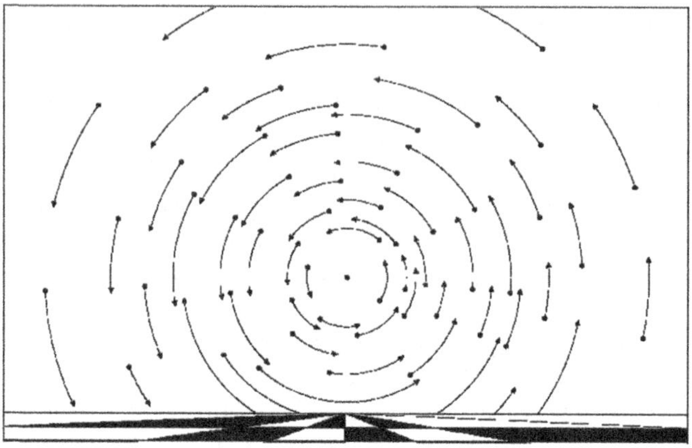

Fig. 12.1 The stars at night

the planets, and the Moon. The representation of the Moon's motion is the most complex element in all the models to be considered, so for brevity and simplicity, we will not treat it here.

Looking at the sky for a few consecutive nights, we see that the stars remain stationary with respect to each other, and, if we stand somewhere in the Northern Hemisphere, we also see that they describe a counterclockwise circular path around a fixed point.[6] The farther a star with respect to the fixed point, the larger the circle it describes (see Fig. 12.1). These observations suggest that the stars lie fixed on a sphere that rotates westwards, with an axis that passes through the fixed point and the center of the Earth (defining the celestial north and south poles), with a period of 23 h and 56 min (the *sidereal* day).[7] The plane perpendicular to the axis is called the celestial equator (see Fig. 12.2). This *celestial sphere* was taken to be the outermost limit of the universe, and this spherical image was adopted in all the astronomical models we will review.

Like the stars, the Sun is also observed to describe a diurnal circular motion. It rises somewhere in the east, describing an arc in the sky until it sets somewhere in the west, to rise again somewhere in the east the next day. As it can be seen in Fig. 12.3, the observed angle of the daily arc with respect to the horizon depends on the terrestrial latitude. The Sun also displays another apparent motion. The point at which it rises on the east and at which it sets in the west is not the same every day.

[6] Another fixed point around which the stars are seen to rotate clockwise is observed from the Southern Hemisphere. The main characters in our story were inhabitants of the Northern Hemisphere, so we will adopt their perspective.

[7] The sidereal day must be distinguished from the *solar* day: the time it takes the Sun to return to the same local meridian (the time between consecutive noons). The solar day thus defined is variable along the year. Using the idealized mean Sun (see below for the distinction between the real and the mean Sun), the solar day can be defined to last 24 h.

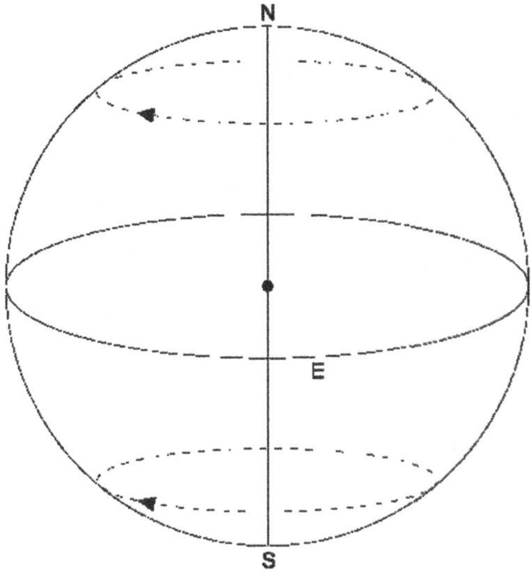

Fig. 12.2 The celestial sphere and the equator E

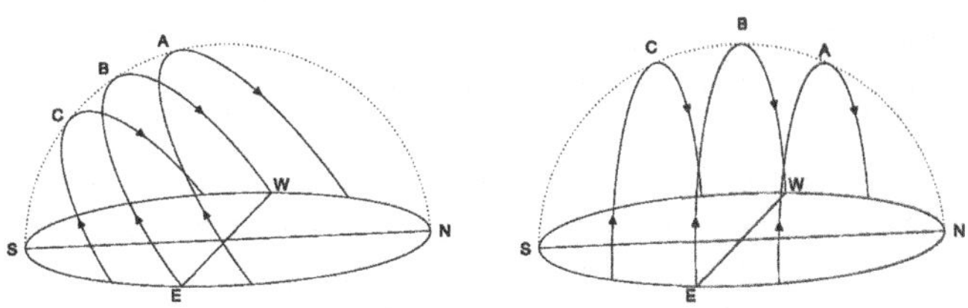

Fig. 12.3 Diurnal motion of the Sun at equinoxes (B) and solstices (A, C), as seen standing about 50° north (left), and at the equator (right)

At two particular days, it rises almost exactly at east and sets almost exactly at west. When that happens, day and night last almost the same everywhere on Earth, which is why we call those days equinoxes (from Latin for *equal night*). Then the Sun rises and sets at points that progress northerly or southerly, reaching the northmost and southmost rising and setting points about 3 months after an equinox. The days when the Sun reaches its north or southmost points for rising and setting are called solstices. At the solstices, the day is the longest or shortest of the year, depending on which hemisphere one is standing on. After a solstice, the same process reverses, reaching an equinox again about 3 months later. The whole cycle lasts a year.

If we consider this yearly motion in connection with the motion of the celestial sphere, we have that the Sun moves with respect to the background stars in a particular way. That is, if we determine the position of the Sun with respect to the celestial sphere over successive sidereal days, we get that it has moved every time a little bit. If we look at a series of such positions, we see that the Sun describes a great circle along the celestial sphere, as represented in Fig. 12.4. The plane

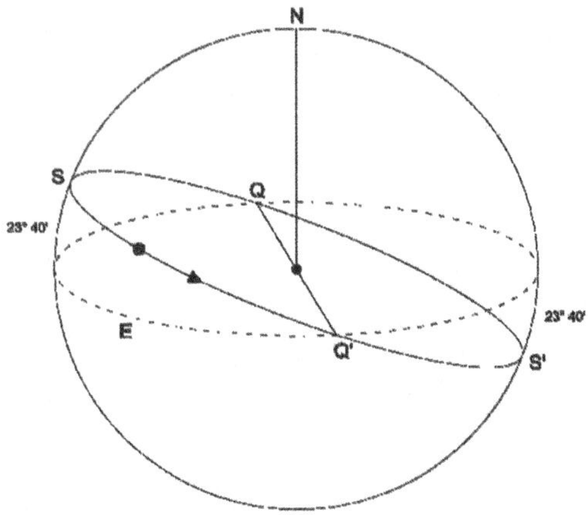

Fig. 12.4 The ecliptic plane SQ′S′Q, determined by the solstices S and S′, and the equinoxes Q and Q′, inclined 23°40′ with respect to the equatorial plane E

described by this great circle is called the *ecliptic*, and it is inclined with respect to the celestial equator by an angle of about 23°40′. This geometric arrangement allows us to define the year, the equinoxes, and the solstices more precisely. The equinoxes are the two points in which the ecliptic intersects the celestial equatorial plane, and the solstices are the points in the ecliptic with maximum and minimum celestial latitude (±23°40′). The year is the time it takes the Sun to complete the ecliptic circle.

The planets that can be seen with the naked eye are Mercury, Venus, Mars, Jupiter and Saturn.[8] They all display a diurnal motion, that is, they rise and set in the horizon every night. They also exhibit a motion similar to the yearly motion of the Sun. If we look at the position of a planet with respect to the background stars over successive sidereal days, it also moves slowly. However, the planets do not describe a great circle in the celestial sphere as the Sun does. Concerning their celestial longitude, the apparent motion of the planets is such that they move eastwards along their trajectory with respect to the background stars, except for periods in which they gradually turn around westwards, to later resume their usual eastward direction. This phenomenon, represented in Fig. 12.5, is called *retrograde motion*. As it can be seen in Fig. 12.6, the planets have a variable celestial latitude, but they are never far from the ecliptic, they always lie within a belt along the celestial sphere, bisected by the ecliptic, of about 16° wide. Each planet has a characteristic zodiacal period, i.e., the average time it takes the planet to make full round around the zodiac, and a characteristic synodic period, i.e., the average time between periods of retrograde motion (see Table 12.1).

[8] Since the observed motion of the Moon, Mercury, Venus, the Sun, Mars, Jupiter and Saturn is not as patently regular as the motion of the stars, the ancient Greeks named them '*planetai*', i.e., wanderers.

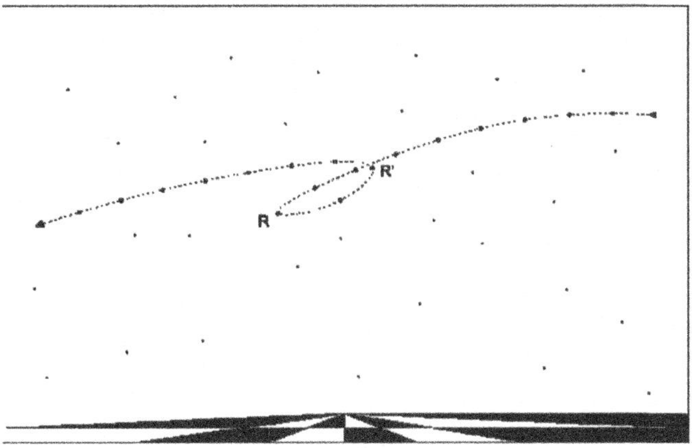

Fig. 12.5 Retrograde motion of a planet, between positions R and R′

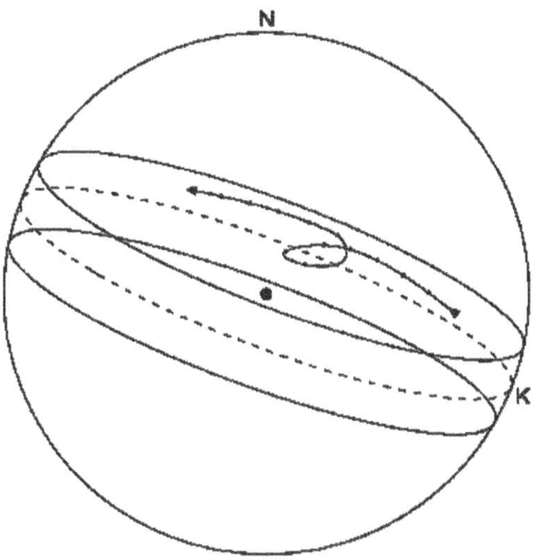

Fig. 12.6 A piece of the motion of planet along the zodiac belt K, including retrograde motion

Table 12.1 Zodiacal and synodic periods of the planets

	Zodiacal period in years	Synodic period in days
Mercury	1	115
Venus	1	584
Mars	1.88	780
Jupiter	11.86	399
Saturn	29.46	378

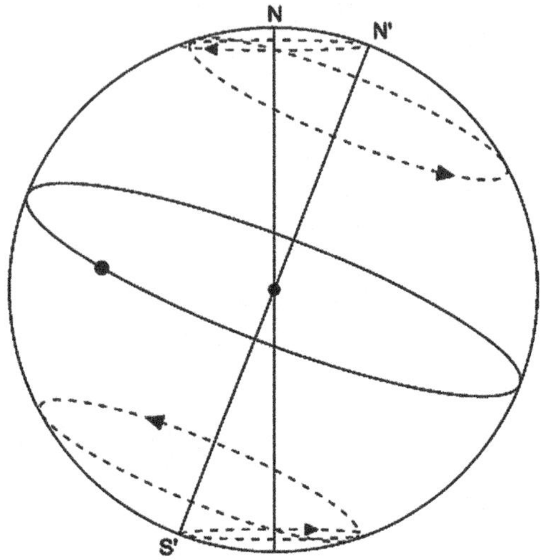

Fig. 12.7 NS is the axis of rotation of the Sun's first sphere, N′S′ the axis of its second sphere

12.2.2 Eudoxus' Concentric Spheres

The first elaborate geometric astronomical model that aims to represent the described phenomena is due to Eudoxus of Cnidus (390 BC – 337 BC). Although his works are lost, we know the essentials of the model through Aristotle's (1924) *Metaphysics* and Simplicius' (2005) commentary (VI century AD) on Aristotle's (1922) *On the Heavens*.

Eudoxus' representation of the universe is a system of spheres, all centered in the Earth. The observable diurnal motion of the stars is represented by a single sphere that rotates westwards around the axis defined by the celestial north and south poles, with a period of a sidereal day—this is just the celestial sphere in Fig. 12.2. The observed motion of the Sun is represented by the combined motion of three spheres. As it is shown in Fig. 12.7, the Sun's first sphere moves exactly as the sphere of the fixed stars. The second sphere rotates eastwards, with a period of a year. Its axis of rotation is inclined 23°40′ with respect to the first sphere's axis, and its endpoints are fixed on two antipodal points in the first sphere.[9] Placing the Sun in a suitable fixed point in the equator of the second sphere, its observed motion is reproduced by the model.[10]

[9] The numerical values for the parameters in Eudoxus' model are contemporary reconstructions. Neither Aristotle nor Simplicius included precise values in their mostly qualitative explanations of the model.

[10] Eudoxus included a third sphere, with its axis inclined a small angle with respect to the second sphere's axis. It is generally affirmed that he wrongly attributed the Sun a small latitudinal motion with respect to the ecliptic. However, Linton (2004, 28) affirms that in Eudoxus' times the ecliptic was vaguely defined as some great circle within the zodiac in the celestial sphere. The definition of the ecliptic in terms of the motion of the Sun, Linton states, was introduced about two

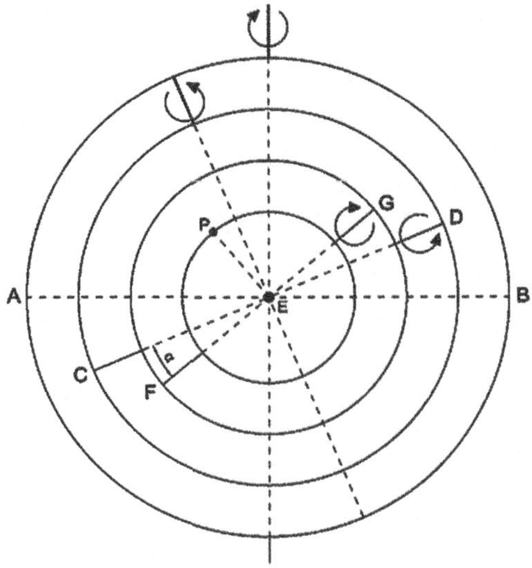

Fig. 12.8 Eudoxus model for a planet. AB is the equator, CD is the ecliptic, the planet is at P

Eudoxus assigned 4 spheres to each planet, as shown in Fig. 12.8. The first sphere reproduces diurnal motion, so it is just like the sphere of the distant stars. The second reproduces the eastward motion of the planet along the zodiac. Just like the Sun's second sphere, its axis is inclined with respect to the axis of the first one by an angle of 23°40′.[11] Its period of rotation is the zodiacal period of the planet. The third and fourth spheres take care of retrograde motion. The third sphere's axis *CD* lies on antipodal points in the equator of the second sphere, that is, it lies on the ecliptic. The axis *FG* of the fourth sphere is slightly inclined with respect to the axis of the third sphere by an angle α, with a specific value for each planet. The third and fourth spheres rotate in opposite directions, both with a period given by the synodic period of the planet. The combined motion of the third and fourth spheres produce loop-shaped figures, called *hippopedes*, representing retrograde motion, and the motion of the second sphere drags the hippopedes along the planet's zodiacal path. Table 12.1 shows the values of zodiacal and synodic periods.

Eudoxus' model says nothing about the order of distances of the planets with respect to the Earth—the radii of the spheres play no role. Actually, for the model to work, we do not need to take them as real physical entities in any sense, but only as geometric-kinematic configurations that represent the motions of the planets (cf. Torretti, 2007, 40). However, ancient astronomers invoked arguments that mixed

centuries later by Hipparchus. If this is correct, Eudoxus' third sphere was justified. Support for this interpretation (see Neugebauer, 1975, 633) lies on the fact that an observational determination of the precise path of the Sun with respect to the background stars was very difficult, so an estimation of a 1/15 part of a circumference, i.e., 24°, naturally suggests. Once established this pseudo-ecliptic plane, more precise observations of the path of the Sun would lead to the introduction of some latitudinal motion with respect to the pseudo-ecliptic.

[11] Or perhaps 24°, see footnotes 9 and 10.

observations and a lucky guess in order to establish that order. Observations clearly show that the Moon is the closest celestial object: its observed size, and the fact that along its trajectory it covers other planets and the Sun (as in solar eclipses). As for the rest of the planets, the distance of a planet to the Earth was given by the zodiacal period—the longer the zodiacal period, the farther the planet from the Earth. Thus, Saturn, Jupiter and Mars are farther than the Sun. For this reason, these were known as the *superior planets*. The Sun, Venus and Mercury, on the other hand, do not differ in their zodiacal periods, so their order was a controversial issue. However, after Ptolemy's work, consensus was reached and Mercury and Venus, in that order, were considered to be closer to the Earth than the Sun—and for this reason they were known as the *inferior planets*.

The kinematic behavior of the inferior planets differs from the superior ones in an important aspect. The maximum elongation (angular distances as seen from the Earth) between the Sun and Mercury and between the Sun and Venus, are, respectively, 29° and 47°. This means that, in terms of their elongation, inferior planets are never far from the Sun, so they can only be seen near sunset or dawn. This also explains why the zodiacal periods of Venus and Mercury are equal to the solar year (see table 12.1). On the other hand, the elongation between the Sun and the superior planets can go up to values near 180°, so the Earth and superior planets can be in *opposition*. Now, although the Earth and inferior planets can never be in opposition, both superior and inferior planets are in *conjunction* when the elongation angle takes minimum values (see Fig. 12.9).[12] Retrograde motion of superior planets is always observed near opposition, whereas for inferior planets it is always observed near conjunction.[13] In Eudoxus' model, both features are coincidences. The system of concentric spheres does not enforce that there is a maximum angle of elongation for inferior planets, nor a connection between retrograde motion and conjunction or opposition.

There are several shortcomings in Eudoxus' model. First, the latitude of planets with respect to the ecliptic cannot be correctly reproduced—according to Simplicius, the representation of celestial latitude was incorrectly taken care of by the width of the hippopedes. Second, the hippopedes resulting from the third and fourth sphere of each planet always have a loop ∞-shape, whereas some of the observed retrograde motions are *z*-shaped. Third, variations in the apparent size of the Moon and some planets could not be accounted for—in Eudoxus' model, a celestial object is always equidistant from the Earth, so its apparent size should not change. Finally,

[12] Elongation at opposition is not exactly 180°, and elongation at conjunction is not exactly 0°, due to the fact that along their orbits of planets show a small latitudinal distance from the ecliptic (although they always stay within the zodiac).

[13] From a heliocentric perspective, for interior planets a distinction between *superior* and *inferior* conjunction can be traced. An interior planet is in inferior conjunction when it lies between the Earth and the Sun, and in superior conjunction when the Sun lies between the planet and the Earth. The retrograde motion of Mercury and Venus is observed near inferior conjunction. From a geocentric perspective, as in Eudoxus' model, this distinction cannot be traced, for the Sun cannot lie between an inferior planet and the Earth.

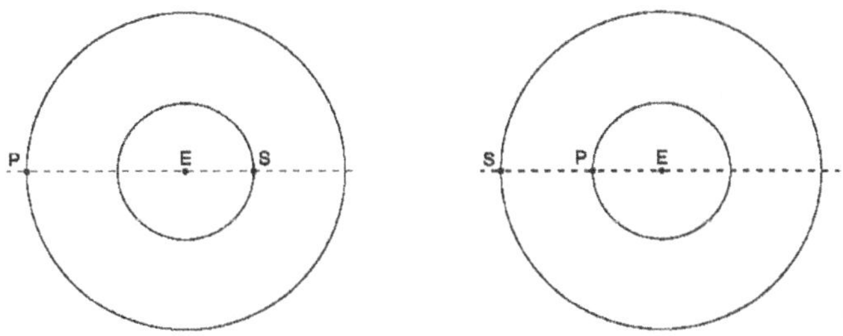

Fig. 12.9 A superior planet P in opposition (left), and an inferior planet in conjunction (right)

and most importantly for the historical development of astronomy, the motion of the Sun along the ecliptic is not strictly uniform. We can divide the ecliptic circle in four equal arcs of 90°, subtended by the solstices and equinoxes. The Sun covers these four arcs in different periods—which means that the seasons of the year are not equally long. A similar feature holds for the planets, for their motion along the zodiac occurs at variable angular velocities. But since in Eudoxus model all spheres rotate uniformly, these variations could not be accounted for.

Despite these issues, Eudoxus' model set the heuristic principles to study the motions of the heavens in subsequent astronomical geometrical models, i.e., the apparent irregularities in celestial motions were to be explained in terms of the combination of several circular uniform motions:

> The details of Eudoxus' theory are not known with any certainty, but we do know that the scheme exerted a profound influence over the development of astronomical thought. [...] This was because it demonstrated the power of geometrical techniques, in that superpositions of simple uniform rotations could be used to model extremely complex behavior. (Linton, 2004, 32)

12.2.3 The Metaphysics of Circular Uniform Motion

The explanation of the motion of celestial objects in terms of circular uniform motion was not just a matter of conceptual and empirical convenience. Since the early days of Greek philosophy, and following a Pythagorean tradition, the idea that this type of motion is essentially appropriate to the heavens was a central metaphysical principle in the examination of nature. In his dialogue *Timaeus*, Plato (2000) claimed that the creation of the heavens responded to the demiurge's decision to introduce time in nature, so circular uniform motion—associated to eternity and strict regularity—was the natural choice for fulfilling this task. This is just an example that metaphysical and aesthetic considerations grounded the value that the ancients assigned to circular uniform motion. Eudoxus' was a member of the Academy, so Simplicius was probably right when he states that his model was a response to the challenge set by Plato of finding a representation of the motion of the planets in terms of circular uniform motion:

> Eudoxus of Cnidus is said to be the first of the Hellenes to have made use of such hypotheses, Plato (as Sosigenes says) having created this problem for those who had concerned themselves with these things: on what hypotheses of uniform and ordered motions could the phenomena concerning the motions of the planets be preserved? (Simplicius, 2005, 33)

In *Physics* and *On the Heavens*, Aristotle (1936, 1922) further elaborated on the importance of circular uniform motion in astronomy. He claimed that the planets (in the ancient sense of the term, see fn. 8) and the stars are made of an element, different of the four elements constituting terrestrial bodies (earth, water, wind and fire), called *ether*. By essence, circular uniform motion is the goal (*télos*) that corresponds to bodies made of this element. Aristotle's conception of celestial objects is thus teleological: the planets and the stars move uniformly in circles not because they are forced to, but because that motion constitutes their essential *télos*.

The Earth is at rest in Aristotle's physics. His central argument against terrestrial motion was that if it rotated, a body thrown straight upwards would fall back to the ground some distance to the west with respect to the place from which it was thrown, against what is observed—and a similar argument can be run against any kind of displacement of the Earth. In Aristotle's terrestrial physics, the *télos* for the elements earth and water is to reach their natural place: the center of the universe. This provides a qualitative-teleological explanation of why the Earth is located precisely at the center of the universe, and why heavy objects fall. Without anything like a concept of inertia, the view of an immobile Earth was the most reasonable, and it remained unchallenged until the sixteenth century.

Eudoxus' model is a nice geometric fit for Aristotle's physics. It is coherent with an immobile central Earth, with the motion of the stars around it, and it provides an explanation of retrograde motion in terms of uniform circular motion—all the concentric spheres rotate uniformly. In Aristotle's interpretation of Eudoxus' model, the spheres are physical and made of ether, not only kinematic configurations.

12.2.4 Ptolemy's Model

The most important astronomers between Eudoxus and Ptolemy are Apollonius of Perga (third century BC) and Hipparchus of Nicaea (ca. 190 BC – ca. 120 BC). Apollonius showed that the problem of the variable angular speed of the Sun along the ecliptic could be solved by a circular uniform orbit not centered on the Earth. In Fig. 12.10, the orbit of the Sun is represented by the dashed circle with radius DS, whose center D is at a distance ED from the Earth E— ED is the *eccentricity* of the Sun's orbit. The angular distance between equinoxes and solstices as seen from the Earth is 90°, but given the eccentricity, they subtend slightly different angles from D, so the determined arcs have slightly different lengths Thus, exactly because the Sun S moves uniformly in a circle centered in D, it covers the seasonal arcs in the ecliptic in different periods.

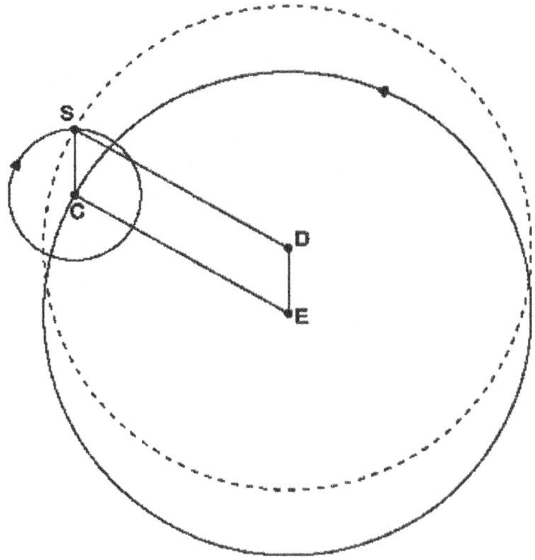

Fig. 12.10 Apollonius' eccentric solar orbit, and its equivalent model of deferent and epicycle

A crucial contribution by Apollonius was that he showed that a different mathematical model can achieve the same result. In Fig. 12.10, consider now the solid circle centered on the Earth with radius *EC*, the *deferent*. Let *C* move uniformly anticlockwise along the deferent, and define a circle centered in *C* with radius *CS*, the *epicycle*. Let *S* move uniformly and clockwise along the epicycle, but with the same angular speed as *C* along the deferent, so that *EDSC* is always a parallelogram. It is clear that the position of *S* in the eccentric orbit model is always the same as the position of *S* in the deferent-epicycle model, so both models reproduce the motion of the Sun equivalently.

The deferent-epicycle model, Apollonius also showed, can be used to represent retrograde motion. As Fig. 12.11 illustrates, by letting the epicycle rotate in the same direction as the deferent, but with a different angular speed, the trajectory of a planet fixed in a point in the epicycle is the dashed line, which contains regular periods of retrograde motion.

In order to put Apollonius ideas to work in a model that saves the phenomena, the value of several parameters must be solved mathematically and determined empirically. In the case of the Sun observed motion (see Fig. 12.10), the direction of *ED* and its relative length with respect to *DS* must be established, through the determination of the angle *DES*. This requires precise observations and mathematical calculations that without modern trigonometry are not trivial.

Hipparchus did just that, and his method is illustrated in Fig. 12.12. P_1, P_2, P_3 and P_4 are the positions of the Sun at the vernal equinox, the summer solstice, the autumnal equinox and the winter solstice, respectively. All four points can be determined observationally, and Hipparchus measured that the Sun travels from P_1 to P_2 in 94 ½ days, and from P_2 to P_3 in 92 ½ days. To determine the direction of the eccentricity *EO*, we must determine the angle $\lambda = P_1EA$, where *A* is the apogee. Apogee and perigee are the *apsides*, i.e., the points in which the

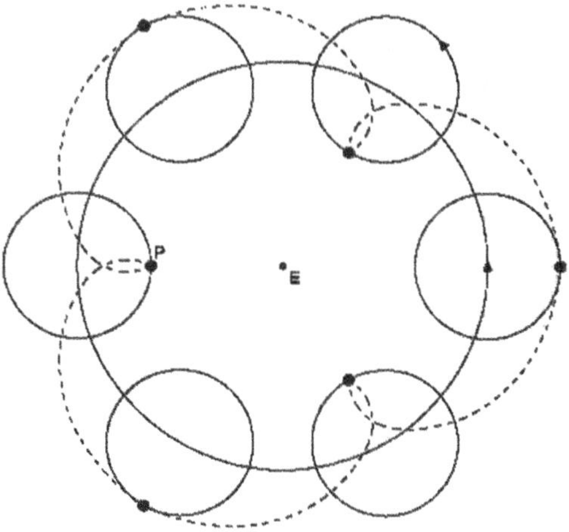

Fig. 12.11 Apollonius' method of deferent and epicycle for retrograde motion

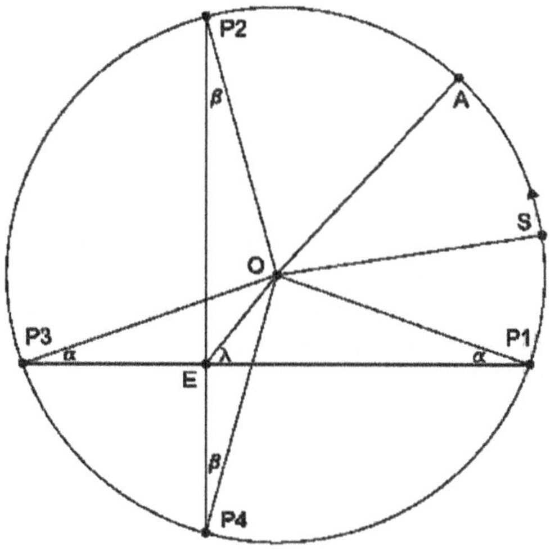

Fig. 12.12 Hipparchus' calibration of Apollonius eccentric model of the Sun

Sun is farthest and closest to the Earth in its eccentric orbit, respectively. If we determine the angle λ, we get the line joining the apsides, and *EO* lies on it. With some modern trigonometry (see Linton, 2004, 56), we obtain $tan\ \lambda = \frac{sin\ \alpha}{sin\ \beta}$ and $\frac{EO}{OS} = \frac{sin\ \alpha}{sin\ \gamma}$. To determine the values of α and β we can use the constant angular sped of the Sun along its orbit, namely, $w{\sim}59'8''$ per day (360° per 365 ¼ days).[14] Then, $\lambda = 65°\ 25'39''$. To calibrate the model to observations, we also need the ratio between the eccentricity and the radius of the Sun's orbit. With modern trigonometry we get $\frac{EO}{OS} = \frac{1}{24.17}$. Without modern trigonometry and following a

[14] $\angle$P1OP2 = $\alpha + \beta + 90°$ and $\angle$P2OP3 = $\alpha - \beta + 90°$, so that $\angle$P1OP3 = $2\alpha + 180°$. The Sun goes from P1 to P2 in 94 ½ = 189/2 days, and from P1 to P3 in 187. Thus, $\alpha + \beta + 90° = 189w/2$, and $2\alpha + 180° = 187w$, solving the last two equations we get α and β.

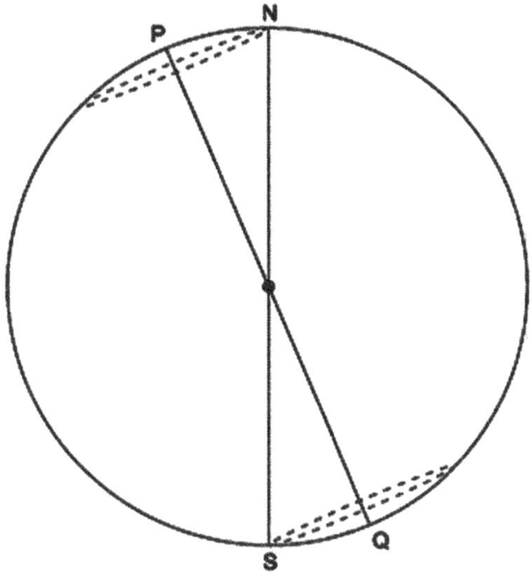

Fig. 12.13 The Ptolemaic representation of equinoxial precession

tortuous mathematical method, Hipparchus obtained $\lambda = 65^\circ\ 30'$ and $\frac{EO}{OS} = \frac{1}{24}$. Notice that these results are necessary to calibrate the deferent epicycle model of the Sun's orbit as well: with $\frac{EO}{OS}$ we get the required ratio between the epicycle radius and the deferent radius, and the direction of *EO* (given by the angle λ) is necessary to build the parallelogram *EDSC* in Fig. 12.10.

Another important contribution by Hipparchus was that the he measured that the time it takes for Sun to return to the same position with respect to the fixed stars (the *sidereal* year), is about 20 minutes longer than the time it takes for the Sun to return to the same solstice or equinox (the *tropical* year). This means that the points at which the celestial equators and the ecliptic plane intersect, the equinoxes, slowly move with respect to the background stars. Hipparchus measured that this *precession of the equinoxes* occurs at a rate of 1° per century—the actual value is 1° every 72 years. This means that a star that at a certain moment coincides with an equinoxial point makes a full eastward circle around the ecliptic to return to the same equinoxial point after about 26,000 years. Ptolemaic astronomers coped with this phenomenon (see Kuhn, 1995, 268) by adding a second sphere to explain the motion of the celestial sphere—and the same method can be used in any geocentric model (see Fig. 12.13). The first sphere rotates with a period of a sidereal day around the celestial north-south axis. The axis of the second sphere is perpendicular to the ecliptic (inclined 23°40′ with respect to the axis of the first sphere), and its endpoints are fixed on antipodal points on the first sphere. The rotation period of the second sphere is 26,000 years. As a result, the position of the celestial poles slowly change with respect to the fixed stars: a star that at some instant coincides with the celestial north pole, makes an eastward circle around the ecliptic north pole, to return the celestial north pole after 26,000 years.

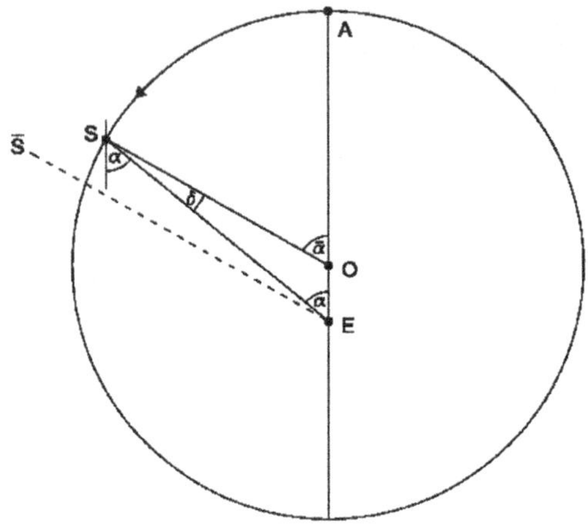

Fig. 12.14 Ptolemy's model of the Sun

Hipparchus did not develop a model of deferents and epicycles for the representation of the motion of the planets. That achievement is due to Ptolemy (ca. 100 – ca. 170 AD), who in his *Mathematical Sintaxis*, commonly known as *The Almagest* (Toomer, 1984), introduced the geometric model of the Universe that dominated astronomy for more than a millennium.

Ptolemy's model for the Sun was basically the same as the one introduced by Hipparchus. An important improvement was that he was able to predict the position of the Sun at any given time along its non-uniform (as seen from the eccentric Earth) path along the ecliptic. In Fig. 12.14, A is the apogee, O the center of the Sun's orbit, and S is the Sun. The angle $\overline{\alpha} = AOS$ increases uniformly. Given O (from Hipparchus model) and the tropical year, $\overline{\alpha}$ can be calculated at any given time. For the angle $\alpha = AES$, it is clear that $\alpha = \overline{\alpha} \pm \delta$. Ptolemy was able to compute the angle δ as a function of $\overline{\alpha}$, allowing thus to predict the position of the Sun at any given time, given by the angle α.

Ptolemy also introduced a crucial concept, the mean sun $\overline{S}$, i.e., the point in the ecliptic that the Sun would occupy at a certain time if it moved with uniform angular speed as seen from the Earth. $\overline{S}$ is a crucial parameter in the calibration of Ptolemy's models for each celestial object. As it can seen in Fig. 12.14, the longitude of $\overline{S}$ at a time t as determined from E, i.e., the angle $AE\overline{S}$, is given by $\overline{\alpha}$.

Ptolemy's representation of the motion of the planets was based on Apollonius' idea of deferent and epicycle—we saw above how this method can represent retrograde motion. Now, to solve the problem of the variable angular speed of the planets along the zodiac, introducing a deferent eccentric with respect to the Earth for each planet seemed the natural strategy. However, Ptolemy noticed that this would not be enough, so he invented a novel geometric device, the *equant*—a point with respect to which the center of the epicycle moves along the deferent with constant angular velocity (see Fig. 12.15). Since the equant is not the center of the deferent, it is clear that the motion of the center of the epicycle is not uniform along the deferent— $DQC = AQB$, so the point P covers the arcs $AB \neq CD$ in the

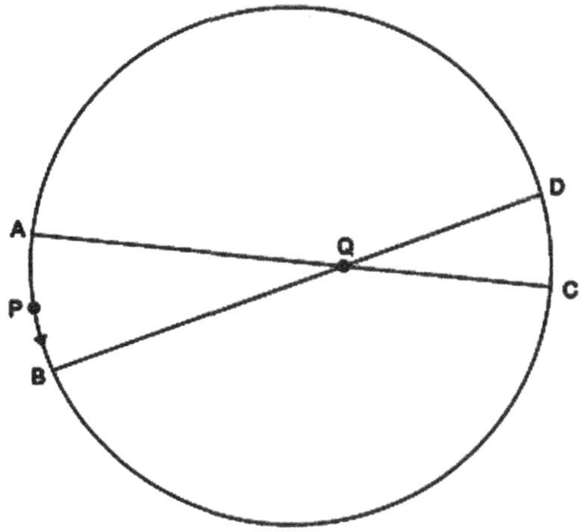

Fig. 12.15 Ptolemy's equant

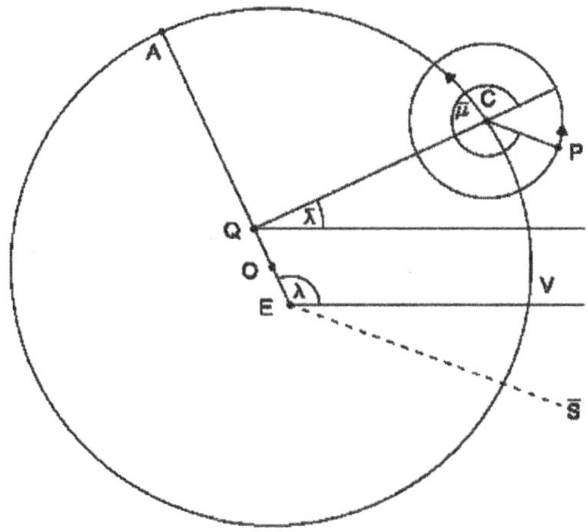

Fig. 12.16 Ptolemy's model of a superior planet

same time. The introduction of the equant, although strongly increases the empirical adequacy of the representation of celestial motions, violates the principle of circular *uniform* motion of celestial bodies.

Figure 12.16 illustrates Ptolemy's model for superior planets. E is the Earth, O is the center of the deferent, and Q the equant—the three points lie on a straight line and $EO = OQ$. C is the center of the epicycle, and it rotates counterclockwise along the deferent. The angle $\overline{\lambda}$ is the *mean longitude* of the planet, and since Q is the equant, $\overline{\lambda}$ increases uniformly with time. The angle $\overline{\lambda}$ for a 'mean planet' is analogous to $\overline{\alpha}$ for the mean Sun $\overline{S}$ in Fig. 12.14, but this time determined from V, the vernal solstice, rather than from the apogee. Thus, $\overline{\lambda}$ is subtended by QC and a straight line from Q parallel to EV. The planet is placed at P, which moves counterclockwise and uniformly along the epicycle, so that the angle $\overline{\mu}$ increases uniformly.

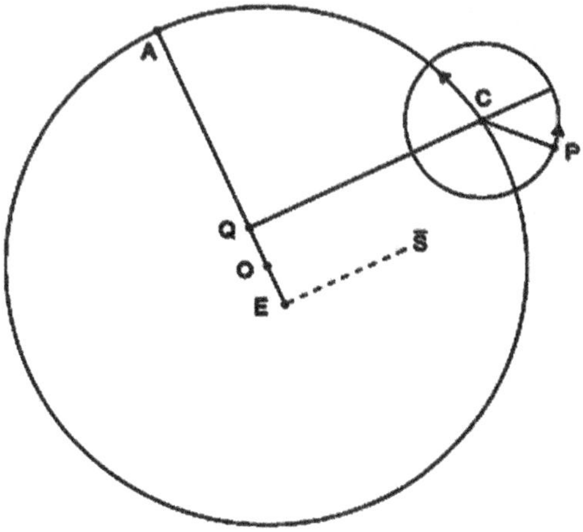

Fig. 12.17 Ptolemy's model of Venus

This gives us the conceptual basis of the model, but then Ptolemy had to calibrate it according to some parameters to save the observed phenomena. A first empirical constraint is that with T_z and T_s the zodiacal and synodic periods of the planet, respectively, and with T the tropical year, it holds that $\frac{1}{T_z} + \frac{1}{T_s} = \frac{1}{T}$. To satisfy this constraint, Ptolemy arranged the model as follows. The period of C around the deferent must be T_z, and the period of P around the epicycle must be T_s. Thus, at a given time t, the angles $\overline{\lambda}$ and $\overline{\mu}$ are given by $\overline{\lambda} = \frac{t}{T_z}$ and $\overline{\mu} = \frac{t}{T_s}$. It follows that the longitude of the mean sun $\overline{S}$ (measured from the solstice V) at t is then $\frac{t}{T} = \overline{\lambda} + \overline{\mu}$, which in turn enforces that CP, the line joining the epicycle center and the planet *is always parallel to* $E\overline{S}$, the line joining the Earth and the mean Sun—this feature will be important below. Finally, by means of rather complicated methods, Ptolemy determined the ratios OA/EQ and OA/CP for each planet, so that he could calculate the true longitude of the planet at any time, given by the angle VEP. An important subtlety in the model was that the angle $\lambda = VEA$—where A is the deferent's apogee, which is assumed to be fixed with respect to the background stars—slowly increases due to the precession of V.

Figure 12.17 represents Ptolemy's model for Venus. For inferior planets, the elongation with respect to the Sun is always small. To represent this, the center of the epicycle C always coincides with the real Sun, so C moves along the deferent around and the equant Q, at a rate such that *QC and $E\overline{S}$ are always parallel*. The remaining features are basically the same as for exterior planets. C moves counterclockwise along the deferent with center in O; E, O and Q lie on a straight line, and the eccentricity EO is equal to OQ; the planet is in P, which rotates around C counterclockwise. The model for Mercury is more complicated, but it retains these basic features—including that QC is always parallel to $E\overline{S}$.

A simplified representation of Ptolemy's model—underscoring the fact that the line joining the center of the epicycle and the position of the planet in the case of superior planets, the line joining the equant and the center of the epicycle in the case of inferior planets, and the line joining the Earth and the mean Sun, are always

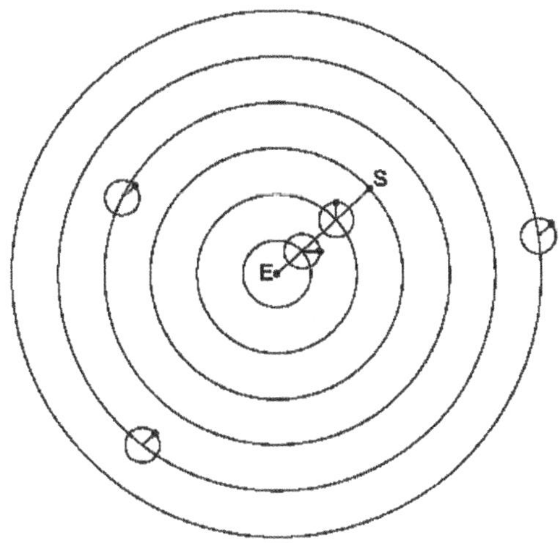

Fig. 12.18 Simplified sketch of Ptolemy's model

parallel—is presented in Fig. 12.18. It is clear that these lines being always parallel is a contingent feature of the model. Actually, it is a rather puzzling feature: of all the possible kinematic configurations of celestial objects, why one in which the mean Sun plays this special role? The question is especially pressing if we consider that there is nothing dynamically special about the Sun in the model, let alone about the mean Sun. Thus, this particular configuration does not get any explanation, and remains literally as a cosmic coincidence in Ptolemy's model.

What we have said so far accounts for the longitude of the planets. In order to represent their latitude, Ptolemy tilted the plane of the deferent with respect to the plane of the ecliptic, and also the plane of the epicycle with respect to the plane of the deferent. The values for these angles were determined by Ptolemy for each planet. As we will see below, from a heliocentric perspective the Ptolemaic epicycle of exterior planets models the motion of the Earth around and the Sun, so the plane of the epicycle should be parallel to the ecliptic. But since Ptolemy had no way to know this, he had to determine the two tilting angles separately. Furthermore, he arranged that the inclination of the deferent with respect to the ecliptic varies as a function of time. As a result, the Ptolemaic theory of planetary latitudes is rather cumbersome.

As for the physical interpretation of the model, Ptolemy does not present explicit statements about it in the *Almagest*. However, in a later work entitled *Planetary Hypotheses*, he did assume a realist view. That is, he claimed that the circles of deferents and epicycles are determined by physical ethereal spheres, arranged as represented in Fig. 12.19. E is the Earth, the center of the universe, and O is the center of the deferent corresponding to the planet P. The 'thick ethereal sphere' of planet P is then situated in the annulus between S_2 and S_3. If we take P to be Jupiter, we can locate Mars within S_4 and determine its own 'thick sphere' in an analogous way, and the same arrangement can be set for Saturn outside S_1.

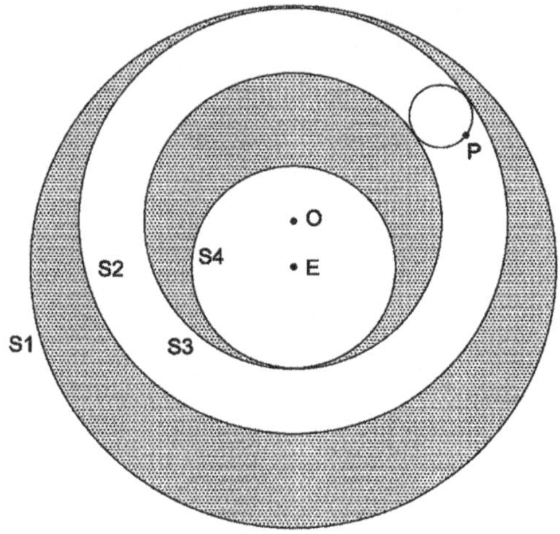

Fig. 12.19 Ptolemy's planetary spheres

An interesting consequence of this physical interpretation is that it allows a method to determine planetary distances from the Earth. The principle is rather simple: assuming that the thick spheres are disposed as close together as possible (following a *horror vacui* principle), the maximum distance of a planet is equal to the minimum distance of the next outer planet. As we saw, Ptolemy had determined the ratios between deferents and epicycles radii for each planet, so the physical realist interpretation allows to determine relative distances for all planets and even the celestial sphere. Furthermore, in the *Almagest*, Ptolemy, using parallax and geometrical reasoning, had calculated a maximum distance for the Moon of 64 Earth radii, which corresponded then to the minimum distance of Mercury. The rest of the distances are given in Table 12.2.[15] If we use the *stadion*, the Greek unit that Eratosthenes (ca. 276 BC – ca. 195 BC) employed in his (roughly correct) calculation of the circumference of the Earth—from which its radius can be obtained—we get an estimation of the size of the universe resulting from Ptolemy's method. Eratosthenes obtained a circumference value of 250.000 *stadia*. With some basic geometry, and picking a value of 160 meters for the *stadion*,[16] it follows that radius of the Ptolemaic universe was *ca.* 128.000.000 km.[17]

[15] Interestingly, the distance that Ptolemy obtained for the Sun by this method in the *Planetary Hypotheses* (1079 Earth radii) is very close to the value he had obtained in the *Almagest* by a different and independent method (1160 Earth radii). For a treatment of this issue, see Carman (2010).

[16] The value in meters of the Greek *stadion* is a disputed issue, but the proposed estimations range between 150 and 200 meters.

[17] As Kuhn (1995, 82) reports, the Arabic astronomer Al-Faraghi (800–870) applied the same method as Ptolemy in the *Planetary Hypotheses*. Using a value of 3250 Roman miles for the radius of the Earth, he calculated a universe radius equivalent to ca. 120.000.000 kilometers. That Ptolemy himself used this method was discovered only in 1967, when the relevant passage in the

Table 12.2 Distances to the Earth, in Earth radii, as determined from Ptolemy's model

	Minimum	Maximum
Moon	33	64
Mercury	64	166
Venus	166	1079
Sun	1079	1260
Mars	1260	8820
Jupiter	8820	14,187
Saturn	14,187	19,865
Fixed stars	20,000	

Ptolemy's model did not suffer any significant modification for centuries. Only when the Islamic medieval civilization flourished, Arabic astronomers introduced some amendments and criticisms (see Linton, 2004, Ch. 4). A curious development was related to Ptolemy's measurement of the obliquity of the ecliptic with respect to the celestial equator—he got a value of 23°51′. Thabit Ibn Qurra, an Arabic astronomer of the eighth century, obtained a more accurate value of 23°33′, but he took Ptolemy's measurements as correct, concluding that the obliquity of the ecliptic varies through time. Furthermore, Ibn Qurra calculated 82°45′ for the longitude of the apogee (the angle λ in Fig. 12.12). The change in value from Hipparchus and Ptolemy's times was natural given precession (recall Hipparchus obtained 65°30′), but since he trusted the different estimations of precession rates obtained by preceding astronomers, which differed from the one he calculated, he concluded that precession rate was also a variable function of time. In order to cope with the alleged varying obliquity of the ecliptic and variable rate of precession, Thabit introduced a *trepidation* theory for the ecliptic in Ptolemy's model.[18] Al-Battani, also in the eighth century, measured an ecliptic obliquity of 23°35′, but he did not conclude a varying value, he simply corrected Ptolemy's result and improved the model of the Sun—amending also the mean Sun motion, the precession of the equinoxes rate, the longitude of the apogee, and the eccentricity of the deferent.

By the twelfth century, substantial criticisms based on Ptolemy's model incompatibilities with Aristotelian principles were levelled by astronomers and philosophers (see Goldstein (1980) for an overview of medieval attacks against Ptolemy). As we already mentioned, the equant involves the rejection of *uniform* circular motion. Furthermore, the eccentricity of the deferents with respect to the Earth was hard to swallow from an Aristotelian standpoint—for some planets, the deferent center lies outside the orbit of the Moon. Prominent thinkers like Averroes

Arabic version of the *Planetary Hypotheses* was found and translated (see Carman, 2010 and the references therein).

[18] The obliquity of the ecliptic is indeed variable, and it is due to gravitational perturbations of other planets. However, the real effect is much smaller than the one that Ibn Qurra deduced from the difference between his measurements and Ptolemy's.

(1126–1198) (see Sabra, 1984; Çimen, 2019) and Maimonides (1138–1204) (see Nutkiewicz, 1978) objected to Ptolemy's model on these grounds.

However, attempts to build models that were faithful to Aristotelian physics were unsuccessful. During the twelfth century, Nur ad-Din Al-Bitruji proposed a model reminiscent of Eudoxus' idea of concentric spheres (see Goldstein, 1971; Sabra, 1984; Çimen, 2019), and, as late as the early sixteenth century, the Italian astronomers Girolamo Fracastoro and Gianbattista Amico tried to revive Eudoxus' model (see Dreyer, 1953, 296–304), but, all such attempts were of no avail.

Anyhow, Ibn Al-Shatir (1304–1375) was successful in introducing a model that retained most of the basic features of Ptolemy's system, but eliminating the equant (see Kennedy & Roberts, 1959). Using an arrangement of epicycles on epicycles, Al-Shatir obtained planetary orbits equivalent to Ptolemy's, but without using the equant, restoring thus the uniformity of circular motion. Although Al-Shatir's model did not have a major impact, his method to eliminate the equant was used by Copernicus in his heliocentric model. Al-Shatir is also responsible for another amendment of Ptolemy's model of the Sun. The astronomer Al-Zarqali (1029–1087) had discovered that the apogee slowly moves with respect to the fixed stars, at a rate of 1° every 279 years, or 12,9″ per year.[19] Al-Shatir coped with this phenomenon by embedding his solar model within another circle along which the apsidal line rotates.

12.2.5 Copernicus' Model

The role of Copernicus' model in the historical development of astronomy and in the constitution of modern science is not as revolutionary as the expression *Copernican revolution* suggests. In short, Copernicus' (1992) work entitled *On the Revolutions of the Heavenly Spheres*, published in 1543, more than a novel geometric astronomical model, is a translation of Ptolemy's to a heliocentric perspective. This subtle maneuver, of course, led to revolutionary science—especially when Kepler entered the stage—but evaluated on its own merits, Copernicus' contribution was hardly groundbreaking. Actually, a central motivation for the heliocentric model was rather conservative: to reintroduce the principle of circular uniform motion. As we saw, Ibn Al-Shatir had already achieved that goal, so the main innovation of Copernicus' work was a simple explanation of retrograde motion as an optical effect due to the orbital motion of the Earth. The heliocentric translation of Ptolemy's model also allowed other advantages, like a purely geometric method to determine planetary distances. Hence the appeal of Copernicus' proposal for astronomers of the time.

[19] What Al-Zarqali discovered was the precession of the apsidal line in the *Earth's* orbit. The modern estimation of apsidal precession period for the Earth is 11,6″ per year. Apsidal precession is caused by concomitant factors, including gravitational planetary perturbations. A full explanation can only be given using general relativity: the precession of Mercury's perihelion was a crucial element in Einstein's formulation of his gravitational theory.

Despite this conceptual attractive, a more nuanced evaluation is that, although it was certainly a spark for the revolutionary work of Kepler and Galileo, "Copernicus might well be described as the last of the ancients, a spiritual companion of Aristarchus, Hipparchus, and Ptolemy" (Linton, 2004, 121).

Copernicus' model assigned two basic motions to the Earth. First, a rotation about an axis fixed on the celestial poles, with a period of a sidereal day. This explains the observed diurnal motion of the stars, the Sun and the planets, but now the celestial sphere is at rest. Second, the Earth undergoes a circular translational motion around a fixed Sun, with a period of a sidereal year (roughly 365 ¼ days) describing a plane inclined $23°28'$ with respect to its rotation axis. This implies of course that the ecliptic gets redefined as the orbital plane *of the Earth*.

This simple geometric scheme allows a simpler explanation of the precession of the equinoxes: if the poles *N* and *S* of the axis of terrestrial rotation describe a circle centered on the poles of the ecliptic *P* and *Q* in about 26.000 years, as represented in Fig. 12.20, a star that at an instant coincides with an equinox makes a circle around the ecliptic, advancing 1° every 72 years.

Copernicus measured a roughly correct period of precession of about 26.000 years. However, his actual account of this phenomenon was rather complicated. Since he thought that the Earth is carried around the Sun by a sphere (here we find yet another conservative principle), he concluded that the axis of rotation should change direction, unless a third compensating motion, that Copernicus named *motion of the inclination*, is added to rotation and translation. Now, this third motion is not exactly opposite to the axial direction change induced by the sphere that carries the Earth, so it was the difference between both motions that accounted for precession. Besides, just like Thabit Ibn Qurra, Copernicus mistakenly believed that the rate of precession is variable in time, and that so is the obliquity of the ecliptic. Thus, Copernicus included a trepidation theory. The Copernican account of the Earth's motion was complicated by yet another factor:

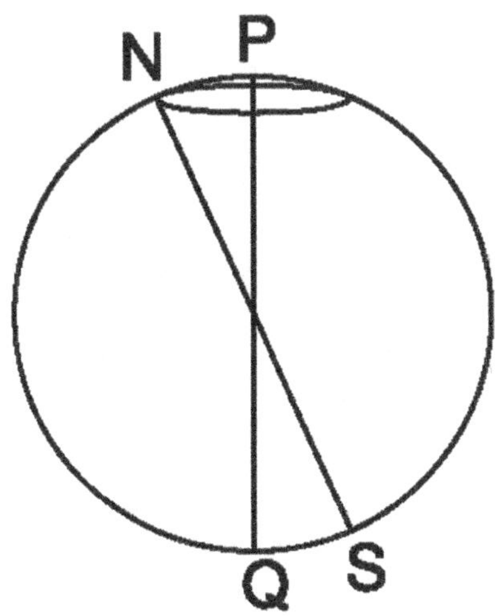

Fig. 12.20 Precession of the equinoxes, heliocentric perspective

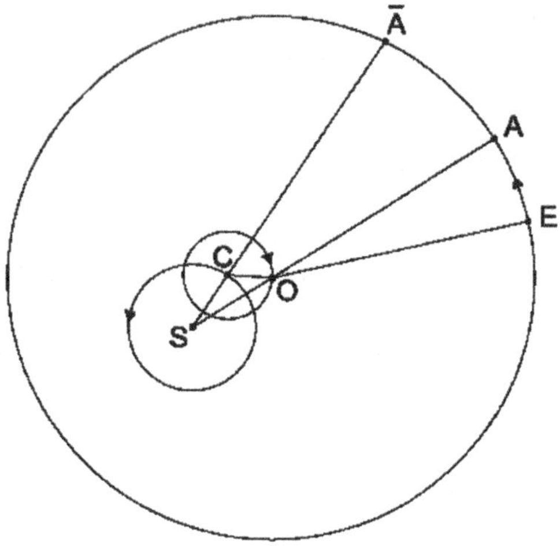

Fig. 12.21 Copernicus' model of the Earth

the model also copes with apsidal precession (discovered by Al-Zarqali), that is, the apsidal line joining aphelion and perihelion (the points at which the Earth is farthest and closest from the Sun) has a variable direction with respect to the fixed stars.

Figure 12.21 depicts Copernicus' model of the Earth's orbit. It moves uniformly and counterclockwise along the circle centered in O with a period of a sidereal year. In turn, O moves uniformly and clockwise along a circle centered in C, with a period of 3.434 years, and C moves uniformly and counterclockwise along a circle centered in the Sun S, with a period of 53.000 years. This complicated arrangement is connected to trepidation, and also to apsidal precession. For example, it can be seen that as a result of the motion of O around C, the eccentricity of the Earth's orbit varies, from which it follows that the aphelion A oscillates around the mean aphelion $\overline{A}$ with a period of 3.434 years. In turn, due to the motion of C around S, $\overline{A}$ describes a full circle around S in 53.000 years. This geometric arrangement guarantees that the point O, the center of the terrestrial orbit, corresponds to (the heliocentric counterpart of) the mean Sun $\overline{S}$.

As we said, a central motivation for Copernicus's model was a simple explanation of retrograde motion, illustrated in Figs. 12.22 and 12.23. The position of the planet against the backdrop of the fixed stars is given by the direction of a line from the Earth E passing through the planet P. Retrograde motion is then nothing but an optical effect due to the orbits of E and P around the Sun. Figure 12.21 also shows why for superior (exterior) planets retrograde motion occurs always in opposition, and Fig. 12.22 shows why for inferior (interior) planets it always occurs in conjunction.

A challenge imposed by planetary motion was that their speed is variable along their path through the zodiac. To deal with it, Ptolemy invented the equant, and added it to the eccentric deferent mechanism. Copernicus was more conservative than Ptolemy in this point, so in the name of the principle of circular uniform motion, he avoided the equant and coped with this irregularity by using a deferent-

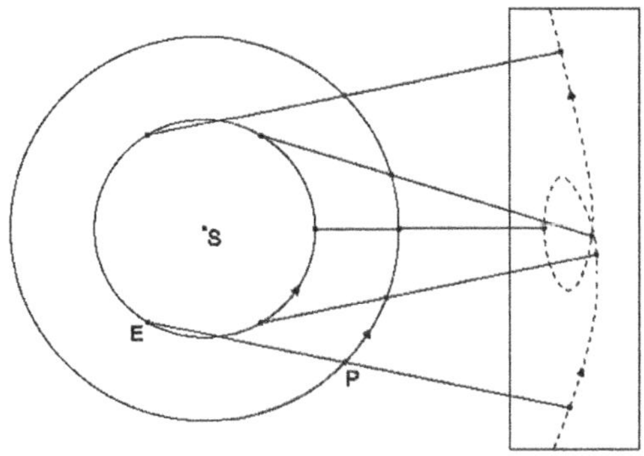

Fig. 12.22 Retrograde motion, exterior planet

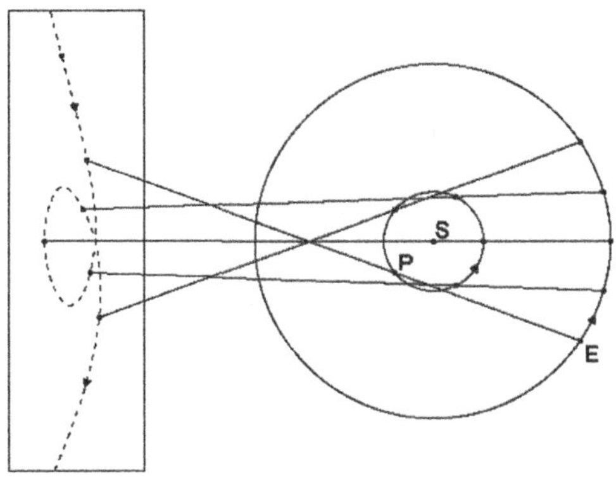

Fig. 12.23 Retrograde motion, interior planet

epicycle system. Figure 12.24 displays Copernicus' model for an exterior planet. O, which is not the Sun, but lies nearby it, is the center of the deferent. The center of the epicycle C moves along the deferent with a period equal to the sidereal period of the planet. The planet at P rotates around C, in the same direction and with the same period. The motion is arranged in such a way that the angles PCO and AOC are equal, where A is the orbit's aphelion. Notice that the deferent-epicycle construction plays a different role than in Ptolemy's planetary models—now it takes care of the observed variable orbital speed of the planet, not of its retrograde motion.

The geometric translation to a geocentric standpoint is illustrated in Fig. 12.25. In Copernicus' model (dashed lines), the Earth E rotates in a circular orbit centered in the mean Sun $\overline{S}$. The vector that gives the position of the planet as seen from the Earth is $E\overline{S} + \overline{S}O + OC + CP$. The figure superimposes Copernicus' model with a geocentric model (solid lines) with two epicycles. O' is the center of the deferent with radius $O'C'$, C' is the center of epicycle with radius $C'C''$, and C'' is the center of the epicycle with radius $C''P$. The position vector of the planet is now $EO' + O'C' + C'C'' + C''P$. It is clear that EO' is parallel to $\overline{S}O$, and that $|EO'| = |\overline{S}O|$. The same holds for $O'C'$ and OC, for $C'C''$ and CP, and for $C''P$ and $E\overline{S}$. It follows then that the planet position vector in the heliocentric model is the same as the planet position vector in the geocentric model.

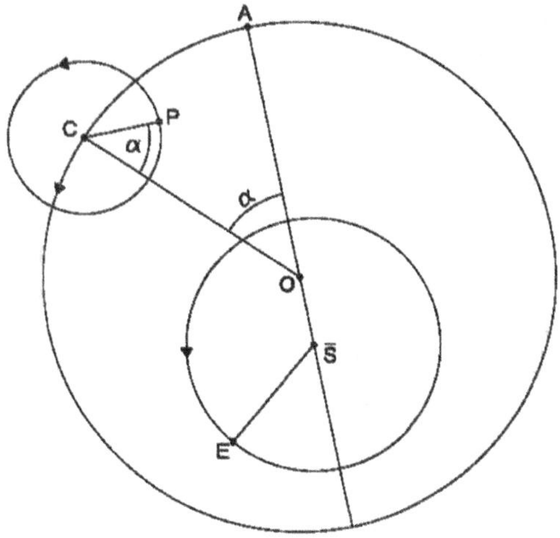

Fig. 12.24 Copernicus' model of an exterior planet

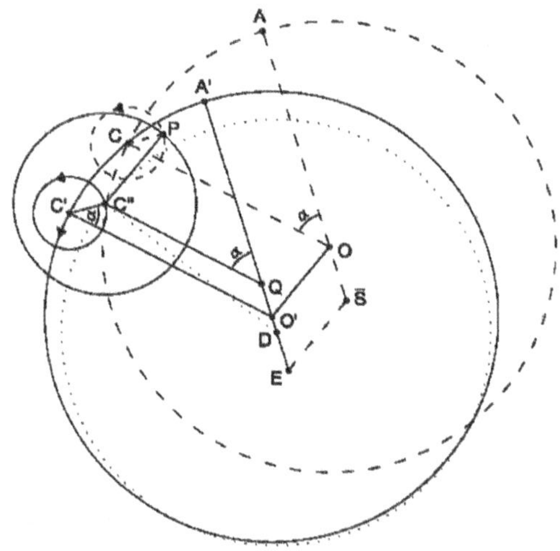

Fig. 12.25 Geometric inter-translability between Copernicus' model for an exterior planet, Ptolemy's, and a geocentric model without equant

To show the translatability to Ptolemy's model, we draw the line $C''Q$, parallel to $O'C'$, so that the angles AOC and $A'QC''$ are equal and increase uniformly. Now, $ED = DQ$, so the center of Ptolemy's deferent must be D. If it holds that $|\overline{S}O| = 3|CP|$, then $|OC|$, the radius of Copernicus' deferent, and $|DC''|$ are equal (see Linton, 2004, 140–141). Copernicus' devised his model so that this condition is met, so Q is the equant of the Ptolemaic deferent (dotted line) with radius DC'', and the radius of the Ptolemaic epicycle is then $C'P$, so C'' rotates along the Ptolemaic deferent according to the equant rule. Then, in the Ptolemaic model the planet position vector is $ED + DC'' + C'P = EQ + QC'' + C'P$, and since $EQ + QC'' = EO' + O'C' + C'C''$, all three planet position vectors are the same. We have illustrated then how the equant can be dispensed with in a geocentric model by using epicycles on epicycles, as Al-Shatir did (solid lines), and how this can be translated into a heliocentric model, as Copernicus did.

Our exposition of the geometric translation between Copernicus' and Ptolemy's model for an exterior/superior planet allows us to see that since the position vector in the heliocentric case is $E\overline{S} + \overline{S}O + OC + CP$, whereas in Ptolemy's model it is $ED + DC'' + C''P$, then in the geocentric case it must be the case that the line joining the Earth and the mean Sun and the line joining the epicycles center and the planet, that is, $E\overline{S}$ and $C''P$, must be always parallel. As we saw, in Ptolemy's model this is a cosmic coincidence, but considering it from the point of view of the inter-tranlsability with Copernicus' model, it gets explained away.

Figure 12.26 shows Copernicus model for Venus. The Earth rotates on a circle centered in $\overline{S}$, C rotates on a deferent centered in O with a period given by Venus' sidereal period, whereas the planet rotates along the epicycle centered in C at twice the rate that the Earth rotates around $\overline{S}$, so that the angle $CO\overline{S}$ is twice the angle $E\overline{S}A$, where A is the point at which the Earth is farthest from O.

The heliocentric configuration of interior planets also explains away the coincidental features of Ptolemy's model. Since an inferior planet in the geocentric model is an interior planet in the Copernicus' model, then it must be the case that its maximum elongation is a small angle (Fig. 12.27)—this is just a contingent feature in Ptolemy's model, but if we take it as the geocentric translation of Copernicus' model, then it must be arranged in this way.

Furthermore, the apparent motion of a planet depends on two factors: the motion of the planet around the Sun, and the motion of the Earth around the Sun. In the Ptolemaic models for the superior planets, the first factor is taken care of by the deferent, and the second factor by the epicycle. This remark offers an intuitive explanation of the Ptolemaic coincidence that for superior planets the line joining the center of the epicycle and the planet must be always parallel to the line joining the Earth and the mean Sun (cf. Fig. 12.25). In the case of the Ptolemaic models

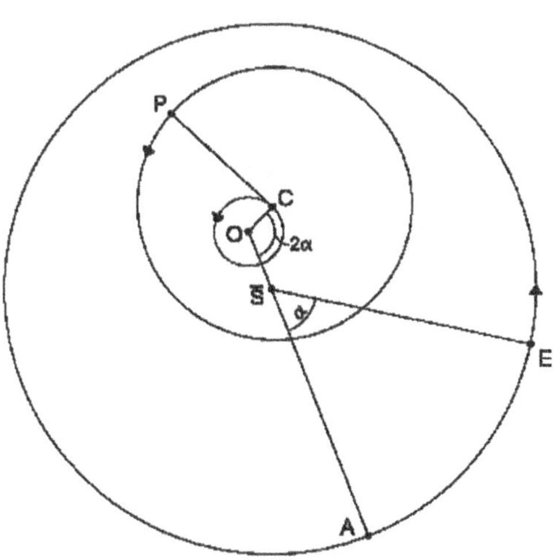

Fig. 12.26 Copernicus' model of Venus

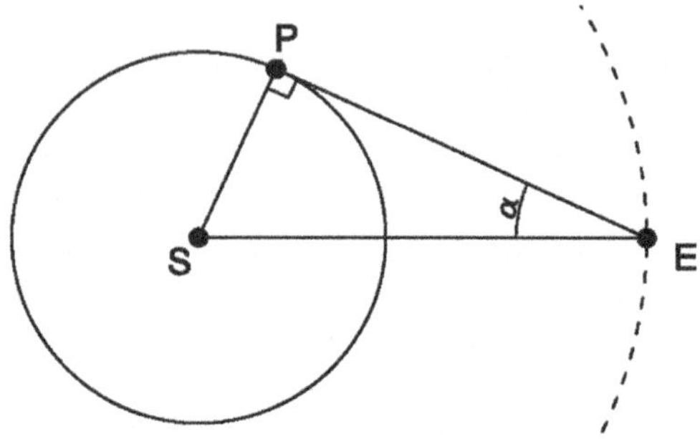

Fig. 12.27 The distance of an interior planet to the Sun in Copernicus' model

for inferior planets, the roles of deferent and epicycle get inverted: the deferent represents the second factor, and the epicycle represents the first one. As a result, the line joining the equant and the center of the epicycle must be always parallel to the line joining the Earth and the mean Sun (cf. Figs. 12.16 and 12.17). That is, this twofold coincidence in Ptolemy's model is explained away in the heliocentric translation.[20]

Another source of appeal of the Copernican model is a purely geometric method to estimate planetary distances to the Sun. The method is very simple for interior planets. When the planet is at maximum elongation, the Earth, the Sun and the planet form a triangle rectangle in the planet. Taking the Earth-Sun distance as our unit of length, it is clear that the planet-Sun distance PS is $sin\ \alpha$ (see Fig. 12.27). The distance to the Sun for exterior planets can also be obtained by a simple trigonometric reasoning (see Jacobsen, 1999, 125). Copernicus' trigonometric method is better than the one formulated by Ptolemy. It can be easily seen that it does not rely on physical assumptions like spheres arrangement.

Furthermore, in the heliocentric model the order of the planets can be established in a more systematic way. In Ptolemy's model, the period used to organize planetary order was the zodiacal period (the average time it takes a planet to return to the same position respect to the background stars as seen from the Earth). But since the zodiacal periods of Mercury and Venus are equal to the sidereal year of the Sun, they could not be invoked to univocally settle the order for these three bodies. In Copernicus' model, the relevant period to determine planetary order is the sidereal one (the period it takes the planet to complete one orbit around the Sun). An intuitive comparison of the geometry of the models shows that for an exterior/superior planet its sidereal and zodiacal periods are the same (cf. Kuhn, 1995, 167), but that is not the case for interior planets. However, the orbital periods of interior planets in Copernicus' model can be easily calculated. After the Copernican explanation of retrograde motion, for an interior planet its synodic period T_p gives the time between inferior conjunctions with the Earth (cf. Fig. 12.23 and fn. 13). Measuring time in years, during T_p years the Earth obviously makes T_p terrestrial orbits, and

---

[20] For an illustration and a rigorous treatment of the inter-translability between the Ptolemaic and Copernican models of Venus, see Neugebauer (1986, 497) and Barbour (2001, 239).

the interior planet makes $1 + T_p$ revolutions around the Sun in that same time. Therefore, in 1 year an interior planet makes $1 + \frac{1}{T_p}$ revolutions around the Sun. Thus, an interior planet makes one revolution around the Sun in $\frac{1}{\left(1+\frac{1}{T_p}\right)}$ years. As it can be seen from Table 12.1, for Mercury, $T_p = 0{,}315$ years, so its sidereal period is 87, 5 days; whereas for Venus, $T_p = 1{,}599$ years, so its sidereal period is 224, 8 days. Sidereal period can be then applied to determine the order of all planets.

On the negative side, although Copernicus' model is *conceptually* simpler—recall the explanation of retrograde motion, precession of the equinoxes, and the cosmic coincidences of Ptolemy's model—in geometric technical terms the model is at least as complicated as Ptolemy's. For example, a complex use of deferents and epicycles is still present, although for purposes different than in Ptolemy's model. This is hardly surprising when we consider that Copernicus' model is, bottom line, a heliocentric geometric translation of Ptolemy's. The Polish astronomer did not introduce novel mathematical methods that could simplify the representation of the phenomena.

This can be seen quite clearly in the latitude theory. Copernicus' aim was to reproduce Ptolemy's results in this matter, so he did not realize the simple representation of planetary latitudes that a heliocentric model allows—as we will see below, Kepler did. In the Copernican model, the inclination of the orbital planes that account for observed planetary latitudes are defined with respect to the mean Sun rather than to the real Sun, and the inclinations are variable in time (as in Ptolemy's model) and connected in a highly complicated way to the longitude of the Earth. Copernicus' latitudinal theory is at least as complicated as Ptolemy's.

A second shortcoming was that an Earth in motion implies that a stellar parallax angle is predicted. As the Earth moves along its orbit, the direction of the apparent position of a star should change night after night at different locations of the Earth along its orbit, but no such variation was observed. In Fig. 12.28, E_1 and E_2 represent two positions of the Earth along its orbit, separated by an interval of 6 months, the direction of the apparent position of a star T should change from E_1 to E_2, according to a parallax angle α. This means either that the thesis of an Earth in motion is wrong, or that the distance to the fixed stars is too large for parallax to be detectable—the value of α is inversely proportional to the distance *ST*. This was especially pressing since the method to determine planetary distance in Copernicus'

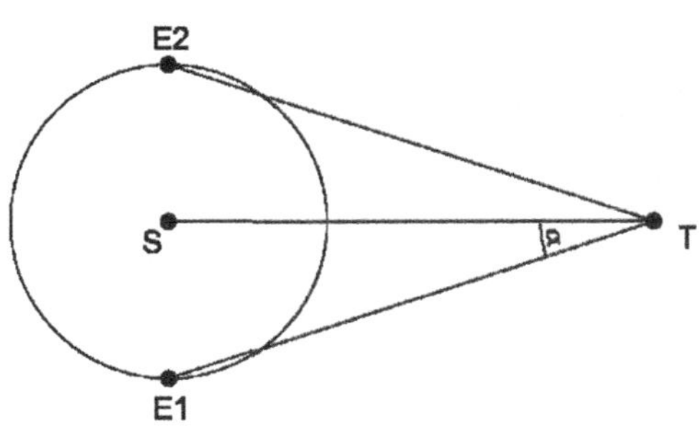

Fig. 12.28 Stellar parallax in a heliocentric model

model gives a solar system smaller than the one determined from Ptolemy's method in the *Planetary Hypotheses*. Using astronomical units—the Earth-Sun distance—in the geocentric model the mean distance to Saturn is 14, whereas the (correct) value obtained from Copernicus' method is 9,6. For the angle of parallax to be undetectable, the radius of the celestial sphere should be of at least a million Earth radii. Recall that the radius of the celestial sphere calculated by Ptolemy was ca. 20.000 Earth radii. A gap of such an unimaginable size between Saturn's orbit and the celestial sphere, devoid of any objects, was very difficult to accept given the ruling aesthetic and metaphysical principles of the time.

Finally, although Copernicus' managed to get rid of the equant and reinstall the principle of regular circular motion, the heliocentric model is in obvious conflict with the then prevalent physics of a fixed Earth. Copernicus argued that given its spherical shape, the natural motion of the Earth, shared by all terrestrial objects, is a circle. To the modern reader this may sound like a small step towards a concept of inertia, but taken at face value is only a qualitative speculation far from being able to allow a satisfactory explanation of terrestrial phenomena for astronomers of the day. Thus, the immediate reaction to Copernicus' model was rather cautious. Although its virtues were valued, it was usually read through instrumentalist glasses.

12.2.6 Tycho's Model

In the second volume of his *Introduction to New Astronomy*, published in 1588, Tycho Brahe presented a geocentric model that grasps the virtues of Copernicus' model over Ptolemy's.[21] The basic idea is simple: the Earth is the center of the universe, the Sun describes a circular orbit around the Earth, and the rest of the planets describe circular orbits around the Sun. In other words, the Sun moves along a deferent, and each planet moves along an epicycle centered in the Sun. The disparity between superior/exterior and inferior/interior and planets (maximum elongation angles, retrograde motion at opposition or conjunction) is dealt with in the Tychonic model by the fact that the radii of the orbits of Mercury and Venus around the Sun are smaller than the radius of the orbit of the Sun around the Earth, as it can be seen in Fig. 12.29.[22]

Tycho's model can be also put in geometric translation with Ptolemy's and Copernicus', as it can be seen in Fig. 12.30 (eccentricities, Ptolemy's equant and Copernican epicycles are ignored). In Tycho's model, the Earth E is fixed, the Sun S rotates around it along the dotted circle, and a generic exterior planet P rotates around S along the larger dashed circle. In the Copernican system, S is fixed, E rotates around S along the smaller dashed circle, and P rotates around S along the larger dashed circle. In Ptolemy's model, E is fixed, S rotates around E along the

[21] For a comprehensive treatment of Tycho's work, see (Dreyer, 2014).

[22] The figure is a simplification. The actual model includes eccentricities and equants.

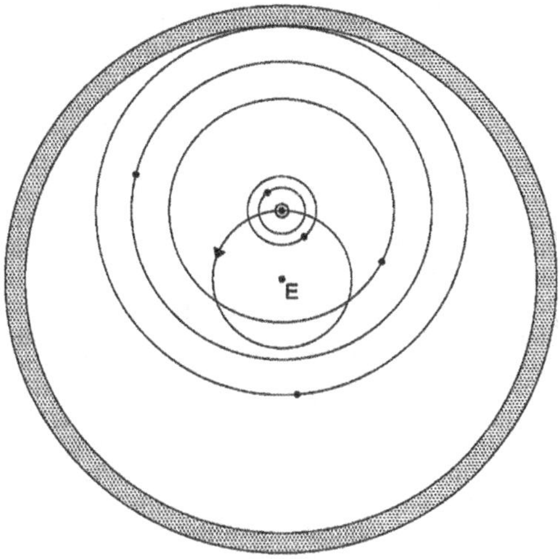

Fig. 12.29 The Tychonic model

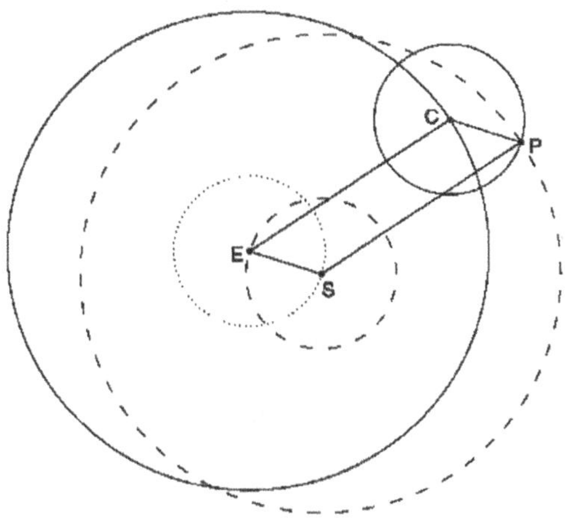

Fig. 12.30 The inter-translability between Ptolemy's, Copernicus and Tycho's model

dotted circle, *C* rotates along a deferent of radius *EC*, the larger solid circle, and *P* rotates along the epicycle centered in *C*. That the observed trajectory of *P* is the same in all three models is guaranteed by the fact that *ESPC* is always a parallelogram.

A quick inspection of Fig. 12.29 shows that Copernicus' trigonometric method to determine planetary distances can be run in the Tychonic model too. On the other hand, as in Ptolemy's model and unlike Copernicus', retrograde motion is explained in terms of the deferent-epicycle method. However, Tycho's model manages to explain why for inferior planets it occurs at conjunction, whereas for superior planets it occurs at opposition (cf. Fig. 12.11). Furthermore, recall that in Ptolemy's system the effect of the motion of the Earth around the Sun is taken care by the epicycle for superior planets, and by the deferent for inferior planets. In Tycho's model it is the deferent that plays that role for all planets, and since the Sun is in the center of the deferent of each planet, there is no need to fine-tune the model with respect to the (mean) Sun.[23]

The Tychonic model also avoids the two most important shortcomings of Copernicus'. Since it is geocentric, the problems for terrestrial physics related to an Earth in motion do not come up. Besides, no stellar parallax is expected, so Tycho could simply assume that the sphere of the fixed stars lies just after Saturn's orbit. As a result, the size of the universe determined in Tycho's model is smaller than the Ptolemaic universe.

After this outline of the Tychonic model, and in evaluative comparison with Copernicus' and Ptolemy's models, it is clear that for contemporary astronomers it was the best of both worlds. As Kuhn reports:

> The remarkable and historically significant feature of the Tychonic system is its adequacy as a compromise solution of the problems raised by the *De Revolutionibus*. Since the Earth is stationary and at the center, all the main arguments against Copernicus' proposal vanish. Scripture, the laws of motion, and the absence of stellar parallax, all are reconciled by Brahe's proposal, and this reconciliation is effected without sacrificing any of Copernicus' major mathematical harmonies. (Kuhn, 1995, 202)

12.2.7 Kepler's Vicarious Hypothesis

Tycho Brahe is a very important character in our story also for his meticulous and abundant observations of the behavior of celestial objects. He was able to obtain observational data for the positions of planets, at different times and in various kinematic configurations, with a margin of error of an angle of $2'$, whereas the margin of error the observations that Ptolemy used was of about $10'$. Johannes Kepler (1571–1630) worked as Brahe's assistant for a short period, and after Brahe's death in 1601, he inherited the invaluable collection of observational data. Armed with it, he introduced subtle but crucial amendments in Copernicus' model. However, the result was a demonstration that at best, the Copernican model would lead to an observational error of $8'$, and given the geometric inter-translation between the models, the same holds for Ptolemy's and Brahe's models.

In his first major work, the *Cosmographic Secret*, published in 1597, Kepler (1981) defended Copernicus' model. This book is mostly known for the thesis that the order and distances of the planets can be represented using the five platonic regular solids, following a Pythagorean aesthetic-metaphysical spirit. This is normally presented as an example of Kepler's extravagant commitment to ancient and conservative views. However, although he never gave up this Pythagorean spirit, in that same work Kepler took a stance towards the Copernican model that is one of the catalysts for the scientific revolution. He noticed that, comparing planets from closest to farthest to the Sun, the orbital periods increase at a rate greater than the

[23] For the precession of equinoxes, Tycho's system must go back to an explanation like the one illustrated in Fig. 12.13.

rate at which the distance increases. This means that the farther the planet, the slower its speed. Kepler concluded that a force emanated from the Sun, which gets weaker with distance, governs all planetary orbits. That is, he arrived to the view that a dynamical mechanism underlies the Copernican model. Before Kepler, nobody had understood an astronomical model in those terms. The Copernican model was either interpreted in instrumentalist terms, or in realist teleological terms. That is, either the model was taken only as a mathematic configuration of the observed phenomena; or as a true description in which the circular motions that determine the orbits are an expression of natural motion, not the outcome of a force. Furthermore, Kepler realized that his dynamical-mechanical thesis implied that the speed of a planet along its orbit is variable, for its distance to the Sun is variable. Thus, he openly rejected the dogma of circular uniform motion for celestial objects.

Kepler (1992) developed this view in his *New Astronomy*, published in 1609. He arranged Copernicus' heliocentric model according to the following principles. First, all planetary orbits are given by a single circle—he fully abandoned the deferent-epicycle method. Second, all orbital planes intersect in the real Sun. Third, there is an equant for each orbit, collinear with the Sun and the orbit's center, and such that the center of the orbit lies between the Sun and the equant—from a heliocentric point of view, the equant corresponds to the mean Sun, of course.[24] The last two principles capture Kepler's mechanical-dynamical stance, for they clearly suggest that the Sun is the driving force determining the orbits, and that the force decreases with distance. Given the collinear arrangement of the equant, orbital center, and the Sun, the velocity of the planet is maximum and minimum at perihelion and aphelion, respectively. As it is clear, this is much simpler than Copernicus' original plan. Epicycles are abandoned, and the latitude of a planet as observed from the Earth is simply a function of the inclination between the planet's orbital plane and the ecliptic.

Given this amended heliocentric configuration, Kepler set himself to the task of determining where exactly the orbit center lies between the equant and the Sun, starting with the most challenging case, Mars. He approached this task using both Mars' longitude and latitude. In Fig. 12.31, *A* and *P* are Mars' aphelion and perihelion, respectively, *S* is the Sun, *O* the orbit center and *Q* is the equant. The angle $\overline{\alpha}$, which increases uniformly and can be obtained from the orbital period, is the mean anomaly, and α the true anomaly. The latter angle can be expressed as a function of the former by the formula $\alpha = \overline{\alpha} - (f+g)\,sin\,\overline{\alpha} + \frac{1}{2}g\,(f+g)\,sin\,2\overline{\alpha}$, which matches Brahe's data if $\frac{f}{g} \approx 0,64$.[25]

Figure 12.32 displays one of the methods by which Kepler determined the inclination of Mars' orbital plane with respect to the ecliptic. *E* is the Earth, *S*

---

[24] Copernicus' made extensive use of the mean Sun, as we saw above, but never as an equant.

[25] This formula for α is obtained using mathematical techniques that were not available to Kepler. The calculations he had to make are tortuous and tedious. Furthermore, the data he had to obtain $\overline{\alpha}$ the angle were observations for M in kinematic configurations with respect to the *mean* Sun, rather than to the real Sun, so Kepler had to extrapolate them. (see Linton 2004, 179).

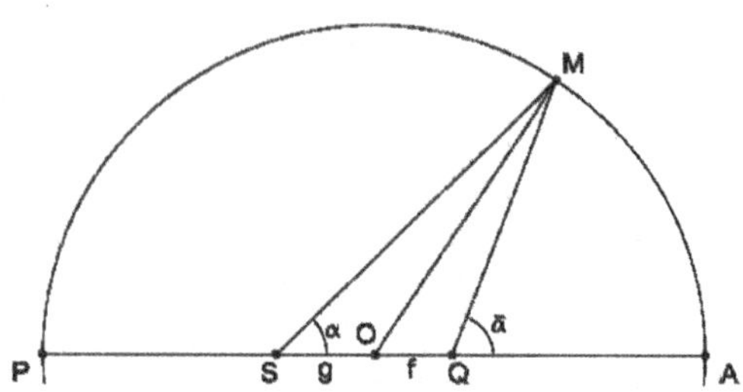

Fig. 12.31 Kepler's method to calculate the ratio from Mar's longitude

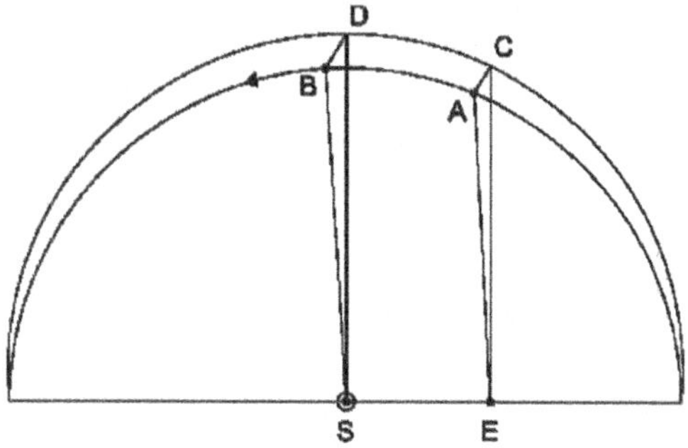

Fig. 12.32 Kepler's method to calculate the inclination between the ecliptic and Mars' orbital plane

is the Sun, and C and D lie on the ecliptic plane. Using again Brahe's data, Kepler extrapolated the position of Mars at A, that is, as seen from the Earth in a configuration such that E lies on the line of the nodes of Mars orbit (the points in which Mars orbital plane intersects the ecliptic plane), so that CAE is a right angle. In this configuration, the angle CEA is equal to DSB, where B is one of the *limits* in Mars' orbit—the points in which Mars' ecliptic latitude is maximum and minimum. It is clear that $CEA = DSB$ is the angle of inclination of Mars orbit, for which Kepler got a value of $1°50'$.[26]

With this result, Kepler was able to determine the ratio $\frac{f}{g}$ once again, for its value has an impact on the predicted maximum latitude of a planet as seen from the Earth. For Mars, the latitude as seen from the Earth is maximum when the planet is in opposition, as in Fig. 12.33. The angle β is determined observationally, α is known to be $1°50'$, the center of Mars orbit, the equant Q and the distance SE are known, so it is clear that $\frac{f}{g}$ can be obtained. The problem was that Kepler found a value of 1 this time. But with this value, the model fails for planetary longitude. When $\overline{\alpha} = \pm 45°$ (cf. Fig. 12.31), the predicted and observed positions of Mars differ by $8'$—before Brahe's data, this error would have been negligible, but with an accuracy of $2'$, it was unacceptable.

[26] A remarkably accurate result, for the true value of the inclination of Mars' orbit is $1°51'$. Kepler determined the nodes of the orbit of Mars from Brahe's data. He found that the red planet returns to the same node every 687 days, which is exactly its sidereal period, and that the Sun lies on the line joining both nodes. This is evidence for Kepler's assumption that the Sun lies on the orbital plane of each planet.

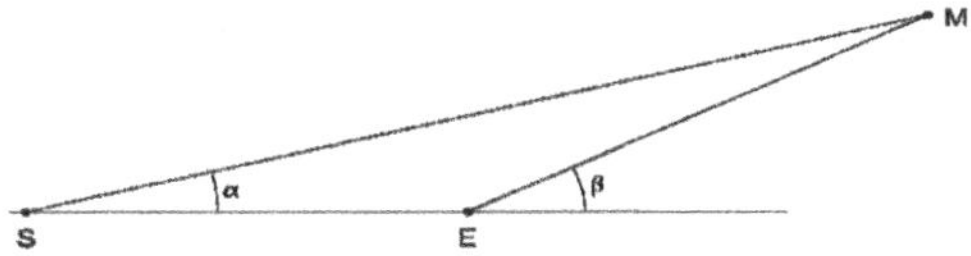

Fig. 12.33 Kepler's method to calculate the ratio from Mar's latitude

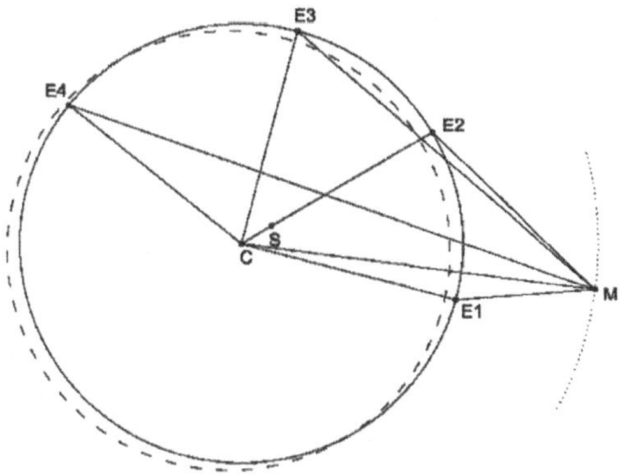

Fig. 12.34 Kepler's method to determine the Earth's orbit

Kepler considered the possibility that the discrepancy could be rooted in an imprecise knowledge of the Earth circular orbit—as it is clear from Fig. 12.33, the calculation of $\frac{f}{g}$ crucially depends on an accurate knowledge of the position of *E*. Kepler devised an ingenious method to determine the circle of the Earth's orbit, shown in Fig. 12.34. He used observations of Mars every 687 days, so that it is the same position with respect to the Sun *S* and the distant stars—Kepler chose moments when Mars crosses the ecliptic, to avoid complications concerning latitude. Since the orbital period of the Earth is different than Mars', the former will be located at different positions at each observation. With three positions, a unique orbital circle can be determined. With a fourth position it can be checked if it lies in the same circle. The center *C* of the found orbit can thus be established, and also its distance to the equant. The solid circle is Kepler's improved terrestrial orbit, whereas the dashed circle is Copernicus'. The result confirmed Kepler's mechanical-dynamical interpretation of the Copernican model: the Sun, the orbital center and the mean Sun are collinear, with the orbital center between the Sun and the equant, so that the Earth's speed is fastest at perihelion and slowest at aphelion. However, for $\frac{f}{g}$ in the terrestrial orbit he also obtained a value 1, so he concluded that the same should hold for Mars, and that the observational error of $8'$ could not be corrected.

Kepler was fully aware of the inter-translability of the models, so the results of his amendments to Copernicus' model—that made it as empirically adequate as possible, but that also showed that it is false—can be introduced, *mutatis mutandis*, in Ptolemy's and Brahe's models as well. Therefore, a choice between the three models would have to be done in terms of extra-empirical considerations. As Barbour states, "by giving the demonstration in all three systems, Kepler highlighted their equivalence at the kinematic level and emphasized that the choice between the rival systems must be assessed primarily on physical and dynamical arguments" (Barbour, 2001, 294).

Kepler demonstrated thus that the three models are equally false. However, he continued using his amended version of the Copernican system as a *vicarious hypothesis*, as he called it, i.e., as a false surrogate model that would lead him to correct principles. The strategy certainly paid off. He derived the second and first laws working on the vicarious hypothesis, which resulted in a new heliocentric model with elliptical orbits.[27] Kepler's vicarious hypothesis is then the last episode in circular motion astronomy.

Kepler took issue with the choice between the three models. He first considered Ptolemy. We know that in Ptolemy's system, for superior planets the role of the epicycle is to take care of the effect of the motion of the Earth around the (mean) Sun. After his correction of the orbit of the Earth, the translation to the Ptolemaic model should be such that the same geometric arrangement must be mirrored by the Ptolemaic epicycle. That is, in the amended Ptolemaic model for superior planets, there should now be an equant for the epicycle, collinear with a *punctum affixionis* (corresponding to the real Sun), and the epicyclic center lies exactly midway between these points. This means that in Ptolemy's system the models of the three superior planets and the model of the Sun must all be constructed on the basis of the terrestrial motion in Kepler's amended version of the Copernican model. Thus, Kepler concluded that "when a comparison of hypotheses has been made, and it has appeared that four theories of the sun [...] can be generated from a single theory of the earth, like many images from one substantial face, the sun itself, the clearest of truth, will melt all this Ptolemaic apparatus like butter, and will disperse the followers of Ptolemy, some to Copernicus' camp, and some to Brahe's" (Kepler, 1992, 337).

This argument does not hold in the Tychonic model. However, in Kepler's vicarious model, leaving the Earth aside for a moment, the apsidal lines of the five ancient planets all intersect in a single point: the real Sun. Furthermore, for the five planets it also holds that the orbital speed is minimum at aphelion and maximum at perihelion. In the translation to Tycho's model, both these features must hold too. However, if the Sun moves around the Earth, the speed of the former in its orbit is fastest at apogee and slowest at perigee. Kepler realizes that a much simpler and unified arrangement obtains in the vicarious model: the apsidal line of the terrestrial orbit meets all the other planetary apsidal lines in the real Sun too, and its orbital speed is also maximum and minimum at perihelion and aphelion, respectively.

This general pattern in the vicarious is of course quite coherent with Kepler's mechanical picture of the solar system, and his discovery of the second law provides

[27] Kepler's first law states that the planets describe elliptical orbits, with the Sun in one of the *foci*. His second law tells us that a straight line from the planet to the Sun sweeps out equal areas in equal times. The first two laws were formulated in the *New Astronomy*. The third law states that the ratio D^3/T^2 has the same value for all planets, where D is the planet's average distance to the Sun, and T is its orbital period. Kepler obtained his third law in *The Harmony of the Universe*, published in 1619.

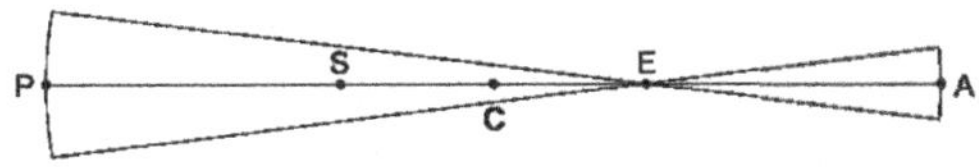

Fig. 12.35 At the apsides, the velocity of the planet is inversely proportional to the distance to the Sun

further support for it, so let us briefly see how the postulation of a solar force helped Kepler to find it. He first derived that at perihelion and aphelion, the linear velocity of the planet is inversely proportional to the distance to the Sun. A reconstruction of Kepler's reasoning, in modern terms, is offered by Barbour (2001, 303). In Fig. 12.35, A is the aphelion, P the perihelion, C the orbit center, S is the Sun and Q is the equant, and $SC = CQ$. By the definition of the equant, in a time t the planet sweeps an angle wt, where w is the (constant) angular speed. Thus, the linear velocity of the planet at A is $v_A = w \bullet EA$, and at P it is $v_P = w \bullet EP$. Now, since $EA = SP$ and $EP = SA$, $(v_A/v_P) = SP/SA$, which in turn implies that both at A and P the linear velocity of the planet is inversely proportional to the distance of the planet to the Sun. The inverse proportionality holds strictly only at perihelion at aphelion, but in agreement with his solar force hypothesis, Kepler postulated it as correct for the whole orbit, so the orbit with an equant became a good approximation.

From a modern perspective, Kepler's "inverse distance law" can be expressed as $\frac{d\theta}{dt} \propto \frac{1}{r}$, where θ is the angle covered by the planet (as determined from the center of the orbit) as a function of time, and r is the distance to the Sun. Thus, to predict the orbit of a planet according to this law, an integration problem must be solved, for $t \propto \int r d\theta$. Kepler could not use calculus, of course, so he devised the following method. He divided the orbital circle in 360 equal arcs of length $\frac{\pi R}{180}$, where R is the orbital radius. The distance to the eccentric Sun in each arc is variable, but taking the average distance as the constant distance r of each arc, then the velocity of the planet along each arc is inversely proportional to r. This method was rather tedious, for in order to calculate the position of a planet at a certain time t, it was necessary to calculate and add the traversed distances in all the previous arcs up to t.

To simplify the task, Kepler reasoned rather mysteriously that "since I knew that the points of the eccentric are infinite, and their distances are infinite, it struck me that all these distances are contained in the plane of the eccentric" (1992, 418). He applied the same principle to arcs of the eccentric circle, that is, the area enclosed by an arc and straight lines from its endpoints to the eccentric Sun 'contains' all the infinitely many different distances from the points in the arc to the Sun. This slapdash reasoning gave him the clue that by calculating that area, a much easier task, he could obtain the velocity and time of the planet along the corresponding arc, according to the inverse distance law. That is, the inverse distance law could be formulated as an *area law* that the velocity of the planet along the arc is inversely proportional (and thus the time directly proportional) to the corresponding area, *if the area is proportional to the distance to the Sun.* This condition holds for triangles whose base is an infinitesimal arc and whose height is the distance to the Sun, but the distance to the Sun is perpendicular to the arc-base only at perihelion and aphelion.

However, with small eccentricity, the area law is a good approximation of the inverse distance law. That is, Kepler obtained his revolutionary second law that a straight line from the planet to the Sun sweeps out equal areas in equal times using the vicarious model of circular orbits, and as an approximation of the inverse distance law, which he formulated on the basis of his dynamical assumption that the Sun governs planetary orbits by means of a force.

12.3 Understanding and Explanation in Circular-Motion Astronomy

After our review of the rise and fall of circular-motion astronomy from Eudoxus to Kepler, we can use this historical episode as a case-study to evaluate the pragmatic account of understanding proposed by de Regt. The role of scientific intelligibility (UT), the variability of standards of intelligibility, the relevance of metaphysics for scientific understanding, and the possibility of getting understanding through false models, are all features that can be recognized in the development of astronomy from Eudoxus to Kepler.

12.3.1 Circular (Uniform) Motion and Scientific Intelligibility

The ancient Greek view that the motion of celestial bodies is circular and uniform was an aesthetic-metaphysical principle that operated as the theoretical basis for the development of astronomy up to the times of Kepler. As we saw, an important point to underscore about this principle is that it is essentially teleological. The idea that nature behaves the way it does because there is a goal given by the essence of all natural things was developed by Aristotle, but it can certainly found in earlier thinkers like Plato and some pre-Socratic philosophers. The aesthetic perfection that the ancient Greeks attributed to circles was naturally associated with the stars and planets, given the degree of uniformity observed in their motion. Thus, the stars and the planets were considered to move uniformly in circles not due to mechanical factors, but simply because that is their *télos*. Barbour characterizes Greek astronomy as *motionic* (as opposed to dynamic):

> the fundamental law of ancient Greek astronomy stated that all celestial bodies move in perfect circles at a uniform (perfectly constant) speed. In accordance with this law, the motion as such *is entirely independent of all the other bodies and matter in the universe*. (Barbour, 2001, 52; my emphasis)

This law of Greek astronomy, if it is to provide any scientific explanation of astronomical phenomena, and not only a qualitative-metaphysical account, must be made intelligible, in the sense of de Regt's UT. That is, from the law of circular uniform motion, models that obey the mentioned law must be built to represent the target phenomena—the motion of celestial objects. Simplicius' anecdote that Plato assigned the task of building a geometrical model that obeys the principle of circular uniform celestial motion, can be certainly understood from this point of view. For the ancient Greek thinkers, something very substantial would be gained in the understanding of nature if the metaphysical principle of circular uniform model got incarnated in what we nowadays call a scientific model.

Eudoxus' method of concentric spheres was the first toolkit of intelligibility with the use of which a scientific model could be built out of the law of ancient astronomy. The method works as the mediator between the theoretical principle of circular uniform motion and the observed phenomena. It is also clear that the task of formulating models out of this basic law was not algorithmic. The idea that the combined motion of concentric spheres can reproduce regular observed motions which are not circular is due to Eudoxus' ingenuity, and to his knowledge of geometry. Thus, making the principle of circular uniform motion scientifically intelligible required skills on Eudoxus' part, which in turn depended on subjective and contextual factors. The success or failure of the model, though, depended on its empirical adequacy, and it soon became clear that the system of concentric spheres is not able to represent the phenomena with acceptable accuracy.

Apollonius introduced a second toolkit of scientific intelligibility for the same law of circular motion. Now, in order to build accurate models following his basic idea of deferents and epicycles, observational and mathematical problems needed to be solved, and their solution also required ingenuity, and also a technical refinement of the basic toolkit was needed. The development of astronomy from Hipparchus to Copernicus can actually be understood as an improvement of the toolkit for understanding provided by the technique of deferents and epicycles, which quite clearly illustrates the virtuously circular connection between understanding and explanations that de Regt defends. With ingenuity and geometric knowledge (once again the subjective-pragmatic factors), Hipparchus was able to determine the mathematical and observational parameters needed to build an accurate solar model from Apollonius' insight.[28] The empirical success of this model motivated the quest for a comprehensive model for all celestial objects using deferents and epicycles. That is, the empirical success of the scientific explanation built using a toolkit of understanding started the process its canonization as a toolkit for UT and UP. A comprehensive model for the planets required further refinement of the tool, and this task was accomplished by Ptolemy. He developed the mathematical structure introduced by Hipparchus, significantly enhancing the scope of empirical adequacy of the model. Given the empirical success of Ptolemy's model, the toolkit of deferents and epicycles gets fully canonized as a standard of intelligibility.

The development of medieval astronomy consists mostly in amendments of parameters in Ptolemy's model in order to cope with newly found phenomena—

[28] An interesting contextual difference is that the deferent-epicycle model is intelligible to us in a simpler and easier way because we have modern trigonometry, to which Hipparchus did not have access. His calibration of the solar model was much more tortuous than for us.

actually a mixture of real and false effects like trepidation and apogee precession—but always using the same geometric-kinematic toolkit of intelligibility. As we saw, in the eighth century, Al-Battani amended some empirical and geometric parameters in the model in order to make it fit better with observations, and Thabit Ibn Qurra introduced a theory of trepidations—in both cases the toolkit of intelligibility is still the method of deferents and epicycles, of course. We also saw that Ibn Al-Shatir in the fourteenth century introduced a model in which he showed that the equant can be dispensed with using a method of epicycles on epicycles. Besides, he also amended Ptolemy's model in order to cope with solar apogee precession. That is, Al-Shatir, further improved the empirical adequacy of the deferent-epicycles method, and he brought it back to coherency with circular-*uniform* motion.

Even Copernicus' heliocentric model can be understood in this way. Although the Polish astronomer explained away retrograde motion, the whole heliocentric model is still built on the method of deferents and epicycles—recall that the observed variable angular velocities of the planets are accounted for by epicycles, and that complex arrangement of epicycles on epicycles manages to avoid the introduction of the equant, retaining coherence with the principle of circular uniform motion. Something similar holds for Brahe's model. The Danish astronomer, exploding the inter-translatability between the Ptolemaic and the Copernican models, formulated yet another astronomical system—that captured the best of both worlds—using the deferent-epicycle method as his toolkit for model construction.

Summarizing, Eudoxus' method of concentric spheres was the invention of a toolkit of intelligibility, insofar as it made possible the construction of a model out of the law of circular uniform motion. Apollonius introduced a second set of tools of intelligibility for the same law, which led to the construction of several models by subsequent astronomers. The increasing empirical success of all the deferent-epicycles models from Hipparchus to Brahe is a clear example of the mutual feedback between understanding and explanation. The toolkit of intelligibility given by the method of deferents and epicycles made possible the construction of several explanatory models, the increasing empirical success of which led to its canonization as a source of understanding. Conversely, the refinement of the toolkit motivated by its fruitfulness in model construction led to the development of better explanatory models.

The relevance of Kepler's work for toolkits of understanding in the historical development of astronomy can be understood in a twofold way. First, he introduced yet another tool of understanding. Although the vicarious hypothesis resembles the previous models, it is certainly the outcome of a novel way to do astronomy. Despite the retention of circular motion, deferents and epicycles are fully abandoned. Besides, the choice of the real Sun as the anchor of the whole model was a revolutionary maneuver. Kepler's new method to build the vicarious model led to important advances in scientific intelligibility: the theory of latitudes could be greatly simplified, it allowed more precise determinations of planetary orbits (recall the improvement in the description of the terrestrial orbit), it allowed several methods to determine the inclination between planetary orbital planes and the ecliptic, and it allowed different ways to estimate the way in which the Sun, the center of the orbit and the equant were arranged. In short, Kepler's new astronomical method resulted in a better model, in the sense that the model scored higher in terms of intelligibility. The novelty of Kepler's approach can also be seen in that even though it retained circular motion, the underlying view of nature that it made scientifically intelligible was not the ancient Greek law, but his revolutionary conception of a mechanical universe. That is, Kepler, based on a sharp and insightful analysis of Copernicus' model, came up with a new dynamical-mechanical principle for the understanding of natural phenomena (UP), and he developed a new approach to make that principle scientifically intelligible through the vicarious model (UT).

On the other hand, despite all the improvement in intelligibility that the vicarious model allowed, he soon noticed that the model was false. Now, since all the models were geometrically inter-translatable, the falsity of the vicarious hypothesis was also a demonstration that all the previous models are equally false. Thus, Kepler's work signified the ruin of the metaphysical bases and geometric methods of ancient astronomy. Kepler showed that both the law of Greek motion and the toolkit of deferents and epicycles were dead ends. A most interesting aspect is that despite this, his new metaphysical-theoretical mechanical principle for the understanding of astronomical phenomena was not falsified by the empirical failure of the vicarious model. On the contrary, further elaboration on the false vicarious model led him to the formulation of the area law, a crucial landmark in the history of science. That is, although his first attempt to make the mechanical principle of understanding the phenomena scientifically intelligible led to a false model, the false model nevertheless led him to spectacular success, a success that in the long run led to the canonization of Kepler's way to do astronomy as a toolkit for scientific understanding.

12.3.2 Differing Standards of Scientific Understanding

We have seen how the deferent-epicycle method raised and fell as a canonical toolkit for understanding in model construction. Our case-study also illustrates another aspect that is central in de Regt's proposal. Recall that the criterion for intelligibility of theories states that a theory is intelligible to scientists *in a context*. That is, there are further pragmatic criteria, apart from usefulness for the construction of empirically adequate models, which determine the evaluation of a theory or model as a source of scientific understanding. Given the interplay between UT and UP, we saw that the canonization of a toolkit for model construction also leads to its canonization as a tool for understanding *the phenomena*. In our case-study, the evaluation of astronomical models in terms of the UP they offered clearly illustrates the relevance of pragmatic factors. Although all the models after Eudoxus and before Kepler were crafted using the toolkit of deferents and epicycles, the resulting models conveyed different forms of intelligibility *of the phenomena*, and there was significant divergence in the evaluation of each of those forms of intelligibility.

Although Ptolemy's model was universally accepted as the most accurate representation of the observed phenomena, there were important critical voices that considered the model unsatisfactory. The main targets of criticism were the introduction of several centers of circular motion, and the use of the equant. These features involved friction with principles of Aristotelian physics, and even with the aesthetic-metaphysical foundations of the Greek law of circular uniform motion. The teleological nature of circular motion of celestial bodies made essential reference to the center *of the universe*, but Ptolemy's model included several different centers for circular motion: what was special about those empty points that defined natural circular motion around them? On the other hand, the equant, as we saw, was in open conflict with the uniformity of motion. Based on these two problems, prominent medieval astronomers and philosophers did not accept Ptolemy's model as a source of intelligibility *of the phenomena*, at most, they took it as an instrument for prediction (see Goldstein, 1980; Sabra, 1984; Nutkiewicz, 1978; Saliba, 1991).

In the eleventh century, Ibn al-Haytham, in his *Doubts Concerning Ptolemy*, stated that Ptolemy's model was plainly false, given its inclusion of the equant and its resulting friction with the Greek law of motion and with Aristotelian physics, and in his *Models of the Motions of Each of the Seven Planets* he attempted a model that got rid of the equant (see Rashed, 2007). In the same spirit, in the twelfth century Averroes expressed the following evaluation of the Ptolemaic system:

> For to assert the existence of an eccentric sphere or an epicyclic sphere is contrary to nature. As for the epicyclic sphere, this is not at all possible; for a body that moves in a circle must move about the center of the universe, not aside from it ... It is similarly the case with the eccentric sphere proposed by Ptolemy [...]. For nothing of the [true] science of astronomy exists in our time, the astronomy of our time being only in agreement with calculations and not with what exists. (Quoted in Çimen, 2019, 141)

Also during the twelfth century, Al-Bitruji, motivated by his dissatisfaction with Ptolemy's model given its friction with circular uniform motion, formulated a model of concentric spheres—similar to Eudoxus' (see Çimen, 2019; Goldstein, 1971). This model, though, was not able to match the empirical accuracy of Ptolemy's system. As we saw above, even as late as the sixteenth century, just before the introduction of Copernicus' model, Girolamo Fracastoro (in his *Homocentrica*, published in 1538) and Giovanni Batista Amici (in his *On the Motion of Celestial Bodies*, published in 1536) attempted to revive the system of concentric spheres, motivated by a strict observance of Aristotelian physics (see Dreyer, 1953, 296–304). Finally, al-Shatir in the fourteenth century, also based on his dissatisfaction with the friction between Ptolemy's model and the principle of circular uniform motion, invented a method to get rid of the equant by a method of epicycles on epicycles (see Kennedy & Roberts, 1959).

All of these criticisms to Ptolemy that motivated attempts to construct alternative models, quite clearly show that during medieval times there were diverging views regarding the value attributed to Ptolemy's model (and even to the method of deferents and epicycles itself, in the case of Al-Bitruji, Averroes, Fracastoro and Amici) as a source of understanding of celestial phenomena. Actually, the *Maragha*

School—an Arabic school of astronomy founded around the Maragha observatory, of which al-Shatir was a member—was a whole scientific sub-community that challenged Ptolemy's system as a canon of understanding, based on the objections concerning its friction with uniform circular motion (see Saliba, 1991). This clearly confirms that other factors apart from empirical adequacy—such as the commitment to uniform motion, or to strict Aristotelian physics—were crucial, *in some contexts*, for whether a certain theory or model is considered as conveying understanding of the phenomena or not. Actually, that these criticisms appeared in the heyday of scholasticism, whereas during Ptolemy's own time and the subsequent centuries there are no records of similar complaints, helps to explain that the criticisms respond to the contextual factor of adherence to Aristotelian views:

> The great conflict between Ptolemaic astronomy and Aristotelian cosmology—which continued right up until the sixteenth century—did not exist when originally Ptolemy wrote the *Almagest*. Ptolemy wrote his masterpiece 500 years after Aristotle, and the whole issue of the physical interpretation of his theory clearly was of secondary importance to him, and presumably to his contemporaries. However, to those rediscovering these works, the time between Aristotle and Ptolemy was not significant—they were both part of the 'ancient learning' that was being resurrected—and taken together they appeared full of contradictions. (2004, 100)

A similar pattern can be recognized in the rivalry between Copernicus and Ptolemy, and later between Copernicus and Brahe. A choice between the Copernican and Ptolemaic systems by, say, 1560, certainly depended on the different evaluations that different astronomers made of the conceptual advantages of one model over the other. That is, since the models were predictively equivalent (at least until the introduction of the telescope), which of the models ranked higher in terms of understanding of the phenomena clearly depended on the value attributed to their clusters of virtues and problems. As it was natural, for most astronomers of the time the price of an Earth in motion was too high—for there were no conceptual basis on which to make sense of it, and no serious contenders for Aristotelian terrestrial physics. The conceptual virtues of the Copernican model, though, were too important to be ignored, so the general attitude was to use the heliocentric model for calculations, but still adhering to Ptolemy as the true model. An illustrative example of this stance is given by the astronomer Thomas Blandeville, who in 1594 wrote "Copernicus [...] affirmeth that the earth turneth about and that the Sun standeth still in the midst of the heavens, *by help of which false assumption* he hath made truer demonstrations of the motions and revolutions of the celestial spheres, than ever were made before" (quoted in Kuhn, 1995, 186; my emphasis).

However, there were astronomers that endorsed the heliocentric model. Soon after the publication of *On the Revolutions*, in the trade between the natural explanation of retrograde motion and the trigonometric method to determine planetary distances in Copernicus' model, on the one hand, and the problems associated to an Earth in motion, on the other, some prominent astronomers ventured to value the former features highly enough as to tolerate the latter. Georg Joachim Rheticus, an Austrian pupil of Copernicus who became a professor in Leipzig and in Wittenberg, was an early advocate of a literal interpretation of the heliocentric system. Rheticus was a driving force behind the publication of *On the Revolutions*, and in 1540 he published an introduction to Copernicus' model entitled *First Account of Copernicus Book on the Revolutions*, in which he defended the model both in mathematical and physical terms (see Danielson, 2006). Michael Mästlin, a German astronomer that worked in Tubingen, was another important early Copernican. He taught and defended the model in Tubingen, and it was under the mentoring of Mästlin that Kepler became a Copernican too (see Methuen, 1996). In England, in 1576 Thomas Digges wrote a review of the heliocentric model in which he defended it as a true model.

This of course illustrates that there is synchronic variation of criteria of understanding. The majority of astronomers of the time, for good reasons, included a geocentrism constraint in the assumed standard of intelligibility of celestial phenomena. The dominance of the commitment to static Earth physics can also be seen in that after Brahe's work most of the supporters of Ptolemy converted to the Tychonic hybrid model—which, as we saw, was able to grasp most of the conceptual appeal of Copernicus' model, but in a geocentric setup. For scientists like Rheticus, Mästlin, Digges, Kepler and Galileo, though, the conceptual appeal of the Copernican model ranked higher than coherence with then accepted terrestrial physics. These astronomers considered that true understanding of the phenomena was conveyed by the Copernican model, even to the extent of tolerating the absence of a dynamical account of terrestrial phenomena compatible with a moving Earth. Actually, Kepler and Galileo were crucial in taking important steps towards a new physics of an Earth in motion. This illustrates that the context-dependency of understanding, with the resulting synchronic variation of standards of intelligibility of the phenomena, can be an engine for scientific progress.

12.3.3 Understanding and Metaphysics

An interesting subtlety in de Regt's account of scientific understanding is given by its connection with metaphysics. He introduces the concept of *metaphysical intelligibility*: "a theory is metaphysically intelligible if it harmonizes with extant, or preferred, metaphysics" (2017, 160). This notion differs from the scientific intelligibility involved in UT. However, he argues that metaphysical and scientific intelligibility interact, for the former can provide tools to render a theory scientifically intelligible in the sense of UT.

Our case-study also provides support for this view. The ancient Greek law of circular uniform motion is actually rooted in metaphysics. As we saw, Plato argued that this type of motion essentially corresponds to celestial objects because the demiurge decided to introduce time in nature, and circular uniform motion is appropriate for that task. Furthermore, Aristotle, based on aesthetic properties of circular motion and teleological considerations, concluded that this must be the form of natural motion corresponding to the ethereal element. That is, the two greatest ancient Greek philosophers attributed circular uniform motion to celestial bodies based on metaphysical views, and these views led to the construction of astronomical models. A clear indication of this is given by Simplicius comment that Eudoxus' model was developed as a response to Plato's challenge of representing celestial motion by a geometric system coherent with the described metaphysical views. Thus, the metaphysics of nature endorsed by Plato and Aristotle prompted the invention of toolkits of scientific intelligibility: the method of concentric spheres and the method of deferents and epicycles.

An even clearer example of the interplay between metaphysical and scientific intelligibility is given by Kepler's vicarious model. Recall that already in the *Secret of the Universe*, published in 1597, Kepler was already convinced that the Sun was responsible for the choreography of the planets by the action of a force. That is, he abandoned the teleological picture associated to the doctrine of circular uniform motion, and embraced a proto-mechanical metaphysical picture of the universe. In 1609, in the *New Astronomy*, Kepler introduced the vicarious model by amending Copernicus' model with his revolutionary metaphysics as a heuristic principle. As we saw, he fully dismissed the method of deferents and epicicyles, and assigned a circular orbit to each planet. All orbital planes intersect in the Sun, and the center of each orbit lies between the Sun and the equant, which implied that the planet's speed is fastest at perihelion and slowest at aphelion—in coherence with the view that the strength of the solar force diminishes with distance. We also saw that this arrangement was a huge improvement in model construction. In the vicarious model, the latitude of the planets is simply a function of the inclination of the orbital planes with respect to the ecliptic, whereas in the previous models the latitudinal theory was rather contrived and inexact. Furthermore, Kepler's model allowed geometric methods to determine such inclination, to determine the distances between the Sun, the orbit center, and the equant, and to determine (and correct) the circle of the Earth's orbit. That is, Kepler's revolutionary conviction to a principle of metaphysical intelligibility—a mechanical universe—led to an improvement in scientific intelligibility in astronomical modelling.

Now, the failure of Kepler's vicarious model, which as we saw was built on the basis of a proto-mechanical metaphysics as heuristic principle, led in turn to further scientific progress in theory and model construction. The unavoidable empirical inadequacy of $8'$ in the vicarious model—and consequently in all the extant models of the time—signified the end of the principles of ancient astronomy. Kepler fully abandoned circular motion, but he remained fully committed to his proto-mechanic metaphysics. Exploiting the fact the orbital motion is fastest at perihelion and slowest and aphelion, in connection with the thesis of a Solar force proportional to distance that he took as the foundation of such fact, Kepler obtained his second law (and also the first one, see Barbour, 2001, Ch. 6; Torretti, 2007, Ch. 4).

The specific details of Kepler's proto-mechanical picture were wrong, of course. The nature of the solar force he conceived was analogous to fan blades: the Sun emits a force similar to light rays, and given a solar rotation that Kepler postulated, the rays of force drag the planets along their circular orbits. Besides, Kepler conceived the force as inversely proportional to the distance rather than to the squared distance, and as producing a velocity rather than an acceleration. His efforts to find a mathematical expression for the solar force he envisioned were thus fruitless. However, Kepler's proto-mechanical metaphysical picture was crucial in the construction of the vicarious model, and in the formulation of his first two laws. Furthermore, the proto-mechanical picture that Kepler introduced had a crucial influence on Descartes' systematic exposition of the mechanical philosophy—which in turn had a crucial relevance in the constitution of modern physics—and on Newton's formulation of the universal law of gravitation.

Despite its success as an engine of scientific understanding, Kepler's proto-mechanical metaphysics was not immediately accepted. A clear example of this is given by Ismael Boulliau's *Astronomia Philolaica*, one of the most important treatises in astronomy between Kepler and Newton, published in 1645. There Boulliau openly acknowledged that Kepler's laws allow a geometrical model that is empirically superior to all the preceding proposals. However, he rejected Kepler's underlying physics, and he introduced a curious model in which the elliptical orbits result from combinations of circular motions. Furthermore, Boulliau proposed that the ultimate foundation of planetary motion was not an external force, but an internal principle given by the essence of celestial objects, pretty much in the spirit of ancient astronomy and Greek teleological metaphysics (for an analysis of Boulliau's work, see Wilson, 1970). Although Descartes' (1979) vortex theory in *The World*, published in 1664, became an influential attempt of a theory of the solar system based on mechanistic principles, the adoption of mechanistic metaphysics in scientific practice, and the full abandonment of teleology was gradual, and culminated with Newton's work.

All of this offers further support for de Regt's account of understanding. The diverging attitudes of astronomers of the time towards the mechanistic framework as a source of metaphysical intelligibility illustrates that this form of intelligibility is also pragmatic and context dependent, just like scientific intelligibility. Besides, the rise and fall of teleological and circular motion metaphysics, and its replacement with the mechanical picture, illustrates the interplay between metaphysical and scientific intelligibility. The success of a certain canon of metaphysical intelligibility as a source of scientific intelligibility can lead to the canonization of the metaphysical picture, but this canonical status is conditional on the coherence of the metaphysical picture with successful science. The teleological circular motion metaphysical framework was dominant for about two millennia, until the mechanistic picture showed to be much more fruitful as a metaphysical background for the construction of scientific knowledge.

12.3.4 Understanding from False Models

There is one last very interesting sense in which our narrative of the history of astronomy from Eudoxus to Kepler gives support for the account of understanding defended by de Regt: scientific understanding of natural phenomena can be obtained from false models. Let us recall de Regt's criterion for understanding a phenomenon

(CUP): "a phenomenon P is understood scientifically if and only if there is an explanation of P that is based on an intelligible theory T and conforms to the basic epistemic values of empirical adequacy and internal consistency" (2017, 92). So far we have paid close attention to the first condition for UP, namely, an intelligible theory from which an explanation for P can be built. The second condition of internal consistency and empirical adequacy of T, at face value at least, seems to amount to a condition of veridicality, i.e., that the explanatory model built out of T must be (approximately) true in order to convey scientific understanding.

However, de Regt and Gijsbers (2017) reformulate the second condition in such a way that it is clearly coherent with the view that scientific understanding can be obtained from explanatory models that are (even wildly) false. Instead of in terms of empirical adequacy and consistency, in the updated formulation of the CUP de Regt & Gijsbers introduce the second condition in terms of the wider concept of *reliable success*. That is, P is scientifically understood iff there is an explanatory model of P based on an intelligible T, and such that the model is reliably successful.

The concept of reliable success is defined in terms of three tasks: making correct predictions, guiding practical applications, and developing better science. The goal of making correct predictions captures the idea of empirical adequacy. Now, reliable success does not require that this goal is attained by means of strict veridicality. That is, explanatory models that, although patently false, have a certain degree of predictive success, can still be reliably successful. In the definition of reliable success none of the three tasks has a privileged status over the other two. Most naturally, scientists will always prefer explanatory models that rank high in all three tasks, but in the absence of such models, they can settle for less. Thus, false theories or models that in a certain range of phenomena predict reasonably well, and which have important practical applications, or that can open paths to new and better science, can comply with the second constraint for providing UP.

Just like scientific intelligibility, reliable success is a pragmatic notion. The assessment of how a certain model ranks in all three tasks is clearly sensitive to the context of evaluation. For example, the degree of success in the first task depends on the realm of phenomena considered—this is why I stated that false models that predict *reasonable well* can still be reliably successful: which degree of predictive success counts as reasonably good is a pragmatic issue. For example, Newtonian mechanics is empirically successful in terrestrial phenomena: it works well for mid-sized objects and velocities much lower than c, but it fails in other contexts. Accordingly, de Regt & Gijsbers (2017, 57) state that a theory T is reliably successful *for a scientist S iff S can use a theory in her scientific work* in such a way that T performs well in at least one of the three mentioned tasks. The reference to S and to her scientific work captures the context-dependent element in judgments about reliable success.

Kepler's vicarious hypothesis constitutes a clear vindication of the thesis that scientific understanding of phenomena can be obtained from a wildly false model. As we saw, the heuristic principle in Kepler's construction of the vicarious model was his proto-mechanical framework for the physics of the solar system. Now, as soon as he attempted the task of determining the parameters in the model, he noticed that the ratio between the distance from the center of the orbit to the equant and the distance from the center of the orbit to the Sun yielded different values depending on whether it was calculated from the latitude or the longitude of Mars, a conflict which in turn led to an unavoidable empirical error of $8'$. That is, as soon as he formulated it, Kepler knew that the model was false.

However, he was explicit about the crucial scientific value that the false model had. The title of chapter 21 of the *New Astronomy* is entitled *Why, and to what extent, may a false hypothesis yield the truth?* There Kepler stated that despite its falsity, the vicarious model does capture some true aspects in the kinematic configuration of the solar system. For example, the model does predict the right longitude at some points in the Martian orbital trajectory. In Kepler's own words:

> There are, however, occasions upon which a false hypothesis can simulate truth, within the limits of observational precision, with respect to the longitude. (Kepler, 1992, 294)

> It is at least now clear to what extent and in what manner the truth may follow from false principles: whatever is false in these hypotheses is peculiar to them and can be absent, while whatever endows truth with necessity is in general aspect wholly true and nothing else.
>
> Further, as these false principles are fitted only to certain positions throughout the whole circle, it follows that they will not be entirely correct outside those positions, except to the extent [...] that the difference can no longer be appraised by the acuteness of the senses. (1992, 298)

In these passages, Kepler is explicit in that the model, despite being false, has a degree of empirical adequacy that is enough to assign it a scientific value. Now, such value does not reside only in the degree of predictive success. If that were the case, the same value could be attributed to the three rival models (Ptolemy's, Copernicus', and Brahe's), given their geometric intertranslability.

The scientific value that Kepler assigned to his vicarious model, which we can read in terms of de Regt & Gijsbers' characterization of UP from false models, relied in its scientific fruitfulness. Unlike its contenders, the vicarious model was able to open an avenue to better science. As we saw above, Kepler formulated his crucial second law in the context of the vicarious model: in Kepler's own formulation, the law does not refer to the elliptic character of the orbits. Actually, the first law of the elliptic orbits was found by Kepler from the empirical failure of the vicarious model and the orbital predictions from the area law. As it is clear from our exposition of the different astronomical models of circular motion, this trail to better science could only be blazed by the vicarious model. Thus, Kepler's vicarious hypothesis yields an UP that cannot be obtained from the preceding models. Despite the fact that they are all equally false, the area law could be formulated from the vicarious hypothesis, but not from the Ptolemaic, the Copernican or the Tychonic model.

References

Aristotle. (1922). *On the heavens*. Clarendon Press.
Aristotle. (1924). *Metaphysics*. Clarendon Press.

Aristotle. (1936). *Physics*. Clarendon Press.
Barbour, J. (2001). *The discovery of dynamics*. Oxford University Press.
Carman, C. (2010). On the determination of planetary distances in the Ptolemaic system. *International Studies in the Philosophy of Science*, 257–265.
Cartwright, N. (1983). *How the laws of physics lie*. Clarendon Press.
Çimen, Ü. (2019). On saving the astronomical phenomena: Physical realism in struggle with mathematical realism in Francis Bacon, al-Bitruji, and Averroës. *HOPOS, 9*, 135–151.
Copernicus, N. (1992). *On the revolutions of the heavenly spheres*. The Johns Hopkins University Press.
Crowe, M. J. (2001). *Theories of the world from antiquity to the Copernican revolution*. Dover.
Danielson, D. (2006). *The first Copernican: Georg Joachim Rheticus and the rise of the Copernican revolution*. Walker & Company.
de Regt, H. (2017). *Understanding scientific understanding*. Oxford University Press.
de Regt, H., & Baumberger, C. (2020). What is scientific understanding and how can it be achieved? In K. McCain & K. Kampourakis (Eds.), *What is scientific knowledge? An introduction to contemporary epistemology of science* (pp. 66–81). Routledge.
de Regt, H., & Gijsbers, V. (2017). How false theories can yield genuine understanding. In S. Grimm, C. Baumberger, & S. Ammon (Eds.), *Explaining understanding* (pp. 50–75). Routledge.
de Regt, H., Leonelli, S., & Eigner, K. (Eds.). (2009). *Scientific understanding: Philosophical perspectives*. Pittsburgh University Press.
Descartes, R. (1979). *The world*. Abaris Books.
Dijksterhuis, E. J. (1986). *The mechanization of the world picture: Pythagoras to Newton*. Princeton University Press.
Douglas, H. (2009). *Science, policy, and the value-free ideal*. Pittsburgh University Press.
Dreyer, J. L. (1953). *A history of astronomy from Thales to Kepler*. Dover.
Dreyer, J. L. (2014). *Tycho Brahe*. Cambridge University Press.
Evans, J. (1998). *The history & practice of ancient astronomy*. Oxfotrd University Press.
Frisch, M. (2014). Models and scientific representations or: Who is afraid of inconsistency? *Synthese, 191*, 3027–3040.
Goldstein, B. R. (1971). *Al-Bitruji: On the principles of astronomy*. Yale University Press.
Goldstein, B. R. (1980). The status of models in ancient and medieval astronomy. *Centaurus, 24*, 132–147.
Grimm, S., Baumberger, C., & Ammon, S. (Eds.). (2017). *Explaining understanding: New perspectives from epistemology and philosophy of science*. Routledge.
Hempel, C. (1965). Aspects of scientific explanation. In *Aspects of scientific explanation and other essays in the philosophy of science* (pp. 333–488). The Free Press.
Jacobsen, T. S. (1999). *Planetary systems from the ancient Greeks to Kepler*. The University of Washington Press.
Kennedy, E. S., & Roberts, V. (1959). The planetary theory of Ibn al-Shatir. *Isis, 50*, 227–235.
Kepler, J. (1981). *Mysterium Cosmographicum: The secret of the universe*. Abaris Books.
Kepler, J. (1992). *New astronomy*. Cambridge University Press.
Koyré, A. (1973). *The astronomical revolution: Copernicus-Kepler-Borelli*. Dover.
Kuhn, T. S. (1995). *The Copernican revolution*. Harvard Universuty Press.
Linton, C. M. (2004). *From Eudoxus to Einstein: A history of mathematical astronomy*. Cambridge University Press.
Longino, H. (1990). *Science as social knowledge: Values and objecitvity in scientific inquiry*. Princeton University Press.
Methuen, C. (1996). Maestlin's teaching of Copernicus: The evidence of his university textbook and disputations. *Isis, 87*, 230–247.
Morgan, M., & Morrison, M. (1999). Models as autonomous agents. In M. Morgan & M. Morrison (Eds.), *Models as mediators: Perspectives on natural and social science* (pp. 38–65). Cambridge University Press.
Neugebauer, O. (1975). *A history of ancient mathematical astronomy*. Springer.

Neugebauer, O. (1986). *Astronomy and history: Selected essays*. Springer.
Nutkiewicz, M. (1978). Maimonides on the Ptolemaic system: The limits of our knowledge. *Comitatus: A Journal of Medieval and Renaissance, 9*, 63–72.
Plato. (2000). *Timaeus*. Hackett Publishing Co.
Rashed, R. (2007). The celestial kinematics of Ibn al-Haytham. *Arabic Sciences and Philosophy, 17*, 3–5.
Sabra, I. (1984). The Andalusian Revolt against Ptolemaic astronomy: Averroes and Al-Bitruji. In E. Mendelsohn (Ed.), *Transformation and tradition in the sciences: Essays in honour of I. Bernard Cohen* (pp. 133–154). Cambridge University Press.
Saliba, G. (1991). The astronomical tradition of Maragha: A historical survey and prospects for future research. *Arabic Sciences and Philosophy, 1*, 67–99.
Simplicius. (2005). *On Aristotle's on the heavens*. Bristol Classical Press.
Toomer, G. J. (1984). *Ptolemy's almagest: Translated and annotated*. Princeton University Press.
Torretti, R. (1990). *Creative understanding: Philosophical reflections on physics*. The University of Chicago Press.
Torretti, R. (2007). *De Eudoxo a Newton: modelos matemáticos en la filosofía natural*. Universidad Diego Portales.
van Fraassen, B. (1980). *The scientific image*. Oxford University Press.
Vickers, P. (2013). *Understanding inconsistent science*. Oxford University Press.
Wilson, C. A. (1970). From Kepler's Laws, so called, to universal gravitation: Empirical factors. *Archive for History of Exact Sciences, 6*, 89–170.
Woodward, J. (2019). Scientific explanation. In E. Zalta (Ed.), *The Stanford encyclopedia of philosophy*. URL = https://plato.stanford.edu/archives/win2019/entries/scientific-explanation/

Chapter 13
Reinterpreting Crucial Experiments

Alejandro Cassini

Abstract Crucial experiments have been largely neglected by philosophers of science. The main reason for this predicament is that Duhem's criticism of that kind of experiment has been accepted as sound and definitive. In this article, I start by revisiting the main argument against the possibility of crucial experiments, which is based on epistemological holism. I contend that the argument rests on the confusion between crucial and decisive experiments. When crucial experiments are deprived of their supposed decisive character, the argument loses its bite. Epistemological holism applies to any experiment, whether crucial or not, but it does not imply that experiments are not possible or that they do not have any epistemological import. This variety of holism simply shows that any evidence has to be interpreted and assessed within a theoretical context that includes many auxiliary hypotheses and presupposed theories, which are regarded as accepted background knowledge. This knowledge is not put to the test in a given experiment, but it is rather employed in describing the experimental result and interpreting its theoretical consequences. The meaning of any crucial experiment has then to be extracted from the theoretical context in which the experimental result is interpreted. When the background of accepted knowledge undergoes a drastic change, a crucial experiment may be reinterpreted in such a way that confirms or refutes hypotheses or theories not available at the moment in which it was performed. I will illustrate this kind of reinterpretation with the historical cases of Fizeau's 1851 experiment, the Michelson and Morley 1887 experiment, and Eddington's 1919 experiment. I will conclude by vindicating crucial experiments.

A. Cassini (✉)
CONICET-Universidad de Buenos Aires, Buenos Aires, Argentina

C. Soto (ed.), *Current Debates in Philosophy of Science*, Synthese Library 477,
https://doi.org/10.1007/978-3-031-32375-1_13

13.1 The Neglect of Crucial Experiments

In a popular talk on the nature of science, the Nobel laureate physicist Richard Feynman described the scientific procedure in the following terms:

> You have, we suppose, two theories about the way something is going to happen, which I will call "Theory *A*" and "Theory *B*". Now it gets complicated. Theory *A* and Theory *B*. Before you make any observations, for some reason or other, that is, your past experiences and other observations and intuition and so on, suppose that you are very much more certain of Theory *A* than of Theory *B*-much more sure. But suppose that the thing that you are going to observe is a test. According to Theory *A*, nothing should happen. According to Theory *B*, it should turn blue. Well, you make the observation, and it turns sort of a greenish. Then you look at Theory *A*, and you say, "It's very unlikely," and you turn to Theory *B*, and you say, "Well, it should have turned sort of blue, but it wasn't impossible that it should turn sort of greenish color." So the result of this observation, then, is that Theory *A* is getting weaker, and Theory *B* is getting stronger. (Feynman, 1999, pp. 67–68)

Feynman was trying to point out how uncertain is the scientific activity and how approximate are the results of scientific tests. Nonetheless, it is significant that he chose a crucial experiment between two rival theories as a typical example of the procedures of science.

In turn, the philosopher of science Ronal Giere has written that

> The idea of a "crucial experiment" as expounded, for example, by Francis Bacon, was a major cultural achievement of the Scientific Revolution. It deserves to be ranked along with such other achievements as the calculus, the telescope, and the air pump. [. . .] Of course with the hindsight of three centuries, we know that the role of crucial experiments has often been exaggerated and the designation of an experiment as "crucial" often comes long after the fact. But this only shows that the idea of a crucial experiment can play a rhetorical as well as an operational role in science. It does not show that, properly understood, it plays no operational role. (Giere, 1999, pp. 123–124)

Despite the historical significance that is attributed to the invention of crucial experiments, it is rather surprising to discover that the topic is rarely discussed by contemporary philosophers of science. The main reason for this predicament seems to be that most philosophers have endorsed Pierre Duhem's stance, according to which "an *experimentum crucis* is impossible in Physics" (Duhem, 1914, p. 285). Duhem's criticism of crucial experiments dates back to his first publication on this topic in an article published in 1894, whose text was the basis of his well-known 1906 book *La théorie physique*.[1] Duhem arrived at this conclusion as a consequence of his epistemological holism, or, as we call it nowadays, confirmational holism. The very essence of his position is contained in the following passage:

> In sum, the physicists can never submit an isolated hypothesis to the control of experience, but only a whole set of hypotheses; when the experience is in disagreement with his

[1] A second augmented edition of the book was published in 1914 and reprinted in 1981. The book was translated into German in 1908 and, much later, into English in 1954. The section on crucial experiments was almost entirely that of the 1894 article. My quotations come from the 1981 reprinted edition.

> predictions, what he learns is that at least one of the hypotheses constituting this set is unacceptable and ought to be modified; but the experience does not designate which one should be changed. (Duhem, 1914, p. 284)

I will argue here that, contrary to Duhem's claim, crucial experiments are perfectly possible in science, as possible as any non-crucial experiment. Epistemological holism certainly places limitations on the reach of experimental results and the conclusions we are entitled to draw from them, but these limitations are the same for crucial and non-crucial experiments alike. Moreover, the theoretical interpretation of every crucial experiment is certainly dependent upon a host of presupposed theories and hypotheses that form the background knowledge against which the experimental results acquire their meaning. But, again, this fact is common to any scientific experiment, as I will show by employing several case studies in which crucial experiments were reinterpreted within the framework of new theories. Finally, a crucial experiment does not need to be decisive in any sense of the term to be crucial.

13.2 Epistemological Holism

Duhem's epistemological holism rests on these two fundamental assumptions: (i) scientific hypotheses are tested by means of their observational consequences, that is, by deducing predictions from them; and (ii) theoretical hypotheses, such as physical laws, do not imply any predictions by themselves. From these premises, Duhem drew the conclusion that theoretical hypotheses cannot be put to the test of experience in isolation. In order to imply a definite prediction, a theoretical hypothesis has to be conjoined with many other hypotheses. When an experiment is performed, a whole system of hypotheses is put to the test. Duhem was very vague concerning the question of which is exactly the group of hypotheses that are experimentally tested. On some occasions, he said that the unity of testing is "a whole set of hypotheses" (*tout un ensemble d'hypothèses*, Duhem, 1914, p. 284), or "a whole theory" (*la théorie tout entière*, 1914, p. 283), or "a whole set of theories" (*tout un ensemble de théories*, 1914, p. 310), or even "the whole system of physical theory" (*le système entier de la théorie physique*, 1914, p. 304).

A very common misinterpretation of Duhem's confirmational holism (and of confirmational holism in general) consists in claiming that an experiment in physics (or in science generally) does not test an isolated hypothesis, but the entire theory to which that hypothesis belongs.[2] Despite being vague on the unity of testing,

[2] For instance, Psillos (2007, p. 109) defines confirmational holism as "the view that theories are confirmed as wholes). In the foreword to the English translation of Duhem's book, Louis de Broglie wrote that "according to Duhem, there are no genuine crucial experiments because it is the ensemble of a theory forming an indivisible whole which has to be compared to experiments" (de Broglie, 1954, p. XI).

Duhem stated clearly that this is not the case: the system of hypotheses put to the test in any experiment in physics always contains a variety of hypotheses that belong to *different theories*. For instance, when he discussed Weber's experiments designed to test Ampère's electrodynamics, he claimed that those experiments "do not confirm this or that isolated proposition of Ampère's theory, but the whole set of *electrodynamical, mechanical, and optical hypotheses* that must be invoked in order to interpret each of Weber's experiments" (Duhem, 1914, p. 303, my emphasis). And, more generally, he claimed that "any experimental control puts into play the most diverse parts of physics and appeals to innumerable hypotheses" (1914, p. 334).

As is well-known, Quine extended confirmational holism up to include portions of mathematics, metaphysical assumptions, and common sense statements into the system of hypotheses that are put to the test in any scientific experiment.[3] For the sake of our argument, this extension does not add anything significant to the consequences of epistemological holism. It suffices to say that the system that is put to the experimental test is not a specific theory but a heterogeneous system composed of different kinds of hypotheses. That system is generally so complex that no experimenter is able to list all the hypotheses that compose it.

Let us recall once again the elementary logic of crucial experiments. We call H the specific hypothesis we are interested in testing; this can be a single hypothesis, a finite set of hypotheses, or a whole theory.[4] Given that H does not imply by itself any observational consequence, in order to derive a testable prediction it is necessary to employ a set of different auxiliary hypotheses A. If H is a theory, by definition, A does not belong to H. The system composed of the conjunction of H and A is the one that permits the deduction of a given prediction O (sometimes called a test implication), which is a proposition of the conditional form $C \rightarrow E$ (where C is a set of propositions that state initial conditions and E is a proposition that describes an observable event). If experience shows that O is false, this implies that at least one proposition belonging to A or to H is false, but we cannot determine which one or how many. In particular, it does not follow that H has been refuted by our experimental results. This is a simple consequence of conformational holism.

A crucial experiment between two different hypotheses H_1 and H_2 is possible when we can derive two incompatible predictions such that O_1 is of the form $C \rightarrow E_1$ and O_2 is of the form $C \rightarrow E_2$ (where E_1 and E_2 are incompatible

[3] See Gillies (1993), chapter 5, for a comparison between Duhem's and Quine's varieties of holism.

[4] From now on, I will assume the classical conception of theories, according to which a theory is a logically closed set of propositions. All discussions of conformational holism are based on that account of theories, which was endorsed by Duhem and Quine themselves. Duhem and Quine also endorsed a deductivist conception of confirmation, according to which theories and hypotheses are tested exclusively by the evidence they imply. Formally, they subscribed to these two conditions: (a) E confirms $H \leftrightarrow H \vDash E$ and (b) E refutes $H \leftrightarrow H \vDash$ not E (where H is a given hypothesis and E is any piece of evidence). These conceptions of theories and testing form the basis of the so-called hypothetico-deductive method. See Quine & Ullian (1978), chapter 8, for an explicit account of deductive confirmation.

propositions, that is, they cannot be true at the same time). Those predictions are deduced from two different sets of hypotheses, such that $(H_1 \& A_1) \vDash O_1$ and $(H_2 \& A_2) \vDash O_2$. The sets of auxiliary hypotheses A_1 and A_2 may be identical in some cases, but generally, they are not. In principle, a crucial experiment aims at refuting one of the two rival hypotheses, either H_1 or H_2, and at the same time, at confirming the other. When H_1 and H_2 are contradictory, the refutation of one of them implies the verification of the other. This is not a very usual situation in real science; more frequently, those hypotheses will be simply contraries, that is, propositions that cannot be true at the same time but can be both false. If, because of the fact that those hypotheses cannot be tested in isolation, no experiment is able to refute one of them, it follows that crucial experiments are not possible for purely logical reasons. That was precisely Duhem's conclusion.

The so-called Duhem-Quine thesis was extensively debated by philosophers who asked whether we can ascertain the falsity of one scientific hypothesis and, if so, on which conditions.[5] I will not follow this strategy here. For the sake of the argument, I will grant that, because of confirmational holism, scientific hypotheses generally cannot be refuted from a strictly logical point of view. Even so, I will argue that crucial experiments are possible in a reasonable sense of the word "crucial".

13.3 Crucial vs. Decisive Experiments

In its common usage, the English word "crucial" means that something is particularly important, significant, or even decisive, such as a crucial decision, for instance. This is also true in other languages, such as French or Spanish.[6] Duhem himself understood the word in this strong sense. According to him, a crucial experiment (to which he referred always by means of the Latin expression *experimentum crucis)* is a decisive experiment, that is, an experiment that proves the truth of a given hypothesis, and, consequently, transforms it into a proposition endowed with total certainty. In his own words, a crucial experiment is, supposedly, "an irrefutable procedure for transforming one of the two hypotheses before us into a demonstrated truth" (Duhem, 1914, p. 288). In more general terms, a crucial experiment is one that establishes the truth of a given hypothesis by refuting all its rivals. Duhem addressed his criticism against this conception of crucial experiments, but his argument missed the point. His criticisms only apply to decisive experiments, in the above sense.

Duhem offered this formulation of crucial experiments:

> Do you want to obtain from a group of phenomena a theoretically certain and indisputable explanation? Enumerate all the hypotheses that can be made to account for this group of

---

[5] See, for instance, Harding (1976) and the articles included in that work.

[6] Standard dictionaries of those languages always mention the word "decisive" as synonymous with "crucial", or as part of its definition.

> phenomena; then by experimental contradiction eliminate all except one; the latter will no longer be a hypothesis, but will become a certainty. (1914, p. 286)

He remarked that in this case the hypothesis that was not refuted by the experience will be "indisputable" and will become "a new truth acquired by science" (1914, p. 286).

We know that in many cases experience cannot verify scientific hypotheses. In the first place, no experiment is able to demonstrate the truth of a general universal proposition, such as those by which the laws of nature are stated. If the experimental results verify a given prediction of conditional form deduced from a universal proposition, this fact, at most, can be taken as confirming that proposition. And this is true for purely logical reasons. On the other hand, if two hypotheses are contradictory (such as, for instance, that the speed of light in a vacuum is finite or infinite), if one of them is refuted by the experimental results, the other is verified by the same evidence. And, again, this is true for logical reasons. In third place, if two rival hypotheses are not contradictory (such as, for instance, that the speed of light in transparent material media is either c/n or cn, where c is the speed of light in a vacuum and n is the refractive index of the medium), if one of them is refuted by the experimental evidence, it does not follow that the other is verified. A given hypothesis would be verified if and only if all its possible rivals were refuted, but this is not possible even in principle, not just in practice, because usually the number of possible rivals is infinite. All this follows from reasons of elementary logic and has nothing to do with epistemological holism. Even in the case that physical hypotheses could be tested in isolation, they could not be verified by experience. No experiment that tests two or more rival hypotheses can verify one of them if they are not contradictory, or if the set of rivals is not exhaustive of all possibilities.

The impact of epistemological holism is not on the possibility of verifying (or not) scientific hypotheses but on the possibility of refuting them. If no theoretical hypotheses can be tested in isolation, and the experimental evidence only refutes a large system of hypotheses, but no one in particular, then, crucial experiments, as characterized by Duhem, cannot even start. By contrast, if an experiment verifies a system of hypotheses, it also verifies each one of the hypotheses that compose that system (otherwise, the system could not be true). However, if the system is just confirmed by experience, by principle, it is not possible to distribute the degrees of confirmation or confidence to the component hypotheses. In this respect, epistemological holism has consequences also for the confirmation of scientific hypotheses. To put it shortly, if we assume epistemological holism, no single theoretical hypothesis can be refuted or confirmed by evidence.

In order to advance our account of crucial experiments, I will introduce the distinction between *decisive experiments* and *crucial experiments*.[7] We know that for purely logical reasons no amount of experimental evidence is able to verify or refute some kinds of scientific hypotheses. Universal hypotheses cannot be verified, and existential hypotheses cannot be refuted. Moreover, statistical or probabilistic

[7] I introduced this distinction in Cassini (2015), but I will elaborate on it in more detail here.

hypotheses can be neither verified nor refuted. If we were to require that an experiment conclusively verified or refuted a theory or a hypothesis, no experiment would be possible, except in some special cases (for instance, existential hypotheses could be verified and universal hypotheses could be refuted, in case they could be tested in isolation). It is not reasonable to demand verification from scientific experiments in general; the most we can expect is to confirm hypotheses, theories, or systems of hypotheses through evidence.

I will call *decisive* those experiments which are regarded by the scientific community in a given moment as providing clear and strong evidence in favor (or against) a given hypothesis or a theory, an amount of evidence that is considered sufficient for the acceptance (or the rejection) of such hypothesis or theory. A decisive experiment is not necessarily conclusive from the logical point of view, but it is one that is regarded as strongly confirmatory (or disconfirmatory) for a given hypothesis. The judgment of a scientific community is rarely unanimous but, in certain circumstances, there is a robust consensus concerning the acceptance of some theories and hypotheses. The existence of decisive experiments is a historical and sociological fact that can be revealed by the history of science. There can be no doubt that some experimental results have been taken as decisive, although the formation of consensus around this fact is not instantaneous; on many occasions, it required debates and provoked controversies.

The concept of a decisive experiment can be explained in terms of *decisive evidence*. A given experience, whether observational or experimental (I will not distinguish between them), delivers a piece of decisive evidence for some hypothesis or theory when that evidence is regarded by a scientific community as providing strong confirmation of that hypothesis or theory. When a piece of evidence is truly decisive, it is usually acknowledged that it produced a breakthrough in the corresponding field of research. For instance, it cannot be doubted that the discovery of the cosmic microwave background in 1965 was regarded almost immediately by the community of cosmologists as decisive evidence in favor of the hot big bang theory. This fact did not force the instantaneous acceptance of that theory or the rejection of its rival, the steady-state theory. Nonetheless, almost all cosmologists agreed on the fact that the new evidence confirmed one of those theories but not the other. Since the discovery of the cosmic microwave background, the steady-state theory waned because it had no natural way of accommodating that evidence, whereas the big bang theory predicted that evidence in advance.[8]

Because the acceptance or rejection of theories is a complex process that involves many criteria besides empirical adequacy, it may be more advisable to define a weaker concept of decisive evidence. In this weaker sense, a piece of evidence is decisive when it is regarded as providing strong confirmation for one theory or

[8] Kragh (1996) is a detailed study of this episode, which was presented here in a very simplified form. The original 1940s articles that predicted the existence of the cosmic microwave background are collected in Bernstein & Feinberg (1986).

hypothesis in such a way that, after the discovery of that evidence, the theory in question qualifies as a serious candidate for acceptance.

In turn, I call *crucial* those experiments that are designed to test simultaneously two or more rival hypotheses or theories. A crucial experiment is *successful* if it provides evidence that confirms one of the rivals and disconfirms the others. Special cases of successful crucial experiments are those in which the evidence verifies one of the rivals and refutes the others or those in which the evidence confirms one of the rivals and refutes the others. It follows from these definitions that a crucial experiment does not need to refute one or more rival hypotheses to be considered successful; it is enough that it disconfirms them, in a sense that is dependent upon a particular theory of confirmation. For instance, it can lower the prior probability of a hypothesis, or undermine our degree of confidence in it. A crucial experiment between more than two rival hypotheses is *partially successful* if it disconfirms at least one of the rivals and confirms two or more of them. Finally, a crucial experiment is *unsuccessful* if it confirms or disconfirms all the rivals at the same time.

From the above considerations, it follows that a successful crucial experiment may be decisive or not, either in the strong or the weak meaning of the expression we have just defined. In turn, a decisive experiment may be crucial or not. Cruciality and decisiveness are independent concepts. In the weaker sense of the expression, a crucial experiment is decisive when it is regarded as one that provides evidence that strongly confirms a given theory and strongly disconfirms its rivals in such a way that they, although not immediately rejected, start to look less appealing to an entire community.

We can find in the history of science some examples of crucial experiments that were regarded as decisive for the acceptance or rejection of a given theory, although they are not as frequent as one would have expected. One of the best-known examples is precisely the one Duhem invoked in his discussion of crucial experiments: the 1850 experiment by Leon Foucault and Armand Fizeau concerning the relative speeds of light in air and water. I do not intend to provide a detailed analysis of the experiment here, but a general description will be sufficient for my purposes.

According to the emission theory, light is composed of small material corpuscles; by contrast, the undulatory theory considers that light is a wave in a subtle material medium, the luminiferous ether. Since the late seventeenth century, it was acknowledged that these rival theories implied different crucial predictions concerning the propagation of light through transparent material media. The emission theory implied that the speed of light is higher in water than in air, whereas the wave theory implied the inverse consequence. Both predictions were clearly stated by Newton and Huygens, who deduced values for the speed of light in the water of *cn* and *c*/*n*, respectively.[9] A crucial experiment between the two theories was then possible in principle, although, apparently, nobody conceived of it until the nineteenth century. Arago (1838) was the first in suggesting the design of a definite crucial experiment, which, significantly, he called a "decisive experience". Arago's design was the following. If two rays of light fall on a rotating mirror, which is put in rotation from right to left, they will be reflected towards the left with an angle equal to the angle of their incidence. Nonetheless, the two reflected rays will remain perfectly aligned, so that they will produce two spots on the same line. But if one of the rays traverses a tube filled with water, the two rays will be not reflected at the same time; one of them, the slower, will be reflected later than the other, and the spots produced by the reflected rays will not appear on the same line anymore; the slower ray will be displaced to the left with respect to the faster one. Arago (1838, p. 963) calculated that a mirror rotating at a thousand revolutions per second will produce a deviation of one arc minute between the two reflected rays when one of them traversed a 28 meters tube filled with water.

The experimental arrangement was delicate and the effect to be observed was very small, but viable using the technology available by the middle of the nineteenth century. The experiment was independently performed between May and June 1850 by Leon Foucault and by Armand Fizeau (together with Louis Breguet). A detailed description of the experimental arrangement, which included a clever improvement of Arago's design, is not relevant here. The experimental results in both experiments provided unambiguous evidence in favor of the prediction according to which the speed of light in water was lesser than its speed in air. On June 17, 1850, Foucault reported to the Academy of Sciences in Paris that "taking into account the longitudes of the air and the water traversed by light, the deviations are shown themselves noticeably proportional to the refractive indexes" (Foucault 1850: 556). In the same session of the Academy, Fizeau concluded that "the observed phenomena are completely in agreement with the theory of undulations and in manifest opposition to the emission theory" (Fizeau & Breguet, 1850, p. 773).

Based on his epistemological holism, Duhem remarked that these experiments did not adjudicate "between two hypotheses, the emission and the wave hypotheses", but rather they "decide between two theoretical sets, each of which has to be taken as a whole, between two complete systems: Newton's optics and Huyghens' optics" (Duhem 1914, p. 287). But this is precisely how the experiments were interpreted by their performers. As Fizeau's quotation shows, his experiment was regarded as crucial between the wave and the emission theories of light, not just between the two specific hypotheses concerning the microscopic structure of light. The wave theory of light was confirmed as a whole and the emission theory was greatly discredited by the experiments. After 1850, the emission theory was at bay, although it was not immediately abandoned by its supporters. The experiments by

[9] Newton (1687, p. 224); Huygens (1690, p. 48). Newton's and Huygens' proofs rested on several auxiliary hypotheses concerning the existence of refractive forces and the microscopic structure of the ether, respectively.

Foucault and Fizeau were decisive in their contribution to the process of the rejection of the corpuscular theory of light. It is generally true in the history of science that no theory or research program is instantaneously abandoned when it faces an adverse experimental result; nonetheless, some experiments play a decisive role against a given theory, when the theory in question is not capable of accommodating the recalcitrant evidence in spite of the efforts of its supporters. That was exactly what happened with the emission theory of light after the 1850 experiments.

13.4 Interpreting Experiments

No experimental result speaks for itself or points to a necessary conclusion. The result of every experiment, whether crucial or not, has to be interpreted against a large background of accepted knowledge. This background includes many established theories, as well as a wealth of different kinds of auxiliary hypotheses, some of which may be regarded as not very well established, or even as insecure. This is not the place to elaborate on auxiliaries, but some remarks on them are necessary to understand the problem of interpreting crucial experiments.

To begin with, an experiment in physics usually assumes some portions of classical mathematics -such as the arithmetic of the real numbers and the differential calculus- which are employed as tools for calculating the empirical prediction that is going to be put to the test. The uses of these formal resources, as well as that of the underlying logic, are regarded as purely instrumental, and for that reason, they are not subject to empirical tests. The geometry of the space or the spacetime seems to be an exception, but it is not; we should regard any hypothesis concerning the geometrical structure of the physical space as an empirical hypothesis that does not belong to pure mathematics. In many experiments in physics, some specific assumptions concerning the local or global structure of the spacetime do play the role of auxiliaries that belong to the accepted background knowledge. However, this is not the case with the underlying logic or the purely mathematical theories, which are used as mere instruments of inference or calculation, not as empirical assumptions. For that reason, I will exclude formal hypotheses of the background knowledge with which experimental results are interpreted.

This background knowledge never includes the whole of science or an entire discipline, as both Duhem and Quine seem to suggest in some passages of their works. No experiment in physics, for instance, presupposes all available physical theories in order to interpret its result. Just some portions of the physical knowledge are relevant to do that but, as Duhem and Quine have rightly asserted, it is by no means easy to determine exactly which portions. In principle, we can never be sure that we have discovered and listed all the assumptions at work in a given experiment. Quine, in his characteristic style, depicts this predicament in the following terms:

> The scientist does not tabulate in advance the whole fund of theoretical tenets and technical assumptions, much less the commonsense platitudes and mathematical laws, that are needed in addition to his currently targeted hypothesis in order to imply the observation categorical

of his experiment. It would be a Herculean labor, not to say Augean, to sort out all the premisses and logical strands of implication that ultimately link theory with observation, if or insofar as linked they be. (Quine, 1992, p. 17)

Nonetheless, there are some specific kinds of assumptions whose inclusion is quite clearly justified. I do not intend to be exhaustive at this point, but I will mention some of the more important ones.

In the first place, there are *presupposed theories*, that is, theories that are assumed in the formulation or the application of other theories. For instance, classical mechanics presupposes that Euclidean geometry provides the structure of physical space and, in turn, statistical physics presupposes that classical mechanics provides the kinematics and the dynamics of the molecules that compose all material bodies. The existence of presupposed theories is manifest not only in the appeal to the laws of such theories but also in the vocabulary employed by a given theory. For example, thermodynamics makes extensive use of concepts such as mass, momentum, or energy, which come from mechanics, as well as of the Newtonian laws of motion for pointlike particles. As a consequence of all this, when we test a given physical theory, we necessarily have to assume other physical theories as part of the background knowledge. Those theories are often older, better established, and better confirmed than the one we want to test. But it is not necessarily so, as we will see later.

In the second place, there are many different hypotheses presupposed in the *experimental design*, the *construction and calibration of the measuring instruments*, and the *use of the materials* with which those instruments interact. For instance, if we were to test a mechanical hypothesis by means of an optical experiment, some hypotheses concerning the nature and the behavior of light would be unavoidably presupposed. On the other hand, if an optical hypothesis were put to the test using a material target with which a beam of light interacts, we should assume some hypotheses concerning the properties of that material target, say, a quartz crystal, as different from the properties of a different target, say, a fluorescent screen.

Finally, any experiment assumes a *ceteris paribus* clause according to which there are no hidden or unknown physical influences that could disturb the experimental outcome. As is well known, these kinds of clauses are mostly vague and frequently cannot be submitted to independent tests. Nonetheless, a given *ceteris paribus* clause might be false, as we sometimes retrospectively discover.

It is clear enough that any experiment in physics assumes a fairly large amount of background knowledge and that experimenters cannot take into account, or even acknowledge, all the auxiliary hypotheses and theories they assume when they perform a single experiment. In order to describe an experimental result, the members of a given scientific community must agree on the use of the concepts they employ in their observational reports. Otherwise, they could not even recognize that they were talking about the same result. For instance, Foucault and Fizeau's 1850 experiment on the speed of light in water was described by means of some mechanical concepts, above all, velocity, which the endorsers of both the emission and the wave theory of light shared. They all agreed on Newtonian mechanics, a presupposed theory that was unquestioned at the time and was universally regarded as a basic part of the accepted, non-problematic, background knowledge. No optical experiment on the nature of light could then be considered as testing mechanical hypotheses. Newtonian mechanics was not put to the test in those experiments.

From a purely logical point of view, the refutation of a given prediction deduced from a set of different hypotheses has a bearing on any member of that set, including all auxiliaries and presupposed theories. In practice, however, the background knowledge is not put to the test, at least in the first instance. The background is safeguarded from refutation because without it the experimental result could hardly be interpreted. The specific function of the background is, precisely, to provide the laws and concepts through which an experimental result acquires a definite meaning. Quine (1992, p. 15) claimed that the fact that logic and mathematics are usually exempted from the system of hypotheses put to the test in the natural sciences is just the product of a decision and not a consequence of the necessity of logical or mathematical truths. I will extend this idea to all background knowledge when an experiment is performed for the first time. Experiments are generally designed to test specific hypotheses or theories (I leave apart here exploratory experiments) and for that reason, all the auxiliaries employed in the deduction of a prediction are conventionally excluded from the body of knowledge that is put to the test. That is why, for instance, scientists do not think they are testing mechanical theories when they perform an optical experiment. Of course, as we will see very soon, this decision can be revised in light of an unexpected experimental result.

In the specific case of crucial experiments, some conditions ought to be fulfilled, if the experiment is to count as crucial between two rival theories or hypotheses. First, the experimental result must be described by employing concepts that do not belong to the rival theories that are put to the test. In the above example, Foucault and Fizeau's results were reported in terms of mechanical concepts as well as in terms of optics, but the latter did not include specific concepts of the undulatory or corpuscular theories of light. They could not be described without vicious circularity by means of such concepts as wavelength or energy of the light particles. Second, the experimental result ought to be interpreted by appealing to shared background knowledge, such as, in our case study, Newtonian mechanics or geometrical optics. Third, the background must be logically independent of the rival theories, as is the case of geometrical optics with respect to theories about the microscopic structure of light. Of course, the endorsers of different rival theories will never agree on the whole set of hypotheses from which the crucial predictions are deduced because each set includes just one of the rival theories or hypotheses. Moreover, for the reasons stated above, both systems of hypotheses are never entirely well-defined and, consequently, scientists cannot ever be sure about how much background knowledge they share.

Despite all the aforementioned limitations, crucial experiments are possible when scientists decide (in most cases, by default) to exempt their background knowledge from being put to the test in some specific cases, that is, when they want to perform an experiment designed to test two or more rival theories or hypotheses.

13.5 Reinterpreting Crucial Experiments

On many occasions, an experiment can give rise to a long-standing controversy concerning its meaning and the bearing of the evidence it delivers. Sometimes, it is not clear at all which hypotheses or theories are confirmed or refuted by the newly acquired evidence. This is often the case, for instance, when some auxiliary hypotheses or some other portions of the background knowledge are questioned. As I expect to show here, there are many different reasons for questioning the background. In principle, any presupposed theory or hypothesis, except for those belonging to logic or mathematics, is open to revision in light of an experimental result. This includes the initial and boundary conditions, which are assumed to obtain when the experiment is performed. I will call the background knowledge and the experimental conditions, taken together, the *theoretical context* of a given experiment. Usually, this context is *fixed* in the very moment in which the experiment is performed and is the one that permits the description of the experimental results and the interpretation of the meaning that is assigned to them.

When the theoretical context of an experiment is disputed, the meaning of that experiment starts to become dubious and open to different interpretations. If the background knowledge undergoes drastic revisions, an experimental result may be reinterpreted in such a way that it acquires a completely different meaning. For instance, an experiment that in a given moment was regarded as providing confirmatory evidence in favor of a given theory later could be regarded as confirming (or refuting) a different theory. This includes the special case in which the experiment seemed to confirm some existential hypothesis; in a different theoretical context it could be interpreted as providing evidence against that hypothesis, that is, instead of proving that entities of some kind exist, it becomes a proof that this very kind of entities does not exist. Moreover, an experiment that was regarded as crucial between two rival theories can be regarded later as crucial between two new different theories and without any bearing on the older ones.

There are well-known historical examples of crucial experiments that were reinterpreted in light of new theories. Fizeau's 1851 experiment on the speed of light in running water is one of the most startling examples of such reinterpretation. There are also examples of experiments whose results were debated and received different interpretations in their own time but later were reinterpreted in light of a new theory. The Michelson-Morley 1887 experiment is a classical example. Finally, there are experiments that were regarded as crucial, and even as decisive, in their time but that would have received a different interpretation in a different theoretical context. Eddington's 1919 eclipse experiment is an interesting example, as we will see soon. I will briefly examine those historical cases.

13.5.1 *Fizeau's 1851 Experiment*

The wave theory of light included as a fundamental hypothesis the one according to which light was a vibrational motion of an all-pervasive substance that filled the whole space: the luminiferous ether. Given that light propagates through fluid and solid transparent media, such as water or glass, the ether had to penetrate those material media. If the ether itself is a material substance, in principle it should be able to interact mechanically with material bodies. When a transparent medium is put in motion, the ether inside it may be disturbed in some way or another, for instance, its velocity could be affected in case it was dragged along by the moving medium. For example, the velocity of the medium might be transferred completely to the ether itself, or, perhaps, the ether might be entirely unaffected.

In order to explain the null result of Fançois Arago's 1810 experiment, Augustin Fresnel proposed in 1818 the hypothesis of the partial convection of the ether by the moving media.[10] Fresnel conjectured that the speed of light inside moving media should be $W = V \pm v\ (1 - 1/n^2)$ (where V is the speed of light in the medium at rest, v is the speed of the moving medium, and n is the refractive index of the medium). This was known as Fresnel's hypothesis and the expression $f = (1 - 1/n^2)$ was called Fresnel's dragging coefficient, or simply Fresnel's coefficient. It implied that the speed of the moving media is partially transferred to the ether according to a proportion that is dependent upon the refractive index of each medium (which seems a rather surprising fact). Foucault and Fizeau's 1850 experiments confirmed that the speed of light in transparent media was c/n when those media were at rest with respect to the laboratory. From that result, nothing definite followed concerning the speed of light when those media were put in motion. Three different hypotheses were at stake at that moment. If the ether were not affected by the motion of the media, that is, not dragged along by them, the speed of light in a moving medium should be equal to its speed in the medium at rest. The first hypothesis was then that $W = c/n$. On the other hand, if the ether were entirely dragged along by the moving media, the speed of light should be $W = c/n \pm v$. Finally, Fresnel's hypothesis of partial convection implied that the speed of light in moving media should be $W = c/n \pm vf$ (where f is the Fresnel coefficient).

Fizeau managed to perform an extremely ingenuous crucial experiment to test these three rival hypotheses concerning the interaction between ether and matter.[11] It was the very first interference experiment, a kind of experiment that was decisive for the future development of optics. It consisted in sending two rays of light that traversed a closed circuit of tubes filled with running water. When recombined, the rays produced interference fringes, which were expected to shift when the water was put in motion and the speed of the water was communicated to the ether. The hypothesis according to which there was no convention of ether by matter implied

[10] Arago (1853) is the delayed publication of the experimental result. A discussion of the experiment can be found in Eisenstaedt (2005), chapter 10.

[11] For a more detailed account of the experiment see Cassini & Levinas (2019).

that no shift of the interference fringes should be observed. In turn, the hypotheses of total and partial convection implied different amounts of shifting in the interference fringes (which Fizeau was able to calculate with high accuracy).

The experiment confirmed Fresnel's prediction within a reasonably low margin of error (of about 15% of the measured value). Fizeau calculated that if water was accelerated up to 7 meters per second, the hypotheses of total convection implied a shift in the interference fringes of almost half a fringe ($N = 0.4597$), whereas Fresnel's hypothesis implied a shift of one-fifth of a fringe ($N = 0.2022$). After performing 19 measurements, Fizeau obtained a mean value of $N = 0.23$, in good agreement with the value predicted by Fresnel's coefficient. It was then clear that this experimental result confirmed the hypothesis of partial convection of ether by matter and refuted both the hypotheses of total and no convection.

Fizeau's report of his experiment concluded that

> It seems to me that the success of this experience must lead to the adoption of Fresnel's hypothesis, or at least to that of the law he has found in order to express the change in the velocity of light as an effect of the motion of the bodies; for although the fact of this law being found to be true constitutes a strong proof in favor of the hypothesis of which it is a mere consequence, yet to many the conception of Fresnel will doubtless still appear both extraordinary and, in some respects, very difficult to be accepted; and before it can be adopted as the expression of the real state of things, additional proofs will be demanded from the physicist, as well as a thorough examination of the subject from the mathematician. (Fizeau, 1851, p. 355)

Notice that the experiment did not measure the speed of light in running water; it merely measured a change in the speed of light when water was put in motion. That change was in agreement with the one predicted by Fresnel's formula but it did not follow from it that the cause of the observed change in the speed of light was the fact that the ether is partially dragged along by the running water. Fizeau's cautious disclaimer seemed justified. It should be remarked that the wave theory of light was not put to the test in the experiment; on the contrary, that theory was presupposed in Fizeau's calculations, which appealed to the wavelength of the solar light used to produce the incident rays.[12]

The theoretical interpretation of Fizeau's result was very much debated by nineteenth-century physicists.[13] The fact that the convection of the ether by the matter was dependent upon the refractive index of the medium seemed perplexing. All physicists accepted that the experiment provided evidence in favor of Fresnel's coefficient, and consequently, that it confirmed Fresnel's law for the composition of the velocities of light and the moving medium. However, it was unclear whether it supported the hypothesis concerning the partial convection of ether by matter. In any case, some decades later, that hypothesis was accepted. The experiment was replicated in 1886 by Michelson and Morley, who provided a much more precise measurement of Fresnel's coefficient. Fizeau's experimental result was then

---

[12] Fizeau (1851) was a short qualitative report of his experiment. He later published his detailed calculations and his measured values in Fizeau (1859).

[13] See Stachel (2005) for a brief account of the controversies.

well established. On the other hand, it seemed very convincing regarding the very existence of the ether, because it was a genuine positive result. As late as 1902, Poincaré wrote that Fizeau's experiment "seems to show us two different media penetrating each other, and yet displacing with respect to each other". His startling conclusion was that in this experiment "one believes to touch the ether with a finger" (Poincaré, 1902, p. 181).

Two years after the discovery of the special theory of relativity in 1905, Fizeau's experimental result received an amazing reinterpretation. In 1907 Max Laue acknowledged that Fresnel's coefficient was a straightforward consequence of the relativistic transformation of velocities, which provided a first-order approximation to it in v/c. In this way, special relativity was able to offer a purely kinematical explanation of Fizeau's experimental result without having recourse to any hypothesis concerning the interaction of ether and matter. According to this reinterpretation, what Fizeau actually measured was not the partial convection of the ether by the running water but just a simple composition of the velocities of the light and the water. Laue concluded that this fact was evidence against the ether hypothesis or, in his terms, that "we are dispensed with the necessity of introducing within the optics an 'ether' that penetrates the bodies without taking part in their motion" (Laue, 1907, p. 990). In a few years, Fizeau's experiment was reinterpreted as a crucial experiment in favor of the special theory of relativity and against a variety of rival theories of mechanics and electrodynamics.[14] Nowadays, different versions of this experiment are performed to test special relativity but it is by no means regarded as relevant concerning the existence of the ether. Its meaning was completely transformed within a relativistic theoretical context.

13.5.2 The Michelson-Morley 1887 Experiment

If there exists a sort of cosmic ether at rest (at least with respect to the Solar System), the speed of light has to be different if measured in a referential that is in motion or at rest with respect to that ether. Within the framework of Newtonian mechanics, the speed of light in a vacuum must be c in a referential that is at rest but it must be $c \pm v$ in a referential that is moving with velocity v relative to the ether. The orbital motion of the Earth provided a moving framework in which, in principle, this kind of absolute motion could be detected. However, a complication arises if we assume that the ether is totally or partially dragged along by the surface of the Earth. For instance, if the ether is completely dragged, as Stokes (1846, 1848) conjectured, the speed of light should not show any variation when it is measured in the referential

[14] Since 1910 Einstein regarded Fizeau's experiment as a crucial experiment that confirmed his special theory of relativity (as well as Lorentz's electrodynamics) and refuted the electrodynamical theories of Hertz and Ritz (besides Galilei mechanics). See an analysis of all the available sources in Cassini & Levinas (2019).

of the moving Earth. On the other hand, if the ether is partially dragged along by the Earth, as follows from Fresnel's hypothesis, the speed of light must exhibit some changes when it is measured in different directions (precisely because of the Earth's orbital motion). Michelson and Morley intended his famous 1887 experiment as a crucial test of these two hypotheses. They expected a straightforward confirmation of Fresnel's hypothesis and a clear refutation of Stokes' hypothesis.

The details of the experimental arrangement are so well-known that we do not need to describe them here.[15] It was an interference experiment in which a shift in the interfering fringes was expected when the speed of the Earth was composed with the speed of the light in different directions. Michelson and Morley predicted, based on Fresnel's hypothesis, that the shift in the fringes should be $N = 0.4$ when they rotated the interferometer by 90°, but they observed just a negligible shift.[16] The (almost) null result surprised the entire community of physicists because it seemed to confirm Stokes' hypothesis, which was regarded as untenable because it could not explain the well-established phenomenon of starlight aberration.

Different possible explanations of the null result were put forward, such as the Lorentz-FitzGerald contraction, which I want not to address here. As everybody knows, the special theory of relativity predicted a null result for every ether-drift experiment as a consequence of the relativistic composition of velocities (which implies that in any inertial referential $c \pm v = c$, for any velocity $v \leq c$). Consequently, the Michelson-Morley null experimental result was explained by the fact that the speed of light is invariant, that is, the same in any referential. Within the relativistic framework, this experiment was reinterpreted as confirmatory evidence for the special theory and against any ether theory that predicted a positive result.

What I am interested in here is not the relativistic reinterpretation of this experiment but rather other different possible pre-relativistic interpretations. Those interpretations were not taken into account by the end of the nineteenth century because they were ruled out by the background of accepted knowledge. Nonetheless, it is enlightening to think about how the Michelson-Morley experiment could have been interpreted within different theoretical contexts.

The first interpretation concerns the orbital motion of the Earth. Michelson and Morley performed their calculations on the expected shift in the interference fringes assuming that the Earth was in motion around the Sun (and consequently with respect to the cosmic ether) with a speed of approximately 30 km per second. The hypothesis of the Earth's motion was not put to the test in that experiment because everybody assumed it as established knowledge beyond any reasonable doubt. Nonetheless, if the experiment would have been performed some centuries

---

[15] See Cassini & Levinas (2005) for a detailed account. See also the classic article by Holton (1969). Swenson (1972) is a comprehensive history of most ether-drift experiments. Michelson made a previous attempt in 1881, which was a failure because of a mistake in his calculations (see Michelson, 1881, 1882).

[16] Given that the refractive index of the air is almost equal to the one of the void, in practice Fresnel's coefficient could be set as null and, as a consequence, the ether could be regarded as being in relative motion with respect to the laboratory with the orbital speed of the Earth.

earlier when the Copernican system was still debated, its interpretation could have been entirely different. It could have been interpreted as providing direct evidence that the Earth was at rest with respect to the Sun and the planets. Instead of an optical experiment, it could have been regarded as a purely mechanical experiment with straightforward implications for the debate on the Earth's motion.

The second interpretation concerns the emission theory of light. Michelson and Morley assumed that light was a wave that propagated through the luminiferous ether. The entire wave theory of light was assumed as a background that was not put to the test in the experiment. By the end of the 1880s, the emission theory of light was almost completely discredited and the wave theory of Maxwell and Lorentz was accepted by the overwhelming majority of the physicists because of its undeniable success in explaining a wealth of optical phenomena, such as color, interference, diffraction, and many others which could not be easily explained by its rival. However, the null result of the experiment might have been easily explained in the framework of the emission theory. The source of light employed by Michelson and Morley was a lamp fixed to the interferometer, which for that reason was affected by the Earth's motion. If we were to assume that light is composed of material corpuscles, those light particles would share the orbital speed of the Earth, which should be composed with their inertial speed. A simple calculation shows that the null result follows from those assumptions without the resource to any auxiliary hypothesis. The emission theory of light was then confirmed by the experiment.

Michelson and Morley, as well as most physicists of their time, were so strongly committed to the wave theory of light that they did not even take into account the possibility of explaining their experimental result by resourcing to the hypothesis according to which light is composed of particles. This would imply the abandonment of the ether hypothesis. In a different theoretical context, however, when the wave theory of light was not well-entrenched as an accepted background, the null experiment could have been regarded as a piece of direct evidence in favor of the emission theory. In the 1880s the attitude of the scientific community was to point out that the emission theory predicted a positive result of a Michelson-Morley type of experiment in which the source of light was in relative motion with respect to the Earth (for instance, starlight or moonlight). In this regard, they simply expressed their confidence in the null result of that kind of experiment. The experiment in question was performed by Rudolf Tomaschek, as late as 1923; its null result actually refuted the prediction of the emission theory but by those times Tomaschek's (1924) experimental result was not regarded anymore as evidence in favor of an ether theory. The experiment was interpreted, and it is nowadays, as one that provided confirmatory evidence for the special theory of relativity.[17]

[17] In 1950 Einstein himself told to Shankland (1963, p. 49) that he considered Tomaschek's null results "really decisive in establishing the speed of light to be independent of the motion of the source".

13.5.3 Eddington's 1919 Experiment

Einstein's general theory of relativity predicted that, due to the spacetime curvature, a ray of light that passes near a massive star should be deflected so that its path could not be a straight line. Einstein derived that prediction in 1911, five years before finding his equation for the gravitational field, solely from the principle of equivalence between inertial and gravitational mass. He calculated that near the surface of the Sun, starlight would be deflected by 0.83 arcseconds (Einstein, 1911, p. 908). He later corrected this calculation taking into account the full general theory of relativity, which implied the existence of an additional deflection caused by the curvature of spacetime produced by the Sun itself. The corrected value was 1.7 arcseconds, twice the 1911 value (Einstein, 1916, p. 822). Einstein himself acknowledged in his 1911 article that the angular distance of a star from the center of the Sun should be increased by 0.83 arcseconds because of the bending of the rays. He also suggested a way of testing this consequence, "since the fixed stars in the portions of the sky that are adjacent to the sun become visible during total solar eclipses" (Einstein, 1911, p. 908).

Eddington's famous 1919 eclipse expedition confirmed Einstein's prediction and was soon hailed as a crucial experiment between the general theory of relativity and the Newtonian theory of gravitation.[18] Eddington and his collaborators stated that the aim of the experiment was "to determine what effect, if any, is produced by a gravitational field on the path of a ray of light traversing it" (Dyson et al., 1920, p. 291). Actually, they did not present it as a crucial experiment between Newton's and Einstein's theories of gravitation, but rather as a test intended to "discriminate" between these three hypotheses:

> (1) The path is uninfluenced by gravitation.
>
> (2) The energy or mass of light is subject to gravitation in the same way as ordinary matter. If the law of gravitation is strictly the Newtonian law, this leads to an apparent displacement of a star close to the sun's limb amounting to 0.87" outwards.
>
> (3) The course of a ray of light is in accordance with Einstein's generalised relativity theory. This leads to an apparent displacement of a star at the limb amounting to 1.75" outwards. (Dyson et al., 1920, p. 291)

Two different sequences of observations, made at Sobral and Principe, reported measured values of $1.98''$ (with an error of $\pm 0.12''$) and $1.61''$ (with a probable error of about $\pm 0.30''$). From these data, Eddington concluded that there was "little doubt that a deflection of light takes place in the neighborhood of the sun and that it is of the amount demanded by Einstein's generalized theory of relativity as attributable to the sun's gravitational field" (Dyson et al., 1920, p. 332).

That experimental result was regarded as a refutation of Newton's theory of gravitation because as early as 1801 Johann von Soldner performed the first calculation of the amount of the deflection of starlight by the Sun and obtained a value of

[18] The details of the experiment are rather complicated and do not concern us here. See the comprehensive account by Kennefick (2019).

0.84 arcseconds (Soldner, 1801, p. 170). This result was derived by assuming that light was made of material corpuscles, which experienced the same gravitational attraction as ordinary matter. It is not obvious how this calculation could have been performed within the framework of the wave theory of light, at least without introducing additional hypotheses concerning the behavior of the luminiferous ether near massive bodies. In any case, it is clear that the emission theory of light unambiguously predicted that rays of light could not propagate in straight lines near massive bodies. Even on the surface of the Earth, the path of light should be parabolic, as is the path of any material projectile, although, due to the high speed of light, the parabolic trajectory would be indistinguishable from a straight line. After the invention of the general theory of relativity, Eddington's experiment was regarded as a crucial result in favor of Einstein's theory and against Newton's law of gravitation. However, had the experiment been performed immediately after Soldner's 1801 prediction, it could have been considered confirmatory evidence for the corpuscular theory of light. In that theoretical context, Newton's law of gravitation was certainly assumed as a non-problematic background, given that it had no rivals. Some years later, the experiment could have been regarded even as a crucial experiment against the wave theory of light, which had no easy way to accommodate its result.

Eddington's 1919 experiment did not convince the astronomic community regarding the acceptance of the general theory of relativity or the rejection of Newton's theory of gravitation. It was not regarded as the decisive experiment that retrospective historical reconstructions often presented. However, it is undeniable that its result gave a significant boost to Einstein's theory as a serious candidate for replacing Newton's theory.

13.6 Crucial Experiments Vindicated

The historian of science Daniel Kennefick has written, when he referred to Eddington's 1919 experiment, that "when interpreting experimental results, context is everything" (Kennefick, 2009, p. 42). This is *a fortiori* true regarding crucial experiments. As we have seen, an experiment qualifies as crucial between two or more rival hypotheses or theories in a given context in which there is a broad agreement among scientists concerning which are the specific hypotheses that are put to the test and which are the theories and auxiliary hypotheses that belong to the background of accepted knowledge. In each context, the background, as a matter of principle, is never put to the test. It may be revised in light of experimental results, but this is a process quite different from testing specific hypotheses or theories.

The theoretical context of an experiment even determines what kind of experiment has been performed and to which aspects of science the experimental results are relevant. Our former examples make the point clear. The Fizeau (1851) experiment and the Michelson and Morley (1886) experiment were conceived of as *optical* experiments designed to test three rival hypotheses concerning the interaction between matter and ether. Nonetheless, when reinterpreted in the framework of the special theory of relativity, they became *mechanical* experiments that tested relativistic kinematics against Newtonian kinematics. In turn, Eddington's 1919 experiment was conceived of as a *gravitational* experiment designed to test three rival hypotheses concerning the effect of a gravitational field on the path of light. In different historical contexts, those experiments could have acquired a very different meaning. The Michelson and Morley experiment could have been interpreted as an *astronomical* experiment designed to test competing systems of the world, whereas the Eddington experiment could have been interpreted as an *optical* crucial test between the emission and the wave theories of light. These cases exemplify a further characteristic of the interpretation of crucial experiments: in a different context, some auxiliary hypothesis and the hypothesis that is regarded as the one put to the test may interchange their roles. The orbital motion of the Earth in the case of the Michelson and Morley experiment is just one outstanding example.

Most crucial experiments are not decisive in the strongest sense of the expression, that is, in being capable of forcing the acceptance of radically new hypotheses or the rejection of well-established theories. Foucault and Fizeau's 1850 experiments did not overthrow the emission theory of light, the special theory of relativity was not accepted solely on the basis of its explanation of the Michelson and Morley 1887 experiment (or even the null results of all ether-drift experiments), and the general theory of relativity was not accepted after the Eddington 1919 experiment (which was not regarded as sufficient for rejecting Newton's theory of gravitation).[19] The acceptance and rejection of theories is a very complicated issue in which several different criteria play a significant role. Empirical adequacy is rarely the decisive reason; internal and external consistency, simplicity, explanatory power, generality, and other epistemic (non-factual) virtues are sometimes regarded as significant as experimental results; finally, metaphysical, political, religious, and ideological considerations have played major roles in the history of science, especially as reasons for rejecting new theories, of which Darwin's theory of evolution is a prominent example. The replacement of a well-entrenched theory is often a slow and gradual process, as the Copernican revolution shows.

The epistemological status of crucial experiments is not essentially different from that of scientific experiments in general. Confirmational holism affects all kinds of experiments in the same way. Every experiment, whether crucial or not, assumes a background knowledge of presupposed theories and hypotheses. Anytime a specific prediction is derived from a theory or hypothesis, some auxiliary hypotheses are employed as explicit or tacit premises of that derivation. All experimental results are interpreted in a given theoretical context on the basis of which the evidence is assessed. Only within a specific context, the experimental evidence acquires a definite meaning that permits to claim that some theory or hypothesis has been confirmed or disconfirmed by that evidence. When the theoretical context

[19] This point has been well argued by Crelinsten (2006). See also Kennefick (2019) for further reflections on the acceptance of the general theory of relativity.

of an experiment undergoes radical changes, the experimental result has to be reinterpreted. In the new context, the same experiment may acquire a different meaning: instead of providing confirmatory evidence for a theory, it may be regarded as providing confirmatory (or refutatory) evidence for new or different theories. Even the evidence that once was considered as "proving" the existence of some kind of entity may be reinterpreted as confirmatory of the existence of another kind of entity.

As a consequence of the former considerations, no experiment establishes the definitive truth or falsehood of any theory or hypothesis, and no experiment proves the existence or non-existence of any kind of entity. For instance, no experiment has verified or falsified the old ether theories nor proved or refuted the existence of the ether. We simply have abandoned the ether hypotheses because we have accepted a theory –special relativity- that was able to explain all known experimental results without resourcing to that hypothesis. On the other hand, the ether hypothesis was discredited, but not refuted, by all the null results of the ether-drift experiments. The special theory of relativity succeeded in reinterpreting all the experiments performed within the framework of the ether theories introducing a new theoretical framework. Nonetheless, we cannot exclude the possibility of a future reinterpretation of those experiments in a completely new theoretical framework that eventually replaces special relativity.

The meaning of an experiment depends entirely upon the theoretical context in which its results are interpreted. Sometimes, there is a fundamental theory that explains that result. Fizeau's 1851 experiment and Eddington's 1919 experiment, for instance, would be meaningless for us, or at least unexplained, without the special and the general theories of relativity, respectively. Our present interpretations of both experimental results are completely dependent upon those theories and would be pointless without them. There is certainly something stable in every experimental result but it is not its theoretical interpretation. What is stable is the description of the result in terms of observations or measurements that presuppose a low degree of theoretical ladenness, such as, for instance, the fact that the interference fringes in the Michelson and Morley experiment do not show any significant shift or the fact that the position of the stars near the Sun appear to be displaced by a certain angle. Those results are easily replicable and the replication of their experiments has shown that Fizeau, Michelson and Morley, and Eddington did not commit any serious experimental mistake, that is, that their measurements were essentially correct. Those experimental results have remained stable until the present day and everything indicates that they will remain stable in the future. However, their theoretical interpretations have experienced radical changes and for that reason, nobody can guarantee that new reinterpretations will not occur in the future.

Once crucial experiments have been freed of the connotation of being decisive and of performing the impossible task of definitively verifying a theory or hypothesis and refuting its rivals, there is no reason to persist in claiming that they are not possible. A crucial experiment is simply a specific kind of scientific experiment, one that is designed to test two or more rival theories and which, in case of being successful, confirms one of them and disconfirms the others. Nobody has argued that confirmational holism implies that scientific experiments, in general, are not possible. On the contrary, everybody thinks that empirical theories should be tested by means of experiments and that our confidence in a given theory depends to a large extent on its experimental basis. Why things should be different regarding crucial experiments? The limitations that confirmational holism imposes on crucial experiments are the same limitations it imposes on every experiment. As a consequence, crucial experiments are as possible as any other kind of experiment.

References

Arago, F. (1810). Mémoire sur la vitesse de la lumière, lu à la prémière classe de l'Institut, le 10 décembre 1810. *Compte Rendue des Séances de l'Académie des Sciences (Paris), 36*(1853), 38–49.

Arago, F. (1838). Sur un système d'experiences à l'aide duquel la théorie de l'émission et celle des ondes seront soumises à des épreuves décisives. *Compte Rendue des Séances de l'Académie des Sciences (Paris), 23*, 954–965.

Ariew, R. (1984). The Duhem thesis. *The British Journal for the Philosophy of Science, 35*, 313–332.

Bernstein, J., & Feinberg, G. (Eds.). (1986). *Cosmological constants: Papers in modern cosmology*. Columbia University Press.

Cassini, A. (2015). Una reivindicación de los experimentos cruciales. *Revista de Filosofía, 40*, 105–137.

Cassini, A., & Levinas, M. L. (2005). La reinterpretación radical del experimento de Michelson-Morley por la relatividad especial. *Scientiae Studia, 3*, 547–581.

Cassini, A., & Levinas, M. L. (2019). Einstein's reinterpretation of the Fizeau experiment: How it turned out to be crucial for special relativity. *Studies in History and Philosophy of Modern Physics, 65*, 55–72.

Crelinsten, J. (2006). *Einstein's jury: The race to test relativity*. Princeton University Press.

De Broglie, L. (1954). Foreword. In Duhem (1954: V–XIII).

Duhem, P. (1894). Quelques réflexions au sujet de la physique expérimentale. *Revue des Questions Scientifiques, 36*, 179–229.

Duhem, P. (1906). *La théorie physique. Son objet et sa structure*. Chevalier et Rivière.

Duhem, P. (1914). *La théorie physique, son objet-sa structure*. Marcel Rivière & Cie. Deuxième édition, revue et augmentée. [Reprinted in Paris: Vrin, 1981].

Duhem, P. (1954). *The aim and structure of physical theory* (Philip P. Wiener, Trans.). Princeton University Press.

Dyson, F. W., Eddington, A. S., & Davidson, C. (1920). A Determination of the deflection of light by the Sun's gravitational field, from observations made at the total eclipse of May 29, 1919. *Philosophical Transactions of the Royal Society of London, Series A, 220*, 291–333.

Einstein, A. (1911). Über den Einfluß der Schwerkraft auf die Ausbreitung des Lichtes. *Annalen der Physik, 35*, 898–908.

Einstein, A. (1916). Die Grundlage der allgemeinen Relativitätstheorie. *Annalen der Physik, 49*, 769–822.

Eisenstaedt, J. (2005). *Avant Einstein. Relativité, lumière, gravitation*. Éditions du Seuil.

Feynman, R. (1999). *The meaning of it all: Thoughts of a citizen-scientist*. Perseus Books.

Fizeau, A. (1851). Sur les hypothèses relatives à l'éther lumineux, et sur une expérience qui paraît démontrer que le mouvement des corps change la vitesse avec laquelle la lumière se propage dans leur intérieur. *Compte Rendue des Séances de l'Académie des Sciences (Paris), 33*, 349–355.

Fizeau, A. (1859). Sur les hypothèses relatives à l'éther lumineux. Et sur une expérience qui paraît démontrer que le mouvement des corps change la vitesse avec laquelle la lumière se propage dans leur intérieur. *Annales de Chimie et de Physique. Troisième Série, 57*, 385–404.

Fizeau, A., & Breguet, L. (1850). Sur l'expérience relative à la vitesse comparative de la lumière dans l'air et dans l'eau. *Compte Rendue des Séances de l'Académie des Sciences (Paris), 30*, 771–774.

Foucault, L. (1850). Méthode générale pour mesurer la vitesse de la lumière dans l'air et les milieux transparents. Vitesses relatives de la lumière dans l'air et dans l'eau. Project d'expérience sur la vitesse de propagation du calorique rayonnant. *Compte Rendue des Séances de l'Académie des Sciences (Paris), 30*, 551–560.

Foucault, L. (1854). Sur les vitesses relatives de la lumière dans l'air et dans l'eau. *Annales de Chimie et de Physique. Troisième Série, 41*, 129–164.

Fresnel, A. (1818). Lettre d'Augustin Fresnel à François Arago sur l'influence du mouvement terrestre dans quelques phénomènes d'optique. *Annales de Chimie et de Physique, 9*, 57–76.

Giere, R. (1999). *Science without laws*. The University of Chicago Press.

Gillies, D. (1993). *Philosophy of science in the twentieth century: Four central themes*. Blackwell.

Harding, S. (Ed.). (1976). *Can theories be refuted? Essays on the Duhem-Quine thesis*. Reidel.

Holton, G. (1969). Einstein, Michelson, and the "crucial" experiment. *Isis, 60*, 132–197.

Huygens, C. (1690). *Traité de la lumière*. Pierre Vander Aa. [Reprinted in Paris: Gauthier-Villars, 1920].

Kennefick, D. (2009). Testing relativity from the 1919 eclipse: A question of bias. *Physics Today, 62*, 37–42.

Kennefick, D. (2019). *No shadow of a doubt: The 1919 eclipse that confirmed Einstein's theory of relativity*. Princeton University Press.

Kragh, H. (1996). *Cosmology and controversy: The historical development of two theories of the universe*. Princeton University Press.

Laue, M. (1907). Die Mitführung des Lichtes durch bewegte Körper nach dem Relativitätsprinzip. *Annalen der Physik, 23*, 989–990.

Michelson, A. (1881). The relative motion of the earth and the luminiferous ether. *American Journal of Science, 22*, 120–129.

Michelson, A. (1882). Sur le mouvement relatif de la Terre et de l'éther. *Comptes Rendues de l'Académie des Sciences (Paris), 94*, 520–523.

Michelson, A., & Morley, E. (1886). Influence of motion of the medium on the velocity of light. *American Journal of Science, 3*, 377–386.

Michelson, A., & Morley, E. (1887). On the relative motion of the earth and the luminiferous ether. *American Journal of Science, 34*, 333–345.

Newton, I. (1687). *Philosophiae naturalis principia mathematica*. S. Pepys.

Poincaré, H. (1902). *La science et l'hypothèse*. Flammarion. [Reprinted 1968].

Psillos, S. (2007). *Philosophy of science A-Z*. Edinburgh University Press.

Quine, W. V. O. (1992). *Pursuit of truth* (Revised ed.). Harvard University Press.

Quine, W. V. O., & Ullian, J. (1978). *The web of belief* (2nd ed.). Random House.

Shankland, R. (1963). Conversations with Albert Einstein. *American Journal of Physics, 31*, 47–57.

Stachel, J. (2005). Fresnel's (dragging) coefficient as a challenge to 19th century optics of moving bodies. In A. J. Knox & J. Eisenstaedt (Eds.), *The universe of general relativity* (pp. 1–13). Birkhäuser.

Stokes, G. (1846). On the constitution of the luminiferous aether viewed with reference to the phenomenon of the aberration of light. *Philosophical Magazine, 29*, 6–10.

Stokes, G. (1848). On the constitution of the luminiferous aether. *Philosophical Magazine, 32*, 343–349.

Swenson, L. (1972). *The ethereal aether: A history of the Michelson-Morley-Miller aether-drift experiments, 1880–1930*. University of Texas Press.

Tomaschek, R. (1924). Über das Verhalten des Lichtes außerirdischer Lichtquellen. *Annalen der Physik, 73*, 105–126.

Von Soldner, J. (1801). Über die Ablenkung eines Lichtstrals von seiner geradlinigen Bewegung, durch die Attraktion eines Weltkörpers, an welchem er nahe vorbei geht. In *Astronomische Jahrbuch Für das Jahr 1804* (pp. 161–172). G. A. Lange.

Chapter 14
Non-reflexive Logics and Their Metaphysics: A Critical Appraisal

Jonas R. Becker Arenhart

Abstract Non-reflexive logics are systems of logic in which the reflexive law of identity is restricted or violated. The most well-known of such systems are Schrödinger logics and quasi-set theory; both are related with the metaphysics of quantum mechanics, aiming at formalizing the idea that quantum entities are non-individuals. We argue in this paper that non-reflexive logics may be seen as attempting to characterize two metaphysically incompatible notions of non-individuals: (i) non-individuals as violating self-identity and (ii) non-individuals as indiscernible entities. The problem is that any choice between these options brings difficult questions, making the understanding of non-individuals through the apparatus of non-reflexive logics rather implausible.

Keywords Non-reflexive logics · Non-individuality · Identity · Individuality · Indiscernibility

14.1 Introduction

A very traditional understanding of quantum mechanics considers that, at least at the metaphysical level, quantum theory deals with *non-individuals*, items for which identity conditions fail; it is said that quantum entities have "lost their identities". This view is now called *the Received View* on quantum non-individuality (from now on, simply RV), given that we inherited it from some of the founding fathers of the theory. The guiding motivation behind the radical suggestion that identity somehow fails in quantum theory came from identity problems issuing from permutation symmetry, manifesting mainly in the quantum statistics for fermions and bosons (see French & Krause, 2006, chap. 3 for historical discussion). In its historical origins, the idea of 'losing identity' was kept at a rather informal level, as a kind of suggestion on what concerns the nature of those entities.

J. R. Becker Arenhart (✉)
Department of Philosophy, Federal University of Santa Catarina, Florianópolis, Brazil

C. Soto (ed.), *Current Debates in Philosophy of Science*, Synthese Library 477,
https://doi.org/10.1007/978-3-031-32375-1_14

Of course, remaining only at the level of a suggestion will not take us very far if we are concerned with a rigorous *metaphysical* characterization of non-individuality. What the founding fathers had suggested, by way of *reading off* a metaphysics from quantum statistics, was that quantum entities somehow had no identity: they could be permuted without giving rise to any distinct quantum state. The traditional illustration of the situation goes as follows: consider as an initial situation two quantum particles labeled *1* and *2*, and boxes *A* and *B*, each box with one of the quantum particles inside it. The issue with identity may be expressed by saying that it makes no sense to inquire whether it is particle *1* or *2* that is in *A* or in *B*. Exchange the particles from their respective boxes and, from the point of view of a quantum mechanical description, you won't get a configuration differing from the one you started with. No identity is available to a quantum system to account for a difference in the system of the boxes *before* the permutation and *after* the permutation of particles. So, the reasonable conclusion seems to be that quantum particles have no identity conditions. Were they to have identity conditions, then these very identity conditions would account for two distinct descriptions, one in which it is *1* that is in *A* and *2* in *B*, and another situation in which it is *2* that is in *A* and *1* that is in *B*. Given that no such distinct descriptions are available, it seems to follow that from a strictly quantum mechanical perspective, the identity of the particles is also not available.

Obviously, that train of thought leading from physics to metaphysics may be resisted in a plurality of ways, and it has indeed been resisted in the literature (see French & Krause, 2006, chap. 4; French, 2019). However, the debate is not settled in favor of any of the sides, and quantum mechanics may still be plausibly interpreted as the main motivation for a metaphysics of non-individuals, in the sense that quantum mechanics may be consistently (and maybe preferably) interpreted as dealing with non-individuals of some kind. This non-individuality has been traditionally cashed in terms of a lack of identity: quantum non-individuals are entities such that the relation of identity makes no sense for them. That seemed to be the main idea to be pursued if we are looking for an appropriate metaphysics of non-individuals, one that is consistent with the 'identity losing' talk present in most of the informal discussion of quantum statistics.

Now, this informal suggestion of a "lack of identity" has found its rigorous formal expression in non-reflexive logics, systems of logic for which identity is restricted in some sense. In particular, in some such systems it makes no sense to say that an entity is identical to itself or different from anything else. That is quite a strong demand, but it helps us making formal sense of something having no identity. As a matter of historical fact, non-reflexive logics were tailored to accommodate such intuitions coming from quantum mechanics. The plan is that, once such a system is advanced, one may grant that the metaphysical proposal is not itself incoherent from the start. As a result, great expectations are put on non-reflexive apparatuses. French (2019, sect. 5), for one, speaking of non-reflexive logics, claims that

> These developments supply the beginnings of a categorial framework for quantum 'non-individuality' which, it is claimed, helps to articulate this notion and, bluntly, make it philosophically respectable.

So, the metaphysical blow of losing identity could be softened by the development of such non-reflexive systems. The idea of characterizing non-individuality through the lack of identity acquires respectability due to its formalization in such systems of logic. Similar hopes are found in French and Krause (2006, p. 244), where it is claimed that quantum mechanics is compatible with a metaphysics of individuals *and* with a metaphysics of non-individuals, and that "these different approaches correspond to different logico-mathematical axiomatizations in terms of standard and non-standard set theories". So, there are high expectations on the job that non-reflexive logics should do in order to help us granting the feasibility of a metaphysics of non-individuals.[1]

However, our worry in this paper will be that in providing a formal system with certain features, we have not yet provided for a fully developed *metaphysical* framework for quantum non-individuality. It must be explained how the lack of identity in the system (a logical notion) accounts for the lack of individuality for the represented entities (a metaphysical notion). In this paper we address precisely this issue; we shall argue that such a metaphysical framework cannot be uniquely developed from the intuitions that guided the formulation of non-reflexive *formal systems*: in fact, such systems are, *prima facie*, compatible with at least two very different metaphysical pictures on non-individuality. Worst: a choice of one of them cannot be performed without leading us to great troubles; some features of non-reflexive logics recommend one of the approaches, other features recommend the other, and the metaphysical approaches seem to be clearly incompatible. As a result, tension arises in the very metaphysical understanding of the system. As we shall see, even if two metaphysical options seem to be on offer, none of them makes for a coherent choice in the stage set by non-reflexive logics.

The paper is structured as follows: in Sect. 14.2 we review the basics of the formalism of non-reflexive systems. In Sect. 14.3 we present what seems to be the standard account of the underlying metaphysics of non-individuality; it concerns the failure of self-identity. In Sect. 14.4 we present a second possible account of a metaphysics of non-individuality, dealing with the failure of the *Principle of the Identity of Indiscernibles* (PII) and an associated failure of a version of a bundle theory of individuality. In Sect. 14.5 we discuss what we take to be the main intuition behind both proposals: lack of individuality must be accompanied by lack of identity. This tenet of the actual formulation of the RV is responsible for most of the troubles of the view. We argue that it is difficult to accommodate all the desiderata of non-individuality in only one of the two metaphysical options presented if the RV is to be formulated as involving a lack of identity. The two metaphysical views are guided by incompatible guiding intuitions, and it seems impossible to coherently choose one of the metaphysical approaches as preferable while keeping

[1] Further claims on the relevance of non-reflexive logics for the metaphysics of non-individuality may be found in French and Krause (2006, chap. 6), for instance, where it is argued that classical logic is inadequate to deal with non-individuals, so that a non-standard system (non-reflexive logic) motivated by quantum mechanics should be developed. See also the discussion in Arenhart (2018).

non-reflexive logics as the underlying logic of the resulting metaphysics. We finish in Sect. 14.6 with prospects of formulations of the RV and its relation with identity.

14.2 Non-reflexive Logics in a Nutshell

Non-reflexive logics, in the specific sense we are considering them in this paper, appeared first with the work of Newton da Costa in (1997, pp. 123–126, originally published in Portuguese in 1980). In this book, by presenting non-reflexive systems, da Costa's original concerns focused on the very possibility of violating the so-called *Principle of Identity*, not with the specific articulation of a formal counterpart for the RV. The general thesis was that there are no inviolable laws of logic: each of the so-called "laws of thought" may be violated if an appropriate system of logic is devised.

Trying to motivate his systems with more than mere formal curiosity, da Costa found his motivation for the development of a non-reflexive system in a quote by Schrödinger, perhaps one of the most famous quotes in the context of the identity debate in quantum mechanics:

> I beg to emphasize this and I beg you to believe it: it is not a question of our being able to ascertain the identity in some instances and not being able to do so in others. It is beyond doubt that the question of 'sameness', of identity, really and truly has no meaning. (Schrödinger, 1996, pp. 121–122)

The idea seems to be that quantum entities are such that identity does not make sense for them. This, as Schrödinger emphasizes, is not a matter of epistemological limitations, of our not being able to ascertain identity in some cases while being able to do so in others, but it is rather due to the features of the entities; the impossibility has an ontological source. According to Schrödinger, that lack of identity is one of the novelties brought about by quantum mechanics (and that kind of claim usually puts Schrödinger among the founding fathers of the RV). Now, while it is very problematic to claim that Schrödinger meant literally that the usual relation of numerical identity does not make sense in some cases, we shall take it as such only for the sake of argument, following the interpretation that the friends of non-reflexive systems have employed to motivate their developments.

Once such exegetical concerns are put aside, let us proceed and check how non-reflexive logics are built over such idea of identity failure. Da Costa formalized that understanding of Schrödinger's claim by introducing the so-called *Schrödinger logics*. These are two-sorted first-order systems in which, besides typical apparatuses such as quantifiers, connectives and non-logical predicate constants, one also introduces *two kinds of individual terms*. One kind of terms, let us call them *first-kind* individual terms, are thought of as representing typical objects, for which identity makes sense. For those terms, it is perfectly legitimate to write formulas of the kind $J = K$, where J and K are both first-kind individual terms. Also, for first-kind terms the reflexive law of identity $\forall X\,(X = X)$ is perfectly valid: for every variable X (identity is a primitive sign of the language). For terms of a *second-kind*, whose intended interpretation concerns quantum entities—the ones Schrödinger seems to be claiming that identity does not make sense for—we cannot have such formulas involving identity. That is, if j or k are second-kind terms, $j = k$ *is not even a formula*. It is in this sense that identity statements cannot even be formulated within the language of Schrödinger logics.

Then, it is in this precise sense that the law of identity is restricted. As a consequence of such a restriction we cannot have $j = j$, for j an individual variable or constant of the second-kind (while we still have $J = J$ for J term of the first kind, recall). It is this restriction of identity that makes Schrödinger logics a kind of *non-reflexive logic*. Notice that it is not the case that some entities are distinct from themselves: the application of identity is restricted and its negation is restricted as well. "Lack of identity" is a matter of not being able to express identity, not of self-identity being false in some model.

While da Costa was the first to advance such a system, there was no immediate metaphysical attempt to interpret the resulting system on what concerns the nature of the entities described. The general kind of view of the role of logic in the development of conceptual systems advanced in da Costa (1997) allows us to consider that a system of logic determines the meaning of some of the most general categories of a scientific theory, such as *objects*, *properties* and *relations*, but that investigation was not further pursued by da Costa in the particular case of Schrödinger logics. Obviously, this form of equating the problem is a nice alternative to typical Quinean views of objecthood, according to which the notion of object is determined by predication and the use of variables in *classical logic*. Allowing for a change of logic allows for a corresponding change in the very notion of object (a thesis that is in fact present in da Costa, 1997), and we may, after all, have "entity without identity": it is enough that we change the logical apparatus to Schrödinger logics. However, pursuing that suggestion would take us far from the route we have stipulated here, and as we have mentioned, da Costa does not pursue those lines either.

Systems of Schrödinger logics were further developed by Krause (1990) into higher-order logic and into a set theory comprising non-individuals, which is now famous as *quasi-set theory* (for further discussion and technical details, see French & Krause, 2006, chaps. 7 and 8). Besides the technical development of those systems, Krause advanced in his 1990 a suggestion of a metaphysical correlate of the system that worked to the effect that non-reflexive systems do indeed work to represent a specific kind of non-individuals. But before going into such a discussion, let us briefly see how quasi-set theory encompasses the idea of a lack of identity.

In quasi-set theory, instead of a two-sorted language, we have a one-sorted language for a theory of sets with atoms. There are two kinds of atoms, the M-atoms and the m-atoms; q-sets, the collections of the theory, are defined as things that are not atoms. The first kind of atoms acts as the formal counterpart of common objects, for which identity does make sense. On the other hand, m-atoms represent quantum entities, for which identity does not make sense. How is such a division achieved at the formal level? Identity is *defined* as a relation holding between M-

atoms belonging to the same q-sets, or else a relation between q-sets having all the same elements. This makes identity a relation not defined for m-atoms. So, we cannot prove in quasi-set theory that the reflexive law of identity holds for m-atoms, while we still can do that for M-atoms and for q-sets.

Obviously, having those apparatuses at our disposal still does not advance much in the sense of promoting a *metaphysics of non-individuals*. One still has to connect, as we mentioned, the very idea of entities without identity with some kind of non-individuality, and develop a detailed account of how to live without identity, at least for some kinds of objects (see Wehmeier, 2012 for advocacy of a *complete* elimination of identity; see also Krause & Arenhart, 2019). In other words, a metaphysics of non-individuals must be developed in order to put some flesh on the formal bones of the non-reflexive systems; otherwise, we have a case of a formal system awaiting for an interpretation. Notice also that this metaphysics we are looking for has already acquired some very specific contours due precisely to its relation with non-reflexive logics: *non-individuality must be cashed in terms of a failure of identity*. In other words: the *metaphysical* notion of non-individuality must be cashed in terms of the failure of the *logical* relation of identity.

As we have already mentioned, non-reflexive systems did not come out of the blue. They were developed motivated by quantum mechanics and a specific interpretation of the message the theory was supposedly sending us about the nature of its objects. It is to this message and its metaphysical counterpart that we shall now turn.

14.3 The Metaphysics of Non-individuality Part 1: Transcendental Individuality

Perhaps one of the main attempts at associating non-reflexive systems with a metaphysical picture of non-individuals was presented by French and Krause (2006, chap. 4). A metaphysical characterization of non-individuals was clearly advanced there and also clearly associated with non-reflexive systems. Curiously, however, this metaphysical articulation is rarely discussed in the context of the metaphysics of quantum mechanics associated with the RV; the focus in those discussions is usually restricted to the formal counterpart, that is, on the idea that identity may somehow fail in non-reflexive systems (see also Sect. 14.5 below).

But how did French and Krause manage to provide for a metaphysical characterization of non-individuals that is compatible with non-reflexive logics? The answer comes from a venerable (albeit highly speculative) tradition on the attribution of *individuality* to particulars, a tradition which French and Krause called *Transcendental Individuality* (the label comes originally from Heinz Post; see French & Krause, 2006, chap. 1 for further discussion). The main idea, as we shall see, is that while individuals may be characterized by possessing (or instantiating) transcendental individuality, *non-individuals may be understood as particular entities that do not possess (or do not instantiate) such transcendental individuality.* Let us check the details.

Roughly speaking, transcendental individuality comprises two distinct packages of individuality principles: it includes principles that account for the individuality of a particular by the introduction of an additional ingredient on the composition of individuals (dealing with compositional theories of particulars), such as substrata and bare particulars, and it also deals with individuality principles that are independent of composition theories of individuality, as for instance non-qualitative features such as primitive thisnesses and haecceities. Notice that substrata and bare particulars are particular items *added* to the composition of a particular; particulars are thought of as being composed by an individual substratum and its properties. On the other hand, primitive thisnesses and haecceities need not be involved in a theory about the composition of a particular (they need not be *parts* or *members* of the individual), but are rather thought of as non-qualitative properties, that is, properties that add no new qualitative feature to the individual.

What is common to both approaches—and what makes them both able to fall under the umbrella of transcendental individuality—is precisely the fact that what confers individuality to a particular in both cases is something that goes over and above any qualitative feature of the particular. In the case of bare particulars and substrata, it is the extra ingredient in the composition of the individual, the bearer of properties, whose identity is unanalyzable, which confers individuality to the particular (see Loux, 1998; Lowe, 2003 for further discussion). Each individual has its own bare particular or substratum, which is responsible for the individuality of the item. Bare particulars and substrata are rather mysterious entities, their characterization is troublesome, to say the least; haecceities and primitive thisnesses may be less problematic to describe, but not less troublesome: they are non-qualitative properties. Socrates, for instance, is individuated, according to this account, by having the non-qualitative and non-shareable feature of "being Socrates". In this case, the idea of a non-qualitative feature is generally couched in terms of a universal uniquely instantiated by that particular for whose individuality it is responsible, but we need not assume this here.

Both kinds of individuality principles play the desired metaphysical role of individuators, that is, they account for an item being precisely what it is. In this sense, individuality goes much beyond granting mere numerical identity. Following Lowe (2003, p. 93), we take it here that an individuality principle has an explanatory role, it accounts for what a particular item is. Transcendental individuality principles, in both forms discussed here, respect that basic requirement.

What is most relevant for our discussion here is that while both kinds of principles (extra ingredients on the one hand, non-qualitative properties on the other) are somehow mysterious, according to French and Krause (2006) both may be characterized formally by the same kind of feature which, not surprisingly by now, is precisely *self-identity*. That is, the possession of a kind of transcendental individuality by Socrates is characterized by Socrates' instantiating or exemplifying his self-identity (see French & Krause, 2006, p. 5; pp. 13–14; p. 140; see also Lowe, 2003, p. 87, and French, 2015 sect. 5; the idea goes back at least to Adams, 1979). In this case, self-identity is enough to grant some kind of transcendental individuality, in one of the versions discussed here.

Obviously, that characterization of individuality fits perfectly well with the underlying motivation for non-reflexive logics: given that the reflexive law of identity can be seen as attributing some form of transcendental individuality, having no individuality (a metaphysical feature) then is to be understood in terms of the lack of such transcendental individuality, which is formally represented by *the lack of self-identity* (a logical feature). The stage is perfectly set for the non-reflexive logics to represent precisely that specific kind of non-individuality. As French and Krause (2006, pp. 13–14) put it:

> … the idea is apparently simple: regarded in haecceistic terms, "Transcendental Individuality" can be understood as the identity of an object with itself; that is, '$a = a$'. We shall then defend the claim that the notion of non-individuality can be captured in the quantum context by formal systems in which self-identity is not always well-defined, so that the reflexive law of identity, namely, $\forall x(x = x)$, is not valid in general.

They also go on to hold that "conceiving of individuality in terms of self-identity will allow us to appropriately represent its denial" (French & Krause, 2006, p. 15), and agree that they "are supposing a strong relationship between individuality and identity … for we have characterized 'non-individuals' as those entities for which the relation of self-identity a=a does not make sense" (French & Krause, 2006, p. 248). Of course, given that it is a kind of Transcendental Individuality that is represented by self-identity, its denial means that no such Transcendental Individuality is present for non-individuals. Notice that even here the main worry concerns formal representation, not the metaphysics *per se*.

So, there we have it. According to *this* first metaphysical dressing of the concept of non-individual, non-reflexive logics are thought of as restricting identity precisely because their aim is to deal appropriately with non-individuals. The latter are thought of as particulars that fail to have individuality according to a kind of transcendental individuality principle, which on its turn is represented by self-identity. In a not very memorable slogan: *no identity, no transcendental individuality*. Of course, this interpretation was not advanced by da Costa, for instance, when he first developed Schrödinger logics, but it makes perfect sense in the context of quantum mechanics and a metaphysics of non-individuals.

However, despite its apparently being made to fit perfectly well with the formal representation already available by non-reflexive logics, this account of non-individuality has an important consequence: if we are to keep perfect symmetry between individuals and non-individuals, then it should be clear that items having identity, in particular those items represented in non-reflexive logics by terms that *do* have identity, will be individuals due to their having identity; more than that, following the present understanding of identity and individuality, *their having identity commits us to a theory of individuality according to transcendental individuality*. In that sense, the idea that self-identity represents transcendental individuality commits identity to playing a metaphysically inflated role: it represents a very important metaphysical role of granting individuality according to a TI form of individuality. We shall discuss this issue soon.

Now, the previous articulation in terms of transcendental individuality is not the only option available for a metaphysics of non-individuality which can be made compatible with non-reflexive logics. As we mentioned, it was not the original interpretation by da Costa, and neither by Krause (1990). It is to the main metaphysical motivations introduced in Krause (1990, 1994), and which is also strongly suggested in French and Krause (2006), that we now turn.

14.4 The Metaphysics of Non-individuality Part 2: Identity of Indiscernibles

Along with the technical innovations which were brought to non-reflexive logics by the development of higher-order Schrödinger logics systems and related systems of quasi-set theory, Krause (first in Krause, 1990, but also discussed in Krause, 1994 and then in French & Krause, 2006, chap. 6 and chap. 8) introduced a crucial novelty which is related to his interpretation of a possible metaphysics underlying the formalism: a relation of indiscernibility. Recall that quantum particles obey permutation symmetry, which means (very roughly) that by permuting such particles no distinct state is obtained. As a result, one cannot perform measurements trying to physically distinguish the particles before or after the permutation: the probability outcomes of any physical magnitude are the same before and after the permutation of the particle labels in the state vector.

The main motivation for the introduction of an indiscernibility relation comes from Heinz Post's (1973) claim that indistinguishability of quantum particles should be present in the formalism "right at the start"; that is, one should not *first label the particles*—thus distinguishing them—and then *later add a symmetry postulate* to the effect that labels play no role (see French & Krause, 2006 chap. 6, Krause, 1990, p. 72, Krause, 1994, p. 400). The indiscernibility relation is introduced in non-reflexive systems to achieve just that purpose. Classical systems of logic and mathematics, such as Zermelo-Fraenkel set theory, always end up collapsing those concepts. According to Krause (1994, p. 402), by using quasi-set theory

> a much more adequate interpretation of the absolute indistinguishability relation would be achieved, and a satisfactory distinction between this relation and the predicate of identity would be established.

Also, French and Krause (2006, p. 240) claim that classical logic

> ... involve a theory of identity which takes the elements of a set (even the *Urelemente*, if they are admitted by the theory) to be *individuals* of a kind. In short, this 'theory of identity' contrasts with the Received View of quantum entities as absolutely or 'strongly' indistinguishable entities, as we have discussed, and cannot provide the grounds for treating 'truly' indistinguishable non-individuals.

So, the problem with classical logic seems to lie in the fact that it makes for distinguishability always available: classical logic always allows for items to be discerned, thence its inadequacy to represent non-individuals (see also French & Krause, 2006, sect. 6.5.2 and 6.5.3 for this specific line of discussion). Along with indiscernibility, as we shall discuss now, comes a metaphysical view differing substantially from the one discussed in the previous section.

From a formal point of view, in non-reflexive logics the relation of indiscernibility is just another binary relation satisfying the postulates of an equivalence relation. Given that in quasi-set theory, for instance, identity is not defined for every item, identity and indiscernibility do not coincide. In particular, *m*-atoms may be indiscernible while not being identical. Something similar happens in Schrödinger logics, respecting the distinction of types of terms: items may be indiscernible and indiscernibility does not collapse on identity (for further details see French & Krause, 2006, chap. 7 and chap. 8).

Now, from a metaphysical point of view, indiscernibility by itself has its own important consequences. If one is attempting to ground individuality in some kind of discernibility relation, then it seems that indiscernible items will not be individuals. That would make a direct link between the formal novelties brought about by an indiscernibility relation and a metaphysics comprising non-individuals. That is precisely the underlying idea forwarded by Krause (1990, 1994, 2011); even French and Krause (2006), although they are explicit on the idea that non-individuals are characterized through the lack of self-identity, oscillate between the approach discussed in the previous section and the idea that failure of discernibility is also characterizing non-individuality. They sometimes seem to imply that non-individuality is a result of indiscernible but non-identical items. Our previous quote of the passage in French and Krause (2006, p. 240) seems to imply precisely that, due to the fact that classical logic cannot legitimately accommodate "'truly' indistinguishable non-individuals".

That is, individuality in classical systems[2] is to be contrasted with the indiscernibility of quantum entities. Even though French and Krause are explicit in other passages that lack of identity represents non-individuality, they sometimes seem to put indiscernibility as the source of non-individuality. This also appears in French and Krause (2006, p. 245), where it is claimed that non-reflexive logics consider non-individuality *right at the start* by legitimately taking them as indiscernible. So, the idea of indiscernibility grounding non-individuality is also present here, sometimes even taking precedence over the lack of identity.

Consider also the following passage, where it is suggested that non-reflexive systems of logic should account for the indiscernibility of quantum entities, something (it is claimed) classical logic cannot do:

> [...] the basic quantum entities of the same kind may be indistinguishable, and so it is a pertinent question to look for the kind of 'logic' they obey. In such a logic, of course, we would be able to talk of indistinguishability, and to consider that some entities may have

[2] Due to discernibility being always available, in the view of French and Krause.

all their relevant properties in common without turning out to be the very same entity, as implied by Leibniz's Law. (French & Krause, 2006, p. 318)

This strongly contributes to a reading of non-individuality, as represented by non-reflexive logics, in terms of indiscernibility. Classical logic deals with individuals because it allows every item to be discernible from every other item; non-reflexive logics allow for non-individuals precisely because it allows that some entities be indiscernible without being identical.

Traditionally, grounding individuality in some kind of discernibility relation is the most natural option for those not willing to endorse one of those forms of TI principle of individuality. This option not only avoids positing metaphysically mysterious ingredients, but also seems to be more in tune with an empiricist foundation for individuality. Individuals are somehow characterized by their properties. Taken in this sense, we have what is traditionally called the *Bundle theory of individuation*, according to which an individual is a bundle of properties. Notice that this is again a theory involving the composition of individuals (although it may be framed without such a commitment, we shall not go into the details here, but see Rodriguez-Pereyra, 2004).

A bundle theory, at least when it is understood as a theory dealing with the composition of individuals, requires some form of the *Principle of Identity of Indiscernibles* to hold: if two entities share all the same properties, then they are the same. Obviously, if individuality is to be granted by the bundles, then no two bundles should be identical. We may follow French (2015) and distinguish three versions of the principle, which differ precisely in the features they allow to count as genuinely distinguishing: (1) any properties and relations whatever are allowed to count as distinguishing features; (2) the same as before, excluding just spatio-temporal properties and relations; (3) only non-relational properties are allowed.

Versions 2 and 3 of the principle are violated already in classical mechanics, that is, classical particles may be such that they share every non-relational property (violating 3) and also may share every property and relation, excluding spatio-temporal properties and relations (violating 2). Quantum mechanics, however, goes one step further and violates 1 too. Due to permutation symmetry, no property whatever could distinguish such particles, and so they are thought to violate even the most general form of the PII.[3]

Given the failure of the PII, the natural thought, then, is that quantum entities are not individuals according to a bundle theory. They may be, as non-reflexive systems represent it, indiscernible without being identical. That is, instead of admitting the failure of the bundle theory, we keep the theory and accept that some entities are not individuals as characterized by the bundle theory. This is a different way of framing the notion of non-individuality than the previous one, grounded on the failure of transcendental individuality. To grant that individuals will obey the underlying intuition of the bundle theory of individuality, it is assured that the Principle of

[3] For those worried about the current discussion on weak discernibility in the context of quantum mechanics, see the last paragraph of this section.

Identity of Indiscernibles will hold for individuals. Notice, that is, that just as in the case of the Transcendental Individuality approach, in which individuals had TI while non-individuals lack it, here *individuals obey PII, while non-individuals fail it*.

From a formal point of view, then, a kind of symmetry in the treatment of individuals and non-individuals is also preserved in this second approach: for items for which identity is defined in quasi-set theory, identity and indiscernibility coincide. For items for which identity is not defined, the *m*-atoms of quasi-set theory, only indiscernibility holds, with identity not being even expressible. This, of course, represents only one way in which the PII fails: one could have indiscernible items that are not identical (that is, that are different), but that option cannot be expressed in the underlying non-reflexive logic.

There we have then, another metaphysical option for the non-reflexive systems. Again, it is curious that this metaphysical approach to non-individuality is rarely discussed in the contexts where the RV is attacked or even thoroughly defended. Notice that this second metaphysical view benefits heavily from the indiscernibility relation that was originally introduced by Krause (1990), while the TI view benefited from the main feature of non-reflexive logics, which is the failure of the reflexive law of identity. In this second approach, failure of the PII must cope with failure of identity, and PII fails not because indiscernible items may be different, but because identity simply does not apply; the principle is somehow restricted. This restriction, on its turn, is inherited from previous versions of non-reflexive logics, namely, the one developed by da Costa in 1997.

Before we leave this section, an important issue must be addressed. Some may be concerned with our discussion of PII. It could be objected that nowadays weaker forms of the PII have been introduced in quantum mechanics appealing to *weakly discerning relations*. Roughly speaking, a relation R weakly discerns x from y if it is irreflexive and symmetric. In this sense, we have xRy and yRx (due to symmetry), but we don't have xRx or yRy (due to irreflexivity). That grants that whenever such a relation is available and holds between x and y, x and y are numerically different, and identity is granted to work just fine for such entities. Quantum entities are said to obey weak discernibility, and then a weak version of the PII also holds (see the pioneering discussion in Muller & Saunders, 2008, and also Caulton & Butterfield, 2012; Huggett & Norton, 2014). That would put a lot of pressure on the RV as presented here, because both major claims by the current articulations of the RV, *viz.*, that quantum entities have no identity and that they are indiscernible, would be under question.

However, even tough weak discernibility could be employed as a weapon against such articulations of the RV, we prefer to leave that kind of criticism aside for this paper. Claiming that the present approach to quantum mechanics is wrong because there are weakly discerning relations granting identity and discernibility would work as a kind of *external criticism*. It may not have full impact on the friends of non-individuality, however. Typically, reactions by friends of non-individuality consist in claiming (i) that weakly discerning relations are not enough to ground numerical distinction (given that a weakly discerning relation requires *two* relata available beforehand), (ii) that proofs that quantum mechanics obeys weak discernibility beg the question against the RV, by employing a formal apparatus (ZFC) that already requires identity available for every entity (and one of the reasons to shift to a non-reflexive system, recall, was to avoid such problem), and (iii) that such relations are not really discerning, given that at best they could establish a numerical difference, but not qualitative difference (for the references and a brief account of the dialectics of the debate, see French, 2019, sect. 4).

So, given that the debate on weak discernibility could lead us to discussions that would distract us from our more immediate goals, our argumentative strategy here follows a different route. We avoid entering into the debate on weak discernibility and its impact on the RV; we shall rather provide for a kind of *internal criticism* of the RV. That is, irrespectively of how such debates between weak discernibility and non-individuality are going to be settled, our aim is to take as a starting point the formalism of the non-reflexive approach to non-individuality, with its two accompanying metaphysical options, as described in this section and in the previous one, and argue that no coherent metaphysical picture has effectively emerged. In this sense, the criticism presented here is independent of the external criticism coming from the weak discernibility debate. So, without further comments, let us check which are the internal difficulties for the metaphysics of non-individuality once a non-reflexive logic is adopted.

14.5 No Individuality, No Identity?

As we have discussed in the two previous sections, non-reflexive systems may be accompanied by at least two distinct and incompatible metaphysical packages.[4] Both benefit from specific formal features of such logics. The existence of two metaphysical packages by itself should not strike us as surprising: given that the concept of an individual may be defined in distinct ways, failing to be an individual may also be defined in distinct ways. What is surprising, perhaps, is that both accounts of non-individuality, in the case in focus, are framed as resulting from a failure of identity; their being so framed means that they may benefit from the lack of identity for some kinds of terms in non-reflexive logics and, consequently, have an appropriate formal representation. As we have seen, it seems in part that the relevant metaphysics ends up being dictated by the formal apparatus that was previously available to represent the idea of a non-individual. We shall turn ourselves now to the idiosyncrasies of the concepts of non-individuals thus framed; we shall discuss

[4] The term 'packages' here refers to the fact that while each metaphysical approach (bundle theory and TI) has very specific general features, there is still a lot of space for fixing the details, so they are not, after all, just two approaches, but rather general starting points for two families of approaches. For instance, the bundle theory has to spell the details of how properties are bundled together, and this may be done in distinct ways, all of them compatible with the general view that goes under the label of 'bundle theory'. The same may be said about TI approaches.

in particular whether each of the metaphysics of non-individuals presented before coheres with our understanding of the formal system itself. We shall argue that some considerable knots remain for both options; in fact, their *prima facie* compatibility with the formal system is threatened by features of the formal system itself, so that tensions remain to be dealt with.

The fact that a system of logic does not have a unique underlying philosophical understanding (to put it in broad terms) is not a novelty. As discussed by da Costa (1997, pp. 85–86), for instance, even traditional Aristotelian logic serves as the formal apparatus for very different such philosophical underpinnings. On the one hand, Aristotle himself saw his own system as grounded in the very nature of things, as reflecting ontological categories. Kant, on the other hand, with his Copernican revolution, saw the very same logic dealing with categories that are rather conditions for our knowledge of things; categories are not thought to follow from things, but things are thought according to the categories.

However, the situation for non-reflexive systems and their accompanying metaphysical interpretations has important differences. Given the importance of non-reflexive logics in the foundation of the RV, an equivalent discussion of the most adequate metaphysical understanding of such systems has not already been made. While non-reflexive systems are very closely tied to the RV nowadays, what the *metaphysics* of non-individuals looks like, is still rarely discussed. The fact that there is one formal system to represent non-individuals seems to imply that there is one metaphysics of non-individuality, that "non-individual" is precisely understood as soon as the formal system is devised. That is not correct, as we have seen. The formal system gives us no unified picture of non-individuality. Furthermore, as we have already mentioned, non-reflexive systems impose a very specific kind of approach to non-individuality, one in which non-individuality must be cashed in terms of lack of identity. Our contention is that when such issues as the proper articulation of the metaphysics of the RV is properly pursued, the disunity inside non-reflexive logics may bring some difficulties that have not yet been addressed, challenging the very coherence of the RV. As we have seen, French (2019, sect. 5) even goes on to claim that the respectability of the non-individuals approach relies, in great measure, on the existence of formal apparatuses to make the idea clear. So, if the metaphysical interpretations available for the system fail to cohere with the system, then, there is something wrong with the view. It is to this claim that we now turn.

Recall that the original reason for formalizing the idea of a "loss of identity" came from da Costa's attempt at capturing Schrödinger's claim about one of the greatest novelties of quantum mechanics: identity makes no sense for quantum particles. It is precisely that lack of identity that characterizes non-reflexive logics. Now, as we have seen, turning to the metaphysical side, according to some of the proponents of such logics, the lack of identity may acquire two distinct metaphysical interpretations in the light of the proponents of the RV:

(i) it may represent the lack of transcendental individuality or,
(ii) when coupled with an indiscernibility relation, it may represent the failure of a version of the principle of identity of indiscernibles, relating the system thus with a bundle theory of individuality.

Notice that there are two major conflicting intuitions behind those two metaphysical projects. One of them, the one originally pursued by da Costa, seems to advise us to keep with Schrödinger and simply forget about identity. That by itself seems to recommend the TI approach to individuals and non-individuals. The other intuition, not completely related to the loss of identity by quantum particles, concerns their indiscernibility and the very idea that it should be formally represented in a mathematical framework allowing us to separate identity from indiscernibility; this comes from Post's demand that indiscernibility should be present right at the start in the formalism (there are also other motivations coming from Yuri Manin's complaint that quantum theory has not its own language, see French & Krause, 2006, chap. 6). Separating identity from indiscernibility seems to recommend that we adhere to the bundle theory of individuality and define non-individuals also according to that package, *i.e.* as indiscernible entities failing PII.

It seems a conflict is ready to take place between the main intuitions behind the two approaches. On the one hand, one could for instance agree with Schrödinger's claim and still not wish to introduce an indiscernibility relation right from the start, leaving indiscernibility to be expressed by the shareability of every property concerned; or on the other hand, one could choose to keep only with discussions about indiscernibility not collapsing in identity, but still maintain identity for whatever reason (disagreeing with the Schrödingerian motivation). What is troubling, perhaps, in the above formulations of the metaphysics of non-individuality, is that the formalism of non-reflexive logics has attempted to accommodate both intuitions. When we turn to the metaphysics of non-individuals, however, those intuitions pull to different directions. As we shall see, the fact that they pull to different directions grants that the packages are indeed distinct (bundle theory and transcendental individuality are deadly enemies!), but it also makes it difficult to keep both intuitions together in the formalism. We end up with no clear idea of what a non-individual should be. Let us check.

14.5.1 Keeping with the TI Package

At first sight, it seems that the most natural metaphysical interpretation for the fact that identity does not make sense for some entities should be cached in terms of a lack of transcendental individuality. However, as we mentioned, there results an accompanying reading of the *presence* of identity, which means that entities having identity have transcendental individuality. That move is inconvenient for it inflates the logical notion of identity with metaphysical content that goes much beyond what most people would wish to accept. Identity (the logical notion), under this particular view, represents individuality (the metaphysical notion) in a metaphysically thick sense as developed by TI. If that is the price of having non-individuals, then not many will buy it. So, in terms of metaphysical economy, the trouble of introducing non-individuals by failure of TI is that one also introduces *individuals* as the items having TI. Typically, empiricists are suspicious of such metaphysical posits, and one would have to face the typical debates concerning the plausibility of the introduction of TI-like individuating devices for individuals.

But that approach is not inconvenient only because of the metaphysical load on individuals and non-individuals, but also because this particular understanding of the lack of identity makes some strong demands that, in the context of current disagreements on the notion of identity in quantum mechanics, seem highly implausible. That is, current debates on the role of identity in quantum mechanics is not framed in terms of having or not having transcendental individuality, so that adding that demand of a metaphysical reading of identity and lack of identity puts a heavy burden on the shoulders of the friend of the RV.

To illustrate what kind of burden we are concerned with, consider the claims by Bueno (2014), according to whom identity is fundamental (something along similar lines could be said of the contentions by Dorato & Morganti, 2013 that identity is primitive). Bueno advanced many theses to the effect that identity cannot be eliminated, so that as a result the RV fails to be intelligible due to its rejection of identity. The main points advanced by him are: (i) we need identity for the correct understanding of the working of concepts, (ii) we need identity for counting and cardinality attribution, (iii) identity is fundamental because it is undefinable, and (iv) we need identity to understand quantification. We shall not discuss the merits of each point advanced by Bueno, but what is relevant for us is that his own understanding of identity is framed based on the assumption that identity can be understood as having a metaphysically deflated meaning. Identity is welcome to empiricists (such as Bueno) because it implies nothing metaphysical. When he argues that identity is fundamental, *it is the deflated identity he has in mind* (see Bueno, 2014, sect. 4).

Now, this deflated understanding of identity must be opposed to the inflated understanding that we are now examining, and must also be taken into account when it comes to debate the role of identity in quantum mechanics and the fundamentality of identity. The RV, articulated as a failure of transcendental individuality, implies that identity has a very substantial content; it is this substantial content that is absent for non-individuals. Bueno wants to keep a rather deflated interpretation of identity, and certainly would also wish to abandon a substantial view on identity. It seems certain that Bueno would not hold that it is a kind of TI that is fundamental when he claims that identity is fundamental. So, in this sense, there would be agreement between Bueno and the friends of the RV on the claim that quantum entities have no transcendental individuality. On the other hand, there is still no agreement on whether the identity relation should also go, because Bueno would clearly want to jettison TI while keeping identity, something implausible if we represent TI through self-identity. So, a first step in the debate is missing: is identity substantial, or a deflated understanding of identity is enough? After answering that question, and only then, we can go on to inquire whether identity is fundamental or not. It seems that it is incumbent of the friend of the RV to show that identity must have the substantial meaning required in order to make sense of the first metaphysical package of non-individuals. That is a rather heavy burden incumbent on them, given that not many are willing to have a metaphysically inflated view on the logical relation of identity.

Furthermore, the problem may go even deeper. Let us concede that Bueno is right and that identity could be interpreted as a deflated relation, with no substantial metaphysical meaning. Then, to stand against Bueno's arguments and keep non-reflexive logics, the friend of the RV would have to grant that not even a deflated identity applies to quantum entities. Perhaps we can do that at the formal level, and identity is really dispensable in some formal systems, but that would be no victory for the version of the RV under analysis here: once identity is understood as metaphysically deflated, we have no longer the possibility of framing the metaphysical understanding of non-individuals required. So, it seems that the non-reflexive formal systems would be able to survive, but not the accompanying metaphysical understanding of the limitation of identity. A metaphysics of non-individuals in which it is a deflated notion of identity that fails is still lacking. In other words: once identity is metaphysically deflated, it is hard to represent non-individuality as lack of identity.

So, on the first approach to non-individuality, the one in which non-individuals lack transcendental individuality, we might end up having many difficulties. It loads the logical relation of identity too much and makes it play the role of a metaphysical individuation principle; while doing this, it prevents that we make complete sense of part of the debate concerning identity in quantum mechanics: it is identity in a rather minimal sense that is at stake (identity, that is, unrelated with the metaphysical problem of individuality). So, even though this seems to be the most natural way to represent metaphysically the lack of identity, it brings us too much trouble, but there is more, which concerns the relation of this metaphysics of non-individuals with the formal systems of non-reflexive logics.

A related problem concerns the introduction of indiscernibility in the formal apparatus of non-reflexive systems. As we mentioned, for objects with identity (the classical objects in non-reflexive logics, the individuals), *indiscernibility and identity coincide*. However, that seems to go strictly against the main motivation to have a transcendental individuality as a principle of individuality: in adhering to that kind of approach, one would like to have at least the possibility of objects being indiscernible but not identical (consider the case of Black's spheres, for instance; see again Loux, 1998; Lowe, 2003). That is, TI approaches to individuality exist precisely to break the tie between identity and indiscernibility. But the collapse of identity and indiscernibility for classical entities in non-reflexive logics makes that case simply impossible; in fact, coinciding identity and indiscernibility for individuals make for transcendental individuality totally unnecessary when it comes to individuals! So, what is the use of adopting a TI package? In this case, the package ends up internally incoherent; while non-individuals lack TI, individuals do have TI, but the fact that for individuals identity and individuality collapse takes away any advantage the positing of TI for individuals could have! We have all the costs of a TI approach, with none of the benefits.

Maybe a bundle approach should be adopted for individuals, leaving the lack of a TI principle to characterize non-individuals? But that last suggestion brings us trouble: it would amount to a hybrid view of individuality and non-individuality. Consider what is involved in that suggestion: we would have non-individuals characterized by the fact that those entities lack self-identity (*i.e.* they lack TI), while individuals would be characterized by their bundles of properties. A double standard would end up being accepted, and we would have the most expensive of the metaphysical options. Worst yet: in this case, non-individuals are not characterized as failing individuality. This is certainly not an option envisaged by anyone, so that to avoid such *ad hoc* solutions we should try to keep uniformity by adhering to the following guiding principle: non-individuals are those entities that somehow lack precisely that which makes individuals the individuals they are. The hybrid view clearly violates this principle.

Another option could be simply to break the link between identity and indiscernibility for individuals, allowing that the TI work just as intended. The individuals would have TI because of the presence of identity, but then they could also be indiscernible without being the same, because now identity and indiscernibility are not equivalent. But that path would leave us with another trouble. Allowing for TI-like individuals that are indiscernible without being the same is just another *possible* metaphysical interpretation for quantum entities (see French & Krause, 2006, chap. 4). In this interpretation, quantum entities are *individuals* according to TI, and this option just rivals the RV. So, by following the recommended strategy, one would be introducing in the formalism of non-reflexive logics all the conceptual tools needed to deal with quantum entities *as individuals*. However, once that is also available, there seems to be no further reason to keep the non-individuals also as an option: economy of posits and uniformity in treatment would recommend us keeping only the individuals, quantum and classical, without the need for non-individuals too. That is, following this strategy seems to render non-individuals less palatable. However, we seem to end up in a dilemma: either the individuals are indiscernible (and then we do not seem to need non-individuals), or else they are not (and then, TI principles of individuality seems a useless addition). The choice is a difficult one, but seems to be forced on us if we are to interpret the formalism of non-reflexive logics, which has both lack of identity and an indiscernibility relation.

As a result of the previous discussion, the trouble with TI approach to non-individuals may be summarized as follows: having no identity makes complete sense as having no transcendental individuality, but then we must also account for the entities having identity. In that case, there are troubles because (i) we unwillingly introduce a TI kind of individuality for individuals, (ii) we frame the debate on the plausibility of non-reflexive logics in terms of the plausibility of having TI or not (which distorts the typical debates, given that they are not about TI), (iii) we have trouble with the relation of indiscernibility for individuals (it is either equivalent to identity, or else it is not; in both cases we are in trouble). Let us now check whether the second approach to non-individuals may perform better.

14.5.2 Keeping with the PII Package

So, given that characterizing non-individuals in terms of a lack of transcendental individuality brings much trouble, the reasonable option seems to be evident: keep only the bundle theory of individuality; individuals are characterized by their properties, identity and indiscernibility collapse for them and the principle of identity of indiscernibles holds. Non-individuals are entities for which the PII fails; identity and indiscernibility are distinct relations for them. This failure of PII, of course, derives its evidence from quantum mechanics.[5] How does that option fare in the context of non-reflexive logics?

There is a major problem here: that view is unable to account for the fact that in non-reflexive logics *some entities have no identity*. That is, on this view, non-individuality could be perfectly well defined and understood *without eliminating identity*, and we would have trouble spelling out why identity is absent for a given kind of entities and how to make sense of the loss of identity. In fact, the loss of identity is no longer understood as the source of non-individuality, but rather it is indiscernibility which accounts for such feature. Schrödinger's intuitions seem not to be vindicated in this case, of course, because they are tied with another, rival, metaphysical package! The loss of identity ends up unmotivated if we allow the bundle theory understanding of individuality and non-individuality.

Krause (1990, p. 58) motivated in part the violation of PII for quantum particles in non-reflexive systems by keeping the lack of identity for quantum systems. There seems to be a technical reason to keep both the lack of identity and the failure of PII. Let us check briefly. Recall that in higher-order languages, when we have $\forall P(P(x)$ *iff* $P(y))$, then obviously we could claim that x and y are indiscernible. However, in higher-order languages that is also precisely the definition of *identity*. In this sense, identity and indiscernibility collapse in higher-order logics (there is a minor issue about the proper semantics for those logics, standard or Henkin semantics, but we shall not concern ourselves with that here; see French & Krause, 2006, chap. 6). In Schrödinger logics that account of indiscernibility does not collapse on identity for quantum entities because *identity is primitive and identity does not produce formulas when the terms involved represent quantum entities*. So, there would be a clearly technical reason to keep the lack of identity while still maintaining a bundle approach to individuality and non-individuality: it is precisely the lack of identity that prevents that indistinguishability collapses in identity.

However, this is such a radical approach to separate indiscernibility and identity. As a first problem, it seems to render the system unable to define indiscernibility in the intended sense, as the sharing of every property, because that definition would also introduce identity. In this sense, we end up also without indiscernibility. As a second problem, one should notice that there are certainly other more economic ways to violate PII while keeping identity in the language (we have no space to

[5] Again, leaving aside the trouble with weak discernibility, as we have already commented.

rehearse the discussion here, but see Caulton & Butterfield, 2012). For instance, one could consider that the relevant version of PII for granting a metaphysically plausible approach to individuation is the one that deals only with monadic properties (absolute indiscernibility), which quantum particles violate, while still assuming that weaker versions obtain but do not grant individuality (for instance, weak discernibility; this is basically the approach in Muller & Saunders, 2008). In this sense we could keep the idea that some items are indiscernible but not identical violating absolute PII, while identity (as a logical relation) still holds overall. Of course, that would not vindicate Schrödinger's claim in our quote above, and we would not have a non-reflexive logic, but we would be able to keep the symmetry between the treatment of individuals and non-individuals when it comes to the metaphysics: individuals obey a strong version of PII, while non-individuals fail it.

Another option would be to allow the introduction of a specific primitive indistinguishability relation $\equiv$ distinct from identity in the language. We would be able to express that items x and y are not individuals by claiming, for instance, that $x \equiv y$ but that $x \neq y$. In this case, non-individual items would fail a version of PII too. However, the lack of identity would play no role in this metaphysical articulation too. In fact, as we mentioned, when non-individuality is framed in terms of indiscernibility, it is hard to motivate the accompanying idea of something 'losing identity' in a literal sense: we have no metaphysical counterpart for the lack of identity in the bundle view of non-individuality. In other words: given that now identity is not tied to individuality as in the TI approach, and that individuality and non-individuality is a matter of discernibility and indiscernibility (respectively), one ends up with no metaphysical role for the lack of identity. In this sense, part of the formalism seems to play no part in the underlying metaphysical view, and is not motivated.

So, there is a clear dilemma that results from the metaphysics of non-reflexive logics if we keep with the two major options put forward in the literature. On the one hand, there is a substantial (and metaphysically expensive) view of individuals and non-individuals with the transcendental individuality account. The problem with this approach, among others, is that we lose something important by the introduction of indiscernibility for individuals (that is, we lose the major reason to stick with transcendental individuality for individuals). On the other hand, we may abandon transcendental individuality and embrace a bundle theory of individuality. This option makes perfect sense of the relation of indiscernibility and how it works for individuals and non-individuals. However, on this option it is difficult to motivate the idea that identity should be abandoned for non-individuals, and it is hard to see that a non-reflexive logic is required. Both metaphysical options are problematic, and they are the most prominent in the literature. Perhaps the articulation of a metaphysics of non-individuals should receive more attention, with non-individuality cashed from tips arising directly from quantum mechanics. That is a task we shall not endeavor here.

14.6 Concluding Remarks

As we have discussed, the metaphysics of non-individuals suggested by non-reflexive logics may come in two flavors: one as resulting from a violation of the TI principle of individuality, the other as resulting from the failure of a bundle theory of individuality that requires the validity of the PII to account for individuality. Those are the two views that may be found in association with non-reflexive logics when it comes to discuss the metaphysical basis of such logics (something that does not happen with the required frequency).

Furthermore, we have pointed that those two metaphysical packages seem to benefit from distinct formal features of non-reflexive systems: the lack of identity and the existence of a distinguished indiscernibility relation that does not coincide with identity. The great trouble lies in answering how to make sense of the relation of indiscernibility for individuals in the first, TI-based metaphysical package. Given that we are adhering to a TI principle, we should expect some individuals to be (at least possibly) indiscernible too, no? So, the coincidence of identity and indiscernibility for individuals causes some troubles. The second greatest trouble seems to make sense of the lack of identity in the second, PII-based, metaphysical approach. Non-individuals may violate PII without requiring a failure of identity: it could just happen that there are distinct non-individuals that share all their features. The lack of identity is not motivated in this case.

Obviously, those are some of the difficulties pointed out through our paper. They are relevant here because they are mainly a result of our wish to tailor our metaphysics to suit a formal system.[6] The formal systems of non-reflexive logics available do have both the lack of identity for some entities and an indiscernibility relation. Accommodating both features in the above metaphysical pictures seems difficult, to say the least. However, some high hopes are usually put in non-reflexive logics; for instance, they seem to be the main reason for us to grant that non-individuals are a live option in the metaphysics of quantum mechanics, that they are "philosophically respectable", as French puts it (see French, 2019, sect. 5). That being so, it is strange that so little interest arouse about the metaphysical counterpart of such systems so far.

There seems to be two major options for those willing to keep with the idea of non-individuals and the main intuition that there is a lack of identity conditions in quantum mechanics. First, keep with non-reflexive systems but try to formulate a less substantial metaphysics, one that could make indiscernibility and lack of identity compatible. Perhaps the idea of a lack of identity conditions could be formulated in a minimal fashion that will not cause such incompatibilities as we

[6] Of course, one could tailor some formal systems to suit those metaphysical options. But that is not what is at stake here. We are focusing on the fact that it is non-reflexive systems that are said to ground the metaphysics of the RV; as a result, we are attempting to check how this metaphysics is to be understood. As we mentioned, while the formal systems are commonly discussed, the metaphysics is rarely discussed; if our arguments are correct, that may be so for good reasons.

have mentioned. The second path would be to keep identity in the formal system, but separate it from any kind of individuality attribution. Identity could be understood then as metaphysically deflated having no essential relation with individuality. That option, of course, has no relation with the fundamental intuition behind non-reflexive logics according to which identity commits us with individuality (see Arenhart, 2017, for further discussion).

Anyway, it should be clear that once we know our problems we are better able to look for answers. The problems presented here do not evidence that non-individuality does not make sense; rather, they seem to indicate that such substantial approaches to individuality, the ones that were favored in the literature, may not be the best starting point for us to look for non-individuality, even in the context of non-reflexive logics.

References

Adams, R. M. (1979). Primitive thisness and primitive identity. *The Journal of Philosophy, 76*(1), 5–26.

Arenhart, J. R. B. (2017). The received view on quantum non-individuality: Formal and metaphysical analysis. *Synthese, 194*, 1323–1347.

Arenhart, J. R. B. (2018). New logics for quantum non-individuals? *Logica Universalis, 12*, 375–395.

Bueno, O. (2014). Why identity is fundamental. *American Philosophical Quarterly, 51*(4), 325–332.

Caulton, A., & Butterfield, J. (2012). On kinds of indiscernibility in logic and metaphysics. *British Journal for the Philosophy of Science, 63*, 27–84.

da Costa, N. C. A. (1997). *Logiques Classiques et Non Classiques. Essai sur les fondements de la logique*. Masson.

Dorato, M., & Morganti, M. (2013). Grades of individuality. A pluralistic view of identity in quantum mechanics and in the sciences. *Philosophical Studies, 163*(3), 591–610.

French, S. (2019). Identity and individuality in quantum theory. In E. N. Zalta (Ed.), *The Stanford encyclopedia of philosophy* (Winter 2019 Edition). https://plato.stanford.edu/archives/win2019/entries/qt-idind/

French, S., & Krause, D. (2006). *Identity in physics. A historical, philosophical and formal analysis*. Oxford University Press.

Huggett, N., & Norton, J. (2014). Weak discernibility for quanta, the right way. *British Journal for the Philosophy of Science, 65*, 39–68.

Krause, D. (1990). *Non-reflexivity, indistinguishability and Weyl aggregates*. Phd thesis, University of São Paulo (in Portuguese).

Krause, D. (1994). Non-reflexive logics and the foundations of physics. In C. Cellucci, M. C. di Maio, & G. Roncaglia (Eds.), *Logica e filosofia della scienza: problemi e prospettive* (pp. 393–405). Edizioni ETS.

Krause, D. (2011). The metaphysics of non-individuality. In D. Krause & A. Videira (Eds.), *Brazilian studies in philosophy and history of science: An account of recent works* (pp. 257–267). Springer.

Krause, D., & Arenhart, J. R. B. (2019). Is identity really so fundamental? *Foundations of Science, 24*, 51–71.

Loux, M. J. (1998). Beyond substrata and bundles: A prolegomenon to a substance ontology. In S. Laurence & C. Macdonald (Eds.), *Contemporary readings in the foundations of metaphysics* (pp. 233–247). Blackwell Publishers.

Lowe, E. J. (2003). Individuation. In M. J. Loux & D. W. Zimmerman (Eds.), *The Oxford handbook of metaphysics* (pp. 75–95). Oxford University Press.

Muller, F. A., & Saunders, S. (2008). Discerning fermions. *British Journal for the Philosophy of Science, 59*, 499–548.

Post, H. (1973). Individuality and physics. *Vedanta for East and West, 32*, 14–22.

Rodriguez-Pereyra, G. (2004). The bundle theory is compatible with distinct but indiscernible particulars. *Analysis, 64*(1), 72–81.

Schrödinger, E. (1996). *Nature and the Greeks and science and humanism, with a foreword by Roger Penrose*. Cambridge University Press.

Wehmeier, K. F. (2012). How to live without identity—And why. *Australasian Journal of Philosophy, 90*(4), 761–777.

Chapter 15
Typicality of Dynamics and Laws of Nature

Aldo Filomeno

Abstract Certain results, most famously in classical statistical mechanics and complex systems, but also in quantum mechanics and high-energy physics, yield a coarse-grained stable statistical pattern in the long run. The explanation of these results shares a common structure: the results hold for a 'typical' dynamics, that is, for most of the underlying dynamics. In this paper I argue that the structure of the explanation of these results might shed some light—a different light—on philosophical debates on the laws of nature. In the explanation of such patterns, the specific form of the underlying dynamics is almost irrelevant. The constraints required, given a free state-space evolution, suffice to account for the coarse-grained lawful behaviour. An analysis of such constraints might thus provide a different account of how regular behaviour can occur. This paper focuses on drawing attention to this type of explanation, outlining it in the diverse areas of physics in which it appears, and discussing its limitations and significance in the tractable setting of classical statistical mechanics.

Keywords Typicality · Statistical mechanics · Stability · Laws of nature · Physical necessity · Non-accidental regularities

15.1 Introduction: Laws, Stability, and Typicality

It is commonly held that there is no satisfactory philosophical account of the notion of physical necessity. While it can be said that philosophers of science have made some progress in proposing candidate accounts of laws of nature, all of these accounts have major flaws, and in particular, the physical necessity of laws of nature

A. Filomeno (✉)
Instituto de Filosofía, Pontificia Universidad Católica de Valparaíso, Valparaíso, Chile
e-mail: aldo.filomeno@pucv.cl

C. Soto (ed.), *Current Debates in Philosophy of Science*, Synthese Library 477,
https://doi.org/10.1007/978-3-031-32375-1_15

is either postulated or left unexplained.[1] In this paper I propose to look at certain branches of physics that might help to improve our understanding of the source of lawful behaviour, at least within their restricted settings. To this end, I examine results of *stability*, that is, results to the effect that a physical system will evolve into a state that is invariant over time—for instance, the state of equilibrium of a classical gas of particles within a closed environment. More specifically, we will look at the approximate results of *emergent stable behaviour* that are acquired for *a typical underlying dynamics*. A typical dynamics is not the same as an arbitrary dynamics, but it is close: roughly stated, a typical dynamics is supposed to cover most of the dynamics, where 'most' is precisely defined.

In this paper I first want to point out that diverse areas in physics have in common (1) a result of the same type, i.e. a coarse-grained stable statistical pattern, *and* (2) that such a result holds for typical dynamics. The examples that I will cite are from classical statistical mechanics, complex systems, quantum mechanics, and diverse projects in high-energy physics. We will reconstruct in detail the case of classical statistical mechanics, in order to critically assess it. Then, the main aim of this paper is to point out that the structure of the explanation of such results, based on the notion of typicality, can be significant for philosophical debates on laws of nature and physical necessity.

The reason for focusing on the idea of typical dynamics is that such emergent stable patterns are explained almost independently of the specific details of the underlying dynamics. The emergence of a stable pattern does not depend on the specific form of the underlying governing laws, unless the form is "ridiculously special" (Goldstein, 2001, 43). Hence I argue that a dynamical system under suitable conditions, with no specific constraint of dynamical laws (i.e. with no deterministic or indeterministic rules of temporal evolution, usually in the form of differential equations), will exhibit a free state-space evolution which will typically display, in the long run, coarse-grained stable patterns.

Thus, the law-like behaviour at the higher level does not require the postulation of the usual underlying guiding rules of temporal evolution. An analysis of the suitable conditions invoked in each particular setting might help us to understand one way in which regular behaviour can occur—a way hitherto unnoticed in the philosophical literature on laws of nature.[2]

[1] I refer to the necessitarian account (Dretske, 1977; Armstrong, 1983; Tooley, 1977), the propensities/dispositional account (Cartwright, 1999; Mumford, 2004; Chakravartty, 2005), the Humean a.k.a. Best System account (Mill, 1884; Lewis, 1999; Earman & Roberts, 2005; Cohen & Callender, 2009), and the primitivist account (Maudlin, 2007; Carroll, 1994).

[2] To avoid confusion, a suitable differential equation can always *describe* the state-space evolution of a physical system. When I say that there is no specific constraint of dynamical laws I refer to the ontological claim that there is no governing dynamical law postulated in the theory, i.e. there is no dynamical law that *governs*, or *guides*, the system's evolution.

This paper is limited to outlining these results in the diverse areas of physics mentioned, and then discussing their limitations and significance in the setting of classical statistical mechanics. In Sect. 15.2 I mention various projects in high-energy physics and quantum mechanics which appeal to the aforementioned dialectics of deriving certain results for most of the underlying dynamics. Then I focus on reconstructing (in Sect. 15.3) and critically assessing (in Sect. 15.4) the approach of typicality in classical statistical mechanics. In Sect. 15.5 I conclude by assessing the philosophical significance that such results may have for philosophical debates on physical necessity.

The suitable conditions aforementioned are standard general constraints that gain a prominent role, for they can be the only modal constraints, that is, the only conditions that play the role of laws. In the literature on complex systems theory it is well known that, besides the underlying laws, the *context* gains an especially prominent role (Frigg & Bishop, 2016). Roughly stated, in the present study it is asked whether sometimes this role is not only prominent but sufficient to account for an otherwise unconstrained motion. [3]

15.2 Overview of Approaches in High-Energy Physics

Let us begin by citing various diverse projects in physics which employ the rationale discussed here. First of all, there are those projects that seek to derive the laws (the standard model interactions) and symmetries of modern physics from what they call a *random dynamics*. According to this hypothesis, *all complex* Lagrangians lead, in the low-energy limit, to the laws of particle physics (Froggatt & Nielsen, 1991; Chadha & Nielsen, 1983; Chkareuli et al., 2011). The authors consider a fundamental level displaying a highly complex behaviour. This level is below the current quantum level, for quantum mechanics does not describe a complex dynamics like the one they assume. The random dynamics is thought to inevitably yield the emergence, within some energy limit, of all current symmetries. The limit is the low energy domain, which corresponds to the experimentally accessible energies below 1TeV. Similar research along these lines includes the work of Mukohyama and Uzan (2013), as well as Jacobson and Wall (2010), both of which are concerned specifically with Lorentz symmetry, drawing an analogy with statistical explanations of the second law of thermodynamics (for an attempt to frame such projects in the philosophy of physics literature, see Smeenk & Hoefer, 2015, §4.1 and references therein).

[3] Besides, the present assessment may influence the plausibility of the physics projects cited: if the explanation of typicality, as we analyse it in the context of classical statistical mechanics, is a successful (or unsuccessful) type of explanation, this tells in favour of (or against) the projects we cite in other fields of physics that employ the same type of explanation.

There are also more speculative projects concerning *entropic forces*. Similarly, according to them, the allegedly fundamental interactions, including gravity, are not fundamental but rather emergent, arising from the statistical behaviour of lower-level degrees of freedom. See Verlinde (2011, 2017) or the more elaborated derivation of the Einstein field equations from thermodynamic assumptions given by Jacobson (1995).

For decades there has also existed research on chaotic cosmologies (see e.g. Misner, 1969; Barrow, 1977; Linde, 1983) that assumes an undetermined fundamental chaotic dynamics. Today such research concerns the instants before inflation, where a chaotic dynamics is assumed as a natural default initial state, and it is then investigated how we arrived from that state to the current standard model with broken symmetries and frozen degrees of freedom. A recent example which recurs to this view is Okon and Sudarsky's (2016) attempt to explain dynamically the Past Hypothesis (the universe's very special initial state of low entropy).

Finally, a similar dialectics is also found in certain projects in the foundations of quantum mechanics: Valentini's (1991) attempt to derive Born's rule with a quantum analogous of Boltzmann's H-theorem, and Nelson's (1966) attempt to derive the Schrödinger equation by presupposing Brownian motion of classical particles. In the same vicinity, in the foundations of Bohmian mechanics, Goldstein et al. (2010a,b), Goldstein and Tumulka (2010) aim to show that *for typical Hamiltonians* with given eigenvalues all initial state vectors evolve in such a way that the wavefunction will be in thermal equilibrium at most times. For more on these projects, which for reasons of space we can only cite here, see Callender (2007) and references therein.

Needless to say, for obvious reasons (dealing with the early universe, at fundamental sub-quantum levels, etc.) most of these approaches are more speculative than usual standard model physics, and thus have difficulties in delivering empirical predictions. In any case, this is irrelevant for our purpose here, as we are interested in the logical form of their common type of explanation and its significance.

So, how reliable is the typicality approach in the widely discussed setting of classical statistical mechanics? Many worries have been raised both to the typicality approach and its predecessors, such as Boltzmann's H-theorem. Those who raise the worries in statistical mechanics may also be skeptic of the dialectics presented in the aforementioned physical theories, in which the underlying dialectics is the same but is not explicitly discussed.

15.3 The Typicality Approach in Statistical Mechanics

Let us delve into the typicality approach as it appears in the foundations of classical statistical mechanics. According to this approach, the tendency towards the equilibrium macrostate occurs for initial conditions that are typical, where 'typical' is spelled out in measure-theoretical terms. After stressing that this is insufficient for explaining the tendency towards thermal equilibrium, the typicality of the dynamics has to be included, which again means that it occurs for the overwhelming majority of them, where 'typical' here is spelled out in topological terms.

15.3.1 Boltzmann's Explanation of the Second Law of Thermodynamics

The point of departure is Ludwig Boltzmann's project of understanding the macroscopic properties and laws of thermodynamics in terms of their microconstituents and their laws. This was the main mission of kinetic theory and statistical mechanics. The latter can be said to be the continuation of the former, after introducing irreducible probabilistic distributions not to the microconstituents but to the states of macroscopic entities (to the state of the whole gas). After Boltzmann, plenty of different paths have been pursued in order to obtain a reductive explanation of the laws of thermodynamics (see Uffink, 2014; Uffink, 2006, Ch. 4; Frigg, 2008; cf. Albert, 2000; Atkins, 2007; Sklar, 1993, II.3; Filomeno, 2019a; for another philosophical assessment of the typicality approach see Lazarovici and Reichert (2015, §2-3)).

In the case of the hard-sphere model of a gas in an isolated container, the macrostate towards which all systems tend is the macrostate in which the gas has spread out all over the box, filling its volume, that is, the 'equilibrium macrostate'.[4] Appealing to combinatoric mathematics, Boltzmann showed that the equilibrium macrostate is compatible with an overwhelmingly higher number of microstates. Consider a gas composed of n particles with two degrees of freedom each. The state of this system is specified in a $4n$-dimensional phase space Γ by a point x. This point is the microstate, which specifies the position q and momentum p of every particle:

$$x = (p_{x_1}, p_{y_1}, p_{x_2}, p_{y_2}, \dots p_{x_n}, p_{y_n}, q_{x_1}, q_{y_1}, q_{x_2}, q_{y_2} \dots q_{x_n}, q_{y_n}).$$

The phase space comes endowed with the natural Lebesgue measure μ.[5] The particles obey the laws of classical Hamiltonian mechanics; they define a phase

[4] We will consider the simple model of hard spheres, which models molecules of a gas closed in a perfectly isolated container. The gas is either ideal or diluted, neglecting long-range forces, with a fixed kinetic energy $T = \frac{p^2}{2m}$. The molecules of the gas, then, are the micro-constituents, and they are modelled not as point-particles but as hard spheres, each with a certain small radius r. The gas molecules interact like billiard balls; they have no effect on one another except when they collide. 'Hard' means that the collisions are elastic, i.e. no kinetic energy is transformed into other forms, for instance none is lost in the form of heat. Also assumed is a large number of microscopic constituents, typically of the order of Avogadro's number $N = 10^{23}$ or more.

[5] Justifying the choice of the "natural" measure is problematic; see e.g. Sklar (2015, Sect. 4), and Werndl (2013).

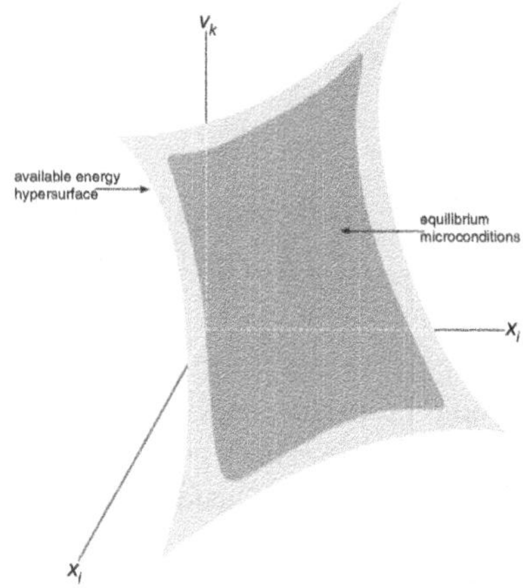

Fig. 15.1 An energy hypersurface, displaying the predominant size of Γ_{Meq}, the region of all the microstates corresponding to the equilibrium macrostate

flow ϕ_t that is measure-preserving, which means that for all regions,

$$R \subseteq \Gamma, \mu(R) = \mu(\phi_t(R))$$

which is known as Liouville's theorem. The system is perfectly isolated from the environment, so the energy is conserved. This restricts the motion of the microstate x over a region of Γ that is the energy hypersurface Γ_E, of $4n - 1$ dimensions. The Lebesgue measure μ restricted to Γ_E, μ_E, is also invariant. From the macroscopic point of view, the gas is characterised by its macrostates, where the equilibrium macrostate is labeled as M_{eq}. $\Gamma_{M_{eq}}$ is the corresponding macroregion in phase space which contains all $x \in \Gamma_E$ for which the system is in M_{eq}. Macrostates M supervene on microstates; a macrostate is compatible with many different microstates.

The main conclusion of Boltzmann's combinatorial argument is that the measure of Γ_{Meq} with respect to μ_E is overwhelmingly larger than any other macroregion. For the details of the proof, see e.g. Uffink (2006, 4.4), or Boltzmann (1877) himself. In fact, this region occupies almost all the energy hypersurface, as Fig. 15.1 conveys.

The entropy is defined as the logarithm of the size of the phase space region of the macrostate:

$$S(M) = k_B \; ln|\Gamma_M|$$

where k_B is the Boltzmann's constant. Given the radical difference between the sizes of the different macro-regions, it was reasonable to think that the non-decrease of entropy stated by the second law and, more generally, the tendency towards equilibrium stated by the 'minus first law' (Brown & Uffink, 2001), will be overwhelmingly more likely to occur. In fact, it follows that

$$S(M_{Eq}) >> S(M_{\neg Eq})$$

where $M_{\neg Eq}$ corresponds to any non-equilibrium macrostate. This would preserve the time-symmetric Newtonian picture of the world while explaining the time-asymmetric behaviour stated by the second law. The second law would not be a strict law but an rather approximation, reflecting the overwhelming likeliness of such behaviour. Hence, there would be no real conflict between reversible microscopic laws and irreversible macroscopic behaviour.

The success of this project, however, was threatened in many ways. A number of obstacles have been showing up ever since, such as

- the reversibility objection,
- the recurrence objection,
- the implausibility of the independence assumptions,
- how to interpret of the various probabilities,
- the status of the past hypothesis,
- the validity of the results outside the simplified models studied,
- the failure of the ergodic hypothesis, as it was proved that ergodicity is not sufficient (nor necessary), etc.

The typicality approach arguably helps in clarifying the status of the probabilities (and we believe that many of the other issues have been correctly responded, but we can leave that aside). Our focus on the typicality approach is however concerned with its potential significance for understanding the general phenomenon of *emergence of lawlike behaviour*, and hence, with its potential significance in the debates in philosophy of science about laws of nature.

15.3.2 Typicality

As with Boltzmann's approach, the main idea of the typicality approach is that a system exhibits entropic behaviour because it is typical for the system to behave in this way. However, as we will see, in contrast to Boltzmann's approach, typicality-based explanations eschew commitment to probabilities.

Maudlin (2011) illustrates the general idea with diverse examples: the toss of a coin, the toss of a die, and the case of a Galton board (depicted in Fig. 15.2). As

Fig. 15.2 The Galton board

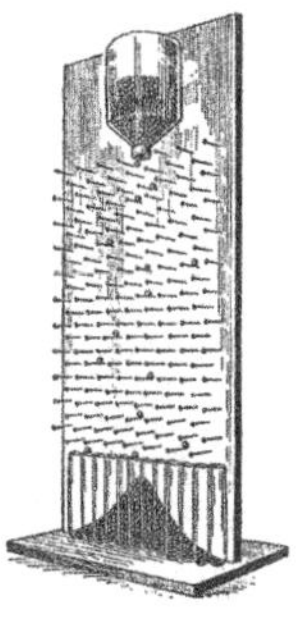

we already know, the Galton board displays, in the long run, a normal distribution centred in the middle basket. How should we understand the nature of this probability distribution, that we take to be neither subjective nor epistemic? The limiting frequency in the middle basket can be explained in terms of typicality, that is, it occurs because most of the possible initial distributions end up with that result. In other words, the typical behaviour of a ball falling in the Galton board is for it to fall in the centre. And this can be explained by focusing on the typical behaviour of a ball hitting a pin, whereby it is deflected to each side half the time. At the core of this phenomenon lies the law of large numbers: if it is typical to be deflected half the time to the left and half to the right, in the long run (i.e. after the balls have hit a large number of pins), it is expected that the number of turns to the left and to the right will be approximately the same, leading the ball to land approximately in the centre.[6] Typicality, then, is understood as follows: "when some specified dynamical behaviour (like passing a single pin to the right, or passing successive pins first to the right and then to the left) has the same limiting frequency in a set of initial states that has measure one, that frequency for the dynamical behaviour is typical" (Maudlin, 2011). In this quote, there is no appeal to probabilities but to measure theory. If the set of states that leads to some outcome has measure one, then it can be defined as typical, where the measure is calculated with a flat Lebesgue measure over the appropriate interval.

We can treat the case of a coin toss similarly. Fair coins typically land heads half the time, because most sequences of fair coin tosses, whatever the initial state, lead to that result in the long run (see Fig. 15.3).

---

[6] The famous 'law of large numbers' (LLN) can be stated thus:

Theorem 1 (The Strong Law of Large Numbers, or Borel Strong Law) *For independent infinite sequences of flips of a fair coin, let B denote the event that the proportion of successes S_n among the first n flips, $\frac{1}{n}S_n$, approaches the limit 1/2 as $n \to \infty$. That is:*

$$B := \left\{ x \in \{0, 1\}^N : \lim_{n\to\infty} \frac{S_n}{n} = \frac{1}{2} \right\}$$

The probability of the event B is 1, i.e. the set B has Lebesgue measure 1 (Dasgupta, 2011, §3.2).

There is also a *weak* version, which permits a small difference between the expected mean value and the effective outcome. The weak version states that the sample average converges in probability towards the expected value. Following Loeve (1977), where each outcome is X_i, the number of trials n, the sample average $\overline{X_n} = (X_1 + X_2 + ... + X_n)/n$, μ the expected value, and ε a small positive number:

$$\lim_{n\to\infty} P\left(|\overline{X_n} - \mu| > \varepsilon\right) = 0$$

The weak version allows for a certain degree of tolerance for departing from the expected value of a finite random sequence, which is quantified by ε. This version leaves open the possibility that $|\overline{X_n} - \mu| > \varepsilon$ happens an infinite number of times, although at infrequent intervals. Instead, in the strong version, for any $\varepsilon > 0$, the inequality holds for all large enough n (Sheldon, 2009). In any case, $X_1, X_2, ...X_n$ are assumed to be an infinite sequence of independent and identically distributed integrable random variables with expected value $E(X1) = E(X2) = ... = \mu$.

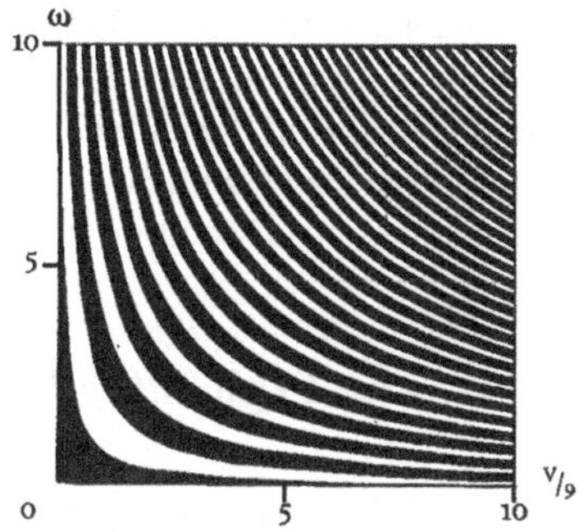

Fig. 15.3 A coin-flip model showing the outputs of heads (black area) and tails in function of the angular speed ω and the vertical velocity V/g (From Diaconis (1998))

In the case of the gas in a box, we can describe more precisely the situation as follows. Following Frigg (2009), an element e of a set Σ is typical if most members of Σ have property P and e is one of them. The element corresponds to a micro-state, the measure employed is the natural Lebesgue measure μ, Σ is the set of all microstates, and P is the property of evolving to equilibrium. 'Most' is thus understood in terms of having measure 1. Conversely, 'atypical' corresponds to having measure 0. (More exactly, the typicality measure is the induced 'microcanonical measure' μ_E restricted to the energy hypersurface.).[7] One of the results of Boltzmann's H-theorem was that, for typical microstates x, the empirical distribution $f_x(q, p)$ will converge to the Maxwell-Boltzmann distribution: $f_x(q, p) \propto e^{-\frac{1}{2}m\beta v^2}$. (For an explanation of the H-theorem as a typicality statement see Lazarovici and Reichert (2015, §3.3).)

More precisely, the definition of typical allows exceptions of subsets of measure zero or even subsets of very small measure. A definition for the crucial notion of 'most' (or 'nearly all') is as follows. Let A_P be the set of elements that exemplify P. Then,

$$\text{“Most of the elements } e \text{ in } \Sigma \text{ exemplify P”} := \frac{\mu(\Sigma \setminus A_P)}{\mu(\Sigma)} < \varepsilon$$

where ε is a small positive number. See Wilhelm (2022) for further details.[8]

[7] The typicality measure is not a probability measure (see e.g. Werndl, 2013). Yet, a link can be established, namely that a typicality measure $\mu(x)$ implies that $P(x) \sim 1$. A suggestion for understanding this link is to interpret the probability P as an objective probability "deriving from ranges in suitably structured initial state spaces" (Rosenthal, 2009, 2012). Thus, the probability of an event x consists in the proportion of initial states that lead to x within the space of all possible initial states. Typicality statements seem to ground this notion of deterministic probability (the first to suggest an interpretation of probabilities in deterministic systems was, as far as we know, Batterman, 1992). Poincaré's method of arbitrary functions lies at the core of Rosenthal's proposal, and thus can be considered to ground typicality statements, as also suggested in Filomeno (2019b).

[8] A minor terminological clarification: for the sake of clarity we (and the authors we cite) talk of 'typical states', 'typical trajectories', etc. although strictly speaking what is defined as typical in a typicality explanation is a property P that is shared by the elements of the set (e.g. the microstates, trajectories, or whatever else). Again, see Wilhelm (2022) for a clear discussion around typicality explanations.

Notably, it turns out that one gets a set of frequencies that are typical as an *analytic consequence* of the deterministic dynamics together with a measure over initial states. And "what we in fact believe as a *purely mathematical fact* is that the set of initial states with 50% limiting frequency for deflections to either side is a set of measure one" (Maudlin, 2011, 286, my italics).

15.4 The Typicality of the Dynamics

So far in the typicality approach, no explicit mention of the dynamics has been made. Yet it has been argued that for the typicality explanation to suffice, something else must be verified, i.e. that the results hold for most variations of the actual Newtonian dynamics. Before analyzing how the typicality of dynamics is defined in Sect. 15.4.2, let us pause to analyse an important aspect regarding how Newtonian dynamics has been treated in statistical mechanics.

15.4.1 A Preliminary Remark on the Actual Newtonian Dynamics

Several assumptions are made in kinetic theory and statistical mechanics, varying depending on the philosophical approach and the system under study. The deterministic laws of classical mechanics are assumed. Now, the assumption that will become relevant concerns the appearance of randomness: in spite of the guiding laws being deterministic Newtonian laws, there is "*an assumption about the erratic nature of the dynamics*" (Uffink, 2006, 5). To quote Sklar (1993), one of the chief assumptions of the Maxwell–Boltzmann programme, and an assumption that runs right through to the ergodic theory of modern physics, is that at some level of description a condition of *independence* must be met "for the theory to properly explain why the correct values of state parameters of gas systems can be obtained from taking the average values of mechanical properties of the individual particles of the system".

There have been various strategies for underpinning this requirement of independence: the assumption of equal initial probabilities, the postulation of dynamical properties as in the ergodic hypothesis (which we can informally phrase as stating that, over long periods of time, the time spent by a system in some region of the phase space of microstates with the same energy is proportional to the volume of this region; for a formal definition see below footnote 14), or rerandomization posits such as hypotheses of molecular chaos, e.g. the Stoßzahlansatz in Boltzmann's theory (the assumption about the lack of correlation of the particles' velocities before they collide). The point is that each of these introduces a condition of randomness or independence at different locations in either the gas model employed or in the method for calculating the properties of the gas (e.g. temperature, pressure, or entropy). Hence a central question in the foundations of statistical mechanics: how can such posits about the erratic wandering nature of the dynamics coexist with the deterministic guiding laws?

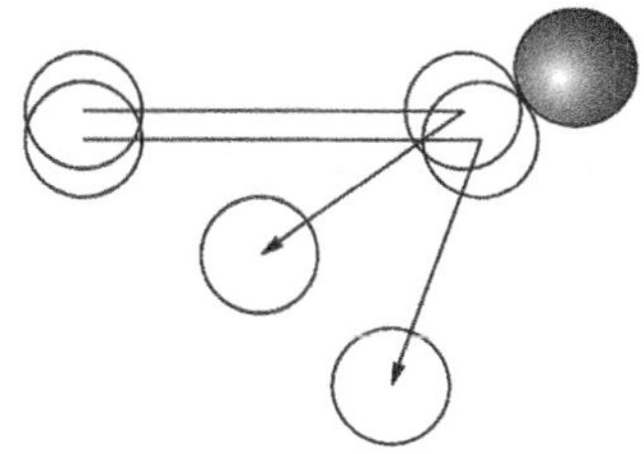

Fig. 15.4 The collision of two classical circular particles is highly sensitive to initial conditions

This apparent randomness is relevant for our purpose, as will become clearer in the next Sects. 15.4.2 and 15.4.3. This apparent randomness is due to the randomizing effect of the actual dynamics, a phenomenon which can be understood by appealing to the relation between determinism and chaos and then between chaos and randomness. Leaving aside the discussions on the definition of these notions, it will suffice for us to understand 'chaos' through its most famous feature, namely high sensitivity to initial conditions, and randomness in terms of a lack of correlations between a state and its past states.[9] As to the relation between deterministic classical mechanics and chaos, we can easily visualize classical mechanics' high sensitivity to initial conditions in the hard-sphere model of Fig. 15.4: the outcome direction after the collision of one sphere with the convex spherical surface of another sphere is highly sensitive. Then the appearance of randomness can be approached via such chaotic property of the Newtonian dynamics, since it washes out the correlations with past states, yielding a random-looking evolution. As it is known in the dynamical systems literature, "interactions in the domain $\Gamma_{M_{ab}}$ will be *so convoluted as to appear uniformly smeared out* in Γ_{M_b}. It is therefore reasonable that the future behaviour of the system, as far as macrostates go, will be *unaffected by their past history*" (Lebowitz, 1999, my italics).[10,11]

[9] 'Randomness' so understood can be embedded in the framework of the Ergodic Hierarchy: a hierarchy of dynamical properties, where the strongest property is the maximum degree of randomness, the Bernoulli property (in which there is no correlation with *any* its past states), and the weakest property is the weakest degree of randomness, the ergodic property (Berkovitz et al., 2006). 'Chaos' can also be defined within such a framework: Werndl (2009) defines it in terms of strong mixing; Belot and Earman (1997), define it in terms of the higher Kolmogorov level; while Berkovitz et al. (2006) advocate seeing it as a matter of degree, quantified according to the position within the hierarchy. For any of these definitions, chaos is a sufficiently strong randomization property, for it will be stronger than ergodicity (which is in turn stronger than ϵ-ergodicity, which will be the required condition for thermodynamic behaviour, as explained later). For definitions of these notions and their interrelation, see e.g. Frigg et al. (2020).

[10] Here, M is the system's macrostate, Γ the phase space, Γ_M the phase space region corresponding to M, M_a the system's initial macrostate, and M_b its later macrostate. $\Gamma_{M_{ab}}$ is the region of Γ_{M_b} that came via Γ_{M_a} (the set of microstates within Γ_{M_b} that are on trajectories that come from Γ_{M_a}).

[11] One of the many details I omit for the sake of space is that the independence assumptions can be of different sorts. One example is Maxwell's assumption, which posits independence in the components of the velocity vector of each gas particle.

Having remarked on this randomizing effect of the actual dynamics, let us spell now out a typicality approach which explicitly includes a specification of the features that the dynamics must verify.

15.4.2 A Typicality Explanation Explicitly Including the Dynamics

Frigg (2009, 5) correctly remarks that *typical states do not automatically attract trajectories*. In other words, before we were implicitly granting that the underlying dynamics leads to equilibrium, but this need not be so. That is, it does not lie in the "nature" of atypical states to evolve into typical ones. It could perfectly well be the case that a phase flow leads to anti-thermodynamic behaviour. Analogously, Frigg points out that the approach to equilibrium will not occur just because the corresponding microstates are more numerous. The disparity of sizes of the macroregion of equilibrium and the other regions does not by itself imply that an entropy decrease will be atypical. Thus, definitions such as those offered by Goldstein and Lebowitz (2004, 57) are unsatisfactory; the dynamics must be explicitly included in the definition of typicality.

This amounts to go one step further and ask whether the second law holds typically in the space of possible Hamiltonians (relative to a measure of typicality over such space, as we will see later), that is, explore whether most possible dynamics, subject to some general constraints, would lead the system to exhibit thermodynamic behaviour.

The scope of this generalization will affect its significance. For instance, it is easier (and in fact Frigg and Werndl (2012) provide some alleged numerical support) to generalize the results in the restricted class of Hamiltonians that describe gases. This would be significant for the project in statistical mechanics of explaining the macroscopic patterns of thermodynamics. Now, as we will discuss later, in this paper we also want to point out that the more these results are generalized in the dynamics space, the more they approach the goal we are aiming at, namely, a typicality explanation of coarse-grained regular behaviour. We want to focus attention on a typicality explanation in which the results would hold almost irrespective of the underlying dynamics. Then, such an explanation could be phrased without postulating any 'dynamics', that is: the results would hold irrespective of the specific physical system's state-space *trajectory*.

The idea amounts to the following. Let a generic dynamical system $(\Gamma, \Sigma, \mu, \phi_t)$ be defined as a probability space $[\Gamma, \Sigma, \mu]$ and a transformation ϕ_t of it.[12] ϕ_t has the generic form of an ordinary differential equation: $\frac{d}{dt}x = f(x,t)$; which is autonomous of the independent variable t which represents time, where x is the position of the system in the n-dimensional state-space and n is the number of degrees of freedom. Then, informally, the idea amounts to seek whether not only an inverse-square guiding law $f(x,t) \propto r^{-2}$ yields evolution to the Maxwell-Boltzmann distribution, but that also a, say, inverse-cube law $f(x,t) \propto r^{-3}$ or an inverse-fifth law $f(x,t) \propto r^{-5}$ would yield the same result in the long term.

If the results were sufficiently typical (sufficiently generic or sufficiently wide in scope), stable behaviour would obtain for almost any underlying dynamics; i.e. it would obtain almost independently of how the dynamics is described. Supported by Maudlin's diagnosis above (p. 400), we would be faced with a non-causal mathematical explanation of some coarse-grained, in the long run, "necessary" behaviour. Notably, the necessity involved would not be physical necessity, but rather something like "mathematical" or "statistical necessity" (along this line, see Filomeno, 2019b).

Thus, together with the statement of the previous subsection that equilibrium states are typical, the Hamiltonian must also be typical. As defined in Sect. 15.3, take Γ_E to be the hypersurface of $6n - 1$ dimensions of the phase space Γ in which the energy is conserved. Taking M_{eq} to be the equilibrium macrostate, take Γ_{Meq} to be the macroregion consisting of all $x \in \Gamma_E$ for which the macroscopic variables assume the values characteristic for M_E. Then, the underlying idea of the typicality of dynamics in statistical mechanics is expressed by Goldstein (2001, 43, my italics):

> Γ_E consists almost entirely of phase points in the equilibrium macrostate Γ_{Meq}, with ridiculously few exceptions whose totality has volume of order $10^{-10^{20}}$ relative to that of Γ_E. For a non-equilibrium phase point x of energy E, the Hamiltonian dynamics governing the motion x(t) would have to be ridiculously special to avoid reasonably quickly carrying x(t) into Γ_{Meq} and keeping it there for an extremely long time—unless, of course, x itself were *ridiculously special*

Frigg and Werndl (2012) formalize which conditions a dynamics has to meet to be typical. A first step is to restrict the discussion to the obtaining of *epsilon-ergodicity*, because they argue that this property is sufficient for thermodynamic behaviour (Vranas, 1998; Frigg & Werndl, 2011).[13] The proposal that the relevant Hamiltonians being ergodic explains thermodynamic behaviour was originally

[12] As previously defined, Γ is a set of elements, interpreted as a state-space (e.g., phase space in classical mechanics); Σ is a σ-algebra of measurable subsets of Γ; and there is a probability measure μ on Σ as usually defined. In our classical case, the natural Lebesgue measure.

[13] In this subsection we will follow the path of these authors for the sake of the exposition, but our main goal does not really depend on its specific actual success.

proposed by Boltzmann, but had since then been subjected to numerous criticisms and has accordingly been left aside by the philosophical and scientific community.[14] ϵ-ergodicity is meant to be a relaxed version of strict ergodicity. It does not require that $\mu_E(B) = 0$, allowing sets of initial conditions to be on non-ergodic solutions with respect to M_{eq}, only if the sets are of small size; that is: $\mu_E(B) \leq \epsilon$ (with ϵ a small number). It is proved in Frigg and Werndl (2012, 9) that, for ϵ-ergodic systems, initial conditions that lie on thermodynamic solutions are typical (cf. Vranas, 1998). This is what they label as 'm-typical' ('m' standing for 'measure'). Then an explicit typicality measure of the dynamics is included. Thus their complete argument has this form (Frigg & Werndl, 2012, 6, 14):

> *Premise 1*: The macrostate structure of the gas is such that equilibrium states are typical in Γ_E.
> *Premise 2*: The Hamiltonian of the gas is typical for the class of all relevant Hamiltonians.
> *Conclusion*: Typical initial conditions lie on solutions exhibiting thermodynamic behaviour.

The task, then, is to ascertain the truth of premise 2, granting their proof in which the relevant property P is being ϵ-ergodic. The Hamiltonian has to be typical with respect to a certain measure. The measure, for a dynamical trajectory, has to be a topological measure; thus, it can be said that it has to be 'topology-typical', or 't-typical'.

The conclusion of Frigg and Werndl (2012), I advance, is the following: the Hamiltonians are t-typical with respect to the so-called Whitney topology, where this is proved for the hard-sphere model and for the Lennard-Jones potentials, a subset of potentials which suffices for a wide class of realistic gases. Here we will also cite recent results that support a wider class of potentials.

[14] Informally, a dynamical system is ergodic if and only if the proportion of time spent in a region A equals the measure of A, that is, the time average is equal to the phase average. Formally, a dynamical system $(\Gamma_E, \mu_E, \phi_t)$ is ergodic if and only if for any measurable subset A and any microstate $x \in \Gamma_E$,

$$T_A(x) = \mu_E(A)$$

except for a set B with $\mu(B) = 0$; where $T_A(x)$ is defined as the time-average of a solution originating in $x \in \Gamma_E$ relative to the measurable set A thus:

$$T_A = \lim_{t\to\infty} \frac{1}{t} \int_0^t \chi_A(\Phi_\tau(x))d\tau$$

where $\chi(A) = 1$ for $x \in A$ and 0 otherwise. Considering B as the set of microstates which lie on non-ergodic solutions with respect to M_{eq}, it can also be said that system is ergodic iff $\mu_E(B) = 0$.

15.4.2.1 Topological Typicality of the Hamiltonians

To measure the typicality of the Hamiltonians in its dynamics space, instead of the Lebesgue measure we must employ a topological notion, namely that of 'generic' (aka 'comeagre'). A set is generic if and only if its complement is a countable union of nowhere-dense sets. Intuitively, the idea is to provide a measure able to inform us whether the other sets are scarce.[15] As we are searching for the typicality of the Hamiltonians, we seek to claim something like the following:

> TYPDYN: ϵ-ergodic Hamiltonians are comeagre/generic in the entire class of gas Hamiltonians G.

TYPDYN is a refined version of premise 2 above. Unfortunately, the class G is too big. G is not easily definable, so Frigg and Werndl restrict their claim to something narrower: first, to the hard-sphere model, then, to the sub-class L of smooth Hamiltonians that are small perturbations of the so-called Lennard-Jones potential. This constraint, the authors argue, is quite acceptable given their aim of explaining the thermodynamic behaviour of gases. Other potentials have also been studied, as we explain below. Thus, first we restrict the Hamiltonians to the smooth ones, namely, those with a fixed kinetic energy $T(p, q) = \frac{p^2}{2m}$. In the hard-sphere model (see footnote 4 in p. 395), the potential V describes the abrupt repulsive forces of elastic collisions between two particles thus: $V(r) = \infty$ for $d < r$, and 0 otherwise; where d is the distance between the particles and r was the radius. In this model, numerical simulations and analytical results support the claim that the motion is ergodic (as originally conjectured by Boltzmann) for almost all parameter values. Sinai (1970) aimed to prove the assumption implicit in Boltzmann and Gibbs that, in the hard-sphere model, the trajectories of the particles are erratic paths wandering freely over the energy surface and spending equal times in equal hyperareas of this surface.[16] Then he proved that the hard-sphere gas is unstable and that this suffices to guarantee ergodicity and mixing—see Frigg and Werndl (2012) for details.

[15] In an equivalent rephrasing, a property is defined as *generic* if and only if it holds for a countable intersection of open dense subsets (Wiggins, 2003, 162). In slight more detail, a property of a vector field is said to be C^k *generic* if and only if the set of vector fields possessing that property contains a residual subset in the C^k topology, where a residual subset contains the intersection of a countable number of sets, each of which is open and dense in the topological space. k denotes the degree of differentiability ($0 \leq k < \infty$).

[16] The proof is limited to 3 spheres; since then many proofs for a larger number of spheres have been sought, for references see Spohn (2012, 149).

Then a potential $V(p, q)$ which has been thoroughly studied is the so-called Lennard-Jones potential, the most notorious and used to describe intermolecular interactions:[17]

> The Lennard-Jones potential is important because there is good evidence that the interaction between many real gas molecules is accurately described by that potential at least to a good degree of approximation. Hence, whatever potentials G comprises, many real gases cluster in a subclass of G, namely L, and so knowing how the members of L behave tells us a lot about how real gases behave. (Frigg & Werndl, 2012, 11)

After that, the choice of the Whitney topology comes from a physically natural way of saying that two Hamiltonians are close; namely, when the difference between the Hamiltonians themselves as well as all their derivatives is small (Frigg & Werndl, 2012, 10). The goal then is to confirm what we summarize as:

> TYPDYN2: ϵ-ergodic Hamiltonians are generic in the sub-class L of smooth gas Hamiltonians with a Lennard-Jones potential with respect to the Whitney topology.
> Now, to prove that TYPDYN2 is the case, they show that the sub-class L is generic by referring to numerical simulations in which Hamiltonians that are ϵ-ergodic for the energy values defined are typical in L. In particular, Ford and Stoddard (1973) improved Sinai's results, extending the proof to systems that have attractive as well as repulsive interparticle forces. For further supportive numerical simulations see the references cited in Frigg and Werndl (2012, 12–14). For still other types of potentials see the references cited in Frigg and Werndl (2012, 18) and the more recent Gallagher et al. (2012) and Pulvirenti et al. (2014).

Before these results, there could have been pessimism about the typicality of dynamics in gases, due to the Kolmogorov–Arnold–Moser (KAM) theorem, which predicts nonergodic regions of stability. These are finite regions in phase space in which there are trajectories confined to perpetual quasiperiodic motion. In spite of the KAM theorem, the simulations of Ford and Stoddard (1973) provide evidence that, as the energies and densities of particles are varied (so even in the allegedly problematic cases of lower energies or higher densities), the expected entropic behaviour obtains without evidence of KAM regions. The KAM regions might be perfectly negligible, given that none has been found in the simulations and "nothing known precludes their being so small as to be physically irrelevant" (Ford and Stoddard, 1973, 1).

[17] Specifically, this potential has the form:

$$V = 4\alpha\left(\left(\frac{\rho}{r}\right)^{12} - \left(\frac{\rho}{r}\right)^{6}\right)$$

where α describes the depth of the potential well, r the distance between two particles, and ρ is the distance at which the inter-particle potential is 0. The potential of the entire system is obtained by summing over all two-particle interactions. The r^{-12} term describes the repulsion forces at short ranges, The r^{-6} term describes the attraction forces at long ranges (van der Waals force or dispersion force).

At the end of the day, the numerical simulations suggest that the Lennard-Jones gas, which has repulsive as well as attractive forces, exhibits thermodynamic behaviour. These results confirm that this type of gases conforms to the properties of ergodicity that were originally demanded by Boltzmann (and Gibbs), and that we demand on behalf of a typicality account that seems to hold for most of the dynamics.

15.4.2.2 The Scope of the Results

The class of Hamiltonians under consideration is not the entire set G, but rather a subset, albeit an important one, i.e. the subset L of Lennard-Jones potentials, together with other potentials, and with the also widely used potential of the hard-sphere model. Frigg and Werndl argue that this is acceptable because L is the most relevant subset for realistic gases and suffices for an approximate proof. While it can be thought that this is quite satisfactory for the long-standing issue in the foundations of statistical mechanics of explaining the tendency of gases towards equilibrium, for our purpose it would be welcome to extend the range of dynamics as far as we can. For ultimately we are exploring to what extent *lawful behaviour can emerge for an almost arbitrary, free state space trajectory under minimal constraints*. The fewer constraints we invoke, the wider the scope of the typicality explanation. Thus, for our purpose, we do not need to stop when typicality is proven for a wide class of *realistic* dynamics of gases. We would also like to know whether typicality is proven for the *unrealistic* dynamics.

Regarding the remaining wide and varied class of unrealistic dynamics, we could try to find further positive or negative results about the typicality of their properties. An exploration of typical properties of the dynamics space has been one of the main tasks in the history of dynamical systems, in the area known as topological dynamics. We want to conclude this section by surveying what are, to our knowledge, its most relevant results.

15.4.3 Further Exploration of the Dynamics Space

In topological dynamics we can find results regarding the typical properties of trajectories, for instance, random-looking behaviour such as ergodicity, mixing and the like. As has often been pointed out, and as we have suggested in Sect. 15.4.1, a dynamics can be random-looking in virtue of its chaotic properties. And such kind of property seems to be the required property, as for instance Frigg (following Dürr and Maudlin) says:

> the Galton Board seems to exhibit random behaviour. Why is this? Dürr's and Maudlin's answer is that the board appears random because random-looking trajectories are typical in the sense that the set of those initial conditions that give rise to nonrandom-looking

> trajectories has measure zero in the set of all possible initial conditions, and this is so because the board's dynamics is chaotic (Dürr, 1998, sec. 2). Translating this idea into the context of SM suggests that the relevant property P is being chaotic. (Frigg, 2009, 1004)

Then, in this section we will outline some key results about whether or not typical trajectories are chaotic, or mixing, or ergodic, or other related properties. We first cite classic results for and against the genericity of chaos, which will be superseded because their domain is too constrained. Then we cite further results concerning wider domains of possible trajectories.

15.4.3.1 Classic but Excessively Constrained Results

There are methods that allow us to rule out the possibility of closed orbits, and thus of periodic motion: the methods of index theory, the existence of Lyapunov functions, and gradient systems (see any textbook on dynamical systems, e.g., Strogatz, 1994). However, these desirable results have to be studied case by case, analysing each dynamical equation, and they say nothing about how frequently such periodic orbits are ruled out. On the negative side, the Poincaré-Bendixson theorem proves the existence of periodic orbits, which is incompatible with the existence of chaotic motion (Strogatz, 1994, 210). However, this result has only been proven for vector fields on a plane, that is, for 2 degrees of freedom.

Similarly, the literature on statistical mechanics has been critical of ergodicity, mainly on the grounds of the KAM theorem. Yet, a careful look at the theorem shows that it can hardly be taken to represent the *general* behaviour of classical mechanical systems; see above (Sect. 15.4.2.1 p. 406), and Frigg (2009, 1005–6), Frigg and Werndl (2011, §5), Berkovitz et al. (2006, §4.1), and Sklar (1993, 174–5).[18]

15.4.3.2 Generic Properties of the Dynamics Space

Non-integrability

Moving to the more general framework of topological dynamics, we find a result that stands for the typicality of chaotic behaviour in Markus and Meyer (1974, §3). They showed that, in the space of all normalized and infinitely differentiable Hamiltonian systems on a compact symplectic manifold, most of the Hamiltonians are non-integrable, a property which is assumed to imply chaos.

[18] See also Frigg and Werndl (2011, §6) for a critique of the limited significance of the result by Markus and Meyer (1974, §4) against the typicality of ergodicity.

Structural Stability

Other results concern the notion of 'structural stability', indirectly related to chaos. Roughly stated, a dynamical system (whether it be a continuous vector field or a discrete map) is said to be structurally stable if nearby systems have qualitatively the same dynamics. Smale (1966) proved that structural stability is *not* a generic property for n-dimensional diffeomorphisms when $n \geq 2$, or n-dimensional vector fields when $n \geq 3$ (Wiggins, 2003, 164; Ruelle, 1989, 43 fn36). The connection between instability and the exponential divergence of nearby trajectories characteristic of chaos is straightforward.

Hyperbolicity

Another result shows that hyperbolic behaviour is generic: the Kupka-Smale theorem. It proves that the hyperbolicity of all periodic points holds for a set of C^r diffeomorphisms that is a dense set in the C^1 topology (Katok & Hasselblatt, 1997, 289,292,295; Ruelle, 1989, 46). 'Hyperbolicity' refers to the presence of expanding and contracting directions of the derivative. The presence of these directions produces exponential behaviour of trajectories on some set and, in the end, the stretching and folding gives rise to complex long-term behaviour. This effectively makes the dynamics appear random.

Ergodicity as a Generic Property

Another positive result comes from Oxtoby and Ulam (1941), who discovered that ergodicity is generic for measure-preserving homeomorphisms on all compact manifolds. More specifically, they discovered that the set of dynamical flows that are not ergodic belongs to the first category in the set of measure-preserving generalized dynamical flows.

Extension of the Results for Ergodicity and weak Mixing

Further, these results have since been extended. The genericity of ergodicity has been extended to automorphisms: Halmos (1944a) shows that ergodicity is generic in the weak topology in the space of all automorphisms. Finally, a further extension of the latter result is that the genericity of the stronger condition of weak mixing has also been proved, both in homeomorphisms and in automorphisms (Katok & Stepin, 1970; Halmos, 1944b).

A discussion of these and other results can be found in Alpern and Prasad (2001). It is beyond the scope and available space of this paper to properly discuss the applicability of these results: the technical notions that appear in the different results impose different constraints, e.g. the results applying to homeomorphisms or to automorphisms, or the scope being restricted to compact manifolds. Thus, the scope of the possible physical systems described requires further discussion (again, see Alpern & Prasad, 2001 for discussion). Then, these results can be studied case by case in different physical theories such as classical statistical mechanics, but also in any of the theories cited in Sect. 15.2. Be that as it may, in what follows we will discuss the type of conditions involved and the corresponding significance that this approach can have.

15.5 Discussion: The Role of the Constraints

We have focused on and assessed a type of explanation, the typicality approach in statistical mechanics which, as we have pointed out, is common in other areas of physics. The motivation for our analysis has been that the typicality approach (1) can be considered an explanation of coarse-grained *stable* behaviour, and (2) such stable, law-like, behaviour is explained almost independently of the specific details of the underlying dynamics.

This is familiar in the literature on scientific explanation, where it has been studied how, in diverse settings, the details of the underlying level are irrelevant (see e.g. Batterman, 2001; Strevens, 2008, 2013; Rohwer & Rice, 2013; Batterman, 2018, 2019). The typicality approach looks like a more general yet less fine-grained explanation in comparison with the renormalization group explanations of the universality of critical phenomena, in which most of the details of the micro-level turn out to be irrelevant. Renormalization group techniques take into account the space of possible Hamiltonians, and in the procedure of coarse-graining and removing irrelevant micro-level details they show that for all Hamiltonians the system's state-space trajectory converges to a fixed point. The typicality approach, as we have seen, can be considered a *mathematical explanation* or, more specifically, a *statistical explanation*, of this sort. This quote by Strevens (2003, 62) clearly illustrates the underlying idea:

> The value of a [...] probability may come out the same on many different, competing stories about fundamental physics. The probability of heads on a tossed coin, for example, is one half in Newtonian physics, quantum physics, and the physics of medieval impetus theory.

One might doubt that the typicality approach does provide a proper explanation. For a defence of the legitimacy of typicality explanations see Wilhelm (2022).[19] We believe that a typicality explanation, although probably not the best explanation, it fulfils central desiderata that one would expect from an explanation. At least in our particular case study it seems that, *if* the kind of explanation we propose were

[19] See Lazarovici and Reichert (2015, §6) for a complementary philosophical assessment of the typicality approach in statistical mechanics, also in connection to the notion of law of nature. See (*ibidem*, §5) for further discussion of common objections to the typicality approach.

successful,[20] we would have gained some insight on the existence of regularities in the world.

Of course, in the cases presented, some specific dynamical law is assumed. The interesting point is that, as we have argued, this does not imply that some specific dynamical law *must* be assumed. In classical statistical mechanics, the facts

1. that the results are supposed to hold for most of the dynamics, and
2. that the relevant typical dynamical property is some randomizing property,

independently support that, merely by imposing prior general constraints, *a free evolution of a point in phase space will display, in the long run, coarse-grained stable patterns.*

Then, the process that has been studied in this paper yields a further reduction of degrees of freedom that occurs in the long run in coarse-grained levels. Now, the significance of the typicality of dynamics is conditional on the prior general constraints that in each case are imposed. For the only law-like assumptions are found in such constraints. We refer for instance to the *boundary conditions*, or to the *kinematical conditions* (spelled out below). A paradigmatic example of a condition that has been implicitly presupposed, both in the results for the actual dynamics in Sect. 15.4.2 as well as in some of the results that widened the range of the dynamics in Sect. 15.4.3, is the principle of *conservation of energy*. (For Hamiltonian dynamical systems are closed measure-preserving systems in which energy is conserved.) In general, the kind of constraints to which we refer are familiar in different physical theories. The point is that they are not the usual differential equations dictating the time-evolution of physical systems, but rather conditions which are still 'lawful' (or 'modal', or 'nomic') in that they constrain the possibility space. An illustrative standard example is that of a marble rolling on a bowl: its motion is constrained to the surface of the bowl. A constraint such as the bowl is different from the guiding dynamics, but it delimits the possible degrees of freedom.[21]

A classification of constraints and assessment of their import in complex dynamical systems is Hooker (2013). An especially basic type of constraint are the holonomic constraints, which means that they can be expressed purely *geometrically*, so that they are *independent of the behaviour of the system*. They can be written as some function of the space-time geometry, satisfying an equation of the form $f(r_1, r_2, ..., r_n, t) = 0$, where the r_i are the system coordinates and t is time (Hooker, 2013, 19).[22] The distinction that we are interested in highlighting, between

[20] In this paper we only propose such kind of explanation, of course we are far from defending its actual success.

[21] And it is possible to also constrain the range of parameters of the system, to constrain the range of quantitative forms its dynamics can take. The effect of such constraints appears as constraints on variables (Hooker, 2013, 2).

[22] More precisely, "A constraint is holonomic if its local (momentary, nearby) differential form is integrable to yield a global constraint relation that is purely a matter of space-time geometry and

constraints and the guiding dynamics, is often neglected in modelling physical systems, because *in practice* it is not necessary to differentiate the constraints from the dynamics (what Hooker calls "the interaction dynamics"). Hence constraints and dynamical laws are conflated in the Lagrangian formalism: the dynamical law that *describes* the motion—a differential equation—encompasses the constraints, it "compresses constraint and interaction information into a flow and in that sense suppresses the explicit details of the interactions and constraints" (Hooker, 2013, 15).

The resulting picture becomes clearer after characterizing the space of the *kinematically and dynamically possible models*. The kinematically possible is a space of functions that represent histories of the system. Define T_s as the class of all the kinematically possible trajectories in state-space. The idea is that T_s delineates the subspace of "metaphysical possibilities consistent with the theory's basic ontological assumptions" (Pooley, 2013, 12). This is generally defined by specifying the independent and dependent variables of the theory and the degree of smoothness of candidate functions, as well as any boundary conditions that they must satisfy (Belot, 2011, 5).[23] In the framework of classical mechanics, consider a model of gravitating point particles with distinct masses, as in Belot (2011, 7). A point in the space of the kinematically possible models of the theory assigns to each of the particles a worldline in spacetime, without worrying about whether the worldlines of each particle jointly satisfy the Newtonian laws of motion. There is then a subset of the kinematically possible models called the set of dynamically possible models. This is a space of solutions which is a $6N$-dimensional submanifold whose points correspond to the motions of particles that obey Newton's laws. Such a submanifold constrains us too much, to a single set of laws, i.e. the actual ones. In more detail, Pooley (2013, 12) (see also ibidem, 40) says:

> "In a coordinate-dependent formulation of Newtonian theory like that so far considered, the [kinematically possible models] might be sets of *inextendible smooth* curves in $\mathbb{R}^4$ which are nowhere tangent to surfaces of constant t (where $(t, \vec{x}) \in \mathbb{R}^4$). The models assign to the curves various parameters (m, ...). Under the intended interpretation, the curves represent possible trajectories of material particles, described with respect to a canonical coordinate system, and the parameters represent various dynamically relevant particle properties, such as mass. The space of [dynamically possible models] consists of those sets of curves that satisfy the standard form of Newton's equations"

The constraints that determine the kinematically possible, then, are *the only modal constraints*, the only conditions that play the role of laws.

---

independent of the system's dynamical states, e.g. the frictionless bowl as constraint for the rolling marble" (Hooker, 2013, 7, fn9).

[23] The whole space of metaphysical possibilities is thought to be too wide: it allows, for instance, that any function from points in space to points in space qualifies as a possible trajectory; thus, T_s follows from assumptions aimed at excluding certain behaviour, i.e. certain conceptual possibilities not ruled out by logic.

An example ubiquitous in all of physics are the global continuous symmetries (which any dynamical law obeys), which are related to the conservation of some quantity. The literature on symmetries abounds in discussions on their status, but a standard view on symmetries is to take them as meta-laws or, quoting Wigner (1960), 'super-principles'. Notice that the symmetries can be exhibited *by the geometrical structure of space* in physical theories (by its topological, affine, and metric structure). In fact, Wigner highlighted that symmetry principles are grounded in the stable properties of the defining structure of spacetime (Martin, 2003, 50). According to such a geometrical interpretation, the symmetries of the laws are interpreted as symmetries of spacetime itself; they codify "the geometrical structure of the physical world" (Brading et al., 2017, §5). To give an example, think of the homogeneity of space, assumed "in the physical description of the world since the beginning of modern science" (ibidem, §2.1). Another example is the time-translation invariance, related with the aforementioned conservation of energy via Noether's theorems.[24,25]

Thus, if in a certain setting the explanation based on the typicality of dynamics is correct, I want to conclude with the moral that, in *any* account of laws of nature (such as those cited in footnote 1 or any other account), in order to explain the high degree of order (the ubiquity and extreme stability of certain regularities) at least in such setting, one does not need to ontologically commit to the usual governing dynamical laws. Instead, one can focus on accounting for conditions of a more general type, the constraints discussed in this section, which would suffice to explain the emergence of law-like behaviour. So, when the primitivist postulates (without much argument, as far as I know) laws of nature, she can appeal to the present moral to at least attenuate the suspicion that the (according to some, mysterious) primitive entity postulated looks like a contrived, highly specific, set of governing differential equations.[26] The deflationist (a Humean, say), according to what I have suggested, is justified in dispensing with some modal entities—the dynamical governing laws—although I do not see how she could dispense with all modal notions, as long as she aims to explain the extremely regular and apparently non-accidental patterns of the Humean mosaic. Some kind of justification should be given by the Humean to the remaining general modal constraints.[27] The propensity theorist might also be interested in picking up our conclusion, in order to suggest that those minimal

[24] However, even if it is a paradigmatic example, we should keep in mind that it might not be true, according to general relativity (see e.g. Maudlin et al., 2019).

[25] I recur to the same examples in an analogous discussion in Filomeno (2019b, §4.2.3). It is interesting to note that both approaches converge at this point, while one starts from a result in chaos theory (Poincaré's method of arbitrary functions) while here we have started from the "neo-Boltzmannian" typicality explanation in statistical mechanics.

[26] In Filomeno (2016) I argue against a widespread satisfaction among philosophers and physicists primitivists about laws, which hope that a future Theory of Everything will be self-explanatory.

[27] I defend the need of the Humean account of laws to say something more in Filomeno (2021). Then, the present study, as well as my Filomeno (2019b), helps the Humean, but is still insufficient.

constraints are to be found in dispositional properties. Although what we have proposed could also be seen as a reductive explanation of such propensities.

To recapitulate, in this paper I have:

1. highlighted the common structure of explanation in certain projects in different fields of physics (Sects. 15.1and 15.2);
2. explicitly seen what is needed for such typicality explanation to be successful in classical statistical mechanics (Sects. 15.3 and 15.4); and
3. argued that, in order to account for the emerging law-like behaviour, the specific form of the underlying dynamical laws is irrelevant, while the constraints gain a prominent role (Sect. 15.5). The fact that in certain contexts a free state-space evolution yields in the long run a stable reduction of further degrees of freedom supports the thesis that stable patterns can occur without needing to postulate—to ontologically commit to—the usual set of governing dynamical equations. The constraints that characterize the context, i.e. that define the kinematically possible states, are left as the only modal constraints. Hence, this shifts the attention to the ontological nature of such kinematical constraints.

Acknowledgments Thanks to Carl Hoefer, John Horden, Ladislav Kvasz, Tim Maudlin, Carlos Romero, the research group on open systems at the department of mathematics of the University of Valparaíso, and audiences at the "Modality in physics" conference at Krakow and the "2nd Chilean conference in philosophy of physics" at Valparaíso.

References

Albert, D. Z. (2000). *Time and chance*. Harvard University Press.

Alpern, S., & Prasad, V. S. (2001). *Typical dynamics of volume preserving homeomorphisms. Cambridge tracts in mathematics*. Cambridge University Press.

Armstrong, D. (1983). *What is a law of nature?* Cambridge University Press.

Atkins, P. (2007). *Four laws that drive the Universe*. Oxford University Press.

Barrow, J. D. (1977). A chaotic cosmology. *Nature, 267*, 117–120.

Batterman, R. W. (1992). Explanatory instability. *Noûs, 26*(3), 325–348.

Batterman, R. W. (2001). *The Devil in the details: Asymptotic reasoning in explanation, reduction, and emergence*. Oxford University Press.

Batterman, R. W. (2018). Autonomy of theories: An explanatory problem. *Noûs, 52*, 858–873.

Batterman, R. W. (2019). Universality and RG explanations. *Perspectives on Science, 27*(1), 26–47.

Belot, G. (2011). Symmetry and equivalence. In R. Batterman (Ed.), *The Oxford handbook of philosophy of physics*. Oxford University Press.

Belot, G., & Earman, J. (1997). Chaos out of order: Quantum mechanics, the correspondence principle and chaos. *Studies in History and Philosophy of Science Part B, 28*(2), 147–182.

Berkovitz, J., Frigg, R., & Kronz, F. (2006). The ergodic hierarchy, randomness and hamiltonian chaos. *Studies in History and Philosophy of Science Part B, 37*(4), 661–691.

Boltzmann, L. (1877). Über die beziehung dem zweiten Haubtsatze der mechanischen Wärmetheorie und der Wahrscheinlichkeitsrechnung respektive den Sätzen über das Wärmegleichgewicht. *Wiener Berichte, 76*, 373–435.

Brading, K., Castellani, E., & Teh, N. (2017). Symmetry and symmetry breaking. In E. N. Zalta (Ed.), *The Stanford encyclopedia of philosophy*. Winter 2017 edition.
Brown, H. R., & Uffink, J. (2001). The origins of time-asymmetry in thermodynamics: The minus first law. *Studies in History and Philosophy of Science Part B, 32*(4), 525–538.
Callender, C. (2007). The emergence and interpretation of probability in Bohmian mechanics. *Studies in History and Philosophy of Science Part B: Studies in History and Philosophy of Modern Physics, 38*(2), 351–370.
Carroll, J. W. (1994). *Laws of nature*. Cambridge University Press.
Cartwright, N. (1999). *The dappled world: A study of the boundaries of science*. Cambridge University Press.
Chadha, S., & Nielsen, H. B. (1983). Lorentz invariance as a low energy phenomenon. *Nuclear Physics B, 217*(1), 125–144.
Chakravartty, A. (2005). Causal realism: Events and processes. *Erkenntnis, 63*(1), 7–31.
Chkareuli, J. L., Froggatt, C. D., & Nielsen, H. B. (2011). Spontaneously generated tensor field gravity. *Nuclear Physics B, 848*, 498–522.
Cohen, J., & Callender, C. (2009). A better best system account of lawhood. *Philosophical Studies, 145*(1), 1–34.
Dasgupta, A. (2011). Mathematical foundations of randomness. In P. Bandyopadhyay & M. Forster (Eds.), *Philosophy of statistics (Handbook of the philosophy of science: Volume 7)*. Elsevier.
Diaconis, P. (1998). A place for philosophy? The rise of modeling in statistics. *Quarterly of Applied Mathematics, 56*, 797–805.
Dretske, F. I. (1977). Laws of nature. *Philosophy of Science, 44*(2), 248–268.
Dürr, D. (1998). *Über den Zufall in der Physik*. Unpublished paper of the talk given at Leopoldina meeting. http://www.mathematik.uni-muenchen.de/~duerr/Zufall/zufall.html
Earman, J., & Roberts, J. (2005). Contact with the nomic: A Challenge for Deniers of Humean Supervenience about laws of nature part I: Humean Supervenience. *Philosophy and Phenomenological Research, 71*(1), 1–22.
Filomeno, A. (2016). Fundamentality, effectiveness and objectivity of gauge symmetries. *International Studies in the Philosophy of Science, 30*(1), 19–37.
Filomeno, A. (2019a). Fundamentos Filosóficos de la Mecánica Estadística. In C. Saborido & M. Martinez (Ed.), *Enciclopedia de la Sociedad Española de Filosofía Analítica*.
Filomeno, A. (2019b). Stable regularities without governing laws? *Studies in History and Philosophy of Modern Physics, 66*, 186–197.
Filomeno, A. (2021). Are non-accidental regularities a cosmic coincidence? Revisiting a central threat to Humean laws. *Synthese, 198*, 5205–5227.
Ford, J., & Stoddard, S. D. (1973). Numerical experiments on the stochastic behaviour of a Lennard-Jones gas system. *Physical Review A, 8*, 1504–1512.
Frigg, R. (2008). A field guide to recent work on the foundations of statistical mechanics. In D. Rickles (Ed.), *The Ashgate Companion to Contemporary Philosophy of Physics* (pp. 99–196). Ashgate.
Frigg, R. (2009). Typicality and the approach to equilibrium in Boltzmannian statistical mechanics. *Philosophy of Science, 76*(5), 997–1008.
Frigg, R., & Bishop, R. (2016). From order to chaos and back again. In N. Cartwright & K. Ward (Eds.), *Rethinking order. After the laws of nature* (pp. 55–75). Bloomsbury.
Frigg, R., & Werndl, C. (2011). Explaining thermodynamic-like behavior in terms of epsilon-ergodicity. *Philosophy of Science, 78*(4), 628–652.
Frigg, R., & Werndl, C. (2012). Demystifying typicality. *Philosophy of Science, 79*(5), 917–929.
Frigg, R., Berkovitz, J., & Kronz, F. (2020). The Ergodic Hierarchy. In E. N. Zalta (ed.), The Stanford Encyclopedia of Philosophy (Fall 2020 Edition), https://plato.stanford.edu/archives/fall2020/entries/ergodic-hierarchy/.
Froggatt, C. D., & Nielsen, H. B. (1991). *Origin of symmetries*. World Scientific.
Gallagher, I., Saint-Raymond, L., & Texier, B. (2012). *From Newton to Boltzmann: hard spheres and short-range potentials*. e-prints. arXiv:1208.5753.

Goldstein, S. (2001). Boltzmann's approach to statistical mechanics. In M. C. Galavotti, G. Ghirardi, F. Petruccione, J. Bricmont, D. Dürr, & N. Zanghi (Eds.), *Chance in physics: Foundations and perspectives* (pp. 39–54). Berlin: Springer-Verlag.
Goldstein, S., & Lebowitz, J. L. (2004). On the (Boltzmann) entropy of non-equilibrium systems. *Physica D Nonlinear Phenomena, 193*, 53–66.
Goldstein, S., Lebowitz, J. L., Mastrodonato, C., Tumulka, R., & Zanghi, N. (2010a). Normal typicality and von Neumann's quantum ergodic theorem. *Royal Society of London Proceedings Series A, 466*, 3203–3224.
Goldstein, S., Lebowitz, J. L., Tumulka, R., & Zanghì, N. (2010b). Long-time behavior of macroscopic quantum systems. Commentary accompanying the English translation of John von Neumann's 1929 article on the quantum ergodic theorem. *European Physical Journal H, 35*, 173–200.
Goldstein, S., & Tumulka, R. (2010). On the approach to thermal equilibrium of macroscopic quantum systems. *Physical Review E, 81*, 011,109. http://link.aps.org/doi/10.1103/PhysRevE.81.011109
Halmos, P. R. (1944a). Approximation theories for measure preserving transformations. *Transactions of the American Mathematical Society, 55*, 1–18.
Halmos, P. R. (1944b). In general a measure preserving transformation is mixing. *Annals of Mathematics (2), 45*, 786–792.
Hooker, C. (2013). On the import of constraints in complex dynamical systems. *Foundations of Science, 18*(4), 757–780.
Jacobson, T. (1995). Thermodynamics of space-time: The Einstein equation of state. *Physical Review Letters, 75*, 1260–1263.
Jacobson, T., & Wall, A. C. (2010). Black hole thermodynamics and Lorentz symmetry. *Foundations of Physics, 40*, 1076–1080.
Katok, A., & Hasselblatt, B. (1997). *Introduction to the modern theory of dynamical systems. Encyclopedia of mathematics and its applications*. Cambridge University Press.
Katok, A. B., & Stepin, A. M. (1970). Metric properties of homeomorphisms that preserve measure. *Uspekhi Matematicheskikh Nauk, 25*, 193–220. English translation in Russian Math. Surveys, 25 (1970), 191–220.
Lazarovici, D., & Reichert, P. (2015). Typicality, irreversibility and the status of macroscopic laws. *Erkenntnis, 80*(4), 689–716.
Lebowitz, J. L. (1999). Statistical mechanics: A selective review of two central issues. *Reviews of Modern Physics, 71*(2), S346.
Lewis, D. (1999). *Papers in metaphysics and epistemology. Cambridge studies in philosophy*. Cambridge University Press.
Linde, A. D. (1983). Chaotic inflation. *Physics Letters B, 129*(3), 177–181.
Loeve, M. (1977). *Probability theory*. Springer Verlag.
Markus, L., & Meyer, K. R. (1974). *Generic Hamiltonian dynamical systems are neither integrable nor ergodic. Number No. 144 in American Mathematical Society: Memoirs of the American Mathematical Society*. American Mathematical Society.
Martin, C. (2003). On continuous symmetries and the foundations of modern physics. In *Symmetries in physics: Philosophical reflections* (pp. 29–60). Cambridge University Press.
Maudlin, T. (2007). *The metaphysics within physics*. Oxford University Press.
Maudlin, T. (2011). Three roads to objective probability. In C. Beisbart & S. Hartmann (Eds.), *Probabilities in physics* (pp. 293–322). Oxford University Press.
Maudlin, T., Okon, E. E., & Sudarsky, D. (2019). On the status of conservation laws in physics: Implications for semiclassical gravity. *Studies in History and Philosophy of Science Part B: Studies in History and Philosophy of Modern Physics, 69*, 67–81.
Mill, J. S. (1884). *A system of logic*. Longman.
Misner, C. W. (1969). Mixmaster Universe. *Physical Review Letters, 22*, 1071–1074.
Mukohyama, S., & Uzan, J.-P. (2013). From configuration to dynamics: Emergence of Lorentz signature in classical field theory. *Physical Review, D87*(6), 065,020.
Mumford, S. (2004). *Laws in nature*. Routledge.

Nelson, E. (1966). Derivation of the Schrödinger equation from Newtonian mechanics. *Physical Review, 150*(4), 1079.
Okon, E., & Sudarsky, D. (2016). A (not so?) novel explanation for the very special initial state of the universe. *Classical and Quantum Gravity, 33*(22), 225,015.
Oxtoby, J. C., & Ulam, S. M. (1941). Measure-preserving homeomorphisms and metrical transitivity. *Annals of Mathematics (2), 42*, 874–920.
Pooley, O. (2013). Substantivalist and relationalist approaches to spacetime. In R. Batterman (Ed.), *The Oxford handbook of philosophy of physics*. Oxford University Press.
Pulvirenti, M., Saffirio, C., & Simonella, S. (2014). On the validity of the Boltzmann equation for short range potentials. *Reviews in Mathematical Physics, 26*, 1450001-294.
Rohwer, Y., & Rice, C. (2013). Hypothetical pattern idealization and explanatory models. *Philosophy of Science, 80*(3), 334–355.
Rosenthal, J. (2009). The natural-range conception of probability. In G. Ernst & A. Hüttemann (Eds.), *Time, chance and reduction: Philosophical aspects of statistical mechanics* (pp. 71–90). Cambridge University Press.
Rosenthal, J. (2012). Probabilities as ratios of ranges in initial-state spaces. *Journal of Logic, Language and Information, 21*(2), 217–236.
Ruelle, D. (1989). *Elements of differentiable dynamics and bifurcation theory*. Academic Press.
Sheldon, R. (2009). *A first course in probability* (8th ed.). Prentice Hall Press.
Sinai, Ya. G. Dynamical systems with elastic reflections. *Russian Mathematical Surveys, 25*, 137.
Sklar, L. (1993). *Physics and chance: Philosophical issues in the foundations of statistical mechanics*. Cambridge University Press.
Sklar, L. (2015). Philosophy of statistical mechanics. In E. N. Zalta (Ed.), *The Stanford Encyclopedia of philosophy*.
Smale, S. (1966). Structurally stable systems are not dense. *American Journal of Mathematics, 88*, 491–496.
Smeenk, C., & Hoefer, C. (2015). Philosophy of the physical sciences. In P. Humphreys *The Oxford handbook of philosophy of science*. Oxford University Press.
Spohn, H. (2012). *Large scale dynamics of interacting particles. Theoretical and mathematical physics*. Springer Berlin Heidelberg. https://books.google.cl/books?id=81joCAAAQBAJ
Strevens, M. (2003). *Bigger than chaos. Understanding complexity through probability*. Harvard University Press.
Strevens, M. (2008). *Depth: An account of scientific explanation*. Harvard University Press.
Strevens, M. (2013). *Tychomancy: Inferring probability from causal structure*. Harvard University Press.
Strogatz, S. H. (1994). *Nonlinear dynamics and chaos: With applications to physics, biology, chemistry, and engineering. Addison-Wesley studies in nonlinearity*. Westview Press.
Tooley, M. (1977). The nature of laws. *Canadian Journal of Philosophy, 7*(4), 667–698.
Uffink, J. (2006). Compendium of the foundations of classical statistical physics. In J. Butterfield & J. Earman (Eds.), *Handbook for philosophy of physics*".
Uffink, J. (2014). Boltzmann's work in statistical physics. In E. N. Zalta (Ed.), *The Stanford encyclopedia of philosophy*. Spring 2014 edition.
Valentini, A. (1991). Signal-locality, uncertainty, and the subquantum H-theorem. I. *Physics Letters A, 156*(1–2), 5–11.
Verlinde, E. P. (2011). On the origin of gravity and the laws of Newton. *JHEP, 1104*, 029.
Verlinde, E. P. (2017). Emergent gravity and the dark universe. *SciPost Physics, 2*(3), 016.
Vranas, P. B. M. (1998). Epsilon-ergodicity and the success of equilibrium statistical mechanics. *Philosophy of Science, 65*(4), 688–708.
Werndl, C. (2009). What are the new implications of chaos for unpredictability? *British Journal for the Philosophy of Science, 60*(1), 195–220.
Werndl, C. (2013). Justifying typicality measures of Boltzmannian statistical mechanics and dynamical systems. *Studies in History and Philosophy of Science Part B: Studies in History and Philosophy of Modern Physics, 44*(4), 470–479.

Wiggins, S. (2003). *Introduction to applied nonlinear dynamical systems and chaos. Texts in applied mathematics*. Springer.

Wigner, E. (1960). The unreasonable effectiveness of mathematics in the natural sciences. *Communications on Pure and Applied Mathematics, 1*(13), 114.

Wilhelm, I. (2022). Typical: A theory of typicality and typicality explanation. *British Journal for the Philosophy of Science, 73*(2), 561–581.

Chapter 16
The Case of Phonons: Explanatory or Ontological Priority

Hernán Lucas Accorinti, Sebastián Fortín, Manuel Herrera, and Jesús Alberto Jaimes Arriaga

Abstract Recent discussions about the microstructure of materials generally focus on the ontological aspects of the molecular structure. However, there are many types of substances that cannot be studied by means of the concept of molecule, for example, salts. For the quantum treatment of these substances, a new particle, called phonon, is introduced. Phonons are generally conceived as a pseudo-particle, that is, a mathematical device necessary to perform calculations but which does not have a "real" existence. In this context, the aim of this paper will be to analyze the ontological status of phonons. For such purposes we will critically analyze the arguments that would account for the presumed nonexistence of phonons. Finally, having already demonstrated that there are not enough reasons to consider phonons as non-existing entities, we will explore some possibilities that allow us to elucidate their ontological status.

Keywords Phonons · Quantum chemistry · Intertheoretical relationships · Ontological commitments · Pluralism

16.1 Introduction

The study of intertheoretical relationships is of particular relevance to the philosophy of chemistry, due to the fact that the close relationships between chemistry and physics affect the independence and autonomy of chemistry as a scientific discipline. The influence that physics has exerted on chemistry from a position of supposed hegemony is manifested, for example, in the modern explanations of the

This work is fully collaborative. The order of the names is alphabetical and therefore does not imply priority.

H. L. Accorinti · S. Fortín (✉) · M. Herrera · J. A. J. Arriaga
CONICET – University of Buenos Aires, Buenos Aires, Argentina
e-mail: sfortin@conicet.gov.ar; 304156359@comunidad.unam.mx

C. Soto (ed.), *Current Debates in Philosophy of Science*, Synthese Library 477,
https://doi.org/10.1007/978-3-031-32375-1_16

chemical bond proposed since the advent of quantum mechanics in the first half of the twentieth century. These explanations led Dirac to declare in 1929 that quantum mechanics already possessed the power to account for the fundamental principles of chemistry. However, the issue of the effective reduction of chemistry to physics has been much discussed recently by authors such as Eric Scerri (2000, 2011, 2012, 2013), Robin Hendry (2010), Hinne Hettema (2012), Lombardi and Labarca (2005) and Lombardi (2014), among many others. The main antireductionist arguments stress that physics is incapable of explaining the structure of chemical substances. In particular, they focus on the incompatibility between the fundamental principles of quantum mechanics and the concept of molecular structure used in chemistry (Woolley, 1978, 1982; Amann, 1992; Sutcliffe & Woolley, 2011, 2012; Bishop, 2005; Fortin et al., 2017). However, it is important to bear in mind that molecules are not the only type of entities in chemistry; in order to enrich the debate, it is essential to consider other types of chemical entities. One example is the case of salts, which are understood not as "large" molecules, but as a *network*; as we will explain in Sect. 16.2, this is a concept completely different from the concept of molecule. As a consequence, although the present work will not specifically delve into the reductionism-anti-reductionist discussion, it will provide new elements for de debate by introducing arguments for the consolidation of the ontological existence of phonons. The discussions around the intertheorical relations and those referred to the ontology are different. Regarding this one can take two paths: to defend a purely intertheorical reductionism without any ontological commitments as Nagel did or practice an analytical metaphysics and postulate, as a product of a conceptual analysis, certain items as fundamental ones. In this paper we will not adhere to any of these attitudes. Our analysis will pretend to be ontological, however such pretension will be based on our best current scientific theories and the philosophical analysis that we will make of them.

Phonons are particles that are appealed to for the quantum treatment of salts. However, phonons are commonly conceived as pseudo-particles. In general, a phonon is considered a mathematical device necessary to perform calculations, but without "real" existence. Some authors classify phonons as quasi-particles, that is, as non-elementary particles that, although can be reduced, on a limited time scale behave as if they were elementary particles (Ladyman, 2015).

In this context, the aim of this paper consists in analyzing the ontological status of phonons in the light of the usual arguments by means of which their real existence is denied. For this purpose, in Sect. 16.2, we will briefly summarize the way in which salts are traditionally studied; we will point out that, in order to account for certain phenomena, mainly thermal acoustic, it is necessary to introduce the concept of phonon. In Sect. 16.3, we will briefly recall the empirical success that underlies the importance of the concept of phonon at an explanatory and predictive level. In Sect. 16.4, we will critically analyze the usual arguments against the real existence of phonons. On this basis, by establishing the analogy between phonons and photons, we will conclude that if the existence of photons is accepted, then the same attitude should be adopted in the case of phonons. Once it is admitted that there are not good reasons to consider phonons as non-existing entities, in Sect. 16.5 we will explore some perspectives that will allow us to elucidate their ontological status. Finally, in Sect. 16.6, we present our conclusions.

16.2 The Internal Structure of Crystalline Solids

Substances in the solid state are commonly studied in a different way from how their liquid or gaseous counterparts are treated, mainly due to the magnitude of their components at the molecular level. For example, a macroscopic crystal can very well be described in terms of a single structural unit that extends through the entire crystal. Crystalline materials, as the salts, are those solid substances which distinctive feature is that their basic structural units have a long-term periodic arrangement.

16.2.1 The Classical Description of a Crystalline Solid

For conciseness, we will introduce two simplifications that will not harm in any way the conclusions we will arrive at in this work, since the generalization to more complex systems without much difficulty is always possible. On the one hand, we will consider only the so-called monatomic crystals, that is, crystalline solids whose units are composed of a single atom. On the other hand, although a real crystal is understood as a perfectly ordered network in three dimensions, here we will consider only models[1] in one dimension.

The simplest model of a crystalline solid is a one-dimensional infinite chain of atoms interconnected by some kind of interaction. Since the chain is infinite and all its links are identical atoms, the chain has a discrete translation symmetry. In turn, by assuming the principle of indifference, the constituent atoms are considered as equidistant in the state of equilibrium. In this model, when an atom is moved away from its equilibrium position, then a restitutive force will appear on that atom that tends to return it to the equilibrium position. The simplest restitutive force is, as in a spring, proportional to the distance, $F \propto d$. If u_s is the variable that represents the position of the particle s, and given that this particle is bound to its neighbors $s+1$ and $s-1$, then the force on it will be

$$F_s = C\left(u_{s+1} - u_s\right) - C\left(u_s - u_{s-1}\right) \tag{16.1}$$

[1] There is a vast amount of literature which presents different versions of the notion of model. We will use a pretheoretical notion of this term. The various perspectives that are assumed from such a notion do not trouble the proposed objectives, not even those positions that attribute a fictional or instrumental character to models (for a detailed analysis of this subject see Gelfert, 2016).

where C is a proportionality constant. Once the force is determined, it is introduced into the equations of classical mechanics, for example, into Newton's second law. In this way, a series of differential equations is obtained. If there are N particles in the chain, then N equations of the following form are obtained:

$$\frac{m}{C}\frac{\partial^2 u_s}{\partial t^2} = u_{s+1} + u_{s-1} - 2u_s \tag{16.2}$$

where m is the mass of particles and $s = 1, 2, \ldots, N$. In principle, if these equations are solved given the initial conditions, the mathematical expression that describes the motion of each atom in time can be computed. Mathematical details are not very important here, but we do want to highlight the picture about the dynamics of solids offered by this model: atoms are linked by interactions that allow them to vibrate about their equilibrium position. In this way, the vibrations in the crystal lattice form waves that propagate through the crystal, by means of which sound and temperature are transported.

16.2.2 Quantum Solids and the Birth of the Concept of Phonon

The problem about the ontological status of phonons arises because the classical description of crystalline solids, offered in the previous section, is not enough to explain the phenomena that can be measured in laboratory. In particular, the dependence of the heat capacity on temperature cannot be explained by the mechanical model. As we will see, in order to account for this kind of phenomena, the introduction of a new entity was required: the phonon.

Let us carefully analyze how phonons are introduced. Since empirical evidence suggests that vibrations are quantized, it is necessary to quantize the model. By assuming that the atoms only interact with their first neighbors, and on the basis of the fact that the inter-atomic forces are equal to those of the harmonic oscillator, quantizing the model is extremely simple and the Hamiltonian $\hat{H}$ can be written as

$$\hat{H} = \sum_{s=1}^{N} \frac{1}{2m}\hat{p}_s^2 + \frac{C}{2}\left(\hat{q}_{s+1} - \hat{q}_s\right)^2 \tag{16.3}$$

where $\hat{p}_s$ and $\hat{q}_s$ are the momentum and position operators from s, respectively, and $\hat{q}_{s+1}$ is the position operator from particle $s + 1$. Since the total number N of particles is a really huge number, of the order of 10^{23}, solving the Hamiltonian as it is expressed is impossible in practice. As a consequence, a strategy that simplifies the calculation is used. The first step is to make a very specific coordinate change, to the so-called *phononic coordinates*:

$$\hat{Q}_k = N^{-1/2} \sum_{s=1}^{N} \hat{q}_s e^{-iksa} \quad \hat{P}_k = N^{-1/2} \sum_{s=1}^{N} \hat{p}_s e^{iksa} \tag{16.4}$$

where $k = 2\pi n N^{-1} a^{-1}$ and $n = 0, \ \pm 1, \ \ldots, \ N/2$. In these coordinates the Hamiltonian is written as a sum of harmonic oscillators that do not interact:

$$\hat{H} = \sum_n \frac{1}{2} \left(\hat{P}_n^2 + \omega_n^2 \hat{Q}_n^2 \right) \tag{16.5}$$

where $\omega_n = (2C/\mathrm{m})^{1/2}(1 - \cos(k_n a))^{1/2}$ is the angular frequency of the oscillator n. The result so obtained shows that the system can be represented by the Hamiltonian corresponding to the sum of many independent harmonic oscillators. Therefore, the formalism based on creation and destruction operators originally developed for the harmonic oscillator can be used (Sakurai, 1994). In this formalism vibrations can be represented as particles that are created and destroyed, where the quantity of particles increases as a function of the increase of the vibrational magnitude. Thus, some states of the system can be represented by means of a state $|n\rangle$, where n is a integer that represents the number of particles in the system. The creation operator $\hat{a}^\dagger$, when applied to a state of the system with n particles, returns another state with $n + 1$ particles, that is, creates a particle. In turn, the destruction operator $\hat{a}$ destroys a particle.

$$a^\dagger |n\rangle = |n+1\rangle \quad a|n\rangle = |n-1\rangle \tag{16.6}$$

On the basis of this representation, the Hamiltonian can be expressed in terms of the creation and the destruction operators as follows:

$$\hat{H} = \sum_k h\omega_k \left(\hat{a}_k^\dagger \hat{a}_k + \frac{1}{2} \right) \tag{16.7}$$

Although this Hamiltonian is the same Hamiltonian of Eq. (16.3), in this representation the atoms originally introduced to constitute chain no longer appear in the model. The new expression represents a new type of particle called *phonon*. Then, the original mechanical picture can be left aside, to move to a model where there are no atoms in motion, but only phonons. From this perspective, the total energy of the crystal is the sum of the energy of the phonons plus a "vacuum" energy associated with the case in which there are no phonons in the network. Indeed,

$$E = \sum_m \varepsilon_m = \sum_m h\omega_m \left(n_m + \frac{1}{2} \right) = \sum_m \varepsilon^m n_m + \varepsilon_{vacuum} \tag{16.8}$$

where ε^m is the energy of a phonon with frequency ω_m, n_m is the number of phonons with this frequency and ε_{vaccum} is the vacuum energy.

16.3 Explanation and Prediction by Means of Phonons

In the previous section we showed that, by means of a series of calculations, it is possible to begin from the Hamiltonian of a system consisting of atoms that vibrate by interacting with their first neighbors, and to arrive to the Hamiltonian of a system composed of phonons. However, the relevance of the model based on phonons, and what makes the question about the ontological issue of phonons so urgent, is the fact that the model is indispensable for the study of the properties of crystalline. In particular, phonons play a central role in explaining the physical and chemical properties of salts. As an example, we present here two simple cases in which the explanatory and predictive power of the concept of phonon can be acknowledged.

16.3.1 The Heat Equation

In the description of crystalline solids, the temperature of the solid is explained microscopically by the presence of phonons; the higher the density of phonons, the higher the temperature. To illustrate this idea, let us consider a metal bar heated at one of its ends and cooled at the other: heat will flow from the hot end to the cold end, producing a temperature gradient. If the phonon model is used for the microscopic explanation, then the situation is described in terms of a source of phonons at the hot end, which travel through the metal bar toward the cold end. In their motion, phonons collide with each other, generating a process of diffusion that ends with the destruction of the phonons at the cold end.

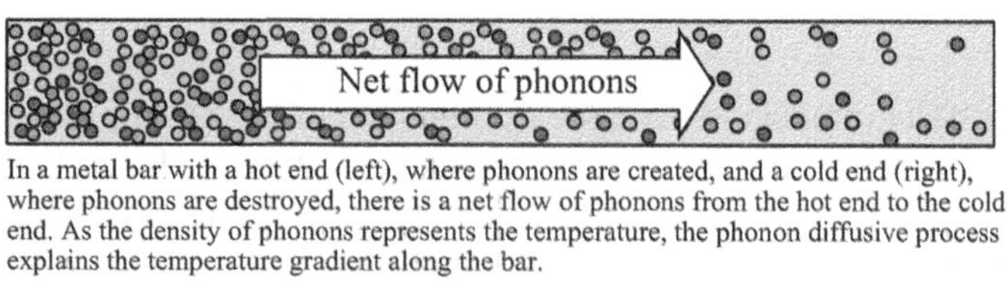

In a metal bar with a hot end (left), where phonons are created, and a cold end (right), where phonons are destroyed, there is a net flow of phonons from the hot end to the cold end. As the density of phonons represents the temperature, the phonon diffusive process explains the temperature gradient along the bar

In this way, phonons offer a microscopic model of heat flow and temperature gradient based on particles. This model is explanatory simple and very effective, since gives an easy picture of thermal phenomena. In fact, it is surprising that the application of quantum mechanics to crystalline solids ends up recovering a conception of heat similar to that developed by Lavoisier in the thirteenth century with his theory of caloric (Best, 2016). In effect, phonons would be a modern and quantized version of the notion of the *substance of heat*.

On the other hand, at the mathematical level, the effectiveness of the model of phonons becomes clear by the fact that it allows deducing the well-known heat equation in a simple way as follows. If collisions are taken into account, the net flow of energy along the longitudinal direction x of the bar is computed as

$$j_U = -\frac{1}{3} C v l \frac{dT}{dx} \tag{16.9}$$

where C is the thermal capacity per unit volume, v is the average velocity of the phonons, and l is the average free path of a phonon between collisions (for a detailed analysis of the deduction of the heat equation, see Ashcroft & Mermin, 1976). Independently of the details of the derivation, it is interesting to notice that the constants appearing in the equation, which have no clear interpretation in the traditional view, acquire a precise meaning in the model of phonons: for example, when interpreted in the light of the phonon model, the heat equation shows that there is a directly proportional relation between the velocity of phonons and the propagation of heat.

16.3.2 The Heat Capacity

Another interesting example that reinforces the richness and epistemic potential of the concept of phonon is the calculation of the heat capacity of solids. In this case, let us consider that phonons are confined within an isolated crystal without sources or sinks. The system can be thus modeled as an ideal gas of phonons confined in a box. On this basis, the total energy of the crystal can be written as the sum of the energy of each of the phonons

$$E = \sum_m \varepsilon^m \langle n_m \rangle \tag{16.10}$$

where $\langle n_m \rangle$ is the average occupation number for phonons of energy ε^m, which is given by the Planck distribution

$$\langle n_m \rangle = \frac{1}{e^{\frac{h\omega_m}{k_B T}} - 1} \tag{16.11}$$

where k_B is the Boltzmann constant and T is the absolute temperature. Then, the specific heat of the crystal $C_{crystal}$ is computed as

$$C_{crystal} = \frac{\partial U}{\partial T} \tag{16.12}$$

Due to the complexity of the calculation, certain approximations are usually applied. In the case of low temperatures, the Debye approximation is used and the result is proportional to the cube of the temperature (for the details of the derivation, see Kittel, 2005)

$$C_{cristal} \propto T^3 \tag{16.13}$$

In the case of high temperatures, the Einstein approach is used, and the thermal capacity results

$$C_{cristal} \propto \frac{1}{T^2} \frac{e^{\frac{h\omega}{k_B T}}}{\left(e^{\frac{h\omega}{k_B T}} - 1\right)^2} \tag{16.14}$$

These equations, deduced from the phononic formalism, fit excellently to the results obtained in the laboratory. For example, the results of the thermal capacity of diamond obtained experimentally are contrasted with the results computed by means of the phononic formalism leading to the above equations. A similar situation occurs with the thermal capacity solid argon. In this case, the agreement between experimental and theoretical results is more than acceptable, it is excellent.

The previous cases are only two of the many applications that show that the phonon formalism is not a mere calculation curiosity; on the contrary, it is a successful model to explain the phenomena associated with solids, which has demonstrated a high empirical success and explanatory power.

16.4 The Ontological Status of Phonons

In the previous sections we have recalled the mathematical derivation of phonons and have stressed their explanatory power. However, such epistemic virtues do not supply sufficient grounds to justify the existence of phonons. Precisely, the anti-realist tradition appeals to the well-known pessimistic meta-induction argument to claim that pragmatic success does not prove the truth of the theory nor, therefore, the effective existence of the entities postulated by it. Aware of this argument, in this section we will try to argue, perhaps not against scientific anti-realism but against a certain type of realism, that if one adopts a realistic attitude in science, there is no reason to consider phonons as non-existing entities. For this purpose, we will follow two strategies: first, we will consider the analogy between phonons and photons, and, second, we will refute the usual arguments by which phonons are commonly considered useful tools but unreal entities.

16.4.1 Phonon, a Cousin of Photon

In Sect. 16.2.2, we have shown how, by means of a series of calculations it is possible to start from the Hamiltonian of vibrating atoms and to arrive to the Hamiltonian of a system composed of phonons in the following way:

$$\hat{H}_{atoms} + \hat{H}_{\text{int}} \xrightarrow{\text{Coordenadas fonónicas}} \sum_m \frac{1}{2}\left(\hat{P}_m^2 + \omega_m^2 \hat{Q}_m^2\right) \xrightarrow{\text{op. creación y destrucción}} \sum_m \hbar\omega_m \left(\hat{a}_m^\dagger \hat{a}_m + \frac{1}{2}\right) \tag{16.15}$$

where $\hat{H}_{atoms} = \sum_{s=1}^{N} \frac{1}{2m} \hat{p}_s^2$ and $\hat{H}_{\text{int}} = \sum_{s=1}^{N} \frac{C}{2} \left(\hat{q}_{s+1} - \hat{q}_s\right)^2$.

This formal operation is highly relevant for the purpose of the present work, due to the complete analogy that can be traced between phonons and photons. In fact, the calculations used to find the phononic Hamiltonian are exactly the same as those used to find the Hamiltonian of photons. Starting from the Hamiltonian of the electric $\hat{H}_E$ and magnetic field $\hat{H}_B$, we formal strategy in the case of photons is:

$$\hat{H}_E + \hat{H}_B \xrightarrow{\text{Coordenadas fotónicas}} \sum_m \frac{1}{2}\left(\hat{P}_m^2 + \omega_m^2 \hat{Q}_m^2\right) \xrightarrow{\text{op. creación y destrucción}} \sum_m \hbar\omega_m \left(\hat{a}_m^\dagger \hat{a}_m + \frac{1}{2}\right) \tag{16.16}$$

As it easy to see, the formal analogy between the derivation of photons and the derivation of phonons is complete; then, it is legitimate to ask why we phonons are endowed with the fictional or degraded status of quasi-particles. If the procedure of the expression (16.16) allows us to understand all forms of electromagnetic radiation in terms of particles, why does the procedure of expression (16.15) not allow us to understand all forms of vibration in a crystal under the same terms? Or, in other words, what are the assumptions that prevent us to say that the temperature and the sound in a crystal are manifestations of the real presence of phonons?

Summing up, in the case of photons, by beginning from the electromagnetic field, certain calculations lead to the expressions that can be easily interpreted in terms of particles named photons. Subsequently, and due to the empirical success of the models based on photons, they acquire the status of real particles. However, in the case of phonons, although exactly the analogous calculations are performed and an equivalent empirical success is reached, the same conclusion is not drawn. In fact, these particles are usually considered as pseudo-particles or quasi-particles (Ladyman, 2015; Franklin & Knox, 2018). In the following subsections we will critically assess the reasons that commonly underlie this different attitude. In particular, we will try to elucidate whether the different arguments usually directed against the non-existence of phonons are powerful enough to draw such a conclusion.

16.4.2 Two Ontologies for Matter

When a picture and a mathematical formalism to describe the composition of chemical substances is needed, definitely the most popular model is that offered by the usual atomic model, according to which matter is composed of atoms in motion, and in whose context chemical phenomena are explained in terms of atoms and their interaction. An atom is classically conceived as a composite particle whose set of properties includes position and velocity. This idea, when applied to the case of a crystalline solid, leads to the model of the chain of interacting atoms as presented in Sect. 16.2. Thus, in order to apply quantum mechanics to a chemical substance under this model, it is necessary to define its Hamiltonian as the Hamiltonian of these component atoms plus the interaction Hamiltonian:

$$\hat{H}_U = \hat{H}_{atoms} + \hat{H}_{\mathrm{int}} \tag{16.17}$$

If some approximations are applied to this Hamiltonian, many natural phenomena can be predicted and explained. For example, the energy levels of a molecule can be computed (Sutcliffe & Woolley, 2012).

On the other hand, in Sect. 16.2 we presented a different ontology for the case of solid state matter based on the phononic model. According to that model, solids are composed of atoms at rest and phonons in motion. In turn, the Hamiltonian of the phonons plus the interaction Hamiltonian can be defined

$$\hat{H}_U = \hat{H}_{phonons} + \hat{H}'_{\mathrm{int}} \tag{16.18}$$

With these Hamiltonians many natural phenomena can also be predicted and explained. For example, the heat equation can be computed.

Although a relevant empirical success is obtained by the application of the two models, the preference for the atomic hypothesis about the constitution of matter prevails over the phononic hypothesis in the scientific community. One may ask for the reasons underlying this choice. In the following paragraphs we critically analyze three arguments that can be used to support the usual atomic hypothesis:

- The "*tool argument*" points out that the atomic hypothesis is originally used to introduce phonons (see Eq. 16.15). Therefore, phonons must be interpreted as a mere tool that simplifies the calculations that cannot be directly solved in the atomic model.

- The "*supercomputer argument*" claims that we are forced to use the phononic model because we do not have yet sufficient computing power to solve the Schrödinger equation of a crystalline network from first principles. The assumption is that, in the future, when we will count with more powerful computers, we will be able to perform the exact calculations with no need to appeal to phonons.
- The "*explanation argument*" asserts that the usual atomic model, unlike the phononic model, can explain a great variety of phenomena. Due to their different explanatory capacity, both models are not at the same level.

The three arguments are intertwined, since the explanation argument only makes sense either by conceiving phonons as mere calculation tools, or by clinging to the hope that in the near future we will have sufficient computing power to do without phonons. However, for reasons of argumentative clarity, we will treat them independently.

16.4.2.1 The Tool Argument

According to the *tool argument*, on the basis of the model of a chain of atoms that vibrate about their equilibrium position, phonons are introduced with the only purpose of simplifying calculations. In fact, starting from the atoms model was the strategy followed in Sect. 16.2 (it is also the way in which phonons appear in the textbooks on the subject; see, e.g., Ashcroft & Mermin, 1976; Kittel, 2005). However, we followed this strategy only for didactic reasons: since the typical reader is familiar with the idea of atoms in motion, it is easier to begin with a model based on them. The relevant point is that those calculations based on the atomic model are not necessary: it is possible to directly introduce phonons with the Hamiltonian

$$\hat{H} = \sum_k h\omega_k \left(\hat{a}_k^\dagger \hat{a}_k + \frac{1}{2} \right) \tag{16.19}$$

In other words, the existence of phonons can be postulated from the very beginning, and, with them, the properties of crystalline solids can be explained. When this fact is kept in mind, the tool argument looses its strength: although the calculations are usually presented from atoms to phonons, they can be presented in the opposite direction, such that phonons appear as the fundamental entities.

Our counterargument is further supported by the fact that the Hamiltonian of the atoms in motion can be obtained by departing from the phononic Hamiltonian. This means that, although there is a transformation that turns the atomic Hamiltonian into the phononic Hamiltonian, there is also an inverse transformation that turns the phononic Hamiltonian into the atomic Hamiltonian. Therefore, one might argue that phonons are the fundamental entities and atoms in motion are useful calculation tools for treating in problems that do not involve thermal phenomena. This shows that, since the situation is symmetric, the tool argument cannot be appealed to for supporting the ontological priority of one description matter over the other. Of course, to prove the formal symmetry is not to prove the existence of phonons. This can not be demonstrated exclusively from this first argument. In this section we only try to reject the idea that the atoms are more fundamental entities. That is, we are not claiming that due to the symmetry we can infer phonon's existence; we affirm instead that symmetry shows that the argument that takes atoms as more fundamental entities due to the supposed primacy of one Hamiltonian over another is not valid. In order to establish the ontological status of phonons, we will need a set of interrelated arguments. All of them as a whole can fulfill our objective. Taken separately they are profusely ineffective. So, it might be taken into account only as a complement in the case that other arguments for the ontological priority of the atoms model were available. If it could be proved, on the basis of other arguments, that the fundamental ontology of matter is one based on atoms in motion, then the tool argument would explain why using phonons is sometimes convenient.

So far we have shown that the two presentation strategies, phonons obtained from atoms and atoms obtained from phonons, are equally legitimate. But this view is valid only in a particular case, the case of non-interacting phonons. In fact, in Sect. 16.2, the phonons appearing at the end of the derivation do not interact with each other. This means that the argument that begins by atoms leads to the very idealized case of phonons that do not interact with each other. But, when phonons are used to describe some phenomenon in a crystal in the general case, a *phononic interaction* term is introduced in the phononic Hamiltonian of Eq. (16.19). For example, the dependence of electrical resistivity with temperature is explained by considering that the electrons that move in a conductive material collide with and are dispersed by the phonons of the crystal lattice: the higher the temperature, the more number of phonons in the material, and consequently more collisions occur. The electron-phonon coupling constant is measured in laboratory, and from this constant the number of phonons surrounding an electron that moves within a crystal is computed: in this way, the type of interaction between the involved particles can be determined (Kittel, 2005). This phenomenon is explained by assuming the existence of both electrons and phonons, and by admitting that particles of those these two kinds interact with each other: the interaction Hamiltonian that accounts for this interaction is built in order to account for the results of experiments, as usual in physics. Summing up, the dependence of electrical resistivity with temperature is explained by taking phonons, not atoms, as the starting point.

The same happens in the experiment carried out by Shinen (1963) on magnesium oxide crystals, where a scattering process between phonons can be observed. Shinen describes a situation in which, as a consequence of the cross-linking between two beams of phonons with frequency ω_1 and ω_2, a third phonon beam is produced with frequency $\omega_1 + \omega_2$. This phenomenon led him to think that, when a phonon with frequency ω_1 collides with a phonon with frequency ω_2 in a scattering process where energy is conserved, a third phonon with frequency $\omega_1 + \omega_2$ is produced. The scattering between the two beams is the consequence of the interaction between phonons; that is, Hamiltonian of phonons must be added an interaction Hamiltonian:

$$\hat{H} = \sum_k h\omega_k \left(\hat{a}_k^\dagger \hat{a}_k + \frac{1}{2} \right) + \hat{H}_{\text{int}} = \hat{H}_{phonons} + \hat{H}_{\text{int}} \tag{16.20}$$

The interaction Hamiltonian has the directly interpretation as the Hamiltonian that accounts for the interaction between phonons. Moreover, it is measured in the experiments in laboratory. Only a posteriori, in certain cases the phononic interaction Hamiltonian can be reformulated in the context of the atomic model, that is, it can be computed by means of the introduction of an extra, non-quadratic term in the Hamiltonian of the atomic model given by Eq. (16.3). But even when such a reformulation is possible, the non-quadratic term in the atomic interaction Hamiltonian lacks a direct interpretation: the interaction between atoms represented by that quadratic term has no clear physical meaning. It might be argued that the atomic model is a mere calculation tool and, consequently, the atomic interaction Hamiltonian turns out to be an artifact derived from the real phononic description. A similar example can be found in Kong et al. (2001), where the phonon-electron coupling in magnesium boride crystals is computed.

The cases described above show that, although in the particular case of non-interacting phonons the position according to which phonons are mere calculation tools might be defended, in the general case such a position becomes untenable. Recall that, according to the tool argument, phonons must be interpreted in an instrumental way since derived from atoms. However, in the case where phonons interact with each other, the roles are reversed: physicists begin with phonons, work with phonons, make experiments with phonons, determine the Hamiltonian interaction of phonons, and finally, just at the end of the process, eventually they translate the obtained results in terms of atoms in motion. In other words, in the generic case, the phonons model is the working methodology, used to pose the problem from the beginning and to design the experiments. Therefore, atoms should be conceived as mere calculation tools without real reference.

As we affirmed at the beginning of this section, there are certain philosophical and scientific positions that may obscure the conclusions that are attempted here. On the one hand, one can (even when the aforementioned criteria were accepted) adopt a radically anti-realist attitude and consider all entities as mere calculation tools. Although this could be an absolutely plausible attitude, we consider that it does not attack our objectives. In short, the audience we target is them who, while having a realistic attitude, nevertheless denies the existence of phonons. Thus, we consider that the analogy between phonons and photons established in Sect. 16.4.1, and the fact that the phononic hypothesis can be considered as a "first principle", undermine the defense of the realists who, accepting the existence of photons, maintain a skeptical attitude toward phonons. Therefore, we are faced to the following dichotomy: either we adopt a general instrumentalist view for all entities, or we accept phonons as existing entities.

However, there is another option. One could adopt a realistic attitude and nevertheless deny the existence of phonons. Indeed, adopting the quantum field theory someone could affirm that fields are the only things that exist at all. From this perspective phonons do not exist because in fact the fundamental reality is not composed by particles: there are no phonons, no photons, no atoms. Without encouragement to argue with this kind of ontology, we could weaken our thesis without it losing strength by affirming the following conditional: if we accept the existence of particles (as most of the not field scientific theories do) then we have no reason to not accept the existence of phonons.

But let's look at the strength of the other two remaining arguments in order to evaluate the legitimacy of the recently proposed conditional. Precisely the weakness of the tool argument will be conclusive when we show the weakness of the other arguments on which it holds.

16.4.2.2 The Supercomputer Argument

This argument insists in that phonons are calculation tools that could be discarded if we had enough computing power to solve the Schrödinger equation of crystals from first principles in an exact way. Let's consider this new argument in more detail.

Given a quantum system U, according to quantum mechanics it has a Hamiltonian associated $\hat{H}_U$, and there are many legitimate ways to divide U into parts. In particular, it is always possible to add and subtract an arbitrary $\hat{H}_X$, so that the original Hamiltonian remains the same:

$$\hat{H}_U = \hat{H}_X - \hat{H}_X + \hat{H}_U = \hat{H}_X + \hat{H}_{\text{int}} \tag{16.21}$$

where $\hat{H}_{\text{int}} = -\hat{H}_X + \hat{H}_U$. This is a highly artificial but correct procedure, which shows that any Hamiltonian can always be written as a given self-Hamiltonian plus some interaction term. The fact that this can always be done does not mean, as we will see later, that it is always useful.

On the basis of the above remark, the Hamiltonian of a crystal can be quantum-mechanically represented as the Hamiltonian of atoms in motion plus an interaction Hamitonian, $\hat{H}_U = \hat{H}_{atoms} + \hat{H}_{\text{int}}$, or as the Hamiltonian of phonons plus an interaction Hamiltonian different from that of the atomic case, $\hat{H}_U = \hat{H}_{phonons} + \hat{H}'_{\text{int}}$. It is important to note that up to now all the Hamiltonians are exact, since no approximation was introduced yet. In the hypothetical case of having a supercomputer that, following the supercomputer argument, allowed us to obtain exact solutions for systems of many equations, then we could use it to solve the equations based on the atomic Hamiltonian and to calculate, for example, the expectation value of a certain observable $\hat{O}$. However, nothing prevents us from using the same supercomputer to solve the equations based on the phononic Hamiltonian, and to calculate the expectation value of the same observable $\hat{O}$. In both cases, because exact Hamiltonians are used and exact solutions are computed, we would obtain exactly the same result.

$$\left\langle \hat{O} \right\rangle_{atoms} = \left\langle \hat{O} \right\rangle_{phonons} \tag{16.22}$$

Summing up, although it is true that with a supercomputer we can exactly describe any problem with the atomic model, it is also true that we can do the same, and with the same effectiveness, with the phononic model: the supercomputer is neutral in this. Therefore, the asymmetric conclusion drawn by the supercomputer argument, according to which atoms exist but phonons do not, is not acceptable, and the argument loses any strength.

16.4.2.3 The Explanation Argument

According to this argument, the ontological priority of atoms in motion over phonons is due to the fact that, by contrast to the latter, the former explain a large number of phenomena.

Let us recall that the Hamiltonian of any crystalline material can be expressed according to the atomic model and to the phononic model:

$$\hat{H}_U = \hat{H}_{atoms} + \hat{H}_{\text{int}} = \hat{H}_{phonons} + \hat{H}'_{\text{int}} \tag{16.23}$$

We have seen that if we had a supercomputer, then we could calculate the exact solutions in both cases, obtaining the same results. But, what happens if we do not have the supercomputer?

If, as it is the case, we do not have the supercomputer, then we are forced to introduce some kind of approximations. The most common approaches are always variations of the traditional perturbation theory (Ballentine, 1990), which requires that the self-Hamiltonian is much larger than the interaction Hamiltonian. Therefore, the requirement to apply the perturbation theory to the Hamiltonian in the atomic model is that $\hat{H}_{atoms} \gg \hat{H}_{\text{int}}$: if this requirement is met, then approximate solutions can be obtained and the expectation value $\left\langle \hat{O} \right\rangle_{atoms}$ of a certain observable $\hat{O}$ can be computed. Analogously, the requirement to apply the perturbation theory to the Hamiltonian in the phononic model is that $\hat{H}_{phonons} \gg \hat{H}'_{\text{int}}$: once again, if this requirement is met, then approximate solutions can be obtained and the expectation value $\left\langle \hat{O} \right\rangle_{phonons}$ of the same observable $\hat{O}$ can be computed. However, unlike what happens in the case of the exact solution presented in the previous section, in this case the results are not the same:

$$\left\langle \hat{O} \right\rangle_{átomos} \neq \left\langle \hat{O} \right\rangle_{fonones} \tag{16.24}$$

This is due to the fact that the two expectation values are obtained by means of approximate solutions. Therefore, one result may be is better than the other, and this can only be decided experimentally. Precisely, in general the perturbation theory cannot be applied to the two Hamiltonians, since the requirements $\hat{H}_{atoms} \gg \hat{H}_{\text{int}}$ and $\hat{H}_{phonons} \gg \hat{H}'_{\text{int}}$ in general cannot be satisfied at the same time. As a consequence, in the generic case the atomic model or the phononic model can be applied, but not both.

The above remark means that, although it is true that without a supercomputer assisting us the atomic model is successful in many cases, it is also true that in many other cases it does not work. In the case of thermal processes, the phononic model must be used, not by whim, but by necessity. It does not matter that the list of applications of the atomic model is longer than that of the phononic model: they complement each other since both are indispensable when the goal is to explain all phenomena. In fact, when one model cannot be applied, the other applies. The phononic model is essential because the atomic model is not universally applicable.

The explanation argument might gain strength if it received additional theoretical support from the other two arguments previously considered. However, as we have shown, this does not happen. So, what do we gain with all three arguments? In particular, what we do is to disassemble the idea that atmos are more fundamental particles than phonons. Could it be possible to continue affirming the opposite, i.e., to continue affirming the ontological priority of atoms over phonons? Of course! But what cannot be affirmed is that such an ontological priority is justified. The refutation of the tool argument (together with the demonstration of formal symmetry), plus the refutation of the arguments based on the explanatory quantity and the hypothesis of the supercomputer, put the two theses on an equal footing. That is, if you want to continue holding the orthodox version, you can do it but being aware that this is done on the basis of certain metaphysical assumptions not sufficiently justified.

16.5 If They Exist, How Is It That They Exist?

Up to this point we have argues that there are good reasons to accept phonons as really existing entities. However, there is still a long way to understand the effective ontological status of phonons. In the current philosophical literature, the prevailing realist position about phonons relies on emergentism. For instance, in the paper "Are there individuals in physics, and if so, what are they?", James Ladyman, conceives phonons as quasi-particles, where quasi-particles are entities with a finite life time and therefore always related to a limited time scale (2015: 202). On the other hand, although in the line of Ladyman's proposal, Alexander Franklin and Eleanor Knox (2018) develop in their paper "Emergence without limits: the case of phonons" develop a purely emergentist proposal to understand the case of phonons. From their perspective, the fundamental entities inhabit a basal level, from which the entities that cannot be explained from the behavior of the basal level emerge. Phonons are entities that emerge from the basal level populated by atoms. In this sense, an emergentist view implies certain kind of ontological dependence of one level to another.

Although the emergentist view is very appealing, is faces a serious difficulty. The particular case of phonons, as Franklin and Knox (2018) themselves admit, is a strange kind of emergence. Usually, at least in physics, the emergence scheme is supported by the fact that the concepts of the involved theories are linked by some singular or asymptotic limit. It is precisely those kinds of limit what introduces the asymmetry indispensable for emergence. In fact, in spite of the many conceptions about emergence, everybody accepts that emergence is an asymmetric relation: if A emerges from B, then B does not emerge from A. In the case of phonons, by contrast, there are no limits relating the atomic model and the phononic model. Furthermore, as explained in detail in the previous section, the two models stand in symmetric relations in several senses. Whereas there is a transformation that turns the atomic Hamiltonian into the phononic Hamiltonian, there is also an inverse transformation that turns the phononic Hamiltonian into the atomic Hamiltonian. Although it is true that a supercomputer could exactly solve any problem with the atomic model, it is also true that it could do the same, and with the same effectiveness, with the phononic model. In turn, the atomic model is successful in many cases but does not work in many others, precisely in those where the phononic model is successful. Given the strong parallelism between the two models, where does the asymmetry required by emergence comes from?

Of course, conceiving phonons as entities emergent from atoms is not a contradictory stance. However, the ontological priority of atoms over phonons is introduced as a metaphysical assumption, which is supported neither by the formal theory of phonons nor by the experimental practice in physics. Of course one could continue thinking phonons as emergentist entities because the symmetry that was shown in all three previous arguments belong to epistemic level while the status attached to the entities concerns to ontological one. However, this does not threaten our arguments because, at least, they shown that this attitude is a consequence of unjustified metaphysical assumptions. However, we want something more. We reaffirm the non-positivist attitude developed at the end of the twentieth century in which the attempt to speak of ontology from our best scientific theories is retaken. Therefore, those who want to avoid those non-scientifically founded metaphysical assumptions may prefer a pluralist perspective that rejects hierarchical relations between levels.

Scientific pluralism, in its ontological version, has been proposed by several authors such as Dupré (1993), Cartwright (1994, 1999), Torretti (2000), and Chakravartty (2011). Despite this, the pluralist perspective is in general very resisted in the scientific community, and even by many philosophers of science. Although scientific practice witnesses an increasing proliferation of models configured from theoretical principles in conflict (Hendry, 1998), or even incompatible models applied to the same system of study, a reductionist, monistic and universalist view prevails in science, as a product of what Scerri (2000) called the "imperialism of physics".

The roots if this situation can be found in the fact that the traditional philosophy of science was constituted as a field of inquiry by taking the evolution of physics as its model, that is, a discipline characterized by trying to provide a unified and integral vision of the world. This may explain the widespread tendency of trying to explain the multiplicity of existing things by means of an economic and simple ontology. However, not only regarding the interdisciplinary relations the reductionist project has faced insurmountable obstacles as a consequence of the impossibility of translating the concepts coming from different scientific disciplines. Also within physics itself similar obstacles appear. It is in this context that pluralism becomes strong because, as Chang asserts, the best reason to be pluralist is that it is not very likely that we will achieve an accurate theory that meets all our needs; and if it is not so, maintaining multiple systems makes much sense (Chang, 2012: 20).

Cartwright does not advocate only for an epistemic pluralism; by taking seriously what scientific practice suggests, she asserts: "as appearances suggest, we live in a dappled world, a world rich in different things, with different natures, behaving in different ways. The laws that describe this world are a patchwork, not a pyramid" (Cartwright, 1999: 01). In agreement with this pluralist stance, although from a slightly different philosophical approach, Olimpia Lombardi and Ana Rosa Pérez Ransanz ascribe to a Kantian-inspired ontological pluralism that retain the realist requirement of a correspondentist conception of truth: "rejecting metaphysical realism does not imply rejecting truth as correspondence. (. . .) a Kantian-rooted realism is able to preserve a notion of truth as correspondence" (Lombardi & Pérez Ransanz, 2012: 50). From this perspective, the simultaneous existence of incompatible entities and/or properties can be asserted whenever it is done from different conceptual schemes. In continuity with certain fundamental aspects of Carnap's position (1950), and inspired by Putnam (1981), the authors assume that the question of what really exists only makes sense in an internal framework, in the context of the ontological domain defined by a scientific successful theory. Thus, the theories of science do not describe a transcendental reality that waits to be discovered as it is in itself. On the contrary, each scientifically successful conceptual scheme constitutes its own ontology, so the problem "is not to establish what actually exists; the problem is to accept that what we call 'object' is constituted within our categorical scheme, thus it is an «object for us»" (Lombardi & Pérez Ransanz, 2012: 23). Whereas a monist and reductionist position only accepts a single reality, the pluralist position implies not only that there is no absolute reality, but that there are multiple realities relative to the conceptual schemes of the different successful theories of science. From a Kantian perspective but with pragmatist nuances Torretti, rejecting God's worldview, proposes a kind of pluralism that depends on the conceptual framework that is imposed by the purpose of scientific practice impose. In this sense he said in "Scientific realism and scientific practice" (2000) that "when I search for my car in a parking lot I bear in mind its visual appearance but do not once think about the photon exchange going on between the sun, the car's surface and my eyes. When I reflect about the origin and evolution of the universe I mix the car and myself and everything else up in a single homogeneous worldwide purée. And these two ways of articulating reality – as well as many others – are quite appropriate for finding our own way in it on different occasions and with different purposes" (Torretti, 2000: 115).

The present discussion about phonons is not the first time that ontological pluralism is considered in the field of chemistry. According to a traditional reductionist view, eminently anti-pluralist, the entities raised by chemistry, although with some explanatory efficiency, are merely apparent since resulting from an approximate description of the only true world described by quantum physics. However, the reductionist position disagrees with the effective scientific practice. For instance, incompatible assumptions coexist in quantum chemistry, coming from structural chemistry and quantum mechanics, and none of them are dispensable. In particular, certain concepts of classical structural chemistry, such as the notion of molecular structure, are so essential for chemical practice that even when a mathematical reduction were possible, the resources of structural chemistry would still be absolutely indispensable. The case of optical isomerism, in turn, shows that even if we could calculate the exact Hamiltonian of a system, from such information we would not be able to obtain the precise structure that allows us to differentiate the dextrorotatory from the levorotatory molecules (for more details see Lombardi, 2014; Fortin et al., 2017, 2018, 2019). Ontological pluralism faces this situation by admitting that quantum chemistry is an autonomous discipline with its own ontology (see Lombardi & Labarca, 2005; Labarca & Lombardi, 2010).

On the basis of the arguments developed in the previous sections, the case of phonons can also be interpreted in a pluralistic philosophical framework, so that both the phononic and the atomic descriptions, even if incompatible with each other, are equally valid in their respective contexts. If a supercomputer existed, any system could be described with both models. And in the actual case in which such a supercomputer does not exist, the two descriptions remain equally valid in their respective fields of application. Therefore, both descriptions of reality must be accepted as equally valid.

From a pluralistic perspective, scientific ontology is context-dependent. In certain contexts, we should admit an ontology of mobile atoms since it allows us to explain a series of phenomena. But assuming such an interpretation in certain explanatory contexts should not limit or cancel other interpretive contexts. The phononic picture of reality, according to which atoms are at rest and many phenomena are the result of the motion of phonons, is another non-exclusive successful interpretation. As Cartwright claims, "all evidence points to the conclusion that I really like – that Nature is not reductive and single-minded. It is rich and diverse" (1994: 361).

16.6 Conclusions

The main aim of this work has been to discuss the ontological status of phonons, usually discredited in physics as mere calculation tools. For this purpose, first we recalled the epistemic virtues of such entities, not only their explanatory power but also their interpretive capability with respect to certain phenomena that remain obscure from other theoretical views.

Then we argued that there are not good reasons to conceive phonons as non-existing entities. To reach this goal we first established an analogy between photons and phonons, so that anyone who accepts photons as existing entities should also accept phonons in the same sense. Second, we examined three arguments that are often used to explain the ontological primacy of atoms over phonons: the *tool argument*, the *supercomputer argument* and the *explanation argument*. By showing the weaknesses of those arguments, we undermined the views of those who, from a realist position, nevertheless deny the real existence of phonons. The first argument shows that to think phonons as tools is not sufficiently justified because formal symmetry reveal that we could start either by the phononic Hamiltonian as by the atomic one. If we consider too the predictive success of phonons, as well as the privilege and the exclusivity power of the phonons to explain some areas, there seems to be no epistemic reason to defend the priority of the atomic hypothesis. This argument is not convincing by itself and the intended conclusions can not be exclusively drawn from it. However, if we consider the second argument, called the supercomputer argument, the new symmetry established reinforces the hypothesis defended in the first argument. The symmetry is manifested when we recognize that supercomputer is neutral with respect to both hypotheses: they could describe the problems with the same degree of effectiveness. Perhaps all our arguments could be refuted by appealing to the great variety of phenomena that the atomic hypothesis can explain. However, the third argument ensures that this does not happen. There are several phenomena that the atomic hypothesis cannot explain, and in this sense the hypothesis that refers to the quantity does not seem to be conclusive. Of course, the three arguments provide epistemic reasons to consider the proposed hypothesis. Being aware that epistemic and ontological issues are different, we consider that we offered good reasons to think that the privilege of the atomic hypothesis can continue to be sustained only as metaphysical assumption. However, we prefer to justify our ontological assertions based on our best scientific theories and an adequate interpretation of them. And that is precisely what we did.

Finally, having already rejected the reductionist position, we considered two possible ways of interpreting the ontological status of phonons. On the one hand, the emergentist perspective proposed by certain authors for the case of phonons is faced to the difficulties derived from the symmetric relations linking the atomic and the phononic models, which turn the asymmetry of emergence into a metaphysical postulate. Indeed, the symmetry really challenges the emergentist perspective since it is founded in an asymmetry of levels. On the other hand, ontological pluralism have the resources to cope with the conceptual problems derived from the case of phonons, with no need to postulate a hierarchical relationship between the different domains of reality.

References

Amann, A. (1992). Must a molecule have a shape? *South African Journal of Chemistry, 45*, 29–38.

Ashcroft, N. W., & Mermin, N. D. (1976). *Solid state physics*. Harcourt College Publishers.

Ballentine, L. E. (1990). *Quantum mechanics*. Prentice Hall.

Best, N. W. (2016). Lavoisier's 'Reflections on Phlogiston' II: On thc nature of heat. *Foundations of Chemistry, 18*, 3–13.

Bishop, R. (2005). Patching physics and chemistry together. *Philosophy of Science, 72*, 716–722.

Carnap, R. (1950). Empiricism, semantics, and ontology. *Revue Internationale de Philosophie, 4*, 20–40. Reprinted in the Supplement to *Meaning and Necessity: A Study in Semantics and Modal Logic*, enlarged edition. Chicago: University of Chicago Press, 1956.

Cartwright, N. (1994). The metaphysics of the disunified world. *Proceedings of the Biennial Meeting of the Philosophy of Science Association, 2*, 357–364.

Cartwright, N. (1999). *The dappled world: A study of the boundaries of science*. CUP.

Chakravartty, A. (2011). Scientific realism and ontological relativity. *The Monist, 94*, 157–180.

Chang, H. (2012). *Is water H2O? Evidence, realism and pluralism*. Springer.

Dupré, J. (1993). *The disorder of things: Metaphysical foundations of the disunity of science*. HUP.

Fortin, S., Lombardi, O., & Martínez González, J. C. (2017). The relationship between chemistry and physics from the perspective of Bohmian mechanics. *Foundations of Chemistry, 19*, 43–59.

Fortin, S., Lombardi, O., & Martínez González, J. C. (2018). A new application of the modal-Hamiltonian interpretation of quantum mechanics: The problem of optical isomerism. *Studies in History and Philosophy of Modern Physics, 62*, 123–135.

Fortin, S., Lombardi, O., & Martínez González, J. C. (2019). Why molecular structure cannot be strictly reduced to quantum mechanics. *Foundations of Chemistry, 21*, 45.

Franklin, A., & Knox, E. (2018). Emergence without limits: The case of phonons. *PhilSci-Archive: Preprints in Philosophy of Science*.http://philsci-archive.pitt.edu/13397

Gelfert, A. (2016). *How to do science with models*. Springer.

Hendry, R. F. (1998). Models and approximation in quantum chemistry. *Poznań Studies in the Philosophy of the Sciences and the Humanities, 63*, 123–142.

Hendry, R. F. (2010). Ontological reduction and molecular structure. *Studies in History and Philosophy of Science Part B: Studies in History and Philosophy of Modern Physics, 41*, 183–191.

Hettema, H. (2012). *Reducing chemistry to physics. Limits, models, consequences*. Rijksuniversiteit Groningen.

Kittel, C. (2005). *Introduction to solid state physics* (8th ed.). Wiley.

Kong, Y., Dolgov, O. V., Jepsen, O., & Andersen, O. K. (2001). Electron-phonon interaction in the normal and superconducting states of MgB_2. *Physical Review B, 64*, 020501.

Labarca, M., & Lombardi, O. (2010). Why orbitals do not exist? *Foundations of Chemistry, 12*, 149–157.

Ladyman, J. (2015). Are there individuals in physics, and if so, what are they? In A. Guay & T. Pradeu (Eds.), *Individuals across the sciences*. Oxford University Press.

Lombardi, O. (2014). Linking chemistry with physics: Arguments and counterarguments. *Foundations of Chemistry, 16*, 181–192.

Lombardi, O., & Labarca, M. (2005). The ontological autonomy of the chemical world. *Foundations of Chemistry, 7*, 125–148.

Lombardi, O., & Pérez Ransanz, A. R. (2012). *Los Múltiples Mundos de la Ciencia. Un Realismo Pluralista y su Aplicación a la Filosofía de la Física*. UNAM-Siglo XXI.

Putnam, H. (1981). *Reason, truth and history*. Cambridge University Press.

Sakurai, J. J. (1994). *Modern quantum mechanics*. (Revised edition. New York.

Scerri, E. (2000). The failure of reduction and how to resist disunity of the sciences in the context of chemical education. *Science & Education, 9*, 405–425.

Scerri, E. R. (2011). Editorial 37. *Foundations of Chemistry, 13*, 1–7.

Scerri, E. R. (2012). Top-down causation regarding the chemistry-physics interface: A sceptical view. *Interface Focus, 2*, 20–25.

Scerri, E. R. (2013). Philosophy of chemistry: Where has it been and where is it going. In J.-P. Llored (Ed.), *The philosophy of chemistry: Practices, methodologies, and concepts*. Cambridge Scholars Publishing.

Shinen, N. S. (1963). Nonlinear acoustic interaction in MgO at 9 Gc/sec. *Physical Review Letters, 11*, 3–6.

Sutcliffe, B. T., & Wolley, R. G. (2011). A comment on editorial 37. *Foundations of Chemistry, 13*, 93–95.

Sutcliffe, B. T., & Wolley, R. G. (2012). Atoms and molecules in classical chemistry and quantum mechanics. In R. F. Hendrya & A. Woody (Eds.), *Handbook of philosophy of science. Vol. 6, philosophy of chemistry*. Elsevier.

Torretti, R. (2000). Scientific realism and scientific practice. In E. Agazzi & M. Pauri (Eds.), *The reality of the unobservable: Observability, unobservability and their impact on the issue of scientific realism* (pp. 113–122). Kluwer.

Woolley, R. G. (1978). Must a molecule have a shape? *Journal of the American Chemical Society, 100*, 1073–1078.

Woolley, R. G. (1982). Natural optical activity and the molecular hypothesis. *Structure and Bonding, 52*, 1–35.

Appendix: Publications by Roberto Torretti

(The following appendix contains a list of publications by Roberto Torretti. The document was kindly provided by Roberto himself.)

16.1 Books

[1] *Manuel Kant. Estudio sobre los fundamentos de la filosofía crítica.* Santiago de Chile: Ediciones de la Universidad de Chile, 1967. 603 pp. Second edition, Buenos Aires: Editorial Charcas, 1980. 605 pp. Third revised edition, Santiago de Chile: Ediciones Universidad Diego Portales, 2005. 1012 pp.

[2] *Filosofía de la Naturaleza. Textos Antiguos y Modernos.* Santiago de Chile: Editorial Universitaria, 1971. 178 pp. Second, enlarged edition, Santiago de Chile: Editorial Universitaria, 1998. 170 pp.

[3] *Problemas de la Filosofía. Textos filosóficos clásicos y contemporáneos.* Río Piedras: Editorial Universitaria, Universidad de Puerto Rico, 1975. 768 pp. Coedited with Luis O. Gómez. Reprinted several times with minor corrections.

[4] *Philosophy of Geometry from Riemann to Poincaré.* Dordrecht: D. Reidel Publishing Co., 1978. xiii + 458 pp. Corrected reprint. Dordrecht: D. Reidel Publishing Co., 1984. xiii+458 pp.

[5] *Relativity and Geometry.* Oxford: Pergamon Press, 1983. xi + 395 pp. Corrected reprint. New York: Dover, 1996. xiv + 395 pp.

[6] *Creative Understanding: Philosophical Reflections on Physics.* Chicago: The University of Chicago Press, 1990. xvi + 369 pp.

[7] *Variedad en la Razón: Ensayos sobre Kant.* Río Piedras: Editorial de la Universidad de Puerto Rico, 1992. x + 248 pp. (Coauthored with CARLA CORDUA.)

[8] *La geometría del universo y otros ensayos de filosofía natural.* Mérida: Consejo de Publicaciones de la Universidad de los Andes, 1994. x + 296 pp.

C. Soto (ed.), *Current Debates in Philosophy of Science*, Synthese Library 477,
https://doi.org/10.1007/978-3-031-32375-1

[9] *Sophocles' Philoctetes*. Text and Commentary. Bryn Mawr, PA: Thomas Library, Bryn Mawr College, 1997. 94 pp. (Bryn Mawr Greek Commentaries).
[10] *El Paraíso de Cantor: La tradición conjuntista en la filosofía matemática*. Santiago de Chile: Editorial Universitaria, 1998. xiv + 589 pp.
[11] *The Philosophy of Physics*. New York: Cambridge University Press, 1999. xv + 512 pp.
[12] *Diccionario de Lógica y Filosofía de la Ciencia*. Madrid: Alianza Editorial. 2002. 670 pp. (Coauthored by JESÚS MOSTERÍN). Second edition, Madrid: Alianza Editorial, 2010. 692 pp.
[13] *Relatividad y espaciotiempo*. Santiago de Chile: RIL editores. 2003. 276 pp.
[14] *Pensar la ciencia*. Madrid: Tecnos. 2004. (Coauthored by MIGUEL ESPINOZA).
[15] *Estudios filosóficos: 1957–1987*. Santiago de Chile: Ediciones Universidad Diego Portales, 2006. 344 pp.
[16] *Estudios filosóficos: 1986–2006*. Santiago de Chile: Ediciones Universidad Diego Portales, 2007. 255 pp.
[17] *De Eudoxo a Newton: Modelos matemáticos en la filosofía natural*. Santiago de Chile: Ediciones Universidad Diego Portales, 2007. 386 pp.
[18] *Crítica filosófica y progreso científico: Cuatro ejemplos*. Santiago de Chile: Ediciones Universidad Diego Portales, 2008. 160 pp.
[19] *Conceptos de gen*. Compilador y traductor: Roberto Torretti. Santiago de Chile: Ediciones Universidad Diego Portales, 2009. 306 pp.
[20] *Estudios filosóficos: 2007–2009*. Santiago de Chile: Ediciones Universidad Diego Portales, 2010. 313 pp.
[21] *Filoctetes de* SÓFOCLES. Texto griego compuesto, anotado y traducido por Roberto Torretti. Santiago de Chile: Ediciones Tácitas, 2011. 207 pp

16.2 Encyclopedia Articles

[22] "Geometry". In Hans Burkhardt and Barry Smith, eds., *Handbook of Metaphysics and Ontology*. Munich: Philosophia Verlag, 1991.
[23] "El método axiomático". In *Enciclopedia Iberoamericana de Filosofía*, Vol. 4, Madrid: Trotta, 1993, pp. 89–110.
[24] "Spazio". In *Dizionario delle Scienze Fisiche*. Roma: Istituto della Enciclopedia Italiana, 1995. Vol. 5, pp. 427–433.
[25] "Space". In *The Routledge Encyclopedia of Philosophy*. London: Routledge, 1998.
[26] "Spacetime". In *The Routledge Encyclopedia of Philosophy*. London: Routledge, 1998.
[27] "Geometry, Nineteenth Century". In *Stanford Encyclopedia of Philosophy*, http://plato.stanford.edu.
[28] "Geometry". In Donald Borchert, ed., *Encyclopedia of Philosophy*, Second Edition. Detroit: Macmillan Reference USA, 2006.

[29] "Callippus of Cyzikus" – "Eratosthenes of Cyrene" – "Friedmann, Alexander Alexandrovich" – "Menaechmus" – "Philolaus of Croton". In Thomas Hockey, ed., *Biographical Encyclopedia of Astronomers*. Berlin: Springer, 2007.

16.3 Other Articles

[30] "Aspectos de la doctrina de Kelsen". *Anales de la Universidad de Chile*, 100: 85–108 (1955).
[31] "Medio geográfico y medio de la conducta en la psicología moderna". *Atenea*, 361–362: 105–119 (1955).
[32] "¿Ha habido progreso de la filosofía en su historia?" *Revista de Filosofía* (Chile), IV.1: 49–56 (1957).
[33] "Causalidad y evolución. Consideraciones sobre el presunto dilema que estas ideas plantearían a la ciencia". *Revista de Filosofía* (Chile), IV.2–3: 38–51 (1957).
[34] "Socialidad del individuo". *Revista de Ciencias Sociales* (Puerto Rico), V.1: 21–29 (1961).
[35] "Kant, filósofo del más acá". *La Torre*, 34: 161–179 (1961).
[36] "Reflexiones sugeridas por el *Tractatus* de Wittgenstein". *Revista de Ciencias Sociales* (Puerto Rico, V.4: 479–490 (1961).
[37] *Hume y la religión*. Concepción: Ediciones Atenea, n.d. (1962), 36 pp.
[38] "Poder político y opresión". *Revista de Filosofía* (Chile), IX.1–2: 35–48 (1962).
[39] "Lecciones sobre el empirismo inglés". *Revista de Filosofía* (Chile), IX.3: 113–155 (1962).
[40] "Para introducir a Heidegger". *La Torre*, 39: 87–102 (1962).
[41] "El pensamiento de José Echeverría". Diario *El Sur* (Concepción, Chile), 22 de enero de 1963.
[42] "Sobre el significado del imperativo categórico". *Revista de Filosofía* (Chile), X.1: 45–66 (1963).
[43] "Finitud del hombre y límites del conocimiento en Descartes y Leibniz". *Anales de la Universidad de Chile*, 128: 33–58 (1963).
[44] "Contrato social y economía dirigida en el pensamiento político de Fichte". *Revista de Ciencias Sociales* (Puerto Rico), VIII.4: 357–375 (1964).
[45] "Unamuno, pensador cristiano". In: *Unamuno*, Santiago: Ediciones de la Universidad de Chile, 1964, pp. 95–112.
[46] "Introducción a un estudio de la deducción trascendental de las categorías en la primera *Crítica* de Kant". *Revista de Filosofía* (Chile), XII: 19–61 (1965).
[47] "Las contrapartidas incongruentes en la gestación de la filosofía crítica de Kant". *Diálogos*, 3: 7–24 (1965).
[48] "Las *Investigaciones* de Wittgenstein y la posibilidad de la filosofía". *Diálogos*, 10: 35–59 (1968).

[49] "Tercer Congreso Internacional de Kant". *Diálogos*, 17: 113–117 (1969).

[50] "Die Frage nach der Einheit der Welt bei Kant". *Kantstudien*, 62: 77–97 (1972).

[51] "On the subjectivity of objective space". In: L. W. Beck, ed., *Proceedings of the Third International Kant Congress*, Dordrecht: D. Reidel Publishing Co., 1972, pp. 535–540.

[52] "Remarks on Salmon's Paradox of Primes". *Philosophy of Science*, 39: 260–262 (1972).

[53] "La filosofía de la aritmética de Husserl". *Studi internazionali di Filosofía*, 4: 183–206 (1972).

[54] "Lógica formal y forma lógica". In: *Überlieferung und Auftrag. Festschrift für Michael de Ferdinandy*, Wiesbaden: Pressler, 1972, pp. 624–633.

[55] "Juicios sintéticos a priori". *Cuadernos de Filosofía* (Buenos Aires), XI,20: 297–320 (1973).

[56] "On the subjectivity of objective space". In: L. W. Beck, ed., *Kant's Theory of Knowledge*, Dordrecht: D. Reidel Publishing Co., 1974, pp. 111–116. (Reprint of [35]).

[57] "On Mr. Kielkopf's not so sober understanding of standard elementary logic". *Mind*, 83: 575–577 (1974).

[58] "La geometría en el pensamiento de Kant". *Anales del Seminario de Metafisica* (Madrid), 9: 9–60 (1974).

[59] "El debate sobre el individualismo metodológico". *Diálogos*, 26: 95–117 (1974).

[60] "Problemas filosóficos del espacio y el tiempo. A propósito de la nueva edición de la obra de Adolf Grünbaum". *Diálogos*, 27: 89–117 (1974).

[61] "Espacio y tiempo: algunos libros recientes". *Diálogos*, 29–30: 255–294 (1977).

[62] "Bedingtes und Unbedingtes in der Mathematik". In: *Transzendenz und Immanenz, Philosophie und Theologie in der veränderten Welt*, hrsg. von D. Papenfuss und J. Söring, Stuttgart: Kohlhammer, 1977, pp. 303–308.

[63] "Tres filósofos de la geometría". *Revista Latinoamericana de Filosofía*, 3: 3–21 (1977).

[64] "Hugo Dingler's philosophy of geometry". *Diálogos*, 32: 85–128 (1978).

[65] "Presencia e idea del mundo". *Cuadernos de la Facultad de Humanidades*, 1: 3–21 (1978).

[66] "Jackson and Pargetter's criterion of distant simultaneity". *Philosophy of Science*, 46: 302–306 (1979).

[67] "Indole y función de los principios de la Teoría General de la Relatividad". *Revista Latinoamericana de Filosofía*, 5: 209–233 (1979).

[68] "Mathematical theories and philosophical insights in cosmology". In: H. Nelkowski *et al.*, eds., *Einstein Symposion Berlin*, Berlin: Springer, 1979, pp. 320–335 (*Springer Lecture Notes in Physics*, 100).

[69] "Three kinds of mathematical fictionalism". In: J. Agassi and R. S. Cohen, eds., *Scientific Philosophy Today*, Dordrecht: D. Reidel Publishing Co., 1981, pp. 399–414.

[70] (With John Stachel) "Einstein's first derivation of mass-energy equivalence", *American Journal of Physics*, 50: 760–763 (1982).
[71] "»Lo que hay«", *Diálogos*, 41: 89–93 (1983).
[72] "Causality and spacetime structure in Relativity". In: R. S. Cohen and L. Laudan, eds., *Physics, Philosophy and Psychoanalysis*, Dordrecht: D. Reidel Publishing Co., 1983, pp. 273–293.
[73] "Kosmologie als ein Zweig der Physik". In: Bernulf Kanitscheider, ed., *Moderne Naturphilosophie*, Würzburg: Königshausen & Neumann, 1984, pp. 183–200.
[74] "La crítica de conceptos en las revoluciones de la física básica". *Revista Latinoamericana de Filosofía*, 10: 25–41 (1984).
[75] "Spacetime physics and the philosophy of science". *British Journal for the Philosophy of Science*, 35: 280–292 (1984).
[76] "Las *Investigaciones* de Wittgenstein y la posibilidad de la filosofía". In: Jorge J.E. Gracia, Eduardo Rabossi, Enrique Villanueva and Marcelo Dascal, eds. *El análisis filosófico en América Latina*, México: Fondo de Cultura Económica, 1985, pp. 536–556. (Reprint of [32]).
[77] "Observation". *British Journal for the Philosophy of Science*, 37: 1–23 (1986).
[78] "Conceptual reform in scientific revolutions". In: Ruth Barcan Marcus *et al.*, eds., *Logic, Methodology and Philosophy of Science VIII*, Amsterdam: North-Holland, 1986, pp. 413–431.
[79] "Physical theories". Part I, *Diálogos*, 48: 183–212 (1986); Part II, *Diálogos*, 49: 147–188 (1987).
[80] "La determinación omnímoda de las cosas y el fenomenismo de Kant". *Revista Latinoamericana de Filosofía*, 13: 132–141 (1987).
[81] "Do conjunctive forks always point to a common cause?" *British Journal for the Philosophy of Science*, 38: 384–387 (1987).
[82] "Probabilidad y determinismo". *Congreso Internacional Extraordinario de Filosofía del 20 al 26 de Setiembre de 1987.* Universidad Nacional de Córdoba, República Argentina. Tomo III, pp. 1201–1207.
[83] "Extractos de una correspondencia" (with C. Ulises Moulines). *Diálogos*, 53: 123–137 (1989).
[84] "The geometric structure of the universe". In: Evandro Agazzi and Alberto Cordero, eds., *Philosophy and the Origin and Evolution of the Universe*, Proceedings of the Conference on the Origin and Evolution of the Universe, held by the International Academy of Philosophy of Science in Lima, Perú, in August 1989. Dordrecht: Kluwer, 1991, pp. 53–73.
[85] "»Y se hizo la luz« . . . Newton y la Ilustración". *La Torre* n.e., 5 [núm. ext.]: 169–178 (1991).
[86] "Mathematical structures and physical necessity". In: Javier Echeverría, Andoni Ibarra y Thomas Mormann, eds., *The Space of Mathematics: Philosophical, Epistemological, and Historical Explorations.* Berlin: Walter de Gruyter, 1992, pp. 132–140.

[87] "La tradición semántica". *Revista Latinoamericana de Filosofía*, 18, 2: 333–340 (1993).
[88] "Una idea feliz". *Revista Latinoamericana de Filosofía*, 19: 289–301 (1993).
[89] "Kitcher on the advancement of science". *Diálogos*, 64: 201–215 (1994).
[90] "El 'observador' en la física del siglo XX". In: Francisco José Ramos, ed., *Hacer: Pensar. Colección de escritos filosóficos*, Río Piedras: Editorial de la Universidad de Puerto Rico, 1994, pp. 581–610.
[91] "Einstein's luckiest thought". In: Jarrett Leplin, ed. *The Creation of Ideas in Physics.* Dordrecht: Kluwer, 1995, pp. 89–96.
[92] "Realismo científico y ciencia real". *Theoria,* XI.26: 29–43 (1996).
[93] "Las analogías de la experiencia de Kant y la filosofía de la física". *Anales de la Universidad de Chile*, Sexta Serie, 4: 77–96 (1996).
[94] "Ruptura y continuidad en la historia de la física". *Revista de Filosofía* (Chile), XLIX-L: 29–44 (1997).
[95] "From Physics to Metaphysics". *Studies in the History and Philosophy of Modern Physics.* 28: 291–298 (1997).
[96] "On Relativity, Time-Reckoning, and the Topology of Time Series", in Jeremy Butterfield, ed., *The Arguments of Time*. Oxford: Published for the British Academy by Oxford University Press, 1999. Pp. 66–82.
[97] "Respuestas a la Encuesta mundial sobre la situación de la filosofía al final del siglo XX". In R. Fornet-Betancourt, ed., *Quo vadis, Philosophie? Antworten der Philosophen. Dokumentation einer Weltumfrage*. Aachen: Wissenschaftsverlag Mainz, 1999. Pp. 314–316.
[98] "Spacetime Models for the World". *Studies in the History and Philosophy of Modern Physics*, **31**: 171–186 (2000).
[99] "Gravity as Spacetime Curvature". *Physics in Perspective*, **2** (2000), 118–134.
[100] "»Scientific Realism« and Scientific Practice". En: Evandro Agazzi and Massimo Pauri, eds., *The Reality of the Unobservable: Observability, Unobservability and their Impact on the Issue of Scientific Realism*, Dordrecht: Kluwer, 2000. Pp. 113–122.
[101] "A Model for Smarandache's Anti-Geometry". In Leonardo Motta y Gheorghe Niculescu, eds., *Proceedings of the Second International Conference of Smarandache Type Notions in Methematics and Quantum Physics*, Rehoboth, NM: American Research Press, 2001, pp. 5–18.
[102] "A Model for Smarandache's Anti-Geometry". *Smarandache Notions Journal*, **12** (2001), 5–18.
[103] "The Friedmann Worlds: Deriving the geometry". Preprint distributed (in English) at the course "¿Se acelera la expansión del Universo? Cosmología y Filosofía", Universidad Internacional Meléndez Pelayo (Santander, España) in July 2000.
[104] "El concepto de probabilidad". *Diálogos*, **81**: 407–447 (2003).
[105] "El cosmos hace noticia". In Álvaro Fischer, ed., *Nuevos paradigmas al comienzo del tercer milenio*. Santiago de Chile: El Mercurio/Aguilar, 2004; Pp. 371–377.

[106] "Intuición pura". In César Ojeda y Alejandro Ramírez, eds., *El sentimiento de lo humano en la ciencia, la filosofía y las artes: Homenaje al Profesor Félix Schwartzmann Turkenich.* Santiago: Editorial Universitaria, 2004. Pp. 111–134.

[107] "Conocimiento discursivo". In *Doctor honoris causa Roberto Torretti*, Barcelona: Universitat Autònoma de Barcelona, 2005. Pp. 13–27.

[108] "Conocimiento discursivo". *RevistaUDP*, **2**: 67–72 (2005). (Reprint of [98]).

[109] "Can science advance effectively through philosophical criticism and reflection?", http://philsci-archive.pitt.edu/archive/00002875/ (2006).

[110] "Causalidad, intervención y determinismo". *Diálogos*, **87**: 217–222 (2006).

[111] "The problem of time's arrow historico-critically reexamined". *Studies in the History and Philosophy of Modern Physics*, **38**: 732–756 (2007).

[112] "'Rod contraction' and 'clock retardation': Two harmless misnomers?" In: Fabio Minazzi, ed., *Filosofia, scienza e bioetica nel dibattito contemporaneo: Studi internazionali in onore di Evandro Agazzi*, Roma: Istituto Poligrafico e Zecca dello Stato, 2007. Pp. 1029–1039.

[113] "Getting rid of the Ether: Could Physics have achieved it sooner, with better assistance from Philosophy?". *Theoria*, **60**: 353–374 (2007).

[114] "Nuevos ensayos de filosofía de la biología". *Revista de Filosofía* (Chile), **LXIV**: 215–229 (2008),

[115] "Objectivity: A Kantian Perspective". In: Michela Massimi, ed., *Kant and Philosophy of Science Today.* Royal Institute of Philosophy Supplement: 63. Cambridge: Cambridge University Press, 2008. Pp. 81–94.

[116] "Diseños y designios". *Estudios Públicos*, **115**: 49–74 (2009).

[117] "Respuestas a mis críticos", en "Simposio: La filosofía de la ciencia de Roberto Torretti". *Teorema*, **XXIX/1**: 147–151 (2010).

[118] "La proliferación de los conceptos de especie en la biología evolucionista". *Theoria*, **69**: 325–377 (2010).

[119] "How Percepts and Concepts Engage the Future". *Foundations of Physics*. **41**: 1717–1728 (2011).

[120] "Fenomenotecnia y conceptualización en la epistemología de Gastón Bachelard". *Theoria*, **72**: 97–114 (2011)

16.4 Book Reviews [1]

[121] Bogumil Jasinowski, *Saber y dialéctica*, Santiago 1957 (*La Torre*, 21: 197–201 (1958)).

[122] Georg Lukacz, *El asalto a la razón*, trad. por Wenceslao Roces, México 1959 (*Revista de Ciencias Sociales*, IV.2: 390–395 (1960)).

[1] All reviews are in Spanish, unless otherwise noted.

[123] Hans Freyer, *Teoría de la época actual*, trad. por Luis Villoro, México 1959 (*Revista de Ciencias Sociales,* V.1: 85–87 (1961)).

[124] Willard van Orman Quine, *From a Logical Point of View*, 2nd ed., Cambridge MA 1961 (*Revista de Filosofía* (Chile), , IX.1–2: 151–154 (1962)).

[125] Alfred J. Ayer, ed., *Logical Positivism*, Glencoe IL 1959 (*Revista de Filosofía* (Chile), , IX.3: 157–164 (1962)).

[126] *Jornadas de Filosofía: Posibilidad de la metafísica*, Tucumán 1961 (*INTER-AMERICAN REVIEW OF BIBLIOGRAPHY*, XIII.1: 82–83 (1963)).

[127] Maine de Biran, *De l'apperception immédiate*, Paris 1963 (*Anales de la Universidad de Chile,* 128: 202–207 (1963)).

[128] Richard B. Braithwaite, *La explicación científica*, trad. por V. Sánchez de Zavala, Madrid 1965 (*Anales de la Universidad de Chile,* 137: 228–236 (1966)).

[129] Mario Bunge, *The Myth of Simplicity*, Englewood Cliffs 1963 (*Anales de la Universidad de Chile,* 138: 250–256 (1966)).

[130] Thomas S. Kuhn, *The Structure of Scientific Revolutions*, Chicago 1962 (*Anales de la Universidad de Chile,* 139: 257–261 (1966)).

[131] Mario Bunge, *Scientific Research*, 2 vols., Berlin 1967 (*Anales de la Universidad de Chile,* 141–144: 346–350 (1967)).

[132] *The Collected Papers of Gerhard Gentzen*, Amsterdam 1969 (*Diálogos,* 21: 104–108 (1970)).

[133] Jaakko Hintikka, ed., *The Philosophy of Mathematics*, New York 1969 (*Diálogos,* 21: 109–114 (1970)).

[134] Robert G. Colodny, ed., *The Nature and Function of Scientific Theories, essays in contemporary science and philosophy*, Pittsburgh 1970 (*Diálogos,* 21: 115–121 (1970)).

[135] *Historisches Wörterbuch der Philosophie*, Band I, hrsg. von Joachim Ritter, Basel 1971 (*Diálogos,* 22: 171–175 (1972)).

[136] Giorgio Tonelli, *A Short-title List of Subject Dictionaries of the 16th, 17th and 18th Centuries as Aids to the History of Ide*as, London 1971 (*Diálogos,* 22: 175–176 (1972)).

[137] Henry E. Kyburg Jr., *Probability and Inductive Logic*, New York 1970 (*Diálogos,* 22: 180–184 (1972)).

[138] Nicholas Rescher, *Scientific Explanation*, New York 1970 (*Diálogos,* 22: 184–188 (1972)).

[139] Wesley Salmon, *Statistical Explanation and Statistical Relevance*, Pittsburgh 1971 (*Diálogos,* 22: 188–192 (1972)).

[140] John R. Searle, ed., *The Philosophy of Language*, New York 1971; Leonard Linsky, ed., *Reference and Modality,* New York 1971 (*Diálogos,* 22: 213–214 (1972)).

[141] Patrick Suppes, *A Probabilistic Theory of Causality*, Amsterdam 1970 (*Diálogos,* 22: 215–216 (1972)).

[142] S. W. P. Steen, *Mathematical Logic with Special Reference to the Natural Numbers*, Cambridge 1972 (*Diálogos,* 23: 219–221 (1972)).

[143] Karel Lambert and Bas C. van Fraassen, *Derivation and Counterexample, an introduction to philosophical logic*, Encino 1972 (*Diálogos,* 23: 221–224 (1972)).
[144] Robert Rogers, *Mathematical Logic and Formalized Theories, a survey of basic concepts and results*, Amsterdam 1971 (*Diálogos,* 23: 224–227 (1972)).
[145] Peter Achinstein, *Law and Explanation, an essay in the philosophy of science*, Oxford 1971 (*Diálogos,* 23: 227–229 (1972)).
[146] *Proceedings of the Third International Kant Congress*, ed. by L.W. Beck, Dordrecht 1972 (*Diálogos,* 23: 229–236 (1972)).
[147] Hilary Putnam, *Philosophy of Logic*, New York 1971; Norwood Russell Hanson, *Observation and Interpretation, a guide to the philosophy of science*, New York 1971; Norman Malcolm, *Problems of Mind, Descartes to Wittgenstein*, New York 1971 (*Diálogos,* 23: 236–239 (1972)).
[148] Bas C. van Fraassen, *Formal Semantics and Logic*, New York 1971 (*Diálogos,* 23: 240 (1972)).
[149] M. Bunge, F. Halbwachs, T.S. Kuhn, J. Piaget et L. Rosenfeld, *Les Théories de la Causalité,* Paris 1971 (*Diálogos,* 23: 241–242 (1972)).
[150] Carlo Borromeo Giannoni, *Conventionalism in Logic, a study in the linguistic foundation of logical reasoning*, The Hague 1971 (*Diálogos,* 24: 167–174 (1973)).
[151] Georg Henrik von Wright, *Explanation and Understanding*, Ithaca 1971 (*Diálogos,* 24: 174–179 (1973)).
[152] Robert G. Colodny, ed., *Paradigms and Paradoxes: The Philosophical Challenge of the Quantum Domain*, Pittsburgh 1972 (*Diálogos,* 24: 196–199, (1973)).
[153] Robert Borger and Frank Cioffi, eds., *Explanation in the Behavioural Sciences*, Cambridge 1970 (*Diálogos,* 24: 200–203 (1973)).
[154] Herbert Meschkowski, ed., *Grundlagen der modernen Mathematik*, Darmstadt 1972; Karl Strubecker, ed., *Geometrie*, Darmstadt 1972 (*Diálogos,* 24: 205–207 (1973)).
[155] Mario Bunge, *Method, Model and Matter*, Dordrecht 1973; *Teoría y realidad*, Barcelona 1972 (*Diálogos,* 25: 156–167 (1973)).
[156] Karl R. Popper, *Objective Knowledge, an evolutionary approach*, Oxford 1972 (*Diálogos,* 25: 168–174 (1973)).
[157] J.M. Jauch, *Are Quanta Real? A Galilean dialogue*, Bloomington 1973 (*Diálogos,* 25: 179–181 (1973)).
[158] Mario Bunge, *Philosophy of Physics*, Dordrecht 1973 (*Diálogos,* 26: 176–181 (1974)).
[159] Ulises Moulines, *La estructura del mundo sensible (sistemas fenomenalistas)*, Barcelona 1973 (*Diálogos,* 26: 181–183 (1974)).
[160] Herbert Feigl, Wilfrid Sellars and Keith Lehrer, eds., *New Readings in Philosophical Analysis*, New York 1972 (*Diálogos,* 26: 183–186 (1974)).

[161] Jürgen Klüver, *Operationalismus: Kritik und Geschichte einer Philosophie der exakten Wissenschaften*, Stuttgart-Bad Cannstatt 1971 (*Diálogos,* 27: 159–165 (1974)).

[162] William A. Wallace, *Causality and Scientific Explanation*, 2 vols., Ann Arbor 1972–1974 (*Diálogos,* 27: 179–181 (1974)).

[163] Fernando Montero Moliner, *El empirismo kantiano*, Valencia 1973 (*Diálogos,* 27: 177–178 (1974)).

[164] José Ferrater Mora, *Cambio de marcha en filosofía*, Madrid 1974 (*Diálogos,* 28: 153–157 (1975)).

[165] Imre Lakatos, *Proofs and Refutations, the logic of mathematical discovery*, Cambridge 1976 (*Diálogos,* 31: 182–186 (1978)).

[166] Jane Bridge, *Beginning Model Theory, the Completeness Theorem and some consequences*, Oxford 1977 (*Diálogos,* 31: 188–192 (1978)).

[167] Farhang Zabeeh, E.D. Klemke and Arthur Jacobson, eds., *Readings in Semantics*, Urbana 1974 (*Diálogos,* 31: 192–193 (1978)).

[168] Bernulf Kanitscheider, *Vom absoluten Raum zur dynamischen Geometrie,* Mannheim 1976 (*Diálogos,* 31: 193–195 (1978)).

[169] Roberto Escobar, *La filosofía en Chile*, Santiago 1976 (*Interamerican Review of Bibliography,* 28: 93–94 (1978)).

[170] Karl R. Popper and John C. Eccles, *The Self and its Brain*, Berlin 1977 (*Diálogos,* 32: 202–206 (1978)).

[171] John Winnie, ed., *Symposium on Space and Time* [special issue of *Noûs,* vol. XI, n°3, September 1977], (*Diálogos,* 32: 207–211 (1978)).

[172] Stephen P. Schwartz, ed., *Naming, Necessity and Natural Kinds*, Ithaca 1977 (*Diálogos,* 32: 211–212 (1978)).

[173] Nelson Goodman, *Ways of Worldmaking*, Indianapolis 1978 (*Diálogos,* 33: 173–174 (1979)).

[174] Arthur W. Burks, *Chance, Cause and Reason, an Inquiry into the Nature of Scientific Evidence*, Chicago 1977 (*Diálogos,* 33: 174–175 (1979)).

[175] George S. Pappas and Marshall Swain, eds., *Essays on Knowledge and Justification*, Ithaca 1978 (*Diálogos,* 33: 175–176 (1979)).

[176] Mario Bunge, *Treatise on Basic Philosophy, Volume 3, Ontology I: The Furniture of the Universe*, Dordrecht 1977 (*Diálogos,* 33: 151–156 (1979)).

[177] John S. Earman, Clark N. Glymour and John J. Stachel, eds., *Foundations of Space-Time Theories*, Minneapolis 1977 [Minnesota Studies in the Philosophy of Science, VIII] (*Diálogos,* 33: 165–171 (1979)).

[178] Robert Geroch, *General Relativity from A to B*, Chicago 1978 (*Diálogos,* 33: 171–173 (1979)).

[179] Wladyslaw Krajewski, *Correspondence Principle and Growth of Science*, Dordrecht 1977 (*Diálogos,* 34: 182–186 (1979)).

[180] George Boolos, *The Unprovability of Consistency, an Essay in Modal Logic*, Cambridge 1979 (*Diálogos,* 35: 183–184 (1980)).

[181] Angel Jorge Casares, *Sobre la esencia del hombre*, Río Piedras 1979 (*Diálogos,* 35: 208–209 (1980)).

[182] José Ferrater Mora, *Diccionario de filosofía*, Sexta Edición, Madrid 1979, 4 vols. (*Diálogos,* 38: 151–152 (1981)).
[183] Ricardo J. Gómez, *Las teorías científicas – desarrollo – estructura – fundamentación*, Tomo I, Buenos Aires 1977 (*Noûs,* 15: 244–246 (1981)). In English.
[184] Nicholas Rescher, *Leibniz's Metaphysics of Nature, a group of essays*, Dordrecht 1981 (*Diálogos,* 39: 160–163 (1982)).
[185] Jill Vance Buroker, *Space and Incongruence: The Origin of Kant's Idealism*, Dordrecht 1981 (*Diálogos,* 39: 163–167 (1982)).
[186] Hartry Field, *Science without Numbers, a Defence of Nominalism*, Princeton 1980 (*Diálogos,* 40: 159–162 (1982)).
[187] Jerzy Giedymin, *Science and Convention, essays on Henri Poincaré's philosophy of science and the conventionalist tradition*, Oxford 1981 (*Diálogos,* 40: 162–167 (1982)).
[188] Richard Swinburne, ed., *Space, Time and Causality*, Dordrecht 1983; D. Mayr and G. Süssmann, eds., *Space, Time and Mechanics: Basic Structures of a Physical Theory*, Dordrecht 1983 (*Diálogos,* 43: 131–136 (1984)).
[189] John R. Searle, *Intentionality, an essay in the philosophy of mind*, Cambridge 1983 (*Diálogos,* 44: 194–200 (1984)).
[190] Dudley Shapere, *Reason and the Search for Knowledge, Investigations in the Philosophy of Science*, Dordrecht 1984 (*Diálogos,* 45: 166–171 (1985)).
[191] Andrew Pickering, *Constructing Quarks, a Sociological History of Particle Physics*, Chicago 1984 (*Diálogos,* 47: 177–183 (1986)).
[192] Isaac Levi, *Decisions and Revisions: Philosophical Essays on Knowledge and Value*, Cambridge 1984 (*Diálogos,* 49: 189–196 (1987)).
[193] Wolfgang Balzer, David A. Pearce and Heinz-Jürgen Schmidt, eds., *Reduction in Science: Structure, Examples, Philosophical Problems*, Dordrecht 1984 (*Diálogos,* 49: 207–213 (1987)).
[194] Arthur Fine, *The Shaky Game. Einstein Realism and the Quantum Theory*, Chicago 1986 (*Diálogos,* 50: 155–160 (1987)).
[195] Joseph Agassi, *Science and Society. Studies in the Sociology of Science*, Dordrecht 1981 (*Philosophia,* 17: 223–223 (1987)). Reviewed in English.
[196] Asim O. Barut, Alwyn van der Merwe and Jean-Pierre Vigier, eds., *Quantum, Space and Time—The Quest Continues. Studies and Essays in Honour of Louis de Broglie, Paul Dirac and Eugene Wigner*, Cambridge 1984. (*Noûs,* 21: 442–444 (1987)). Reviewed in English.
[197] Nicanor Ursua Lezaun *et al.*, *Filosofía de la ciencia y metodología crítica*, Bilbao 1981. (*Noûs,* 22: 327–329 (1988)). Reviewed in English.
[198] *Los filósofos presocráticos.* I. Introducción general por Conrado Eggers Lan. Introducciones, traducciones y notas por Conrado Eggers Lan y Victoria E. Juliá. II. Introducciones, traducciones y notas por Néstor Luis Cordero, Francisco José Olivieri, Ernesto La Croce y Conrado Eggers Lan. III. Introducciones, traducciones y notas por Armando Poratti, Conrado Eggers Lan, María Isabel Santa Cruz de Prunes y Néstor Luis Cordero. Madrid 1978–1980. (*Diálogos,* 51: 221–225 (1988)).

[199] Giovanni Reale, *A History of Ancient Philosophy. I. From the Origins to Socrates. III. The Systems of the Hellenistic Age.* Edited and translated [by] John R. Catan. Albany 1985–1987. (*Diálogos,* 51: 226–229 (1988)).

[200] G. S. Kirk, J. E. Raven, M. Schofield, *The Presocratic Philosophers. A Critical History with a Selection of Texts.* Second edition. Cambridge 1983. (*Diálogos,* 51: 230–232 (1988)).

[201] Peter Galison, *How Experiments End.* Chicago 1987. (*Diálogos,* 52: 155–163 (1988)).

[202] J. V. Field, *Kepler's Geometrical Cosmology.* Chicago 1988. (*Diálogos,* 53: 155–163 (1988))

[203] B. A. Rosenfeld, *A History of Non-Euclidean Geometry: Evolution of the Concept of a Geometric Space.* Translated [from the Russian] by Abe Shenitzer. New York 1988. (*Diálogos,* 54: 250–254 (1989)).

[204] Igal Kvart, *A Theory of Counterfactuals.* Indianapolis 1986. (*Diálogos,* 54: 254–260 (1989)).

[205] Pietro Redondi, *Galileo Heretic (Galileo Eretico).* English translation by Raymond Rosenthal. Princeton 1987. (*Revista Latinoamericana de Filosofía,* 15: 359–361 (1989)).

[206] Ronald Giere, *Explaining Science: A Cognitive Approach.* Chicago 1988. (*Diálogos,* 55: 199–203 (1990)).

[207] Raúl Fornet Betancourt, *Introducción a Sartre.* México 1989. (*Diálogos,* 55: 211 (1990)).

[208] Aristóteles, *Las Categorías.* Edición bilingüe. Introducción, traducción y notas de Humberto Giannini y María Isabel Flisfisch. Santiago de Chile 1988. (*Diálogos,* 55: 212–213 (1990)).

[209] Gottfried Wilhelm Leibniz, *Investigaciones Generales sobre el análisis de las nociones y las verdades.* Introducción, traducción y notas de Mauricio Beuchot y Alejandro Herrera-Ibáñez. México 1986. (*Diálogos,* 55: 213 (1990)).

[210] Immanuel Kant, *Pensamientos sobre la verdadera estimación de las fuerzas vivas.* Traducción y comentario de Juan Arana Cañedo-Argüelles. Berna 1988. (*Diálogos,* 55: 213–214 (1990)).

[211] F. W. J. Schelling, *Sistema del idealismo trascendental.* Traducción, prólogo y notas de Jacinto Rivera Rosales y Virginia López Domínguez. Barcelona 1988. (*Diálogos,* 55: 214 (1990)).

[212] Kurt Gödel, *Obras completas.* Introducción y traducción de Jesús Mosterín. Segunda edición. Madrid 1989. (*Diálogos,* 55: 214–216 (1990)).

[213] Ian Hacking, *The Taming of Chance.* Cambridge 1990. (*Diálogos,* 58: 185–187 (1991)).

[214] Lawrence Sklar, *Philosophy and Spacetime Physics.* (*Noûs,* **25**: 574–578 (1991)). Reviewed in English.

[215] L. T. F. Gamut, *Logic, Language, and Meaning.* Chicago 1991. (*Diálogos,* 59: 191–193 (1992)).

[216] Hans Joachim Krämer, *Plato and the Foundations of Metaphysics: A Work on the Theory of the Principles and Unwritten Doctrines of Plato with a*

Collection of the Fundamental Documents. Albany 1990. (*Diálogos,* 59: 194–200 (1992)).

[217] John Bigelow y Robert Pargetter, *Science and Necessity.* Cambridge 1990. (*Diálogos,* 60: 219–227 (1992)). Reviewed in English.

[218] John Earman, *World Enough and Space-Time: Absolute versus Relational Theories of Space and Time* (*Philosophical Review,* **101**: 723–725 (1992)). Reviewed in English.

[219] Jon Barwise y John Etchemendy. *The Language of First Order Logic.* Chicago 1991. (*Diálogos,* 61: 195–200 (1993)).

[220] Ana María Vicuña N., *Filosofía, poesía y mito a la luz de Eros en el Symposio de Platón.* [Santiago de Chile]: Pontificia Universidad Católica de Chile, 1993. (*Diálogos,* 65: 221–224 (1995)).

[221] Lawrence Sklar. *Physics and Chance: Philosophical Issues in the Foundations of Statistical Mechanics.* Cambridge 1993. (*Diálogos,* 65: 243–249 (1995)).

[222] Kurt Gödel, *Ensayos inéditos.* Edición a cargo de Francisco Rodríguez Consuegra. Prólogo de W. V. Quine. Barcelona: Mondadori, 1994. (*Diálogos,* 66; 185–190 (1995)).

[223] Richard Dedekind, *¿Qué son y para qué sirven los números? y otros escritos sobre los fundamentos de la matemática.* Edición e introducción a cargo de José Ferreirós. Madrid: Alianza, 1998. (*Revista Latinoamericana de Filosofía,* 25: 320–322 (1999)).

[224] Bertrand Russell, *Análisis filosófico.* Introducción y traducción de Francisco Rodríguez Consuegra. Barcelona: Ediciones Paidós, 1999. (*Revista Latinoamericana de Filosofía,* 26: 181–184 (2000)).

[225] Mara Beller, *Quantum Dialogue: The Making of a Revolution*, Chicago: University of Chicago Press, 1999. (*Revista de Filosofía* (Chile), **LV–LVI** 213–216 (2000)).

[226] Miguel Espinoza, *Philosophie de la nature*, Paris: Ellipses, 2000. (*Revista de Filosofía* (Chile), **LVII**: 183–185 (2001)).

[227] Max Jammer, *Concepts of Mass in Contemporary Physics and Philosophy*, Princeton NJ: Princeton University Press, 2000. (*Studies in the History and Philosophy of Modern Physics*, **33B**: 730–735 (2002)).

[228] Steve Fuller, *Thomas Kuhn: A Philosophical History for Our Times*, Chicago: University of Chicago Press, 2000. (*Physics in Perspective*, **4:** 120–123 (2002)).

[229] Lorenzo Magnani, *Philosophy and geometry: Theoretical and historical issues.* Dordrecht: Kluwer, 2001. ((*Studies in the History and Philosophy of Modern Physics*, **34B**: 158–160 (2003)).

[230] Manuel Correia, *La lógica de Aristóteles: Lecciones sobre el origen del pensamiento lógico en la antigüedad*, Santiago: Ediciones Universidad Católica de Chile, 2003. (*Revista de Filosofía* (Chile), **LX:** 165–169 (2004)).

[231] Kurt Gödel, *Collected Works*, vols. IV–V, Oxford: Clarendon Press, 2003 (*Revista de Filosofía* (Chile), **LX**: 169–172 (2004)).

[232] Simon Blackburn, *Truth: A Guide*. Oxford: Oxford University Press, 2005. (*Revista de Filosofía* (Chile), *Revista de Filosofía* (Chile), **LXI**: 229–233 (2005)).

[233] *Vocabulaire européen des philosophies*: *Dictionnaire des intraduisibles*, sous la direction de Barbara Cassin. Paris: Éditions du Seuil/Dictionnaires Le Robert, 2004. (*Revista de Filosofía* (Chile), **LXI**: 233–234 (2005)).

[234] William James, *Psicología para profesores*. Traducción de Paulina Dittborn. Santiago de Chile: Editorial Biblioteca Americana, 2005. (*Revista de Filosofía* (Chile), **LXI**: 235–236 (2005)).

[235] Paul Guyer, ed., *The Cambridge Companion to Kant and Modern Philosophy*. Cambridge: Cambridge University Press, 2006. (*Revista de Filosofía* (Chile), **LXII**: 181–185 (2006)).

[236] Michel Onfray, *Traité d'athéologie: Physique de la métaphysique*, Paris: Bernard Grasset, 2005. (*RevistaUDP*, **3**: 83–84 (2006)).

[237] Talal A. Debs y Michael L.G. Redhead, *Objectivity, Invariance, and Convention: Symmetry in Physical Science*. Cambridge MA: Harvard University Press, 2007. (*Revista de Filosofía* (Chile), **LXIII**: 206–209 (2007)).

[238] Pseudo-Longino, *De lo sublime*. Traducción de Eduardo Molina C. y Pablo Oyarzun R. Noticia Preliminar, Notas e índices de Pablo Oyarzun R. Santiago de Chile: Metales Pesados, 2007. (*Revista de Filosofía* (Chile) **LXIV**: 264–265 (2008)).

16.5 Translations

[239] *Cinco escritos de Leibniz*. Introducción, versión española y notas de Roberto Torretti. *Revista de Filosofía* (Chile), IX.3: 51–81 (1962).

Spanish translation of five Latin texts published by Couturat in *Opuscules et fragments inédits de Leibniz*, Paris 1903, pp. 533–535, 11–16, 518–523, 401–403, 16–24.

[240] Immanuel Kant, *La falsa sutileza de las cuatro figuras del silogismo*. Versión española por Roberto Torretti. *Diálogos*, 19: 7–22 (1970).

Spanish translation of *Die falsche Spitzfindigkeit der vier syllogistischen Figuren erwiesen von M. Immanuel Kant,* Königsberg 1762.

[241] Immanuel Kant, *Sobre el fundamento primero de la diferencia entre las regiones del espacio*. Traducción de Roberto Torretti. *Diálogos*, 22: 139–146 (1972).

Spanish translation of "Von dem ersten Grunde des Unterschiedes der Gegenden im Raume", published in the weekly *Wochentliche Königsbergsche Frag- und Anzeigungsnachrichten,* on February 6th, 13th and 20th, 1768.

[242] Gottfried Wilhelm Leibniz, *Ensayos filosóficos alemanes*. Versión española de Roberto Torretti. *Diálogos*, 23: 139–159 (1972).

Spanish translation of four German texts published by Guhrauer in *Deutsche Schriften*, Berlin 1838–40, vol. II, pp. 48–55, vol. I, pp. 410–413, 420–426, 414–419.

[243] Gottfried Wilhelm Leibniz, *Principios metafísicos de las matemáticas*. Traducción de Roberto Torretti. *Diálogos,* 24: 131–149 (1973).

Spanish translation of *Initia rerum mathematicarum metaphysica*, published by Gerhardt in G. W. Leibniz, *Mathematische Schriften*, Berlin and Halle 1849–63, t. VII, pp. 17–29.

[244] Immanuel Kant, *Sobre la nitidez de los principios de la teología natural y de la moral.* Traducción y notas de Roberto Torretti. *Diálogos*, 27: 57–87 (1974).

Spanish translation of *Untersuchung über die Deutlichkeit der Grundsätze der natürlichen Theologie und der Moral.* Berlin 1764.

[245] Bernhard Riemann, *Sobre las hipótesis que están en la base de la geometría.* Traducción de Roberto Torretti. *Diálogos*, 31: 151–168 (1978).

Spanish translation of *Über die Hypothesen, welche der Geometrie zugrunde liegen*, posthumously published in 1867 by Dedekind in *Abhandlungen der Königlichen Gesellschaft der Wissenschaften zu Göttingen*, vol.13.

[246] Immanuel Kant, *Monadología física.* Traducción del latín por Roberto Torretti. *Diálogos*, 32: 173–190 (1978).

Spanish translation of *Metaphysicae cum geometria junctae usus in philosophia naturalis cuius Specimen I continet Monadologiam Physicam*, Königsberg 1756.

[247] George Berkeley, *Sobre el movimiento o sobre el principio y la naturaleza del movimiento y sobre al causa de la trasmisión de los movimientos.* Traducción del latín por Roberto Torretti. *Diálogos*, 34: 119–141 (1979).

Spanish translation of *De Motu*, London 1721.

[248] Immanuel Kant, *Nuevo concepto del movimiento y el reposo.* Traducción del alemán por Roberto Torretti. *Diálogos*, 34: 143–152 (1979).

Spanish translation of *Neuer Lehrbegriff der Bewegung und Ruhe und der damit verknüpften Folgerungen in den ersten Gründen der Naturwissenschaft*, Königsberg 1758.

[249] Ludwig Lange, *Sobre la ley de la inercia.* Traducción del alemán por Roberto Torretti. *Diálogos*, 34: 153–170 (1979).

Spanish translation of "Über das Beharrungsgesetz", *Leipziger Berichte*, 37: 333–351 (1885).

[250] G. W. Leibniz, *Escritos filosóficos.* Edición de Ezequiel de Olaso. Notas de Ezequiel de Olaso y Roberto Torretti. Traducciones de Roberto Torretti, Tomás E. Zwanck y Ezequiel de Olaso. Buenos Aires: Editorial Charcas, 1982. 666 pp.

Includes all the Leibniz translations listed above and a part of the notes appended to them.

[251] Gottfried Wilhelm Leibniz, *Seis escritos de lógica*. Traducción y notas de Roberto Torretti. *Diálogos*, 51 (1988), 163–215.

Spanish translation of six Latin texts published by Couturat in *Opuscules et fragments inédits de Leibniz*, Paris 1903, pp. 49–57, 77–84, 232–237, 421–423, 193–202, 410–416.

[252] Hilary Putnam, "Los modelos y la realidad". Traducido del inglés por Francisco Rodríguez Consuegra y Roberto Torretti. *Diálogos*, 63: 7–45 (1994).

Annotated translation of "Models and Reality", originally published in *Journal of Symbolic Logic*, 45: 464–482 (1980). The translation includes "Afterthoughts on 'Models and Reality'", written by Putnam for this issue of *Diálogos* and printed, with facing translation, on pp. 41–45.

[253] Bruno de Finetti, "Sobre el significado subjetivo de la probabilidad". Traducido del italiano por Roberto Torretti. *Revista de Filosofía* (Chile), 58: xx–yy (2002).

Spanish translation of "Sul significato soggetivo della probabilità", *Fundamenta Mathematicae*, 17: 298–329 (1931).

The manufacturer's authorised representative in the EU is Springer Nature Customer Service Centre GmbH, Europaplatz 3, 69115 Heidelberg, Germany. If you have any concerns regarding our products, please contact ProductSafety@springernature.com

Printed and bound by CPI Group (UK) Ltd, Croydon, CR0 4YY
28/09/2026
02224061-0006